Lecture Notes
in Control and Information Sciences   244

Editor: M. Thoma

Springer
London
Berlin
Heidelberg
New York
Barcelona
Hong Kong
Milan
Paris
Santa Clara
Singapore
Tokyo

H. Nijmeijer and T.I. Fossen (Eds)

New Directions in Nonlinear Observer Design

Springer

Series Advisory Board

A. Bensoussan · M.J. Grimble · P. Kokotovic · H. Kwakernaak
J.L. Massey · Y.Z. Tsypkin

Editors

H. Nijmeijer
Faculty of Mathematical Sciences, University of Twente, PO Box 217,
7500 AE Enschede, The Netherlands

T.I. Fossen
Department of Engineering Cybernetics, Norwegian University of Science and
Technology, N-7034 Trondheim, Norway

ISBN 1-85233-134-8 Springer-Verlag London Berlin Heidelberg

British Library Cataloguing in Publication Data
New directions in nonlinear observer design. – (Lecture
   notes in control and information sciences ; 224)
   1.Observers (Control theory) 2.Nonlinear control theory
   3.Feedback control systems
   I.Nijmeijer, Henk, 1955- II.Fossen, Thor I.
   629.8'36
   ISBN 1852331348

Library of Congress Cataloging-in-Publication Data
New directions in nonlinear observer design / H. Nijmeijer and
   T.I. Fossen (eds.).
      p.   cm. – (Lecture notes in control and information sciences
   ; 244)
   Includes bibliographical references and index.
   ISBN 1-85233-134-8 (alk. Paper)
      1.Observers (Control theory)--Congresses. 2. Nonlinear control
   Theory--Congresses. I. Nijmeijer, H. (Henk), 1955- .
   II. Fossen, Thor I. III. Series.
   QA402.3.N487 1999   99-12174
   629.8'312--dc21   CIP

Typesetting: Camera ready by contributors
Printed and bound at the Athenæum Press Ltd., Gateshead, Tyne & Wear
69/3830-543210 Printed on acid-free paper

Preface

"New Directions in Nonlinear Observer Design" contains a collection of contributions presented at an international workshop with the same name held from June 24-26, 1999, in Geiranger Fjord, Norway. The workshop included participants from Africa, Asia, Europe and the USA. The subject of the book is nonlinear observer design with focus on general developments in nonlinear observer theory, output feedback control, synchronization of complex dynamical systems, and fault detection and isolation. The book contains the latest achievements in nonlinear observer design and it is a supplement to engineers who are familiar with linear design techniques like the Kalman-filter, Luenberger observer etc. The main advantage of using nonlinear observer theory to linear theory is the improved performance achievement and that global convergence and stability can be established for a large class of nonlinear plants. Traditional state estimators like the Kalman filter must be linearized about a set of pre-defined operating points often in combination with a gain scheduling technique in order to cover the whole state-space. This approach only guarantees local convergence/stability. Therefore, nonlinear observer design is becoming an important field of research with increased international attention.

Outline

The book is organized into four parts each divided into several chapters:

I Nonlinear observer design (9 chapters)

II Output feedback control design (8 chapters)

III Fault detection and isolation (5 chapters)

IV Synchronization and observers (3 chapters)

Acknowledgments

The editors are grateful to:

- **Strategic University Program (SUP) in Marine Cybernetics** at the Norwegian University of Science and Technology (NTNU), Departments of Engineering Cybernetics, Marine Hydrodynamics and Marine Strutures (*Professor Dr.-Ing. Olav Egeland, Program Manager*).

- **ABB** (*Professor Dr.-Ing. Asgeir J. Sørensen, Technology Manager - Business Area Marine and Turbochargers*)

for their financial support.
The authors want to thank all the workshop contributors for contributing to this book project.

Finally, Mrs. Alison Jackson at Springer-Verlag London should be thanked for editorial suggestions and for helping us with general publishing questions.

Trondheim, February 1999   Thor I. Fossen
Enschede, February 1999   Henk Nijmeijer

Contents

Contributors

Alcorta García, E., Department of Measurement and Control, University of Duisburg, Duisburg, Germany.

Ashton, S. A., School of MIS, Coventry University, U.K.

Astolfi, A., Centre for Process Systems Engineering, Imperial College of Science, London, U.K.

Bastin, G., Centre for Systems Engineering and Applied Mechanics, Universite Catholique de Louvain, Louvain–La–Neuve, Belgium.

Battilotti, S., Dipartimento di Informatica e Sistemistica, Università di Roma "La Sapienza", Italy.

Besançon, G., Laboratoire d'Automatique de Grenoble, ENSIEG, Saint-Martin d'Hères, France.

Blanke, M., Department of Automatic Control, Aalborg University, Denmark.

Canudas de Wit, C. Laboratoire d´Automatique de Grenoble, ENSIEG-INPG, ST. Martin d´Hères, France.

Cruz, C., Department of Electronics & Telecom., Scientific Research and Advanced Studies Center of Ensenada (CICESE), México.

Deng, H., Department of Applied Mechanics and Engineering Sciences University of California at San Diego, La Jolla, USA.

Egeland, O., Department of Engineering Cybernetics, Norwegian University of Science and Technology, Trondheim, Norway.

El Bahir, L., Department of Control Engineering, Université Libre de Bruxelles, Brussels, Belgium.

El Yaagoubi, E. H., LCPI ENSEM Cassablanca, Morocco.

Fossen, T. I., [1]Department of Engineering Cybernetics, Norwegian University of Science and Technology, Trondheim, Norway and [2]ABB Industri AS, Marine Division, Oslo, Norway.

Frank, P. M., Department of Measurement and Control, University of Duisburg, Duisburg, Germany.

Glumineau, A. Institut de Recherche en Cybernétique de Nantes, France.

Hammouri, H., LAGEPT University of Lyon, France.

Huijberts, H. J. C., Department of Mathematics and Computing Science, Eindhoven University of Technology, The Netherlands.

Horowitz, R. Department of Mechanical Engineering, University of California, Berkeley, CA, U.S.A.

Isidori, A., [1]Department of Systems Science and Mathematics, Washington University, St. Louis, USA and [2]Dipartimento di Informatica e Sistemistica, Università di Roma "La Sapienza", Italy.

Izadi-Zamanabadi, R., Department of Automatic Control, Aalborg University, Denmark.

Jiang, Z.-P., Department of Electrical Engineering, Polytechnic University, Brooklyn, U.S.A.

Junge, L., Drittes Physikalisches Institut, Universität Göttingen, Germany.

Khalil, H. K., Department of Electrical and Computer Engineering, Michigan State University, USA.

Kinnaert, M., Department of Control Engineering, Université Libre de Bruxelles, Brussels, Belgium.

Kocarev, L., Department of Electrical Engineering, St Cyril and Methodius University, Skopje, Macedonia.

Kristiansen, D., Department of Engineering Cybernetics, Norwegian University of Science and Technology, Trondheim, Norway.

Krstić, M., Department of Applied Mechanics and Engineering Sciences University of California at San Diego, La Jolla, USA.

Lilge, T., Institut für Regelungstechnik, University of Hannover, Hannover, Germany.

Lohmiller, W., Nonlinear Systems Laboratory, Massachusetts Institute of Technology, Cambridge, Massachusetts, USA.

López-Morales, V., Institut de Recherche en Cybernétique de Nantes, France.

Loria, A., Laboratoire d'Automatique de Grenoble, ENSIEG, St. Martin d'Hêres, France.

Nijmeijer, H., [1]Faculty of Mathematical Sciences, Dept. of Systems, Signals and Control, University of Twente and [2]Faculty of Mechanical Engineering, Eindhoven University of Technology, The Netherlands.

Ortega, R., Laboratoire des Signaux et Systèmes, Ecole Supérieure d'Electricité, Paris, France.

Panteley, E., I.N.R.I.A., Rhône Alpes, St. Martin d'Hêres, France.

Parlitz, U., Drittes Physikalisches Institut, Universität Göttingen, Germany.

Pettersen, K. Y., Department of Engineering Cybernetics, Norwegian University of Science and Technology, Trondheim, Norway.

Praly, L., Centre Automatique et Systèmes, École des Mines de Paris, Fontainebleau, France.

Rodrigues–Cortes, H., Laboratoire des Signaux et Systèmes, Ecole Supérieure d'Electricité, Paris, France.

Schaffner, J., Institute for Systems, Informatics and Safety, European Commission Joint Research Centre, Ispra, Italy

Schreier, G., Department of Measurement and Control, University of Duisburg, Duisburg, Germany.

Shields, D. N., School of MIS, Coventry University, U.K.

Shiriaev, A., Department of Engineering Cybernetics, Norwegian University of Science and Technology, Trondheim, Norway.

Slotine, J. J. E., Nonlinear Systems Laboratory, Massachusetts Institute of Technology, Cambridge, Massachusetts, USA.

Strand, J. P., ABB Industri AS, Marine Division, Oslo, Norway.

Teel, A., Department of Electrical and Computer Engineering, University of California, Santa Barbara, USA.

Tsiotras, P., Georgia Institute of Techology, School of Aerospace Eng., Atlanta, Georgia, USA.

Vik, B. Department of Engineering Cybernetics, Norwegian University of Science and Technology, Trondheim, Norway.

Zeitz, M., Institut für Systemdynamik und Regelungstechnik, University of Stuttgart, Germany.

Part I

Nonlinear Observer Design

The general question of determining or reconstructing the state of a system when only some output function, usually a function of the sensor measurements, is referred to as an observer or state estimation problem in the control literature. The observer problem is, except for linear systems, only solved for particular classes of nonlinear systems, and a large variety of open problems concerning nonlinear observers exist. Within this first part of the book, the reader can find a sample of recent advances and directions in the study of nonlinear observer designs and their applications. Contributions in the next chapters range from continuous-time to discrete-time, from ordinary differential equations to partial differential equations, and from mechanical and chemical systems to general nonlinear system models. All contributions treat the observer problem per se, and no attempts included in this part deal with exploiting the state estimate found in further control design. Such contributions are discussed in the next parts of the book.

A Viewpoint on Observability and Observer Design for Nonlinear Systems

Gildas Besançon

Laboratoire d'Automatique de Grenoble
ENSIEG BP46 - 38402 Saint-Martin d'Hères, France

1   Introduction

Given a dynamical system, the observer aims at obtaining an estimate of the current state by only using available measurements. For linear systems, the property of observability, characterized by the Kalman rank condition, guarantees the possibility to indeed design an observer. In the case of non-linear systems, observability is not enough, basically because this property in general depends on the input of the system. In other words, observability of a nonlinear system does not exclude the existence of inputs for which two distinct initial states cannot be distinguished by using the knowledge of the measured output. This results in the fact that in general, observer gains can be expected to depend on the applied input. Moreover, the existing observers generally tightly depend on some specific structure of the considered system.

This chapter discusses such characteristics of observer design for non-linear systems, basically following the recent work of [1]: on the basis of background definitions, the main observability requirements for observer design are first recalled in Section 2, so as to put into relief the various contexts which can be found. In particular, designs which are *non uniform* w.r.t. the input are distinguished from those which are *uniform*, with a special attention paid to the case of *uniform design* for *non uniformly observable* systems in Section 3.

Two directions of extension of available designs are then highlighted and illustrated, namely the possibility of interconnecting sub-observers to still obtain an observer in Section 4, and the issue of state transformation to make some observer design possible in Section 5. Some conclusions are finally given in Section 6.

2   Basic Definitions and Proposed "Classification"

Let us consider a nonlinear system defined by the following representation:

$$(\Sigma_{nl}) \left\{ \begin{array}{lll} \dot{x} & = & f(x,u) \quad x \in I\!\!R^n,\, u \in I\!\!R^m \\ y & = & h(x) \quad y \in I\!\!R^p \end{array} \right. , \qquad (1.1)$$

and let $\chi_u(t, x_0)$ denote its solution at time t, with initial condition x_0 at time $t = 0$ and control $u(t)$. Admissible inputs $u(.)$ are assumed to be taken in some set $\mathcal{U}$ of measurable and bounded functions.

Observability can then be defined by the notion of indistinguishability [22] (see [11] for a synthesis).

Definition 1.1 *Indistinguishability.*
A pair $(x_0, \tilde{x}_0)$ will be said to be indistiguishable by u if $\forall t \geq 0$, $h(\chi_u(t, x_0)) \equiv h(\chi_u(t, \tilde{x}_0))$. The pair is just said to be indistinguishable, if it is so for any u.

From this definition, observability of (1.1) can be defined as follows:

Definition 1.2 *Observability.*
A nonlinear system (1.1) is observable if it does not have any indistinguishable pair of states.

At this point, one can notice that from this definition, observability does not exclude the possible existence of inputs for which some states are indistinguishable.

As an example, the following system:

$$\begin{array}{lll} \dot{x}_1 & = & u x_2 \\ \dot{x}_2 & = & -x_2 \\ y & = & x_1 \end{array} \qquad (1.2)$$

is clearly observable, and yet $u \equiv 0$ makes every pair

$$\left(\begin{array}{c} \bar{x}_1 \\ \bar{x}_2 \end{array} \right) \neq \left(\begin{array}{c} \bar{x}_1 \\ \tilde{x}_2 \end{array} \right)$$

indistinguishable.

This means that in general observability is not enough to be able to design an observer and that the problem of inputs must be taken into account. A particular case of interest, is the case of inputs for which no indistinguishable pair can be found:

Definition 1.3 *Universal inputs.*
An input u is universal on $[0, t]$ if for every pair of distinct states $x_0 \neq \bar{x}_0$, there exists $\tau \in [0, t]$ such that $h(\chi_u(\tau, x_0)) \neq h(\chi_u(\tau, \bar{x}_0))$.
If u is universal on $I\!\!R^+$, it is just said to be universal.

From this definition, the notion of *singular inputs* can be derived:

Definition 1.4 *Singular inputs.*
A non universal input is called singular.

As one can guess from example (1.2) above, a typical class of systems which have singular inputs is the class of so-called *state-affine* systems of the following form:

$$\begin{aligned} \dot{x} &= A(u)x \\ y &= Cx. \end{aligned} \tag{1.3}$$

For such systems, one can define - as for linear time-varying systems - the following quantities:

- the transition matrix $\Phi_u(\tau, t)$ by:

$$\begin{aligned} \frac{d\Phi_u(\tau, t)}{d\tau} &= A(u(\tau))\Phi_u(\tau, t) \\ \Phi_u(t, t) &= Id, \end{aligned} \tag{1.4}$$

- the observability Grammian $\Gamma(t, T, u)$ by:

$$\Gamma(t, T, u) := \int_t^{t+T} \Phi_u^T(\tau, t) C^T C \Phi_u(\tau, t) d\tau), \tag{1.5}$$

- and some universality index $\gamma(t, T, u)$, defined as the smallest eigenvalue of $\Gamma(t, T, u)$.

On this basis one can characterize inputs which are "universal enough" so that an observer design will be possible:

Definition 1.5 *[10] Regularly persistent inputs.*
An admissible input u is said to be regularly persistent for system (1.3) if $\exists T > 0$, $\alpha > 0$, and $t_0 > 0$ such that $\gamma(t, T, u) \geq \alpha$ for $t \geq t_0$.

These remarks show that in general, the observer gain, as well as its stability property, depend on the input.
In view of the above definitions, a particular case of special interest is the one of systems without singular inputs:

Definition 1.6 *Uniformly observable systems.*
A system whose all inputs are universal is called uniformly observable.
If, for every $t > 0$, all inputs are universal on $[0, t]$, the system is locally uniformly observable.

A class of "sufficiently regular" locally uniformly observable control affine systems have been characterized in [14]:

Theorem 1.1 *[14] An observable nonlinear system in the form:*

$$
\begin{aligned}
\dot{\xi} &= f(\xi) + ug(\xi) \\
y &= h(\xi)
\end{aligned}
\tag{1.6}
$$

with $u = 0$ as a universal input, and nonsingular Jacobian of $\Phi(\xi) = (h(\xi), L_f(h(\xi)), \ldots L_f^{n-1}(h(\xi)))$ at ξ_0, is locally uniformly observable at ξ_0 if and only if the change of coordinates $x = \Phi(\xi)$ turns it into the following form:

$$
\begin{aligned}
\dot{x}_1 &= x_2 + \varphi_1(x_1)u \\
\dot{x}_2 &= x_3 + \varphi_2(x_1, x_2)u \\
&\vdots \\
\dot{x}_{n-1} &= x_n + \varphi_{n-1}(x_1, \ldots x_{n-1})u \\
\dot{x}_n &= \bar{\varphi}_n(x) + \varphi_n(x)u \\
y &= x_1
\end{aligned}
\tag{1.7}
$$

$\square$

Extensions to non-control-affine case mono and multi output can be found in [15] and [12] respectively.

For such systems, so-called *high gain observers* may exist [16], which are observers with gain and stability independent of the input. However, one can find several other cases of observers which have been proposed irrespective of the input, although the considered systems are not uniformly observable [21, 29, 13, 2].

The only possible explanation for such a phenomenon is that in these cases, the difference between trajectories resulting from two distinct indistinguishable states *naturally* tends to zero. This is for instance what happens in system (1.2) : for this system $u = 0$ is singular insofar as it cannot distinguish

$$
\begin{pmatrix} \bar{x}_1 \\ \bar{x}_2 \end{pmatrix} \neq \begin{pmatrix} \bar{x}_1 \\ \tilde{x}_2 \end{pmatrix}
$$

However, for such initial conditions and input, the error e_2 on trajectories of x_2 satisfies $\dot{e}_2 = -e_2$, while that of x_1 is identically zero, and thus, it is clear that the difference between the two trajectories asymptotically goes to zero. In this case, an observer irrespective of the input can be designed, simply as $\dot{\hat{x}}_2 = -\hat{x}_2$.

Following the terminology of linear systems, this suggests to define some *detectability* property as follows:

Definition 1.7 *Detectability.*
A nonlinear system (1.1) will be called detectable if for every couple $((\bar{x}_0, \tilde{x}_0), u(.))$ in $(\mathbb{R}^n \times \mathbb{R}^n) \times \mathcal{U}$ such that there exists t_0 for which

$\forall t \geq t_0; h(\chi_u(t, x_0)) = h(\chi_u(t, \tilde{x}_0))$ *then*

$$\|\chi_u(t, x_0) - \chi_u(t, \tilde{x}_0)\| \xrightarrow[t \to \infty]{} 0.$$

In the case of uncontrolled system, we only consider pairs of initial conditions.

As a summary, in view of the above observability properties w.r.t. the inputs, one can distinguish the following cases in observer designs:

- Either the system does not have singular inputs, and in that case, one can hope to be able to design an observer irrespective of the input (*uniform observation of uniformly observable systems*), but may also only find an observer depending on the input (*non uniform observation of uniformly observable systems*);

- Or the system may have singular inputs, and in that case possible observer designs will generally depend on the inputs (*non uniform observation of non uniformly observable systems*), except in special cases of systems which are detectable in the sense of definition 1.7 (*uniform observation of non uniformly observable systems*).

3   Examples of Non Uniform and Uniform Observation

3.1   Non Uniform Observation: the Case of State-Affine Systems

Let us consider here a system described by the following equations:

$$\begin{array}{rclcl} \dot{x} & = & A(u)x + B(u), & x \in \mathbb{R}^n, u \in \mathbb{R}^m \\ y & = & Cx & y \in \mathbb{R}^p. \end{array} \tag{1.8}$$

For such a system, the observability generally depends on the input, and under appropriate excitation, an observer has been proposed with a gain indeed depending on u, as recalled below:

Theorem 1.2 *[10, 17] If u is regularly persistent for (1.8), and $A(u), B(u)$ are uniformly bounded on the set of admissible inputs, then there exists θ_0 s.t. for any $\theta \geq \theta_0$, the following system is an observer for (1.8):*

$$\begin{array}{rcl} \dot{\hat{x}} & = & A(u)\hat{x} - S^{-1}C^T(C\hat{x} - y) + B(u) \\ \dot{S} & = & -\theta S - A(u)^T S - SA(u) + C^T C \\ S_0 & > & 0, \end{array} \tag{1.9}$$

and $\forall \zeta > 0, \exists \theta > 0 : \|\hat{x}(t) - x(t)\| \leq \lambda \exp(-\zeta t)$, for some $\lambda > 0$.

□

This result can clearly be extended to the case of systems in the form $\dot{x} = A(s)x + B(s)$ for any measured signal s which is regularly persistent for $\dot{x} = A(s)x$, for instance $s = (u, y)$ [18].

We present in Section 4 another type of extension, based on interconnections of observers in the form (1.9).

3.2   Uniform Observation: the Case of Uniformly Observable Systems

We consider here a system of the following form:

$$\begin{aligned} \dot{x} &= Ax + \varphi(x, u), \quad x \in \mathbb{R}^n, \, u \in \mathbb{R}^m \\ y &= Cx \in \mathbb{R} \end{aligned} \tag{1.10}$$

with $A = \begin{pmatrix} 0 & 1 & \cdots & 0 \\ \vdots & \ddots & \ddots & 0 \\ 0 & \cdots & 0 & 1 \\ 0 & \cdots & & 0 \end{pmatrix}$ and $C = (\, 1 \quad 0 \cdots \quad 0 \,)$.

Under structure condition as in (1.7) - ensuring uniform observability - and some Lipschitz condition on φ, one can here design an observer with a gain which is uniform w.r.t. u as recalled hereafter (where x_i - resp. φ_i - denotes each component of x - resp. φ):

Theorem 1.3 *[16] If:*

- φ *is globally Lipschitz w.r.t.* x, *uniformly w.r.t.* u;

- $\dfrac{\partial \varphi_i}{\partial x_j} \equiv 0$, *for* $i = 1, \ldots n - 1, j = i + 1, \ldots n$.

then there exists θ_0 *such that for all* $\theta > \theta_0$, *the following system is an asymptotic observer for (1.10):*

$$\begin{aligned} \dot{\hat{x}} &= A\hat{x} - S^{-1} C^T (C\hat{x} - y) + \varphi(\hat{x}, u) \\ 0 &= -\theta S - A^T S - SA + C^T C, \end{aligned} \tag{1.11}$$

and $\forall \zeta > 0, \exists \theta > 0 : \|\hat{x}(t) - x(t)\| \leq \lambda \exp(-\zeta t)$, for some $\lambda > 0$.

□

We will use this result in Section 4 to propose a uniform observer for some non uniformly observable system. But let us first illustrate this phenomenon in next subsection.

3.3   An Example of Uniform Observation of Non-uniformly observable Systems

To be able to find an observer which is non uniform w.r.t. the input for a system which is not uniformly observable, as well as a uniform observer for a system which is uniformly observable, is to some extent quite consistent. More impressive are cases of *uniform* observers for *non uniformly* observable systems.

An illustrative example of the phenomenon is given by the case of systems of the following form:

$$\begin{aligned} \dot{x} &= Ax + f(x,u) \\ y &= Cx \end{aligned}$$ (1.12)

for which one can find matrices K, D such that

$$T = \begin{pmatrix} C \\ D \end{pmatrix}$$

is invertible, $(KC + D)f(x,u) = \varphi(Cx, u)$, and $A_{22} + KA_{12}$ is stable, with

$$\begin{pmatrix} A_{11} & A_{12} \\ A_{21} & A_{22} \end{pmatrix} = TAT^{-1}$$

. Such systems indeed, generally admit singular input (like system (1.2) for instance, for which $D = (0\ 1)$ and $K = 0$ satisfy the above conditions, and $u = 0$ is a singular input), and yet:

$$\begin{aligned} \dot{z} &= (A_{22} + KA_{12})z + (KA_{11} + A_{21} - (A_{22} + KA_{12})K)y + \varphi(y, u) \\ w &= \begin{pmatrix} C \\ KC + D \end{pmatrix}^{-1} \begin{pmatrix} y \\ z \end{pmatrix} \end{aligned}$$

(1.13)

is an observer for (1.12) irrespective of the input [2]. Such systems in fact enjoy the following structure:

$$\dot{x} = \begin{pmatrix} A_{11} & A_{12} \\ A_{21} & A_{22} \end{pmatrix} x + \begin{pmatrix} I_p \\ \Lambda \end{pmatrix} F(x,u) + \begin{pmatrix} 0 \\ \psi(y,u) \end{pmatrix}, \ y = Cx = (I_p, 0)x$$

(1.14)

and as soon as Λ makes $A_{22} - \Lambda A_{12}$ asymptotically stable, then one can design an observer in the form (1.13) for (1.14).

Such a property can in particular be used to design *robust* observers in the sense that the state estimation does not require the knowledge of $F(x,u)$.

For instance, various manufacturing systems admit a representation in the form (1.14), and an example can be found in [28] where the thermal behaviour of some machine-tool spindle-bearing system is considered. Considering temperatures of the various elements involved as state variables, a numerical realization of the systems reads as follows:

$$
\dot{x} = \left(
\begin{array}{c|cc}
-8.000.10^{-5} & 2.982.10^{-6} & 0 \\
\hline
0.0421 & -0.0325 & 0.0104 \\
0 & 2.483.10^{-6} & -1.724.10^{-4}
\end{array}
\right) x
$$

$$
+ \left(
\begin{array}{c}
1.407.10^{-5} \\
0.0080 \\
2.495.10^{-5}
\end{array}
\right) \hat{Q}(u, x) \tag{1.15}
$$

$$
y = (1 \quad 0 \quad 0)x \tag{1.16}
$$

where $\hat{Q}(u, x)$ is some inaccurate model of the friction heat flow.

This system is thus in the form (1.14), with the nonlinear part in the form

$$
\left(
\begin{array}{c}
B_1 \\
B_2
\end{array}
\right) F(x, u)
$$

and one can check that here $\Lambda = \frac{B_2}{B_1}$ leaves $A_{22} - \Lambda A_{12}$ stable. Hence an observer can here be designed with complete decoupling of the uncertain part $\hat{Q}$.

Some more general conditions for such a design to be possible are given in [2], but a general formulation of the idea can be expressed as follows:

Proposition 1.1 *If a nonlinear system:*

$$
\begin{aligned}
\dot{x} &= f(x, u) \\
y &= h(x)
\end{aligned} \tag{1.17}
$$

can be transformed into:

$$
\begin{aligned}
\dot{z}_1 &= \bar{f}_1(z_1, z_2, u) \\
\dot{z}_2 &= \bar{f}_2(z_1, z_2, u) \\
y &= z_1
\end{aligned} \tag{1.18}
$$

by change of coordinates $z = \Phi(x)$, such that for any couple (u, z_1) of admissible functions and any pair of initial conditions $z_2^0 \neq \hat{z}_2^0$ we have $\|\chi_2^{(u, z_1)}(t, z_2^0) - \chi_2^{(u, z_1)}(t, \hat{z}_2^0)\| \to 0$ when $t \to \infty$, then:

$$
\begin{aligned}
\dot{\hat{z}}_2 &= \bar{f}_2(y, \hat{z}_2, u) \\
\hat{x} &= \Phi^{-1}\left(
\begin{array}{c}
y \\
\hat{z}_2
\end{array}
\right)
\end{aligned} \tag{1.19}
$$

is an observer for (1.17).

This illustrates how an observer can be designed on the basis of some state transformation, here in some particular conditions ensuring detectability. Other results on state transformation for observer design are given in section 5.

4   Observer Interconnection

One way to extend the class of systems for which an observer can be designed is to interconnect observers in order to design an observer for some interconnected system, when possible. If indeed a system is not under a form for which an observer is already available, but can be seen as an interconnection between several subsystems each of which would admit an observer if the states of the other subsystems were known, then a candidate observer for the interconnection of these subsystems is given by interconnecting available sub-observers.

Notice that in general, the stability of the interconnected observer is not guaranteed by that of each sub-observer, in the same way as separate designs of observer and controller do not in general result in some stable observer-based controller for nonlinear systems (no *separation principle*). However, Lyapunov-based sufficient conditions can be given so that the existence of sub-observers results in that of an interconnected observer [7]. Consider for instance the case of systems made of two subsystems of the following form:

$$(\Sigma) \begin{cases} \dot{x}_1 &= f_1(x_1, x_2, u), \ u \subset U \subset \mathbb{R}^m; \ f_i \ \mathcal{C}^\infty \text{ function}, \ i = 1, 2; \\ \dot{x}_2 &= f_2(x_2, x_1, u), \ x_i \in X_i \subset \mathbb{R}^{n_i}, \ i = 1, 2; \\ y &= (h_1(x_1), h_2(x_2))^T = (y_1, y_2)^T, \ y_i \in \mathbb{R}^{n_i}, \ i = 1, 2. \end{cases}$$

$$(1.20)$$

Assume also that $u(.) \in \mathcal{U} \subset \mathcal{L}^\infty(\mathbb{R}^+, U)$, and set $\mathcal{X}_i := \mathcal{AC}(\mathbb{R}^+, \mathbb{R}^{n_i})$ the space of absolutely continuous function from $\mathbb{R}^+$ into $\mathbb{R}^{n_i}$. Finally, when $i \in \{1, 2\}$, let $\bar{\imath}$ denote its complementary index in $\{1, 2\}$.

The system (1.20) can be seen as the interconnection of two subsystems (Σ_i) for $i = 1, 2$ given by:

$$(\Sigma_i) \quad \dot{x}_i = f_i(x_i, v_{\bar{\imath}}, u), \quad y_i = h_i(x_i), \quad (v_{\bar{\imath}}, u) \in \mathcal{X}_{\bar{\imath}} \times \mathcal{U}. \qquad (1.21)$$

Assume that for each system (Σ_i), one can design an observer $(\mathcal{O}_i)$ of the following form:

$$(\mathcal{O}_i) \quad \dot{z}_i = f_i(z_i, v_{\bar{\imath}}, u) + k_i(g_i, z_i)(h_i(z_i) - y_i), \quad \dot{g}_i = G_i(z_i, v_{\bar{\imath}}, u, g_i),$$

$$(1.22)$$

for smooth k_i, G_i and $(z_i, g_i) \in (\mathbb{R}^{n_i} \times \mathcal{G}_i), \mathcal{G}_i$ positively invariant by (1.22). The idea is to look for an observer for (1.20) under the form of the following interconnection:

$$(\mathcal{O}) \begin{cases} \dot{\hat{x}}_i &= f_i(\hat{x}_i, \hat{x}_{\bar{i}}, u) + k_i(\hat{g}_i, \hat{x}_i)(h_i(\hat{x}_i) - y_i); \ i = 1, 2; \\ \dot{\hat{g}}_i &= G_i(\hat{x}_i, \hat{x}_{\bar{i}}, u, \hat{g}_i); \ i = 1, 2 \end{cases} \quad (1.23)$$

Set $e_i := z_i - x_i$, and for any $u \in \mathcal{U}, v_i \in \mathcal{X}_i$ consider the following system (where $k_i^{v_{\bar{i}}}(t)$ denotes gain $k_i(g_i, z_i)$ defined in (1.22)) :

$$\mathcal{E}_i^{(u,v_i)} \begin{cases} \dot{e}_i &= f_i(z_i, v_{\bar{i}}, u) - f_i(z_i - e_i, v_{\bar{i}}, u) + k_i^{v_i}(t)(h_i(z_i) - h_i(z_i - e_i)) \\ \dot{z}_i &= f_i(z_i, v_{\bar{i}}, u) + k_i^{v_i}(t)(h_i(z_i) - h_i(z_i - e_i)) \\ \dot{g}_i &= G_i(z_i, v_{\bar{i}}, u, g_i). \end{cases}$$

Then sufficient conditions for (1.23) to be an observer for (1.20) have been expressed in [7] as follows:

Theorem 1.4 *[7] If for $i = 1, 2$, any signal $u \in \mathcal{U}, v_{\bar{i}} \in \mathcal{AC}(\mathbb{R}^+, \mathbb{R}^{n_{\bar{i}}})$, and any initial value $(z_i^0, g_i^0) \in \mathbb{R}^{n_i} \times \mathbb{G}_i$, $\exists V_i(t, e_i), W_i(e_i)$ positive definite functions such that:*

(i) $\forall x_i \in X_i; \forall e_i \in \mathbb{R}^{n_i}; \forall t \geq 0$,

$$\frac{\partial V_i}{\partial t}(t, e_i) + \frac{\partial V_i}{\partial e_i}(t, e_i)[f_i(x_i + e_i, v_{\bar{i}}(t), u(t)) - f_i(x_i, v_{\bar{i}}(t), u(t))$$
$$+ k_i^{v_i}(t)(h_i(x_i + e_i) - h_i(x_i))] \leq -W_i(e_i)$$

(ii) $\exists \alpha_i > 0; \forall x_i \in X_i; \forall x_{\bar{i}} \in \mathbb{R}^{n_{\bar{i}}}; \forall e_i \in \mathbb{R}^{n_i}; \forall e_{\bar{i}} \in \mathbb{R}^{n_{\bar{i}}}; \forall t \geq 0$,

$$\left\| \frac{\partial V_i}{\partial e_i}(t, e_i)[f_i(x_i, x_{\bar{i}} + e_{\bar{i}}, u(t)) - f_i(x_i, x_{\bar{i}}, u(t))] \right\| \leq \alpha_i \sqrt{W_i(e_i)} \sqrt{W_{\bar{i}}(e_{\bar{i}})},$$

(iii) $\alpha_1 + \alpha_2 < 2$,

then (1.23) is an asymptotic observer for (1.20).

$\square$

In the *weaker* case of *cascade* interconnection, namely when $f_1(x_1, x_2, u) = f_1(x_1, u)$ in (1.20), assumptions can be weakened in the following way:

Theorem 1.5 *[7] Assume that:*

I. System $\dot{x}_1 = f_1(x_1, u); \ y_1 = h_1(x_1)$ admits an observer $(\mathcal{O}_1)$ as in (1.22) (without v_2), s.t. $\forall u \in \mathcal{U}$ and $\forall x_1(t)$ admissible trajectory of the system associated to u:

$$\lim_{t \to \infty} e_1(t) = 0 \text{ and } \int_0^{+\infty} \|e_1(t)\| dt < +\infty \quad (\text{with } e_1 := z_1 - x_1);$$
$$(1.24)$$

II. $\exists c > 0; \ \forall u \in U; \ \forall x_2 \in X_2, \|f_2(x_2, x_1, u) - f_2(x_2, x_1', u)\| \leq c\|x_1 - x_1'\|$;

III. $\forall u \in \mathcal{U}, \forall v_1 \in AC(\mathbb{R}^+, \mathbb{R}^{n_1}), \forall z_2^0, g_2^0, \exists v(t, e_2), w(e_2)$ *positive definite functions s.t for every trajectory of* $\mathcal{E}_2^{(u,v_1)}$ *with* $z_2(0) = z_2^0, g_2(0) = g_2^0$:

(i) $\forall x_2 \in X_2, e_2 \in \mathbb{R}^{n_2}, t \geq 0$,

$$\frac{\partial v}{\partial t}(t, e_2) + \frac{\partial v}{\partial e_2}(t, e_2)[f_2(x_2 + e_2, v_1(t), u(t)) - f_2(x_2, v_1(t), u(t))$$
$$+ k_2^{v_1}(t)(h_2(x_2 + e_2) - h_2(x_2))] \leq -w(e_2)$$

(ii) $\forall e_2 \in \mathbb{R}^{n_2}, t \geq 0;$ $v(t, e_2) \geq \bar{w}(e_2)$

(iii) $\forall e_2 \in \mathbb{R}^{n_2} \backslash \mathcal{B}(0, r), t \geq 0;$ $\left\| \frac{\partial v}{\partial e_2}(t, e_2(t)) \right\| \leq \lambda(1 + v(t, e_2(t)))$ *for some constants* $\lambda, r > 0$ *and* $\mathcal{B}(0, r) := \{e_2 : \|e_2\| \leq r\}$.

Then:

$$\begin{array}{rcl}
\dot{\hat{x}}_1 & = & f_1(\hat{x}_1, u) + k_1(\hat{g}_1, \hat{x}_1)(h_1(\hat{x}_1) - h_1(x_1)) \\
\dot{\hat{x}}_2 & = & f_2(\hat{x}_1, \hat{x}_2, u) + k_2(\hat{g}_2, \hat{x}_2)(h_2(\hat{x}_1) - h_2(x_1)) \\
\dot{\hat{g}}_1 & = & G_1(\hat{x}_1, u, \hat{g}_1); \quad \dot{\hat{g}}_2 = G_2(\hat{x}_2, \hat{x}_1, u, \hat{g}_2).
\end{array} \quad (1.25)$$

is an observer for (1.20) where $f_1(x_1, x_2, u) = f_1(x_1, u)$.

$\square$

In view of these conditions, and using available observers for systems in some particular forms, one might be able to design observers for further nonlinear systems.

As an example, one can obtain in this way, and on the basis of observer (1.9) for system (1.8), a non uniform observer for a class of *cascade block state affine* systems of the following form:

$$\Sigma_c \left\{ \begin{array}{rcl}
\dot{x}_1 & = & A_1(u, y)x_1 + B_1(u, y) \\
\dot{x}_2 & = & A_2(u, y, x_1)x_2 + B_2(u, y, x_1) \\
& \vdots & \\
\dot{x}_q & = & A_q(u, y, x_1, \ldots x_{q-1})x_q + B_q(u, y, x_1, \ldots x_{q-1}) \\
y_1 & = & C_1 x_1 \\
& \vdots & \\
y_q & = & C_q x_q
\end{array} \right. \quad (1.26)$$

where $x_i \in \mathbb{R}^{n_i}, y_i \in \mathbb{R}^{n_i}, u \in \mathbb{R}^m, y = (y_1^T, \ldots y_q^T)^T = Cx$ and A_i, φ_i are continuous functions. Here the stability of the interconnected observer can only be guaranteed provided the inputs are "rich enough".

Denoting by $\chi_u^i(t, x_0)$ the projection of the solution onto $\mathbb{R}^{n_1 + \ldots n_i}$ which takes components from 1 to $n_1 + \ldots n_i$ of x, $\omega_i(t, u, x_0)$ the extended input

$$\begin{pmatrix} u \\ C\chi_u(t, x^0) \\ \chi_u^i(t, x^0)) \end{pmatrix},$$

and $\mathcal{B}(u)$ the set $\{x \in \mathbb{R}^n : \|\chi(t, u, x)\| < \infty, \quad t \in [0, +\infty[\ \}$ of initial conditions generating bounded trajectories with u, we define this "richness" as:

Definition 1.8 *Given $E \subset \mathbb{R}^n$, an input u will be said to be $E-$regularly persistent for Σ_c if for any compact K of E such that $K \cap \mathcal{B}(u) \neq \emptyset$ there exist $t_0 > 0, \alpha > 0, T > 0$ such that:*

$$\forall t \geq t_0; \forall x \in K \cap \mathcal{B}(u), \quad \forall i = 1 \text{ to } q-1, \quad \gamma(t, T, \omega_i(t, u, x)) \geq \alpha.$$

Theorem 1.6 *[4] Given $E \subset \mathbb{R}^n$, assume that:*

- *For $i = 2, \ldots q$, A_i, φ_i are globally Lipschitz w.r.t. $(x_1, \ldots x_{i-1})$ uniformly w.r.t. (u, y).*

- *Input u is E-regularly persistent for (1.26).*

- *$x(0) \in E \cap \mathcal{B}(u)$.*

then for any $\zeta > 0$, there exist $\theta_1 > 0, \ldots, \theta_q > 0$ and $\lambda > 0$ such that the following system:

$$\mathcal{O}_c \begin{cases} \dot{\hat{x}}_1 &= A_1(u,y)\hat{x}_1 + \varphi_1(u,y) - \hat{S}_1^{-1}C_1^T(C_1\hat{x}_1 - y_1) \\ \dot{\hat{x}}_2 &= A_2(u,y,\hat{x}_1)\hat{x}_2 + \varphi_2(u,y,\hat{x}_1) - \hat{S}_2^{-1}C_2^T(C_2\hat{x}_2 - y_2) \\ &\vdots \\ \dot{\hat{x}}_q &= A_q(u,y,\hat{x}_1,\ldots,\hat{x}_{q-1})\hat{x}_q + \varphi_q(u,y,\hat{x}_1,\ldots,\hat{x}_{q-1}) - \hat{S}_q^{-1}C_q^T(C_q\hat{x}_q - y_q) \\ \dot{\hat{S}}_1 &= -\theta_1\hat{S}_1 - A_1^T(u,y)\hat{S}_1 - \hat{S}_1 A_1(u,y) + C_1^T C_1, \quad \hat{S}_1(0) > 0 \\ \dot{\hat{S}}_2 &= -\theta_2\hat{S}_2 - A_2^T(u,y,\hat{x}_1)\hat{S}_2 - \hat{S}_2 A_2(u,y,\hat{x}_1) + C_2^T C_2, \quad \hat{S}_2(0) > 0 \\ &\vdots \\ \dot{\hat{S}}_q &= -\theta_q\hat{S}_q - A_q^T(u,y,\hat{x}_1,\ldots,\hat{x}_{q-1})\hat{S}_q - \hat{S}_q A_q(u,y,\hat{x}_1,\ldots,\hat{x}_{q-1}) + C_q^T C_q, \\ & \qquad \hat{S}_q(0) > 0 \end{cases}$$

(1.27)

is an observer for (1.26), with: $\|\hat{x}(t) - x(t)\| \leq \lambda e^{-\zeta t}$.

$\square$

The above design illustrates the case of cascade interconnection.
As an example of "full interconnection" of observers, let us consider a system of the following form:

$$\begin{aligned} \dot{x}_1 &= A_1 x_1 + f_1(x_1, u) + g_1(x_1, x_2, u); & x_1 \in \mathbb{R}^{n_1} \\ \dot{x}_2 &= A_2(u)x_2 + f_2(x_1, u) := \varphi(x_2, x_1, u); & x_2 \in \mathbb{R}^{n_2} \\ y_1 &= C_1 x_1 \in \mathbb{R} \\ y_2 &= C_2 x_2 \in \mathbb{R}^p \end{aligned}$$

(1.28)

with:

(C1) A, C as in (1.10) and $f(x_1, u)$ satisfying uniform observability and Lipschitz assumptions;

(C2) $g(x_1, x_2, u) = (0, \ldots 0, g_n(x_1, x_2, u))^T$ and g_n is globally Lipschitz w.r.t. x_1 (resp. x_2), uniformly w.r.t. (x_2, u) (resp. (x_1, u));

(C3) f_2 is globally Lipschitz w.r.t. x_1, uniformly w.r.t. u.

For such a system, one can easily identify two subsystems for which (1.11) and (1.9) are candidate sub-observers, and on the basis of the associated Lyapunov functions [16, 17], one can check that as soon as u is regularly persistent for $\dot{z} = A_2(u)z$, conditions of theorem 1.4 can be satisfied for θ_1 large enough.

This gives an observer of the following form:

$$\begin{aligned}
\dot{\hat{x}}_1 &= A_1\hat{x}_1 + f_1(\hat{x}_1, u) + g_1(\hat{x}_1, \hat{x}_2, u) - S_1^{-1}C_1^T(C_1\hat{x}_1 - y_1); \\
\dot{\hat{x}}_2 &= A_2(u)\hat{x}_2 + f_2(\hat{x}_1, u) - S_2^{-1}C_2^T(C_2\hat{x}_2 - y_2); \\
0 &= -\theta_1 S_1 - A_1^T S_1 - S_1 A_1 + C_1^T C_1 \\
\dot{S}_2 &= -\theta_2 S_2 - A_2(u)^T S_2 - S_2 A_2(u) + C_2^T C_2; \quad S_2(0) > 0.
\end{aligned}$$

Notice that here the observer gain is non uniform due to the state affine part of the system. But one could imagine a similar case where some uniform gain can be used, provided that detectability is guaranteed.

As an example, consider system (1.28) again, now with $C_2 = 0$ and φ being some function now enjoying the following property:

(C3') φ is globally Lipschitz w.r.t. x_1, uniformly w.r.t. (u, x_2) and there exists V positive definite s.t. $\forall(\xi, e) \in \mathbb{R}^{n_2}, \|\frac{\partial V}{\partial e}\| \leq -\beta_2\|e\|$ and $\frac{\partial V}{\partial t}(t, e) + \frac{\partial V}{\partial e}(t, e)[\varphi(\xi + e, x(t), u(t)) - \varphi(\xi, x(t), u(t))] \leq -\beta_1\|e\|^2$, for every admissible input function u and absolutely continuous function x.

Then one can again check that under conditions (C1), (C2), (C3'), an observer can be obtained as follows [7]:

$$\begin{aligned}
\dot{\hat{x}}_1 &= A_1\hat{x}_1 + f_1(\hat{x}_1, u) + g_1(\hat{x}_1, \hat{x}_2, u) - S_1^{-1}C_1^T(C_1\hat{x}_1 - y_1) \\
\dot{\hat{x}}_2 &= \varphi(\hat{x}_2, \hat{x}_1, u) \\
0 &= -\theta_1 S_1 - A_1^T S_1 - S_1 A_1 + C_1^T C_1.
\end{aligned} \tag{1.29}$$

5   State Transformations and Observer Design

One can notice that observer designs presented till now are all based on a particular structure of the system. The subsequent idea is that these designs also give state observers for systems which can be turned into one of these forms by change of state coordinates.

We will call *equivalent*, two systems related by such a relationship:

Definition 1.9 *Given $x_0 \in \mathbb{R}^n$, a system described by:*

$$\begin{cases} \dot{x} &= f(x, u) = f_u(x) \; x \in \mathbb{R}^n, u \in \mathbb{R}^m \\ y &= h(x) \in \mathbb{R}^p \end{cases} \tag{1.30}$$

will be said to be **equivalent at x_0** *to the system:*

$$\begin{cases} \dot{z} & = & F(z,u) = F_u(z) \\ y & = & H(z) \end{cases}$$

if there exists a diffeomorphism $z = \Phi(x)$ defined on some neighbourhood of x_0 such that:

$$\forall u \in I\!\!R^m, \quad \frac{\partial \Phi}{\partial x} f_u(x) \mid_{x=\Phi^{-1}(z)} = F_u(z) \qquad et \qquad h \circ \Phi^{-1} = H.$$

The interest of such a relationship for observer design is then motivated by the following proposition:

Proposition 1.2 *Given two systems (Σ_1) and (Σ_2) respectively defined by:*

$$(\Sigma_1) \begin{cases} \dot{x} & = & X(x,u) \\ y & = & h(x) \end{cases} \quad and \; (\Sigma_2) \begin{cases} \dot{z} & = & Z(z,u) \\ y & = & H(z) \end{cases}$$

and equivalent by $z = \Phi(x)$,

If:

$$(\mathcal{O}_2) \begin{cases} \dot{\hat{z}} & = & Z(\hat{z},u) + k(w, H(\hat{z}) - y)) \\ \dot{w} & = & F(w,u,y) \end{cases}$$

is an observer for (Σ_2),

Then:

$$(\mathcal{O}_2) \begin{cases} \dot{\hat{x}} & = & X(\hat{x},u) + \left(\dfrac{\partial \Phi}{\partial x}\right)^{-1}_{\mid \hat{x}} k(w, h(\hat{x}) - y) \\ \dot{w} & = & F(w,u,y) \end{cases}$$

is an observer for (Σ_1).

This kind of remark has motivated various works on characterizing (*rank observable*) systems which can be turned into some "canonical form" for observer design, from the linear one up to output injection [23, 8, 24] to several forms of cascade block state affine systems up to nonlinear injections from block to block [3, 26, 4, 5].

Using the formalism of differential forms [9] e.g. used in [19], we can indeed characterize systems equivalent to "special forms" of (1.26) (a general characterization would further require the use of explicit PDE's in its formulation).

With the following notations:

- $d, L_Z, i_Z, \wedge$ to respectively denote usual differentiation, Lie derivative along a vector field Z, inner product with Z and exterior product of differential forms;

- $d^r v := dv_1 \wedge \ldots \wedge dv_r$, and $i_X v := (i_X v_1, \ldots i_X v_r)$ if $v = (v_1, \ldots v_r)$;

- $i_X\Omega := \{i_X\omega, \omega \in \Omega\}$, $\bigwedge^d\Omega := \{\omega_1 \wedge \ldots \omega_d, \omega_i \in \Omega\}$, $d\Omega := \{d\omega, \omega \in \Omega\}$ if Ω is a set of differential forms;

- $\Omega \otimes \Theta$ to denote the set of finite linear combinations of elements of Ω with coefficients in Θ;

- $i_X\omega := i_{X^1}\ldots i_{X^r}\omega$ if ω is an $(r+1)$-differential form, and $X = (X^1, \ldots X^r)$ an r-tuple of vector fields;

- $\mathcal{O}(y)$ to denote the *observability subspace* of the considered system with output y, namely the smallest vectorial subspace of $I\!R^n$ which contains all output functions, and is invariant under Lie derivation along the vector fields of the system, obtained when u describes $I\!R^m$;

let us privilege systems $(\Sigma^{\nu_1 \ldots \nu_q}_{n_1 \ldots n_q})$ of the following form:

$$
(\Sigma^{\nu_1 \ldots \nu_q}_{n_1 \ldots n_q})
\begin{cases}
\dot{z}_1 &= A_1(u, y^1)z_1 + \varphi_1(u, y^1) \\
\dot{z}_2 &= A_2(u, y^2, z_1)z_2 + \varphi_2(u, y^2, z_1) \\
&\vdots \\
\dot{z}_q &= A_q(u, y^q, z_1, \ldots z_{q-1})z_q + \varphi_q(u, y^q, \ldots z_{q-1}) \\
y &= \begin{pmatrix} C_1 z_1 \\ \vdots \\ C_q z_q \end{pmatrix} = \begin{pmatrix} y^1 \\ \vdots \\ y^q \end{pmatrix} \\
& u \in I\!R^m, z_i \in I\!R^{n_i}, y^i \in I\!R^{\nu_i},
\end{cases}
\tag{1.31}
$$

assumed to satisfy the following *cascade rank observability condition* at x_0: for any x in some neighbourhood of x_0,

$$
\begin{aligned}
dim d\mathcal{O}(y^1)(x) &= n_1 \\
dim d\mathcal{O}(y^2) \wedge d^{n_1}\mathcal{O}(y^1)(x) &= n_2 \\
&\vdots \\
dim d\mathcal{O}(y^q) \wedge d^{n_1}\mathcal{O}(y^1) \ldots \wedge d^{n_{q-1}}\mathcal{O}(y^{q-1})(x) &= n_q
\end{aligned}
\tag{1.32}
$$

where $n_i \in I\!N^*$ such that $\sum_{i=1}^q n_i = n$. We will call those integers *cascade observability indices*.

This construction means that in (1.31), variables z_i of each block are exclusively "observed" by output h_i, as soon as $z_1, \ldots z_{i-1}$ are known . The characterization of such systems will use the following tools:

Given $n_m, \nu_m, q \in I\!N^*$, ν_m-tuples of functions $y^m = (y^m_1, \ldots y^m_{\nu_m})$ and ν_m-tuples of vector fields X^m, for $1 \le m \le q$, we define:

- $\mathcal{H}(y^m)$ the space such that $d\mathcal{H}(y^m) \wedge d^{\nu_m}y^m = 0$,

- $\begin{cases} \Omega^{X^m}_1(y^m) = Span_{I\!R}\{dL_{f_u}(y^m_j) \wedge d^{\nu_m}y^m, u \in I\!R^m, 1 \le j \le \nu_m\} \otimes \mathcal{H}(y^m) \\ \Omega^{X^m}_{k+1}(y^m) = Span_{I\!R}\{di_{f_u}i_{X^m}(\omega) \wedge d^{\nu_m}y^m, u \in I\!R^m, \\ \qquad\qquad\qquad\qquad \omega \in \Omega^{X^m}_k(y^m)\} \otimes \mathcal{H}(y^m) \\ \Omega^{X^m}(y^m) = \sum_{k=1}^\infty \Omega^{X^m}_k(y^m) \end{cases}$

- $\Theta_{n_1..n_m}^{y^1...y^{m+1}}$ the space of functions such that:

$$\bigwedge_{j=1}^{m}(d^{\nu_j}y^j \wedge \bigwedge^{n_j-\nu_j} i_{X^j}\Omega^{X^j}(y^j)) \wedge d^{\nu_{m+1}}y^{m+1} \wedge d\Theta_{n_1..n_m}^{y^1...y^{m+1}} = 0,$$

- $\Omega^{y^1...y^m}(\Sigma)$ the space $\bigwedge_{j=1}^{m-1}(d^{\nu_j}y^j \wedge \bigwedge^{n_j-\nu_j} i_{X^j}\Omega^{X^j}(y^j)) \wedge \Omega^{X^m}(y^m) \otimes \Theta_{n_1..n_{m-1}}^{y^1...y^m}$ for $m > 1$ and $\Omega^{X^1}(y^1) \otimes \mathcal{H}(y^1)$ for $m = 1$.

These definitions associate to each block (m) - corresponding to a ν_m-tuple of outputs y^m - the set of functions of these outputs $\mathcal{H}(y^m)$, a space of differential (ν_m+1)−forms Ω^{X^m}, a set of functions $\Theta_{n_1..n_m}^{y^1...y^{m+1}}$ gathering all functions of z_1 to z_{m-1} and a module to characterize the state affine structure of each block.
We can then state:

Theorem 1.7 *A nonlinear system (1.1) cascade observable w.r.t. outputs* $h = (y^1,...y^q)$ *in the sense of (1.32) is equivalent at* x_0 *to a system* $(\Sigma_{n_1...n_q}^{\nu_1...\nu_q})$ *described by (1.31)* **if and only if** $y^i \in \mathbb{R}^{\nu_i}$, $(n_1,...n_q)$ *are cascade observability indices of (1.1) and there exist q ν_m-tuples of vector fields* $X^m = (X_1^m,...X_{\nu_m}^m), 1 \le m \le q$, *such that, for* $1 \le m \le q$:

1. $L_{X_j^m}(y_k^m) = 0$ *if* $j \ne k$ *and 1 otherwise, for* $1 \le j,k \le p$

2. $dim(\Omega^{y^1...y^m}(\Sigma)) = n_m - \nu_m$ *on* $\Theta_{n_1...n_{m-1}}^{y^1...y^m}$

3. $\begin{cases} di_{X^m}\Omega^{X^m}(y^m) \wedge \bigwedge_{j=1}^{m}(\bigwedge^{n_j-\nu_j} i_{X^j}\Omega^{X^j}(y^j) \wedge d^{\nu_j}y^j) = 0 \\ di_{X^m}\Omega^{X^m}(y^m) \wedge \bigwedge_{j=1}^{m-1}(\bigwedge^{n_j-\nu_j} i_{X^j}\Omega^{X^j}(y^j) \wedge d^{\nu_j}y^j) \\ \subset \sum_{l=1}^{\nu_m} i_{X^m}\Omega^{X^m}(y^m) \wedge dh_l \wedge \bigwedge_{j=1}^{m-1}(\bigwedge^{n_j-\nu_j} i_{X^j}\Omega^{X^j}(y^j) \wedge d^{\nu_j}y^j) \otimes \Theta_{n_1...n_{m-1}}^{y^1...y^m} \end{cases}$

4. $\bigwedge_{j=1}^{q}(d^{\nu_j}y^j \wedge \bigwedge^{n_j-\nu_j} i_{X^j}\Omega^{X^j}(y^j))|_{x=x_0} \ne 0$ □

This statement follows previous results of [20], [4] or [6], and can be checked by the same kind of arguments: necessity is obtained by verifying that conditions 1 to 4 are indeed satisfied for a system of the form (1.31) with $X_i^m = \frac{\partial}{\partial y_i^m}$, and sufficiency is established by inductively defining new coordinates under the form $dz_j \wedge dz_{j-1} \wedge ... \wedge dz_1 = M_j(y^j, z_{j-1},...z_1)i_{X^j}\Omega^{X^j}(y^j) \wedge dz_{j-1} \wedge ... \wedge dz_1$ where M can be found on the basis of condition 3 along the same lines as in [6].

The problem in such a characterization is to find appropriate vector fields X^i. Let us sketch a constructive procedure giving such vector fields in the case of systems equivalent to (1.31) where each block takes the following form:

$$
\begin{aligned}
\dot{z}_{i1} &= A_{11}^i(u,\underline{z}_{i-1},y_i)+A_{12}^i(u,\underline{z}_{i-1})z_{i2}\\
\dot{z}_{i2} &= A_{21}^i(u,\underline{z}_{i-1},y_i)+A_{22}^i(u,\underline{z}_{i-1},y_i)z_{i2}+A_{23}^i(u,\underline{z}_{i-1})z_{i3}\\
\dot{z}_{i3} &= A_{31}^i(u,\underline{z}_{i-1},y_i)+A_{32}^i(u,\underline{z}_{i-1},y_i)z_{i2}+A_{33}^i(u,\underline{z}_{i-1},y_i)z_{i3}\\
&\quad +A_{34}^i(u,\underline{z}_{i-1})z_{i4}
\end{aligned}
$$

$$\vdots$$

$$
\begin{aligned}
\dot{z}_{in_i-1} &= A_{n_i-1,1}^i(u,\underline{z}_{i-1},y_i)+A_{n_i-1,2}^i(u,\underline{z}_{i-1},y_i)z_{i2}+\\
&\quad A_{n_i-1,3}^i(u,\underline{z}_{i-1},y_i)z_{i3}+A_{n_i-1,4}^i(u,\underline{z}_{i-1})z_{i4}\\
&\quad +...+A_{n_i-1,n_i}^i(u,\underline{z}_{i-1})z_{in_i}\\
\dot{z}_{in_i} &= A_{n_i1}^i(u,\underline{z}_{i-1},y_i)+A_{n_i2}^i(u,\underline{z}_{i-1},y_i)z_{i2}+\\
&\quad A_{n_i3}^i(u,\underline{z}_{i-1},y_i)z_{i3}+A_{n_i4}^i(u,\underline{z}_{i-1})z_{i4}+..+A_{n_in_i}^i(u,\underline{z}_{i-1})z_{in_i}\\
y_i &= C_iz_i=z_{i1}\in I\!R
\end{aligned}
$$

$$(1.33)$$

where $(z_{i1},\dots,z_{in_i})^T=z_i$ and $\underline{z}_{i-1}=(z_1^T,z_2^T,\dots z_{i-1}^T)^T$ for $i\geq 2$ and is empty otherwise.

Notice that any system equivalent to a form (1.31) where $A_i(u,\underline{z}_{i-1},y^i)=A_i(u,\underline{z}_{i-1})$ - as characterized in [4] - is equivalent to a form where each block has this triangular structure (1.33).

Notice also that, as in the case of state-affine equivalent systems where $A(u,y)$ does not depend on y, e.g. considered in [20], one can compute sets of constant control sequences I_k^i in the form $\{(u_{11}^i,\dots u_{1k}^i),\dots(u_{\nu_j1}^i,\dots u_{\nu_jk}^i)\}$ such that: $\{dL_{f_{I_k^1}}(h_1),\ k=0\dots r_1\}$ spans $d\mathcal{O}(h_1)$ and inductively, $\{dL_{f_{I_k^i}}(h_i),\ k=0\dots r_i\}\wedge\bigwedge_{l=1}^{i-1}d\mathcal{O}(h_l)$ spans $d\mathcal{O}(h_i)\wedge\bigwedge_{l=1}^{i-1}d\mathcal{O}(h_l)$ (where h_i denotes the output function for y_i and $L_{f_{I_k^i}}(h_i)$ is the vector of components $L_{f_{u_{l1}^i}}\dots L_{f_{u_{lk}^i}}(h_i)$).

On this basis, one can inductively compute candidates for X^1 to X^q on the same pattern, given hereafter for X^1:

- Compute Y (uniquely) defined by:

$$
\begin{aligned}
L_Y(h_1) &= 0,\quad \text{and for } j=1 \text{ to } p & (1.34)\\
L_YL_{f_{I_j^1}}(h_1) &= 0 \text{ if } j\neq r_1,\ 1 \text{ otherwise}; & (1.35)\\
L_YL_{f_{I_j^l}}(h_l) &= 0 \quad \text{for } j=1,..r_l,\ l=2,...q.
\end{aligned}
$$

- By successive Lie Brackets, compute $Y_{v_{p-1}\dots v_1}:=[f_{v_{p-1}},\dots[f_{v_1},Y]\dots]$ and $Y_{v_p\dots v_1}:=[f_{v_p},Y_{v_{p-1}\dots v_1}]$ for some constant v_i's, and set:

$$
Z:=Y_{v_p\dots v_1}+\frac{L_{f_{v_p}}(h_1)}{2}[Y_{v_{p-1}\dots v_1},Y_{v_p\dots v_1}]
$$

- Check: $dL_Z(h_1)\equiv 0$; $L_Z(h_1)\neq 0$ and finally set: $X^1:=\dfrac{1}{L_Z(h_1)}Z$.

One can check by inspection that such a construction necessarily gives a candidate for X^1. The same construction can be used to find X^2 to X^n and finally, verification of conditions of Theorem 1.7 reduces to differentiations and tests of linear dependencies. In this way, the particular structure (1.33) can be fully intrinsically characterized, as this is done for several cases in [6].

6   Conclusions

The purpose of this chapter was to draw some lines of recent advances in the problem of observer design for nonlinear systems and highlight several further directions of research. In particular, the problem of the input has been underlined for the observability properties of the systems, and several aspects of observer designs based on interconnection of sub-observers as well as state transformations have been discussed.

In terms of the technique used for the design, obviously further methods can be thought of, for instance including optimization [25], sliding modes [27] etc.

Acknowledgement

The author would like to thank Professors Hassan Hammouri and Guy Bornard for having awoken and fed his interest in nonlinear observers.

7   REFERENCES

[1] G. Besançon. *Contributions à l'Etude et à l'Observation des Systèmes Non Linéaires avec Recours au Calcul Formel.* PhD thesis, Institut National Polytechnique de Grenoble, 1996. Laboratoire d'Automatique de Grenoble.

[2] G. Besançon and H. Hammouri, "On uniform observation of non-uniformly observable systems," *Systems & Control Letters*, vol. 33, no. 1, pp. 1–11, 1996.

[3] G. Besançon and G. Bornard. "A condition for cascade time-varying linearization," in *IFAC Proc., Nonlinear Control Systems Design Symposium, Tahoe City, CA, USA*, pp. 684–689, 1995.

[4] G. Besançon, G. Bornard, and H. Hammouri. "Observer synthesis for a class of nonlinear control systems," *Europ. Journal of Control*, vol. 3, no. 1, pp. 176–193, 1996.

[5] G. Besançon and G. Bornard. "State equivalence based observer design for nonlinear control systems," in *Proc. IFAC World Congress, San Francisco, CA, USA*, pp. 287–292, 1996.

[6] G. Besançon and G. Bornard. "On characterizing classes of observer forms for nonlinear systems," in *Proc. 4th European Control Conf., Brussels, Belgium*, 1997.

[7] G. Besançon and H. Hammouri. "On observer design for interconnected systems," *Journal of Mathematical Systems, Estimation, & Control*, vol. 8, no. 3, 1998.

[8] D. Bestle and M. Zeitz. "Canonical form observer design for nonlinear time-variable systems," *Int. Journal of Control*, vol. 38, no. 2, pp. 419–431, 1983.

[9] W. M. Boothby. *An Introduction to Differentiable Manifolds and Riemannian Geometry*. Academic Press, New York, 1975.

[10] G. Bornard, N. Couenne and F. Celle. "Regularly persistent observer for bilinear systems," in *Proc. of the Colloque International en Automatique Non Linéaire, Nantes*, June 1988.

[11] G. Bornard, F. Celle-Couenne and G. Gilles. "Observability and observers," in *Nonlinear Systems - T.1, Modeling and Estimation*, pp. 173–216, Chapman & Hall, London, 1995.

[12] K. Busawon. *Sur les Observateurs pour des Systèmes Non Linéaires et le Principe de Séparation*. PhD thesis, Université Claude Bernard, Lyon I, 1996.

[13] D. Dawson, Z. Qu and J. Carroll. "On the state observation and output feedback problems for nonlinear uncertain systems," *Systems & Control Letters*, vol. 18, pp. 217–222, 1992.

[14] J.P. Gauthier and G. Bornard. "Observability for any $u(t)$ of a class of nonlinear systems," *IEEE Trans. on Automatic Control*, vol. 26, no. 4, pp. 922–926, 1981.

[15] J.P. Gauthier and A. Kupka. "Observability and observers for nonlinear systems," *Siam Journal on Control and Optimization*, vol. 32, no. 4, pp. 975–994, 1994.

[16] J.P. Gauthier, H. Hammouri and S. Othman. "A simple observer for nonlinear systems - applications to bioreactors," *IEEE Trans. on Automatic Control*, vol. 37, no. 6, pp. 875–880, 1992.

[17] H. Hammouri and J. D. L. Morales. "Observer synthesis for state-affine systems," in *Proc. 29th IEEE Conf. on Decision and Control, Honolulu, Hawaii*, pp. 784–785, 1990.

[18] H. Hammouri and F. Celle. "Some results about nonlinear systems equivalence for the observer synthesis," in *New Trends in Systems Theory*, pp. 332–339, Birkhäuser, 1991.

[19] H. Hammouri and J.P. Gauthier. "Bilinearization up to output injection," *Systems & Control Letters*, vol. 11, pp. 139–149, 1988.

[20] H. Hammouri and M. Kinnaert. "A new formulation for time-varying linearization up to output injection," *Systems & Control Letters*, vol. 28, pp. 151–157, 1996.

[21] S. Hara and K. Furuta. "Minimal order state observers for bilinear systems," *Int. Journal of Control*, vol. 24, no. 5, pp. 705–718, 1976.

[22] R. Hermann and A. Krener. "Nonlinear controllability and observability," *IEEE Trans. on Automatic Control*, vol. 22, no. 5, pp. 728–740, 1977.

[23] A. J. Krener and A. Isidori. "Linearization by output injection and nonlinear observers," *Systems & Control Letters*, vol. 3, pp. 47–52, 1983.

[24] A. J. Krener and W. Respondek. "Nonlinear observers with linearizable error dynamics," *Siam Journal on Control and Optimization*, vol. 23, no. 2, pp. 197–216, 1985.

[25] H. Michalska and D. Mayne. "Moving horizon observers," in *IFAC Proc., Nonlinear Control Systems Design Symposium, Bordeaux, France*, pp. 576–581, June 1992.

[26] J. Rudolph and M. Zeitz. "A block triangular nonlinear observer normal form," *Systems & Control Letters*, vol. 23, pp. 1–8, 1994.

[27] J. J. E. Slotine, J. Hedrick and E. Misawa. "On sliding observers for nonlinear systems," *Journal of Dynamic Systems, Measurements, and Control*, vol. 109, pp. 245–252, 1987.

[28] J. Tu and J. Stein. "Model error compensation and robust observer design - part 2: Bearing temperature and preload estimation," in *Proc. American Control Conference, Baltimore, Maryland, USA*, pp. 3308–3312, June 1994.

[29] B. Walcott and S. Żak. "State observation of nonlinear uncertain dynamical systems," *IEEE Trans. on Automatic Control*, vol. 32, no. 2, pp. 166–170, 1987.

Model-Based Observers for Tire/Road Contact Friction Prediction

Carlos Canudas de Wit[1], Roberto Horowitz[2] and P. Tsiotras[3]

[1]Laboratoire d´Automatique de Grenoble, UMR CNRS 5528
ENSIEG-INPG, ST. Martin d´Hères, France.
[2]Department of Mechanical Engineering, University of California
Berkeley, CA 94720-1740, U.S.A.
[3]Georgia Institute of Techology, School of Aerospace Eng. Atlanta,
Georgia 30332-0150, U.S.A.

1 Introduction

This contribution is devoted to the problem of tire-road friction estimation. The need for such type of studies, steers from the difficulty of direct sensing of tire forces, slip, slip angles and other external factors. Observer algorithms are, in this context, a low cost alternative for sensors. Tire forces information is relevant to problems like: optimization of Anti-look brake systems (ABS), traction system, diagnostic of the road friction conditions, etc.

Literature for tire/road friction estimation is numerous. Bakker *et al* [1] and Burckhardt [4] describe two analytical models for tire/road behavior that are intensively used by researchers in the field. In these two models the coefficient of friction, μ, or more precisely, the normalized friction force, i.e.

$$\mu = \frac{F}{F_n} = \frac{\text{Friction force}}{\text{Normal force}}$$

is mainly determined based on the wheel slip s and some other parameters like speed and normal load. Fig. 1 shows two curves, obtained from Harned *et al* [9], that represent typical μ versus s behavior.

It is current practice to name the ratio between the friction and the normal forces, μ, as being the "coefficient" of friction. Under constant normal force conditions, μ, is a constant if and only if the Coulomb model is used to describe friction. Nevertheless, the Coulomb model is too simplistic to suitable represent forces between the rubber tire and the road, which are dominated by the elesto-plastic force/displacement characteristics. Therefore, to consiser μ as a constant is a pure idealistic view. μ should thus

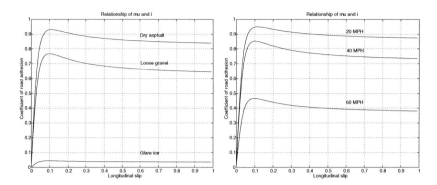

FIGURE 1. a) Variations between coefficient of road adhesion μ and longitudinal slip s for different road surface conditions (left). b) Variations between coefficient of road adhesion μ and longitudinal slip s for different vehicle velocities (right).

be viewed more as the ratio between friction and normal forces (i.e. the normalized force), which is indeed a (static or dynamic) function of the system state variables.

The expression given by Bakker *et al* [1], and Paceijka and Sharp [14], also known as "magic formula" is derived heuristically from experimental data to produce a good fit. It provides the tire/road coefficient of friction μ as a function of the slip s. The expression in Burckhardt [4] is derived with a similar methodology. The final map expresses μ as a function of s, the vehicle velocity, v and the normal load on the tire F_n.

Kiencke [10] presents a procedure for real-time estimation of μ. A simplification to the analytical model by Burckhardt [4] is introduced in such a way that the relation between μ and s is linear in the parameters. Kiencke [10] uses a two stages identification algorithm. In the first stage, the value of μ is estimated. This estimate of μ is used in the second stage to obtain the parameters for the simplified μ versus s curve.

The paper by Gustafsson [8] derives an scheme to identify different classes of roads. He assumes that by combining the slip and the initial slope of the μ versus s curve it is possible to distinguish between different road surfaces. The author tests for asphalt, wet asphalt, snow and ice and identifies the actual value of the slope with a Kalman filter and a least square algorithm.

Ray [16] estimates μ based on a different approach. Instead of using the slip information to derive a characteristic curve, Ray [16] estimates the forces on the tires with an extended Kalman filter. Using a tire model introduced by Szostak *et al* [17], that expresses the tire forces as a function of μ, the author tries this model for different values of μ. A Bayesian approach is used to determine the value of μ that is most likely to produce the forces estimated with the extended Kalman filter.

The works of Kiencke [10], Gustafsson [8], and Ray [16] do not consider any velocity dependence in the derivation of μ, as suggested by Burckhardt

[4] and Harned *et al* [9]. An attempt to consider the velocity dependence for ABS control is presented in Liu and Sun [13]. The authors assume the tire/road characteristics to be known. Due to the limitations in the available data, the authors are not able to compare their algorithm with other methods.

There are other works related to the on line identification of the tire/road friction, as for example Lee and Tomizuka [12], and Yi and Jeong [18]. However, in these papers only the instantaneous coefficient of friction is identified.

The coefficient of tire/road friction, or coefficient of road adhesion, μ is mainly a function of the longitudinal slip, the velocity of the vehicle and the normal load.

The estimators proposed in the literature depends very much on the type of used models, and verification of the hypothesis used for the model derivation. As shown by the figures above, the relation of the curves $\mu - s$, depends very much on system operating conditions, such as the vehicle velocity. It is clear that parameters describing a curve like the one in Fig .1-(a), will not be invariant, as shown in Fig .1-(b). It is thus interesting to introduce models described by parameters that are more likely to be invariant and have physical significance. Theory never exactly matches reality, but some times closely resembles it.

To achieve this goal, we propose in this paper to use a dynamical tire/road friction model, together with a nonlinear observer specifically designed for this application. This paper is organized as follows: The next section reviews some of the existing tire/road friction models, and also introduces lumped and distributed dynamic representations. In Section 3 we set-up the observation problem, using the particular case of a one-wheel system with lumped contact friction. Inspired from previous works by Canudas-de-Wit and Lischinsky [6] on adaptive friction estimation and compensation, Section 4 presents a general framework for the design of nonlinear observers for the on-line estimation of the road conditions. In Section 5 we apply this design to the case study case set in Section 3. Finally, Section 6 presents simulation results.

2   Tire-road Friction Models

This section reviews some friction models that can be used for the study of the on-line identification of the friction force (or coefficient, if we consider normalized force). We first present the some of the pseudo steady-state models proposed in the literature, then we discuss some alternative dynamic (lumped and a distributed) models.

The sep up for this study is the simple case of an one-wheel model with tire-road contact friction, shown schematically in Fig. 2. In this study we

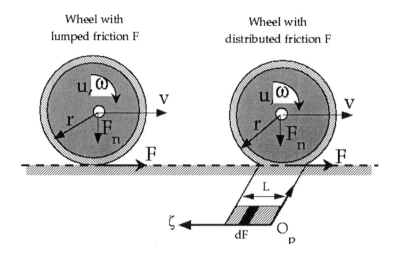

FIGURE 2. One-wheel system with: lumped friction (left), distributed friction (right)

will thus consider a system of the form

$$m\dot{v} = F \tag{2.1}$$

$$J\dot{\omega} = -rF + u_\tau - \sigma_\omega\omega, \tag{2.2}$$

where:

m – wheel mass,

J – wheel inertia,

r – wheel radius,

v – linear velocity,

ω – angular velocity,

u_τ – braking/driving torque,

F – tire/road friction force.

Therefore, only longitudinal motion (longitudinal slip) will be considered.

2.1 Pseudo-Steady State Models

This type of models are currently used in the literature. They are defined as one-to-one (memory less) maps between the friction F, and the longitudinal slip rate s, defined as:

$$s = \begin{cases} \frac{v - r\omega}{v} & \text{if } v > r\omega, v \neq 0 \quad \text{breaking} \\ \frac{r\omega - v}{r\omega} & \text{if } v < r\omega, \omega \neq 0 \quad \text{driving} \end{cases} \tag{2.3}$$

The slip rate results from the reduction of the effective circumference of the tire (consequence of the tread deformation due to the elasticity of the tire rubber), which implies that the ground velocity will not be equal to $v = r\omega$. The slip rate is defined in the interval $[0, 1]$. When $s = 0$ there is no sliding (pure rolling), whereas $s = 1$ indicates full sliding.

In opposition to steady-state friction models, the pseudo steady-state (PSS) models aim at describing the shapes shown in Fig.1 via static maps $F(s)$, mapping s to F, and not as a steady-state relation between contact relative velocity and friction forces. They are named pseudo steady-state because some of these models depend on the vehicle velocity v, i.e. $F(s, v)$. The models also depends on the operating conditions, although they are only valid under steady-state conditions.

One of the most well known models of this type is Pacejka's model (see, Pacejka and Sharp [14]), also known by the name of "magic formula". This model has been shown to suitably match experimental data, obtained under particular conditions of constant linear and angular velocity. The Pacejka model has the form

$$F(s) = c_1 \sin(c_2 \arctan(c_3 s - c_4(c_3 s - \arctan(c_3 s)))),$$

where the $c_i's$ are the parameters characterizing this model. The parameters can be identified by matching experimental data, as shown in Bakker et al [1].

The model proposed by Burckhardt [4] for the tire/road friction characteristics is of the form

$$F(s, v) = \left(c_1(1 - e^{-c_2 s}) - c_3 s\right) e^{-c_4 v}, \tag{2.4}$$

where $c_1, \cdots, c_4$ are constants. The normal load at the tire is kept constant in this model. Note also the velocity dependency of this model, seeking to match variations like the one shown in Fig. 1-(b).

Kiencke and Daiss [11] neglect the velocity dependent term in Eq. (2.4) and approximate the curve by

$$F(s) = K_s \frac{s}{c_1 s^2 + c_2 s + 1}, \tag{2.5}$$

where K_s is the slope of the $F(s)$ versus s curve when $s = 0$ and c_1 and c_2 are properly chosen parameters. Notice that Eq. (2.5) is only dependent on the slip s. The value of K_s is assumed to be known. Kiencke and Daiss [11] choose a fixed value of about 30° for it.

Alternative, Burckhardt [3] proposes a simpler three parameters model,

$$F(s) = c_1(1 - e^{-c_2 s}) - c_3 s.$$

Since these models are highly nonlinear in the unknown parameters, they are not well adapted to be used for on-line identification. For this reason, simplified models like

$$F(s) = c_1 \sqrt{s} - c_2 s$$

are used in connection with a linear recursive identification algorithms, has been proposed in the literature.

A part from the nonlinearity in the unknown parameters, the major limitation of this models seems to steer from the fact that the unknown parameters are not really invariant, they may strongly depend on the tire characteristics (such as compound, tread type, tread depth, inflation pressure, temperature), on the road conditions (such as type of surface, texture, drainage, capacity, temperature, lubricant, i.e. water or snow), and on the vehicle operational conditions (velocity, load), see Pasterkamp and Paceijka [15].

As an alternative to the pseudo static models that depend instantaneously only on s (memory-less models), dynamic models based on the preliminary studies on dynamic friction models of Dahl [7], can be adapted to suitably describe the road-tire contact friction. The Dahl's models leads to a friction displacement relation that bears much resemblance with stress-strain relations proposed in classical solid mechanics.

A potential advantage of such models is their ability to describe some of the physical phenomena found in road/tire friction (such as: hysteresis loops, pre-sliding displacement, etc), as well as their dependance on parameters that may have physical meaning. Although the parameters of such models, may also depend on some of the factors mentioned above, other parameters may be more like to be invariant, or be more directly related with the phenomena to be observed, like for instance the change on the road characteristics (i.e. dry wet, etc.). Dynamic models can be formulated as a lumped or distributed models, as shown in Fig. 2. This distinction will be discussed next.

2.2   Lumped Dynamic Models

A lumped friction model assumes punctual tire-road friction contact. An example of such a model can be derived from the LuGre model (see Canudas et al, [5]). This model differs from the one in [5] in the way that the function $g(v)$ is defined. Here we propose to use the term $e^{-|v_r/v_s|^{1/2}}$ instead the term $e^{-(v_r/v_s)^2}$ as in the LuGre model in order to better match the pseudo-stationary characteristic of this model (map $s \mapsto F(s)$) with the shape of the Paceijka's model, as it will be shown later.

The model of Canudas et al, [5] is written:

$$\dot{z} = v_r - \frac{\sigma_0 |v_r|}{g(v_r)} z \tag{2.6}$$

$$F = (\sigma_0 z + \sigma_1 \dot{z} + \sigma_2 v_r) F_n \tag{2.7}$$

with,

$$g(v_r) = \mu_C + (\mu_S - \mu_C)e^{-|v_r/v_s|^{1/2}}$$

where,

σ_0 – rubber longitudinal lumped stiffness,

σ_1 – rubber longitudinal lumped damping,

σ_2 – viscous relative damping,

μ_C – normalized Coulomb friction,

μ_S – normalized Static friction, $\mu_C \leq \mu_S, \in [0, 1]$,

v_S – Stribeck relative velocity,

F_n – normal force,

v_r – relative velocity $= (r\omega - v)$,

z – internal friction state.

Remark: This model, has the following important properties:

(i) if $|z(0)| < \mu_S/\sigma_0$, thus $|z(t)| < \mu_S/\sigma_0$, $\forall t \geq 0$,

(ii) $\infty > \mu_S \geq g(v_r) \geq \mu_C > 0$, $\forall v_r$

(iii) the right hand side of (2.6) is Lipschitz (globally if v_r is assumed bounded, and locally if not).

In particular Property (i) ensures that the internal friction states are bounded and that its upper bound is given by the static friction parameter (Property (ii). Property (iii), provides existence and uniqueness of a solution to (2.6).

2.3   Distributed Dynamic Models

Distributed models assume the existence of an area of contact (or patch) between the tire and the road, as shown in Fig. 2. This patch represent the projection of the part of the tire that is in contact with the road. The contact patch is associated to the frame O_p, with ζ as the axis coordinate. The patch length is L.

Distributed dynamical models, as well as their relation with the pseudo-static models, has been studied previously in works of Bliman *et al* [2]. They propose second order rate independent model (similar to the Dahl ones), and have shown that, under constant v and ω, there exist a choice of parameters that closely match a curve similar to the one characterizing the magic formula.

Similar results can be obtained by using a model based in the first-order LuGre friction model, i.e.

$$\frac{d\,\delta z}{dt}(\zeta, t) = v_r - \frac{\sigma_0|v_r|}{Lg(v_r)}\delta z \tag{2.8}$$

$$F = \int_0^L dF(\zeta, t)d\zeta, \tag{2.9}$$

with $g(v_r)$ defined as before and

$$dF = \left(\frac{\sigma_0}{L}\delta z + \sigma_1\,\delta\dot{z} + \sigma_2 v_r\right)dF_n,$$

where,

σ_0/L – rubber longitudinal distributed stiffness per length,

dF – differential friction force,

dF_n – distributed normal force $[F_n/L]$,

v_r – relative velocity $= (r\omega - v)$,

δz – differential internal friction state.

Note that in this formulation the differential internal friction state $\delta z(\zeta, t)$, depends on both time t, and space ζ. Indeed, Eq. (2.8) describes a partial differential equation.

$$\frac{d\,\delta z}{dt}(\zeta, t) = \frac{\partial\,\delta z}{\partial\zeta}(\zeta, t) + \frac{\partial\,\delta z}{\partial t}(\zeta, t) = v_r - \frac{\sigma_0|v_r|}{Lg(v_r)}\delta z$$

2.4 Relation Between Distributed Dynamical Model and the Magic Formula

The linear motion of the differential dF in the patch frame O_p is $\dot{\zeta} = r\omega$, for positive ω, and $\dot{\zeta} = -r\omega$, for negative ω (the frame origin change location when the wheel velocity reverses). Hence $\dot{\zeta} = r|\omega|$. We can thus rewrite (2.8) in the ζ coordinates as

$$\frac{d\,\delta z}{dt}(\zeta, t) = \frac{d\,\delta z}{d\zeta}\frac{d\zeta}{dt}(\zeta, t) = \frac{d\,\delta z}{d\zeta}|r\omega| = v_r - \frac{\sigma_0|v_r|}{Lg(v_r)}\delta z \tag{2.10}$$

$$\frac{d\,\delta z}{d\zeta} = -\frac{\sigma_0|s|}{Lg(v_r)}\delta z + s\,\mathrm{sgn}(r\omega - v). \tag{2.11}$$

Assuming that v, and ω are constant (hence also v_r, and s), the above equation describes a linear space invariant system having the sign of the relative velocity as its input.

Considering a positive value for $\text{sgn}(r\omega - v)$ over the space interval $[\zeta(t_0), \zeta(t_1)]$, or equivalent over $[\zeta_0, \zeta_1]$, we have that the solution of the above equation is:

$$\delta z(\zeta_1) = \delta z(\zeta_0) e^{-\frac{\sigma_0 |s|}{Lg(v_r)}(\zeta_1 - \zeta_0)} + \frac{Lg(v_r)}{\sigma_0}\left(1 - e^{-\frac{\sigma_0 |s|}{Lg(v_r)}(\zeta_1 - \zeta_0)}\right)$$

Introducing this solution together with Eq. (2.11) in Eq. (2.9), and integrating with $\delta z(\zeta_0) = \zeta_0 = 0$, we have that $F(s)$, is given as:

For the driving case:

$$F(s,\omega) = F_n g(s)\left[1 - (1 - \sigma_1 s)\frac{g(s)}{\sigma_0 s}\left(e^{-\frac{\sigma_0 s}{g(s)}} - 1\right)\right] + \sigma_2 r\omega s \qquad (2.12)$$

with

$$g(s) = \mu_C + (\mu_S - \mu_C)e^{-|r\omega s/v_s|^{1/2}}$$

for some constant ω, and $s \in [0, 1]$.

For the breaking case:

$$F(s,v) = F_n g(s)\left[1 - (1 - \sigma_1 s)\frac{g(s)}{\sigma_0 s}\left(e^{-\frac{\sigma_0 s}{g(s)}} - 1\right)\right] + \sigma_2 vs \qquad (2.13)$$

with

$$g(s) = \mu_C + (\mu_S - \mu_C)e^{-|vs/v_s|^{1/2}}$$

for some constant v, and $s \in [0, 1]$.

Figure 3 shows the plot of $F(s)$ with the parameters shown in the table 2.1.

Parameter	Value	Unit
σ_0	40	[N/m]
σ_1	4.9487	[N $\cdot$ s/m]
σ_2	0.0018	[N$\cdot$ s/m]
μ_C	0.5	[-]
μ_S	0.9	[-]
v_s	12.5	[m/s]

TABLE 2.1. Data used for the plot shown in Fig. 3 and Fig. 4

Uncertainty in the knowledge of the function $g(v_r)$, can be modeled by introducing the parameter θ, as

$$g(v_r) = \frac{\tilde{g}(v_r)}{\theta},$$

where $\tilde{g}(v_r)$ is some nominal known value for $g(v_r)$. Computation of the function $F(s, \theta)$, from Eq. (2.13), as a function of θ, gives the curves shown

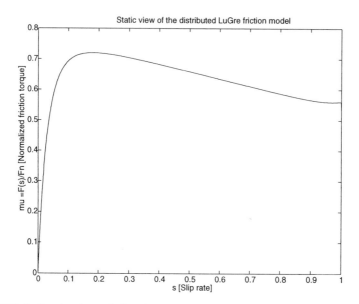

FIGURE 3. Static view of the distributed LuGre model (breaking case, with $v = 20m/s = 72Km/h$). This curve shows the normalized friction $\mu = F(s)/F_n$, as a function of the slip rate s.

in Fig. 4. These curves matches reasonable well the experimental data shown in Fig. 1-(a), for different coefficient of road adhesion. Hence, the parameter θ, suitable describes the changes in the road characteristics.

Note that the pseudo-static representation, Eqs. (2.12) and (2.13), does not depends on the patch length L. Hence, the parameters obtained by feeding this model to experimental data, can also be used in the simpler lumped model. This model will be used in the sequel for the observation problem to be defined next.

3   Problem Formulation

We consider the one-wheel model with the lumped tire/road friction model, i.e.

$$m\dot{v} = F_n(\sigma_0 z + \sigma_1 \dot{z}) + F_n \sigma_2 v_r \qquad (2.14)$$

$$J\dot{\omega} = -rF_n(\sigma_0 z + \sigma_1 \dot{z}) - \sigma_\omega \omega + u_\tau \qquad (2.15)$$

$$\dot{z} = v_r - \theta \frac{\sigma_0 |v_r|}{g(v_r)} z \qquad (2.16)$$

with,

$$g(v_r) = \mu_C + (\mu_S - \mu_C)e^{-|v_r/v_s|^{1/2}},$$

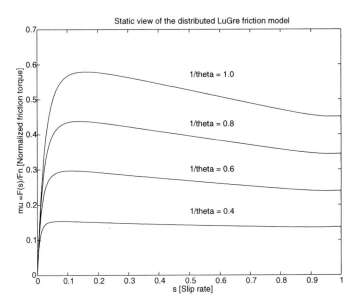

FIGURE 4. Static view of the distributed LuGre model, under different values for $1/\theta$. Breaking case, with $v = 20m/s = 72Km/h$. This curve shows the normalized friction $\mu = F(s)/F_n$, as a function of the slip rate s.

where we have neglected the term σ_2 in the equation (2.15), and introduced the parameter θ to capture variation and uncertainties in the function $g(v_r)$. The observation problem can be now formulated as follows.

Problem formulation: Assume that the lumped friction parameters with $\theta = 1$, has been identified off-line, and assume that the variable w, and v are measurable from some sensors (the need for the measure of v may be relaxed later). The problem is to design an on-line observer for θ, that allows the controller to monitor the eventual changes in the road conditions.

4   General Observer Design

Consider the following system:

$$\dot{x} = Ax + B\left[\theta\varphi(y,u,x)\right] + Ru + Ey \qquad (2.17)$$
$$\dot{\theta} = 0 \qquad (2.18)$$
$$y = C^T x \qquad (2.19)$$

with $y, \theta, \varphi(y,u,x) \in R, x \in R^n$, and $u \in R^m$.

We assume that system states are bounded, and that the following holds:

A1) (A, C) is an observable pair,

A2) One of the following properties holds for $\varphi(y, u, x)$:

There exist a known function $\infty > \rho_0 \geq \rho(y, u) \geq 0$, such that:

(i) $|\varphi(y, u, x_1) - \varphi(y, u, x_2)| \leq \rho(y, u)||x_1 - x_2||, \qquad \forall x_1, x_2,$

(ii) $|\varphi(y, u, x)| \leq \rho(y, u)||x||, \qquad \forall x,$

A3) The map $\psi \mapsto \tilde{y}$ of the system

$$\dot{\tilde{x}} = [A - KC^T]\tilde{x} + B\psi \qquad (2.20)$$
$$\tilde{y} = C^T \tilde{x} \qquad (2.21)$$

is strictly passive, i.e. $\forall Q > 0, \exists P = P^T > 0$, and K, such that

$$P[A - KC^T] + [A - KC^T]^T P = -Q \qquad (2.22)$$
$$PB = C. \qquad (2.23)$$

A4) The trajectories of the system $(y(t), u(t), x(t))$, satisfy:

$$\lim_{t \to \infty} \varphi(y(t), u(t), x(t)) \neq 0$$

Under this hypothesis, we propose the following observer structure:

$$\dot{\hat{x}} = A\hat{x} + B\left[\hat{\theta}\varphi(y, u, \hat{x})\right] + Ru + Ey + K(y - \hat{y}) + B\nu_1 \quad (2.24)$$
$$\dot{\hat{\theta}} = \nu_2 \qquad (2.25)$$
$$\hat{y} = C^T \hat{x} \qquad (2.26)$$

where ν_1, and ν_2, are design variables, which will be defined subsequently. Introducing the error variables:

$$\tilde{x} = x - \hat{x} \qquad (2.27)$$
$$\tilde{\theta} = \theta - \hat{\theta} \qquad (2.28)$$
$$\tilde{y} = y - \hat{y} = C^T \tilde{x}, \qquad (2.29)$$

The error equation becomes

$$\dot{\tilde{x}} = [A - KC^T]\tilde{x} + B\left[\theta\varphi(y, u, x) - \hat{\theta}\varphi(y, u, \hat{x})\right] - B\nu_1 \quad (2.30)$$
$$\dot{\tilde{\theta}} = -\nu_2 \qquad (2.31)$$
$$\tilde{y} = C^T \tilde{x}, \qquad (2.32)$$

where,

$$\theta\varphi(y, u, x) - \hat{\theta}\varphi(y, u, \hat{x}) = \tilde{\theta}\varphi(y, u, \hat{x}) + \theta\left[\varphi(y, u, x) - \varphi(y, u, \hat{x})\right]$$

Now, defining the Lyapunov function

$$V = \tilde{x}^T P \tilde{x} + \frac{1}{\gamma} \tilde{\theta}^2$$

and using properties $A1$, and $A3$, we have

$$
\begin{aligned}
\dot{V} &= -\tilde{x}^T Q \tilde{x} + 2\tilde{\theta} \left[\tilde{y}\varphi(y,u,\hat{x}) - \gamma^{-1}\nu_2 \right] && (2.33) \\
&\quad +2\tilde{y}\theta \left[\varphi(y,u,x) - \varphi(y,u,\hat{x}) \right] - \tilde{y}\nu_1 && (2.34)
\end{aligned}
$$

Defining the adaptation law ν_2 as

$$\nu_2 = \gamma\varphi(y,u,\hat{x})\tilde{y} \qquad (2.35)$$

we obtain

$$\dot{V} \leq -\tilde{x}^T Q \tilde{x} + 2|\tilde{y}||\theta| \left[\varphi(y,u,x) - \varphi(y,u,\hat{x}) \right] - \tilde{y}\nu_1 . \qquad (2.36)$$

If $A2 - (i)$ holds, then we have,

$$
\begin{aligned}
\dot{V} &\leq -\tilde{x}^T Q \tilde{x} + 2|\tilde{y}||\theta|\rho(y,u)||\tilde{x}|| - \tilde{y}\nu_1 && (2.37) \\
&\leq -q||\tilde{x}||^2 + 2||C^T|||\theta|\rho(y,u)||\tilde{x}||^2 - \tilde{y}\nu_1 && (2.38) \\
&\leq -||\tilde{x}||^2(q - 2||C^T|||\theta|\rho_0) - \tilde{y}\nu_1 , && (2.39)
\end{aligned}
$$

where

$$q = \lambda_{min} Q .$$

Since (2.22) holds for any Q, the minimum eigenvalue of Q can be selected such that the term within the parenthesis of the last inequality is positive, i.e.

$$q = 2||C^T|||\theta|\rho_0 + q_0$$

with any $q_0 > 0$, and $\theta_{max} \geq |\theta|$. Note that a value of θ_{max} can be obtained from the knowledge on the road characteristics, as discussed in previous sections.

In this case we can simply set $\nu_1 = 0$, to get

$$\dot{V} \leq -q_0||\tilde{x}||^2 .$$

In the second case, when only $A2 - (ii)$ holds, we have

$$
\begin{aligned}
\dot{V} &\leq -\tilde{x}^T Q \tilde{x} + 2|\tilde{y}||\theta|\rho(y,u)(||x|| - ||\hat{x}||) - \tilde{y}\nu_1 && (2.40) \\
&\leq -\tilde{x}^T Q \tilde{x} - |\tilde{y}| \left[-2\theta_{max}\rho(y,u)(||x||_{max} - ||\hat{x}||) + \mathrm{sgn}(\tilde{y})\nu_1 \right] && (2.41)
\end{aligned}
$$

which suggests that ν_1 should be defined to have a high-gain component, i.e.

$$\nu_1 = 2\theta_{max}\rho(y,u)(||x||_{max} - ||\hat{x}||)\mathrm{sgn}(\tilde{y}) ,$$

where θ_{max}, and $||x||_{max}$, are respectively constant upper bounds of the parameter θ and the state norm $||x||$.

With this choice of ν_1 we have as before that

$$\dot{V} \leq -q||\tilde{x}||^2 .$$

Thus, in both of the cases considered by assumption $A2$ $\tilde{x}$, and $\tilde{\theta}$ are bounded, and $\tilde{x} \rightarrow 0$. Finally, from the error equation (2.30), we have that

$$\lim_{t \rightarrow \infty} \left\{ B\tilde{\theta}\varphi(y, u, x) \right\} = 0 ,$$

which together with assumption $A4$ leads us to conclude that

$$\lim_{t \rightarrow \infty} \hat{\theta} = \theta .$$

We have proved the following theorem:

Theorem 2.1 *Consider the following system*

$$\dot{x} = Ax + B\left[\theta\varphi(y, u, x)\right] + Ru + Ey \qquad (2.42)$$
$$\dot{\theta} = 0 \qquad\qquad\qquad\qquad\qquad\qquad\quad (2.43)$$
$$y = C^T x \qquad\qquad\qquad\qquad\qquad\quad (2.44)$$

under the assumptions $A1 - A4$, with $y, \theta, \varphi(y, u, x) \in R$, $x \in R^n$, and $u \in R^m$. Then the following observer

$$\dot{\hat{x}} = A\hat{x} + B\left[\hat{\theta}\varphi(y, u, \hat{x})\right] + Ru + Ey + K(y - \hat{y}) + B\nu_1 \quad (2.45)$$
$$\dot{\hat{\theta}} = \gamma\varphi(y, u, \hat{x})\tilde{y} \qquad\qquad\qquad\qquad\qquad\qquad\qquad (2.46)$$
$$\hat{y} = C^T\hat{x} , \qquad\qquad\qquad\qquad\qquad\qquad\qquad\qquad\quad (2.47)$$

with

$$\nu_1 = \begin{cases} 0 & if \quad A2\text{-}(i)holds \\ 2\theta_{max}\rho(y, u)(||x||_{max} - ||\hat{x}||)sgn(\tilde{y}) & if \quad A2\text{-}(ii)holds \end{cases} \quad (2.48)$$

ensures (under verification of $A4$), that

$$\lim_{t \rightarrow \infty} \hat{\theta} = \theta .$$

5   Application to the One-Wheel Model

We consider the one-wheel model with lumped friction as described by the equations (2.14)-(2.16). As formulated previously, we assume that both v

and ω are measurable variables. To set our system in the same framework that the structure (2.17)-(2.19), we introduce the new variable

$$\chi = J\omega + rF_n\sigma_1 z \,,$$

from which we get:

$$\dot{\chi} = -\frac{\sigma_0}{\sigma_1}\chi + (J\frac{\sigma_0}{\sigma_1} - \sigma_w)\omega + u_\tau \tag{2.49}$$

$$\dot{z} = (r\omega - v) - \theta\sigma_0\frac{|r\omega - v|}{g(v_r)}z \tag{2.50}$$

$$y = \frac{1}{J}(\chi - rF_n\sigma_1 z) = \omega \,. \tag{2.51}$$

Defining x, u and y respectively as

$$x = \begin{bmatrix} \chi \\ z \end{bmatrix}, \quad u = \begin{bmatrix} u_\tau \\ v \end{bmatrix}, \quad y = \omega \,,$$

we can rewrite the above system as

$$\dot{x} = \begin{bmatrix} -\frac{\sigma_0}{\sigma_1} & 0 \\ 0 & 0 \end{bmatrix} x + \begin{bmatrix} 0 \\ 1 \end{bmatrix} \theta\varphi(y, u, x) + \begin{bmatrix} (J\frac{\sigma_0}{\sigma_1} - \sigma_w) \\ r \end{bmatrix} y + \begin{bmatrix} 1 & 0 \\ 0 & -1 \end{bmatrix} \begin{bmatrix} u_\tau \\ v \end{bmatrix}$$

where $\varphi(y, u, x)$, is defined as

$$\varphi(y, u, x) = -\frac{\sigma_0|ry - v|}{g(ry - v)}z \,.$$

With this representation we shall now verify condition under which the assumptions $A1 - A3$ hold. The last condition $A4$ depends on the operational conditions, in particular on the applied torque u_τ.

Condition $A1$ (linear observability). With A, and C defined as above, i.e

$$A = \begin{bmatrix} -\frac{\sigma_0}{\sigma_1} & 0 \\ 0 & 0 \end{bmatrix} \quad C = \begin{bmatrix} \frac{1}{J} \\ -\frac{rF_n\sigma_1}{J} \end{bmatrix}$$

we have that condition $A1$ holds for any values of the system parameters.

$$\text{rank}\,[C, A^T C] = \text{rank}\begin{bmatrix} \frac{1}{J} & 0 \\ -\frac{rF_n\sigma_1}{J} & \frac{rF_n\sigma_0}{J} \end{bmatrix} = 2$$

This rank condition clearly shows that the existence a non-zero normal force F_n is necessary to build the friction observer.

Condition $A2$ (global Lipschitz condition). With $\varphi(y, u, x) = -\frac{\sigma_0|ry-v|}{g(ry-v)}z$, we have that

$$|\varphi(y, u, x_1) - \varphi(y, u, x_1)| \leq \frac{\sigma_0|ry - v|}{g(ry - v)}|z_1 - z_2| \leq \rho(y, u)|z_1 - z_2| \leq \rho(y, u)\|x_1 - x_2\|$$

where $\rho(y, u) = \sigma_0 \mu_C |ry - v|$.

Condition $A3$ **(Passivity).** Finding a vector K, so that the map $\psi \mapsto \tilde{y}$ of the system description (2.20)-(2.21), is strictly passive, is equivalent to searching for a vector $K = [k_1, k_2]^T$, such that the I/O-map $G(s)$, defined as

$$G(s) = C^T \left[Is - A + KC^T \right]^{-1} B, \qquad (2.52)$$

is strictly positive real (SPR), i.e. $\mathrm{Re}\{G(j\varpi)\} > 0, \forall \varpi \in [0, \infty]$

Computation of $G(s)$ with the corresponding values for A, B, C gives the map

$$G(s) = \frac{s + \beta}{s^2 + \alpha_1 s + \alpha_2} \qquad (2.53)$$

with

$$\beta = \frac{\sigma_0}{\sigma_1} + \frac{k_1}{J}$$

$$\alpha_1 = \beta - k_2 \frac{rF_n \sigma_1}{J}$$

$$\alpha_2 = -\beta k_2 \frac{rF_n \sigma_1}{J} + k_1 k_2 \frac{rF_n \sigma_1}{J^2}$$

A sufficient condition to this function be SPR is that $k_2 < 0$. Then a simple choice for K is thus

$$k_1 = 0$$
$$k_2 = -k,$$

for some $k > 0$. From the Kalman-Yakubovich-Popov lemma, we thus ensure with this choice of K that there exist P satisfying the Lyapunov equation with $PB = C$.

Condition $A4$ **(Persistence of excitation).** To ensure parameter convergence we need to guarantee that

$$\lim_{t \to \infty} \varphi(y(t), u(t), x(t)) = \lim_{t \to \infty} \frac{\sigma_0 |ry(t) - v(t)|}{g(ry - v)} z(t) \neq 0.$$

This implies that the relative velocity should not tend to zero in order for the estimated parameter to converge. This in turn implies that the internal friction state $z(t)$ will not asymptotically converge to zero.

Finally, we have,

Theorem 2.2 *Consider the one-wheel model with lumped dynamic friction (2.49)-(2.51), then the following observer:*

$$\dot{\hat{\chi}} = -\frac{\sigma_0}{\sigma_1}\hat{\chi} + (J\frac{\sigma_0}{\sigma_1} - \sigma_w)\omega + u_\tau \tag{2.54}$$

$$\dot{\hat{z}} = (r\omega - v) - \hat{\theta}\frac{\sigma_0|r\omega - v|}{g(v_r)}\hat{z} - k(\omega - \hat{y}) \tag{2.55}$$

$$\dot{\hat{\theta}} = -\gamma\frac{\sigma_0|r\omega - v|}{g(v_r)}\hat{z}(\omega - \hat{y}) \tag{2.56}$$

$$\hat{y} = \frac{1}{J}(\hat{\chi} - rF_n\sigma_1\hat{z}), \tag{2.57}$$

with positive nonzero k, and γ, ensures that all the estimated states are bounded, and that:

$$\lim_{t\to\infty}\hat{\chi} = \chi \tag{2.58}$$

$$\lim_{t\to\infty}\hat{z} = z. \tag{2.59}$$

If in addition, the relative contact velocity does not vanishes, then we also have that

$$\lim_{t\to\infty}\hat{\theta} = \theta.$$

5.1 Simulation Results

Simulations have been performed with the one-wheel system and the lumped LuGre model. The friction parameters used in the simulations are the ones given in Table 1, with the following additional values for the wheel: $r = 25[cm]$, $m = 5[Kg]$, $J = 0.75 * m * r^2 = 0.2344[Kgm^2]$, $F_n = 14[Kgm^2/s^2]$.

Fig. 5, shows simulation results. Fig. 5-(a) shows the time-profile of the contact friction force resulting form the application of the time torque profile $u_\tau(t)$ shown in Fig. 5-(c). The simulation has first an acceleration phase, and then a breaking phase. From Fig. 5-(a), we can see that about 2 seconds are needed for the friction torque to reach its maximum value.

The observation error of the χ, and z is shown in Fig. 5-(d). According to the theorem these two variables should converge to their true values regardless the profile evolution of the system states. This is verified by this curve showing the exponential convergence of the $||(\tilde{\chi}(t), \tilde{z})||$ to zero.

Since the ultimate goal of this work is to be able to on-line estimate this variation, the simulation was done under variations of the parameter θ, representing the road variation conditions (see Fig. 4). Fig. 5-(b) shows in bold lines the value of θ, which evolves within fourth different conditions: the first quarter of the simulation corresponds to dry asphalt conditions. The second quarter corresponds to a sudden change from dry to wet. During the third quarter, there is a smooth variation from wet to snow. The last

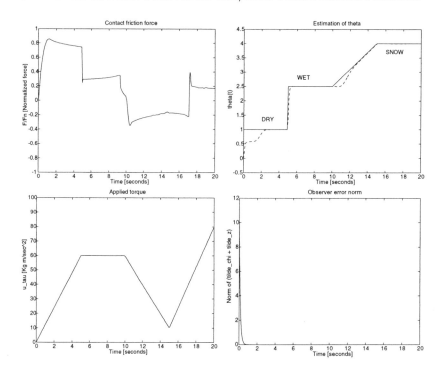

FIGURE 5. a) Contact torque friction $F(t)$ (up left). b)Estimated parameter $\hat{\theta}(t)$, and evolution of θ (up right). c) Applied wheel torque $u_\tau(t)$ (low left). d)Observe error norm of $(\tilde{\chi}(t), \tilde{z})$ (low right).

quarter is keep constant at the snow conditions. In dotted lines we can see the evolution of the estimate $\hat{\theta}(t)$. As we can observe, a good parameter tracking is obtained, as long as the relative contact velocity is different from zero. During the small time-period when this velocity is small or zero, the adaptation law yields a constant $\hat{\theta}(t)$.

6   Conclusions

We have presented a method to estimate on-line the changes in road condition. To achieve this goal we have introduced dynamical friction models that, one hand provide a more accurate description of the contact friction, and one the other hand, allow us to characterize road condition variations via a single parameter.

It has been shown that the distributed parameter version of these model also capture stationary shape profiles between normalized friction and slip rate that are similar to the ones obtained from experimental data (i.e. magic formula).

We have introduced a model-base observer that ensure asymptotic tracking of road condition, under mild conditions implying a non-vanishing evolution of the slip rate. This condition are quite natural in this context (they imply that the vehicle should operate away to the ideal pure rolling condition). Mathematically, this condition correspond to the persistently excitation condition, which is well known in the adaptive control literature. In the context of nonlinear observers, this condition appear as being the characterization of "good " inputs, which are required to recover state observability.

The observer presented here has been derived in a general framework allowing to extend our study to the case where the vehicle velocity is not measurable. In particular, assumption $A2 - (ii)$ will allows for this extension, if it can be shown that the assumption $A3$, also holds. This study and the introduction of other factors like: wheel vertical deformation, and suspension dynamics, are currently under study.

Acknowledgements

The LuGre version of the dynamic friction model presented here, was derived during the first author visit at the Department of Aeronautics at the Georgia Institute of Technology (CNRS/NSF collaboration project). A more complete report on this topic is in preparation.

The first author would like also to thanks M. Sorin and P.A. Bliman for the interesting discussion on distributed friction models.

7 REFERENCES

[1] E. Bakker, L. Nyborg and H. Pacejka. Tyre Modelling for Use in Vehicle Dynamic Studies. Society of Automotive Engineers Paper # 870421, 1987.

[2] P. A. Bliman, T. Bonald and M. Sorine. Hysteresis Operators and tire Friction Models: Application to vehicle dynamic Simulator. *Prof. of ICIAM. 95*, Hamburg, Germany, 3-7 July, 1995.

[3] M. Burckhardt. ABS und ASR, Sicherheitsrelevantes, Radschlupf-Regel System, *Lecture Scriptum.* University of Braunschweig, Germany, 1987.

[4] M. Burckhardt. *Fahrwerktechnik: Radschlupfregelsysteme.* Vogel-Verlag, Germany, 1993.

[5] C. Canudas de Wit, H. Olsson, K. J. Åström and P. Lischinsky. A New Model for Control of Systems with Friction, *IEEE TAC*, Vol. 40, No. 3, pp.419-425, March 1995.

[6] C. Canudas de Wit and P. Lischinsky. Adaptive friction compensation with partially known dynamic friction model, *International Journal of Adaptive Control and Signal Processing*, Vol. 11, pp.65-85, 1997.

[7] P. R. Dahl. Solid Frictioin Damping of Mechanical Vibrations. *AIAA Journal*, 14, No. 12, pp.1675-1682, 1997.

[8] F. Gustafsson. Slip-based Tire-road Friction Estimation. *Automatica*, 33(6):1087–1099, 1997.

[9] J. Harned, L. Johnston and G. Scharpf. Measurement of Tire Brake Force Characteristics as Related to Wheel Slip (Antilock) Control System Design. *SAE Transactions*, 78(690214):909–25, 1969.

[10] U. Kiencke. Realtime Estimation of Adhesion Characteristic Between Tyres and Road. In *Proceedings of the IFAC World Congress*, volume 1, 1993.

[11] U. Kiencke and A. Daiss. Estimation of Tyre Friction for Enhaced ABS-Systems. In *Proceedings of the AVEG'94*, 1994.

[12] H. Lee and M. Tomizuka. Adaptive Traction Control. PATH Technical Report UCB-ITS-PRR-95-32, Institute of Transportation Studies, University of California at Berkeley, 1995.

[13] Y. Liu and J. Sun. Target Slip Tracking Using Gain-Scheduling for Antilock Braking Systems. In *The American Control Conference*, pages 1178–82, Seattle, Washington, 1995.

[14] H. B. Pacejka and R. S. Sharp. Shear Force Developments by Psneumatic tires in Steady-state conditions: A review of Modeling Aspects.. *Vehicle Systems Dynamics*, Vol. 20, pp.121-176, 1991.

[15] W. R. Pasterkamp and H. B. Pacejka. The Tire as a Sensor to Estimate Friction. *Vehicle Systems Dynamics*, Vol. 29,(1997) pp.409-422, 1997.

[16] L. R. Ray. Nonlinear Tire Force Estimation and Road Friction Identification: Simulation and Experiments. *Automatica*, 33(10):1819–1833, 1997.

[17] H. T. Szostak,R. W. Allen and T. J. Rosenthal. Analytical Modeling of Driver Response in Crash Avoidance Manuevering. Volume II: An Interactive Tire Model for Driver/Vehicle Simulation. Report no. DOT HS 807-271, U.S. Department of Transportation, 1988.

[18] K. Yi and T. Jeong. Observer Based Estimation of Tire-road Friction for Collision Warning Algorithm Adaptation. *JSME International Journal*, 41(1):116–124, 1998.

Observer Design for Nonlinear Oscillatory Systems

Dag Kristiansen and Olav Egeland

Department of Engineering Cybernetics
Norwegian University of Science and Technology
Trondheim, Norway

1 Introduction

Numerous vibration phenomena which are theoretically interesting as well
as practically important can only be understood on the basis of nonlinear vibrations. For instance, the wide field of self-excited, parametric and
auto-parametric vibration demands nonlinear treatment from the very beginning. The sources of the nonlinearities may be either geometric, inertial, material, damping or a combination of these things. Nonlinearities
bring a whole range of phenomena that are not found in linear systems.
In single-degree-of-freedom systems these phenomena include multiple solutions, jumps, limit cycles, natural frequency shift, subharmonic and su-
perharmonic resonances, period-multiplying bifurcations, and chaotic motions [12]. Large excitation levels are usually needed to produce period-
multiplying bifurcations and chaotic motions in single-degree-of-freedom
systems.

In addition to the above mentioned phenomena, the response of nonlinear multi-degree-of-freedom systems can exhibit combinations resonances
and modal interactions. The latter may provide a coupling or an energy
exchange between the system's modes and arises if there exists a special
relationship between two or more natural frequencies of the linear modes
and an excitation frequency. This means that the long-time responses of the
system can contain significant contributions in many modes of vibration.
The presence of significant responses in more than one mode increases the
number of modal equations that must be analyzed, and this generally serves
to complicate the dynamics of the system. More importantly, modal interactions can lead to dangerously large responses in modes that are predicted
by linear analysis to have insignificant response amplitudes. The extent of
the interaction and its conditions depend on the linear natural eigenfrequencies w_i and the nonlinearities of the system. More precisely, autopara-
metric resonances in systems with n linear natural frequencies $(\omega_1, \dots, \omega_n)$
and n corresponding modes (the eigenfrequencies are assumed to real and

nonzero) occur whenever two or more eigenfrequencies are *commensurable* or *nearly commensurable* (see e.g. [12, 1]). If a harmonic external excitation of frequency Ω acts on a multi-degree-of-freedom system, then in addition to all primary and secondary resonances ($r\Omega \approx s\omega_i$, where r and s being integers) of a single-degree-of-freedom system, there might exist other resonant combinations of the frequencies in the form $r\Omega \approx s_1\omega_1 + \cdots + s_n\omega_n$, where r and s_i are integers such that $r + \sum_{i=1}^{n} |s_i| = N$, where N is the order of the nonlinearity plus one and n is the number of degrees of freedom. This means for multidegree-of-freedom systems with cubic nonlinearities, to the first approximation, combination resonances may occur if $\Omega \approx |\pm\omega_m \pm \omega_k|/2$, $\Omega \approx |\pm 2\omega_m \pm \omega_k|/2$ or $\Omega \approx |\pm\omega_m \pm \omega_k \pm \omega_l|$. If quadratic nonlinearities are added, additional combination resonances may occur if $\Omega \approx |\pm\omega_m \pm \omega_k|$. Thus, a high-frequency excitation may produce large amplitude responses in low-frequency modes that are involved in the combination resonance and vice versa. Interestingly, the concept of modal interactions can also be utilized in control design, see e.g. [14, 13, 3].

In this chapter we will focus on designing full-state nonlinear observers for systems where *internal resonance* is present. We will assume that we do not have any measurements of the velocities and also that we cannot measure each position separately. Direct applications include e.g. cylinder gyroscopes [4]. As an analysis-tool, we shall use the concept of contraction theory [10]. A short review of this concept is given in Section 2.

2   Contraction Theory

In connection with the observer design, *contraction theory* will play an important role in the analysis. Here we will give a short review of the theory known as contraction theory which was proposed by [10]. The results are based on ideas from fluid mechanics and tools from differential geometry. The basic idea is to view the system differential equations as an n-dimensional "fluid-flow" described by Euler coordinates. By calculating the squared distance between two trajectories in the "flow-field" one ends up with a concept called contraction region. The interested reader is referred to [10] and the references therein for more on this subject. It is also worth mentioning that a thoroughly mathematical treatment of similar ideas can be found in [2].

Given the nonlinear, non-autonomous system

$$\dot{\mathbf{x}} = \mathbf{f}(\mathbf{x}, t) \tag{3.1}$$

where $\mathbf{x} \in \mathbb{R}^n$ and $\mathbf{f} : \mathbb{R}^n \times \mathbb{R}_+ \longrightarrow \mathbb{R}^n$ is assumed to be sufficiently smooth. This equation can be written differentially as

$$\delta\dot{\mathbf{x}} = \frac{\partial \mathbf{f}}{\partial \mathbf{x}}(\mathbf{x}, t)\, \delta\mathbf{x} \tag{3.2}$$

where $\delta\mathbf{x}$ is a virtual displacement.

The squared distance between two neighboring trajectories can be defined as $\delta\mathbf{x}^T\delta\mathbf{x}$, which means that the rate of change is given by

$$\frac{d}{dt}\left(\delta\mathbf{x}^T\delta\mathbf{x}\right) = 2\delta\mathbf{x}^T\delta\dot{\mathbf{x}} = 2\delta\mathbf{x}^T\frac{\partial\mathbf{f}}{\partial\mathbf{x}}\delta\mathbf{x} \tag{3.3}$$

Let $\lambda_{\max}(\mathbf{x},t)$ denote the largest eigenvalue of $\frac{1}{2}\left(\frac{\partial\mathbf{f}}{\partial\mathbf{x}} + \frac{\partial\mathbf{f}}{\partial\mathbf{x}}^T\right)$, then

$$||\delta\mathbf{x}|| \leq ||\delta\mathbf{x_0}||\, e^{\int_0^t \lambda_{\max}(\mathbf{x},\tau)d\tau} \tag{3.4}$$

If $\lambda_{\max}(\mathbf{x},t)$ is uniformly strictly negative, (3.4) shows that $||\delta\mathbf{x}||$ converges exponentially to zero. This implies by path integration that the length of any finite path converges exponentially to zero.

Now consider the differential coordinate transformation

$$\delta\mathbf{z} = \Theta(\mathbf{x},t)\,\delta\mathbf{x} \tag{3.5}$$

where $\Theta(\mathbf{x},t)$ is a square matrix. Then a generalization of the squared length is

$$\delta\mathbf{z}^T\delta\mathbf{z} = \delta\mathbf{x}^T\mathbf{M}(\mathbf{x},t)\,\delta\mathbf{x} \tag{3.6}$$

where $\mathbf{M}(\mathbf{x},t) = \Theta^T(\mathbf{x},t)\Theta(\mathbf{x},t)$ represents a symmetric and continuously differentiable metric. If $\mathbf{M}(\mathbf{x},t)$ is uniformly positive definite, exponential convergence of $\delta\mathbf{z}$ to zero implies exponential convergence of $\delta\mathbf{x}$ to zero. We also have that $\frac{d}{dt}\delta\mathbf{z} = \mathbf{F}\delta\mathbf{z}$, where $\mathbf{F} = \left(\dot{\Theta} + \Theta\frac{\partial\mathbf{f}}{\partial\mathbf{x}}\right)\Theta^{-1}$, and we can state the following definition and theorem:

Definition 3.1 ([10]) *Given the system equations $\dot{\mathbf{x}} = \mathbf{f}(\mathbf{x},t)$, a region of the state space is called a contraction region with respect to a uniformly positive definite metric $\mathbf{M}(\mathbf{x},t) = \Theta^T(\mathbf{x},t)\Theta(\mathbf{x},t)$ if $\mathbf{F}$ is uniformly negative definite in that region.*

Regions where $\mathbf{F}$ is negative semi-definite are called semi-contracting, and regions where $\mathbf{F}$ is skew-symmetric are called indifferent.

Theorem 3.1 ([10]) *Given the system equations $\dot{\mathbf{x}} = \mathbf{f}(\mathbf{x},t)$, any trajectory which starts in a ball of constant radius with respect to the metric $\mathbf{M}(\mathbf{x},t)$, centered at a given trajectory and contained at all times in a contraction region with respect to $\mathbf{M}(\mathbf{x},t)$, remains in that ball and converges exponentially to this trajectory.*

Furthermore global exponential convergence to the given trajectory is guaranteed of the whole state space is a contraction region with respect to the metric $\mathbf{M}(\mathbf{x},t)$.

Remark 3.1 ([7]) *Note that* $\Gamma = \Theta^T \frac{\partial \Theta}{\partial \mathbf{x}}$ *can be written in terms of* Christoffel *symbol of the first kind, i.e., [11]:*

$$\Gamma_{hlk} = \frac{1}{2} \left(\frac{\partial M_{kl}}{\partial x^h} + \frac{\partial M_{lh}}{\partial x^k} - \frac{\partial M_{hk}}{\partial x^l} \right) \tag{3.7}$$

Since $\Gamma_{hlk} = \Gamma_{klh}$ *[11], we have that*

$$\Gamma_{hlk} a^h b^k = \Gamma_{klh} a^k b^h = \Gamma_{hlk} b^h a^k \tag{3.8}$$

where a^k *is the k-th component of the vector* $\mathbf{a}$ *and* b^h *is the h-th component of the vector* $\mathbf{b}$. *This means that for autonomous* $\Theta (\mathbf{x})$, *we can analyze the dynamics* $\dot{\mathbf{z}} = \Theta \dot{\mathbf{x}}$ *in place of* $\mathbf{F} = \left(\dot{\Theta} + \Theta \frac{\partial \mathbf{f}}{\partial \mathbf{x}} \right) \Theta^{-1}$, *i.e.,*

$$\frac{d}{dt} \left(\delta \mathbf{z}^T \delta \mathbf{z} \right) = 2 \delta \mathbf{z}^T \delta \dot{\mathbf{z}} \tag{3.9}$$

since

$$\delta \dot{\mathbf{z}} = \delta \left(\Theta \dot{\mathbf{x}} \right) = \frac{\partial \Theta}{\partial \mathbf{x}} \delta \mathbf{x} \dot{\mathbf{x}} + \Theta \frac{\partial \mathbf{f}}{\partial \mathbf{x}} \delta \mathbf{x} = \frac{\partial \Theta}{\partial \mathbf{x}} \dot{\mathbf{x}} \delta \mathbf{x} + \Theta \frac{\partial \mathbf{f}}{\partial \mathbf{x}} \delta \mathbf{x} \tag{3.10}$$

$$2 \delta \mathbf{z}^T \delta \dot{\mathbf{z}} = 2 \delta \mathbf{x}^T \Theta^T \left(\frac{\partial \Theta}{\partial \mathbf{x}} \dot{\mathbf{x}} + \Theta \frac{\partial \mathbf{f}}{\partial \mathbf{x}} \right) \delta \mathbf{x} = \frac{d}{dt} \left(\delta \mathbf{z}^T \delta \mathbf{z} \right) \tag{3.11}$$

A last result which will be used in the observer design is the following [10]: Consider a smooth virtual dynamics of the form

$$\frac{d}{dt} \begin{pmatrix} \delta \mathbf{z}_1 \\ \delta \mathbf{z}_2 \end{pmatrix} = \begin{pmatrix} \mathbf{F}_{11} & \mathbf{0} \\ \mathbf{F}_{21} & \mathbf{F}_{22} \end{pmatrix} \begin{pmatrix} \delta \mathbf{z}_1 \\ \delta \mathbf{z}_2 \end{pmatrix} \tag{3.12}$$

and assume that $\mathbf{F}_{21}$ is bounded. Exponential convergence of $\delta \mathbf{z}_1$ can be concluded for uniformly negative definite $\mathbf{F}_{11}$. Also, if $\mathbf{F}_{22}$ is uniformly negative definite, this implies exponential convergence of the whole system to a single trajectory since $\mathbf{F}_{21} \delta \mathbf{z}_1$ represents an exponentially decaying disturbance in the second equation. We can think of the dynamics of $\delta \mathbf{z}_1$ as the plant, and the dynamics of $\delta \mathbf{z}_2$ as the observer. By designing the observer such that the system trajectory is contained in the "flow-field" of the observer, this means that the observer is exponential convergent if $\mathbf{F}_{11}$ and $\mathbf{F}_{22}$ are uniformly negative definite, and $\mathbf{F}_{21}$ is bounded.

3   System Equations

We will assume that our system is given as n nonlinearly coupled oscillators with constant mass $\mathbf{M} = \text{diag}\{m_i\} > \mathbf{0}$ and linear viscous damping $\mathbf{C} = \text{diag}\{\mu_i\} > \mathbf{0}$, i.e. [12],

$$\mathbf{M}\dot{\mathbf{v}} = -\mathbf{C}\mathbf{v} - \frac{\partial V}{\partial \mathbf{q}} + \mathbf{F} \tag{3.13}$$

$$\dot{\mathbf{q}} = \mathbf{v} \tag{3.14}$$

$$y = \sum_i q_i \tag{3.15}$$

where $\mathbf{F}$ is an external forcing and $V = V(\mathbf{q})$ will decide what types of nonlinearities which are present in the system (e.g. quadratic, cubic or both). More precisely, V can be written as $V = V_1 + V_2$ where V_1 is due to the linear spring constants and is assumed to be positive, and V_2 reflects the nonlinear coupling terms. Note that our measurement y, given by (3.15), makes this problem in some sense different from e.g. robotics where one usually can measure each position separately. We will assume that the system parameters (mass, damping etc.) are known.

3.1 Analysis

Contraction analysis of mechanical systems in Hamiltonian form were investigated in [9, 8, 6], while systems in Lagrangian form were considered in [7]. Here we will give an alternative analysis, which is a direct consequence of energy considerations. We will assume that

1. V can be written as $V = \mathbf{q}^T \mathbf{P}(\mathbf{q}) \mathbf{q}$ where $\mathbf{P}$ is positive definite.

2. $\frac{\partial V}{\partial \mathbf{q}}$ can be written as $\frac{\partial V}{\partial \mathbf{q}} = \mathbf{K}(\mathbf{q}) \mathbf{q}$ where $\mathbf{K}$ is a square matrix.

Now (3.13) can be written differentially as $(\mathbf{F} = \mathbf{0})$

$$\mathbf{M}d\mathbf{v} = -\mathbf{C}d\mathbf{q} - \frac{\partial V}{\partial \mathbf{q}} dt \tag{3.16}$$

Using $\frac{\partial V}{\partial \mathbf{q}} = \mathbf{K}(\mathbf{q}) \mathbf{q}$, (3.16) can be written as

$$\mathbf{M}d\mathbf{v} = -\mathbf{C}d\mathbf{q} - \mathbf{K}(\mathbf{q}) \mathbf{q}dt \tag{3.17}$$

Since $\mathbf{P} > 0$, there exists a matrix $\mathbf{W}(\mathbf{q})$ such that

$$\mathbf{W}^T(\mathbf{q}) \mathbf{W}(\mathbf{q}) = \mathbf{P}(\mathbf{q}) \tag{3.18}$$

Define

$$\dot{\mathbf{z}} = \begin{pmatrix} \sqrt{\mathbf{M}} & 0 \\ 0 & \sqrt{2}\mathbf{W}(\mathbf{q}) \end{pmatrix} \begin{pmatrix} \mathbf{v} \\ \mathbf{q} \end{pmatrix} \tag{3.19}$$

and introduce

$$\dot{\phi} = \mathbf{q} \tag{3.20}$$

then

$$dz = \begin{pmatrix} \sqrt{\mathbf{M}} & \mathbf{0} \\ \mathbf{0} & \sqrt{2}\mathbf{W}\left(\mathbf{q}\right) \end{pmatrix} \begin{pmatrix} d\mathbf{q} \\ d\phi \end{pmatrix} \tag{3.21}$$

$$\Updownarrow$$

$$\delta z = \begin{pmatrix} \sqrt{\mathbf{M}} & \mathbf{0} \\ \mathbf{0} & \sqrt{2}\mathbf{W}\left(\mathbf{q}\right) \end{pmatrix} \begin{pmatrix} \delta\mathbf{q} \\ \delta\phi \end{pmatrix} \tag{3.22}$$

Also from (3.19)

$$\delta\dot{z} = \begin{pmatrix} \sqrt{\mathbf{M}} & \mathbf{0} \\ \mathbf{0} & \sqrt{2}\frac{\partial(\mathbf{W}(\mathbf{q})\mathbf{q})}{\partial\mathbf{q}} \end{pmatrix} \begin{pmatrix} \delta\mathbf{v} \\ \delta\mathbf{q} \end{pmatrix} \tag{3.23}$$

Then, using (3.9)

$$\begin{aligned} \frac{d}{dt}\left(\frac{1}{2}\delta z^T \delta z\right) &= \delta z^T \delta\dot{z} \\ &= \left(\delta\mathbf{q}^T\sqrt{\mathbf{M}} \quad \sqrt{2}\delta\phi^T\mathbf{W}^T\left(\mathbf{q}\right) \right) \begin{pmatrix} \sqrt{\mathbf{M}}\delta\mathbf{v} \\ \sqrt{2}\frac{\partial(\mathbf{W}(\mathbf{q})\mathbf{q})}{\partial\mathbf{q}}\delta\mathbf{q} \end{pmatrix} \\ &= -\delta\mathbf{q}^T\mathbf{C}\delta\mathbf{q} - \delta\mathbf{q}^T\mathbf{K}\left(\mathbf{q}\right)\delta\phi \\ &\quad +2\delta\phi^T\mathbf{W}^T\left(\mathbf{q}\right)\frac{\partial\left(\mathbf{W}\left(\mathbf{q}\right)\mathbf{q}\right)}{\partial\mathbf{q}}\delta\mathbf{q} \end{aligned} \tag{3.24}$$

Note that

$$\frac{\partial V}{\partial\mathbf{q}} = \frac{\partial\left(\mathbf{q}^T\mathbf{W}^T\left(\mathbf{q}\right)\mathbf{W}\left(\mathbf{q}\right)\mathbf{q}\right)}{\partial\mathbf{q}} = 2\frac{\partial\left(\mathbf{W}\left(\mathbf{q}\right)\mathbf{q}\right)^T}{\partial\mathbf{q}}\mathbf{W}\left(\mathbf{q}\right)\mathbf{q} = \mathbf{K}\left(\mathbf{q}\right)\mathbf{q} \tag{3.25}$$

i.e.,

$$\mathbf{K}^T\left(\mathbf{q}\right) = 2\mathbf{W}^T\left(\mathbf{q}\right)\frac{\partial\left(\mathbf{W}\left(\mathbf{q}\right)\mathbf{q}\right)}{\partial\mathbf{q}} \tag{3.26}$$

such that

$$\frac{d}{dt}\left(\frac{1}{2}\delta z^T \delta z\right) = -\delta\mathbf{q}^T\mathbf{C}\delta\mathbf{q} \tag{3.27}$$

which means that the "flow-field" is semi-contracting. Bounded $\delta\mathbf{q}$ and $\delta\phi$ and by assuming that $\mathbf{K}\left(\mathbf{q}\right)$ is bounded, leads to bounded $\delta\mathbf{v}$ (using $\mathbf{M}\delta\mathbf{v} = -\mathbf{C}\delta\mathbf{q} - \mathbf{K}\left(\mathbf{q}\right)\delta\phi$). Assuming bounded $\frac{\partial^2 V}{\partial\mathbf{q}^2}$ means that $\delta\dot{\mathbf{v}}$ is bounded since $\mathbf{M}\delta\dot{\mathbf{v}} = -\mathbf{C}\delta\mathbf{v} - \frac{\partial^2 V}{\partial\mathbf{q}^2}\delta\mathbf{q}$. This means that $\delta\mathbf{q}$ and $\delta\mathbf{v}$ converges asymptotically to zero.

Remark 3.2 *As in [7], the above analysis can be regarded as a generalization of the energy conservation since*

$$\frac{d}{dt}\left(\frac{1}{2}\delta\mathbf{q}^T\mathbf{M}\delta\mathbf{q} + \delta\phi^T\mathbf{W}^T(\mathbf{q})\,\mathbf{W}(\mathbf{q})\,\delta\phi\right) = -\delta\mathbf{q}^T\mathbf{C}\delta\mathbf{q} \quad (3.28)$$

$$\Updownarrow$$

$$\frac{d}{dt}\left(\frac{1}{2}d\mathbf{q}^T\mathbf{M}d\mathbf{q} + d\phi^T\mathbf{W}^T(\mathbf{q})\,\mathbf{W}(\mathbf{q})\,d\phi\right) = -d\mathbf{q}^T\mathbf{C}d\mathbf{q} \quad (3.29)$$

Multiplying with $\frac{1}{dt^2}$:

$$\frac{d}{dt}\left(\frac{1}{2}\mathbf{v}^T\mathbf{M}\mathbf{v} + \mathbf{q}^T\mathbf{W}^T(\mathbf{q})\,\mathbf{W}(\mathbf{q})\,\mathbf{q}\right) = -\mathbf{v}^T\mathbf{C}\mathbf{v} \quad (3.30)$$

$$\Updownarrow$$

$$\frac{d}{dt}\left(\frac{1}{2}\mathbf{v}^T\mathbf{M}\mathbf{v} + V\right) = -\mathbf{v}^T\mathbf{C}\mathbf{v} \quad (3.31)$$

4   Observer Design

Since we do not have any measurements of the velocities, we will take advantage of the following result due to [5]: Given the system

$$\dot{\mathbf{x}} = \mathbf{f}(\mathbf{x}, \mathbf{t}) \quad (3.32)$$

with measurement

$$\mathbf{y} = \mathbf{h}(\mathbf{x}) \quad (3.33)$$

and the following general observer

$$\dot{\hat{\mathbf{x}}} = \mathbf{g}(\hat{\mathbf{x}}, \mathbf{y}, t) \quad (3.34)$$
$$\hat{\mathbf{y}} = \mathbf{h}(\hat{\mathbf{x}}) \quad (3.35)$$

where $\mathbf{g}, \mathbf{h}$ are assumed to be smooth functions.

We can state the following result:

Proposition 3.1 ([5]) *Given a smooth coordinate transformation of the observer dynamics $\bar{\mathbf{x}} = \bar{\mathbf{x}}(\hat{\mathbf{x}}, \hat{\mathbf{y}})$, where for each $\hat{\mathbf{y}}$ the mapping $\hat{\mathbf{x}} \mapsto \bar{\mathbf{x}}$ is invertible, and given the n-dimensional system equations and m-dimensional measurements*

$$\dot{\mathbf{x}} = \mathbf{f}(\mathbf{x}, t) \quad (3.36)$$
$$\mathbf{y} = \mathbf{h}(\mathbf{x}) \quad (3.37)$$

then the observer equations

$$\dot{\hat{x}} = g(\hat{x}, y, t) \tag{3.38}$$

$$\hat{y} = h(\hat{x}) \tag{3.39}$$

transform to

$$\dot{\hat{x}} = g(\hat{x}, y, t) + \frac{\partial \hat{x}}{\partial y}(\dot{y} - \hat{y}) \tag{3.40}$$

if

$$\dot{\bar{x}} = \frac{\partial \bar{x}}{\partial \hat{x}}(\hat{x}(\bar{x}, y), y) g(\hat{x}(\bar{x}, y), y, t)$$
$$+ \frac{\partial \bar{x}}{\partial y}(\hat{x}(\bar{x}, y), y) \frac{\partial h}{\partial \hat{x}}(\hat{x}(\bar{x}, y)) g(\hat{x}(\bar{x}, y), y, t) \tag{3.41}$$

is integrated instead of (3.38) and (3.39).

The proof can be found in [5].

We now propose the following observer for (3.13)-(3.15):

$$M\dot{\hat{v}} = -C\hat{v} - \frac{\partial V(\hat{q})}{\partial \hat{q}} + F \tag{3.42}$$

$$\dot{\hat{q}} = \hat{v} \tag{3.43}$$

Introduce $M\bar{v} = M\hat{v} - \begin{pmatrix} \gamma_1 \\ \vdots \\ \gamma_n \end{pmatrix} \hat{y}$, then $(F = 0)$

$$M\dot{\bar{v}} = -C\hat{v} - \frac{\partial V(\hat{q})}{\partial \hat{q}} - H\hat{v} \tag{3.44}$$

where $H = \begin{pmatrix} \gamma_1 & \cdots & \gamma_1 \\ \vdots & \vdots & \vdots \\ \gamma_n & \cdots & \gamma_n \end{pmatrix}$ and

$$M\dot{\hat{v}} = M\dot{\bar{v}} + \begin{pmatrix} \gamma_1 \\ \vdots \\ \gamma_n \end{pmatrix} \dot{y} = -C\hat{v} - \frac{\partial V(\hat{q})}{\partial \hat{q}} - H(\hat{v} - v) \tag{3.45}$$

We can now view (3.13), (3.14), (3.43) and (3.45) as a hierarchical combination as in (3.12). Since H is bounded, this means that under the assumption that $K(\hat{q})$ and $\frac{\partial^2 V}{\partial \hat{q}^2}$ are bounded, $\delta\hat{q}$ and $\delta\hat{v}$ converge asymptotically to zero.

Remark 3.3 *Note that due to our measurement (3.15), if the observer equation (3.43) is changed to $\dot{\hat{\mathbf{q}}} = \hat{\mathbf{v}} - \mathbf{H}(\hat{\mathbf{q}} - \mathbf{q})$, this implies that (using the relation $\mathbf{K}^T(\hat{\mathbf{q}}) = 2\mathbf{W}^T(\hat{\mathbf{q}}) \frac{\partial(\mathbf{W}(\hat{\mathbf{q}})\hat{\mathbf{q}})}{\partial \hat{\mathbf{q}}})$*

$$\frac{d}{dt}\left(\frac{1}{2}\delta\mathbf{z}^T\delta\mathbf{z}\right) = \delta\mathbf{z}^T\delta\dot{\mathbf{z}}$$

$$= \left(\ \delta\hat{\Psi}^T\sqrt{\mathbf{M}} - \delta\hat{\phi}^T\mathbf{H}^T\sqrt{\mathbf{M}}\ \ \sqrt{2}\delta\hat{\phi}^T\mathbf{W}^T(\hat{\mathbf{q}})\ \right)\left(\begin{array}{c}\sqrt{\mathbf{M}}\delta\hat{\mathbf{v}}\\\sqrt{2}\frac{\partial(\mathbf{W}(\hat{\mathbf{q}})\hat{\mathbf{q}})}{\partial\hat{\mathbf{q}}}\delta\hat{\mathbf{q}}\end{array}\right)$$

$$= -\delta\hat{\Psi}^T(\mathbf{C}+\mathbf{H})\delta\hat{\Psi} + \delta\hat{\phi}^T\mathbf{H}^T(\mathbf{C}+\mathbf{H})\delta\hat{\Psi}$$

$$-\delta\hat{\phi}^T\left(\mathbf{H}^T\mathbf{K}(\hat{\mathbf{q}}) + 2\mathbf{W}^T(\hat{\mathbf{q}})\frac{\partial(\mathbf{W}(\hat{\mathbf{q}})\hat{\mathbf{q}})}{\partial\hat{\mathbf{q}}}\mathbf{H}\right)\delta\hat{\phi} \qquad (3.46)$$

where $\dot{\hat{\Psi}} = \hat{\mathbf{v}}$, $\dot{\hat{\phi}} = \hat{\mathbf{q}}$, $\dot{\mathbf{z}} = \left(\begin{array}{cc}\sqrt{\mathbf{M}} & 0\\0 & \sqrt{2}\mathbf{W}(\hat{\mathbf{q}})\end{array}\right)\left(\begin{array}{c}\hat{\mathbf{v}}\\\hat{\mathbf{q}}\end{array}\right)$,

$$\delta\mathbf{z} = \left(\begin{array}{c}\sqrt{\mathbf{M}}\delta\hat{\Psi} - \sqrt{\mathbf{M}}\mathbf{H}\delta\hat{\phi}\\\sqrt{2}\mathbf{W}(\hat{\mathbf{q}})\delta\hat{\phi}\end{array}\right), \ \delta\dot{\mathbf{z}} = \left(\begin{array}{cc}\sqrt{\mathbf{M}} & 0\\0 & \sqrt{2}\frac{\partial(\mathbf{W}(\hat{\mathbf{q}})\hat{\mathbf{q}})}{\partial\hat{\mathbf{q}}}\end{array}\right)\left(\begin{array}{c}\delta\hat{\mathbf{v}}\\\delta\hat{\mathbf{q}}\end{array}\right).$$

Generally, there seems to be no conclusion about the contraction behaviour of this observer design. However from (3.46) we see that when the gains in the observer (γ_i) are in some sense small, this observer "behaves" in the same way as (3.42) and (3.43).

5   Simulations

5.1   Example 1: 2-DOF Oscillatory System with Cubic Nonlinearities

Nonlinear oscillations in multi-degree-of-freedom systems with cubic non-linearities can be found in many physical systems such as the vibration of strings, beams, membranes, and plates for which stretching is significant, the motion of spherical, centripetal, and double pendulums, and the motion of masses connected with nonlinear springs [12]. For a 2-dof mechanical system with cubic nonlinearities, V is given by

$$V = \frac{k_1}{2}q_1^2 + \frac{k_2}{2}q_2^2 + \alpha_1 q_1^4 + \alpha_2 q_1^3 q_2 + \alpha_3 q_1^2 q_2^2 + \alpha_4 q_1 q_2^3 + \alpha_5 q_2^4 \qquad (3.47)$$

where $k_i > 0$ are the linear spring constants and α_i are constants. Note that V can be written as

$$V = \left(\ q_1\ \ q_2\ \right)\left(\begin{array}{cc}\frac{k_1}{2}+\alpha_1 q_1^2 + \frac{1}{2}\alpha_3 q_2^2 & \frac{1}{2}\left(\alpha_2 q_1^2 + \alpha_4 q_2^2\right)\\\frac{1}{2}\left(\alpha_2 q_1^2 + \alpha_4 q_2^2\right) & \frac{k_2}{2}+\alpha_5 q_2^2 + \frac{1}{2}\alpha_3 q_1^2\end{array}\right)\left(\begin{array}{c}q_1\\q_2\end{array}\right)$$

$$= \left(\ q_1\ \ q_2\ \right)\mathbf{P}(\mathbf{q})\left(\begin{array}{c}q_1\\q_2\end{array}\right) \qquad (3.48)$$

Also

$$\frac{\partial V}{\partial \mathbf{q}} = \begin{pmatrix} k_1 + 4\alpha_1 q_1^2 + 2\alpha_3 q_2^2 & 3\alpha_2 q_1^2 + \alpha_4 q_2^2 \\ \alpha_2 q_1^2 + 3\alpha_4 q_2^2 & k_2 + 2\alpha_3 q_1^2 + 4\alpha_5 q_2^2 \end{pmatrix} \begin{pmatrix} q_1 \\ q_2 \end{pmatrix}$$
$$= \mathbf{K}(\mathbf{q})\,\mathbf{q} \tag{3.49}$$

The following parameters were used in the simulations: $m_1 = m_2 = 1$, $\alpha_1 = 1$, $\alpha_2 = 0.9$, $\alpha_3 = 0.8$, $\alpha_4 = 0.6$, $\alpha_5 = 0.5$, $\mu_1 = \mu_2 = 0.001$, $k_1 = 1$, $k_2 = 9$, and $\mathbf{F} = \begin{pmatrix} \sin(k_1 t) \\ 0 \end{pmatrix}$. V is positive for $\mathbf{q} \neq \mathbf{0}$ since $\alpha_i > 0$ and

$$\frac{\alpha_1 \alpha_3}{2} - \frac{\alpha_2^2}{4} \geq 0 \tag{3.50}$$

$$\frac{\alpha_3 \alpha_5}{2} - \frac{\alpha_4^2}{4} \geq 0 \tag{3.51}$$

$$\alpha_1 \alpha_5 + \frac{\alpha_3^2}{4} + \frac{\alpha_2 \alpha_4}{2} \geq 0 \tag{3.52}$$

The initial conditions of the plant were: $q_1(0) = q_2(0) = v_1(0) = v_2(0) = 0$, while the initial conditions of the observer were $\hat{q}_1(0) = 1$, $\hat{q}_2(0) = 1$, $\bar{v}_1(0) = 0.5$, $\bar{v}_2(0) = -0.4$. The results using $\gamma_1 = \gamma_2 = -1$ is shown in Figures 1–5

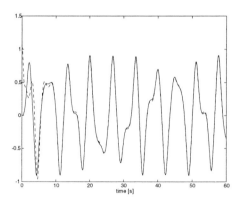

FIGURE 1. $q_1(t)$ [solid line] and $\hat{q}_1(t)$ [dotted line].

5.2   Example 2: Cylinder Gyroscope

The nonlinear dynamics of a cylinder gyroscope was modelled and analyzed by [4]. The model included geometric nonlinearities, and it was shown that V is given by

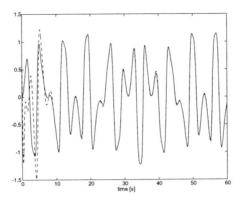

FIGURE 2. $v_1(t)$ [solid line] and $\hat{v}_1(t)$ [dotted line].

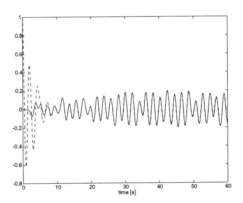

FIGURE 3. $q_2(t)$ [solid line] and $\hat{q}_2(t)$ [dotted line].

$$V = \begin{pmatrix} q_1 & q_2 & q_3 \end{pmatrix} \begin{pmatrix} \frac{1}{2}k_1 + a_1 q_1^2 + \frac{1}{2}a_4 q_2^2 + \frac{1}{2}a_5 q_3^2 & \frac{1}{2}a_8 q_1 + \frac{1}{2}a_{10} q_3 & \frac{1}{2}a_7 q_1^2 + \frac{1}{2}a_9 q_2^2 \\ \frac{1}{2}a_8 q_1 + \frac{1}{2}a_{10} q_3 & \frac{1}{2}k_2 + a_2 q_2^2 + \frac{1}{2}a_4 q_1^2 + \frac{1}{2}a_6 q_3^2 & 0 \\ \frac{1}{2}a_7 q_1^2 + \frac{1}{2}a_9 q_2^2 & 0 & \frac{1}{2}k_3 + a_3 q_3^2 + \frac{1}{2}a_5 q_1^2 + \frac{1}{2}a_6 q_2^2 \end{pmatrix} \begin{pmatrix} q_1 \\ q_2 \\ q_3 \end{pmatrix}$$

where k_i are positive constants and a_j are constants depending on the linear axial mode shapes of the gyroscope. Straightforward calculations show

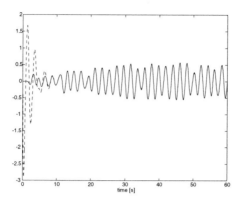

FIGURE 4. $v_2(t)$ [solid line] and $\hat{v}_2(t)$ [dotted line].

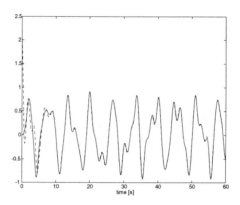

FIGURE 5. $y(t)$ [solid line] and $\hat{y}(t)$ [dotted line].

that

$$\frac{\partial V}{\partial \mathbf{q}} = \left(\begin{array}{c} k_1 + 4a_1 q_1^2 + 2a_4 q_2^2 + 2a_5 q_3^2 + 2a_8 q_2 \\ a_8 q_1 + a_{10} q_3 \\ a_7 q_1^2 + a_9 q_2^2 + a_{10} q_2 \end{array} \right.$$
$$\left. \begin{array}{c} 0 \\ k_2 + 4a_2 q_2^2 + 2a_4 q_1^2 + 2a_6 q_3^2 + 2a_9 q_1 q_3 \\ 0 \end{array} \right.$$
$$\left. \begin{array}{c} 3a_7 q_1^2 + a_9 q_2^2 + a_{10} q_2 \\ 0 \\ k_3 + 4a_3 q_3^2 + 2a_5 q_1^2 + 2a_6 q_2^2 \end{array} \right) \left(\begin{array}{c} q_1 \\ q_2 \\ q_3 \end{array} \right)$$
$$= \mathbf{K}(\mathbf{q})\,\mathbf{q}$$

The following parameters were used in the simulations: $m_1 = m_2 = m_3 = 1$, $a_1 = 1$, $a_2 = 1$, $a_3 = 1$, $a_4 = 0.3$, $a_5 = 0.4$, $a_6 = 0.3$, $a_7 = 0.5$, $a_8 = 0.5$,

$a_9 = 0.7$, $a_{10} = 0.3$, $\mu_1 = \mu_2 = \mu_3 = 0.001$, $k_1 = 1$, $k_2 = 9$, $k_3 = 25$, and $\mathbf{F} = \begin{pmatrix} \sin(k_1 t) \\ 0 \\ 0 \end{pmatrix}$. Note that with these data, V can be shown to be positive for $\mathbf{q} \neq \mathbf{0}$. The initial conditions of the plant were: $q_1(0) = q_2(0) = q_3(0) = v_1(0) = v_2(0) = v_3(0) = 0$, while the initial conditions of the observer were $\hat{q}_1(0) = 1$, $\hat{q}_2(0) = 1$, $\hat{q}_3(0) = -0.2$, $\bar{v}_1(0) = -0.5$, $\bar{v}_2(0) = -0.4$, $\bar{v}_3(0) = -0.1$. The results using $\gamma_1 = \gamma_2 = \gamma_3 = -1$ is shown in Figures 6–12.

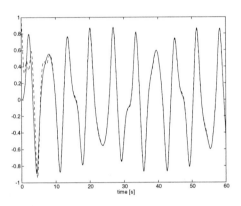

FIGURE 6. $q_1(t)$ [solid line] and $\hat{q}_1(t)$ [dotted line].

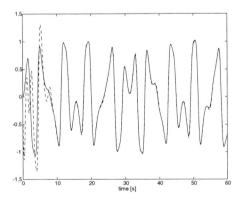

FIGURE 7. $v_1(t)$ [solid line] and $\hat{v}_1(t)$ [dotted line].

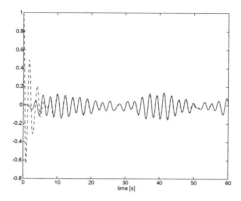

FIGURE 8. $q_2(t)$ [solid line] and $\hat{q}_2(t)$ [dotted line].

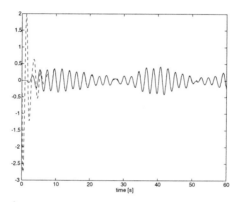

FIGURE 9. $v_2(t)$ [solid line] and $\hat{v}_2(t)$ [dotted line].

6   Conclusions

We have proposed an observer for nonlinear oscillatory systems in Lagrangian form with a single measurement given by $y = \sum_{i=1}^{n} q_i$. The analysis was mainly based on contraction theory which can be found in the
papers by Lohmiller and Slotine [5]- [10]. It was shown that the proposed
observer was asymptotically convergent.

The observer was simulated first on a 2-dof system with cubic nonlinearities, and then on a model of a cylinder gyroscope. The simulations showed
agreement with the theoretical analysis.

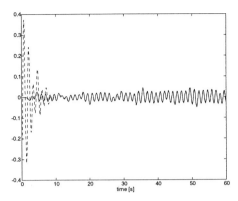

FIGURE 10. $q_3(t)$ [solid line] and $\hat{q}_3(t)$ [dotted line].

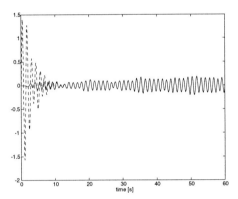

FIGURE 11. $v_3(t)$ [solid line] and $\hat{v}_3(t)$ [dotted line].

7 REFERENCES

[1] R. Evan-Iwanowski *Resonance Oscillations in Mechanical Systems*, Elsevier, New York, 1976.

[2] P. Hartman. *Ordinary Differential Equations*, Birkhauser Verlag, Boston, 1982.

[3] A. Khajepour, F. Golnaraghi and K. A. Morris. "Modal Coupling Controller Design Using a Normal Form Method, Part 1 & 2," *Journal of Sound and Vibration*, vol. 205, pp. 657-688, 1997.

[4] D. Kristiansen and O. Egeland. "Nonlinear Oscillations in Coriolis Based Gyroscopes," Accepted for publication in *Nonlinear Dynamics*.

[5] W. Lohmiller and J. J.-E. Slotine "On Metric Observers for Nonlinear Systems," *Proceedings IEEE International Conference on Control*

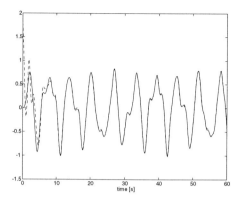

FIGURE 12. $y(t)$ [solid line] and $y(t)$ [dotted line].

Applications, Dearborn, MI, pp. 320-326, 1996.

[6] W. Lohmiller and J.-J.E. Slotine. "On Metric Controllers and Observers for Nonlinear Systems," *Proceedings 35th IEEE Conference on Decision and Control*, Kobe, Japan, pp. 1477-1482, 1996.

[7] W. Lohmiller and J.-J.E. Slotine. "Applications of Contraction Analysis," *Proceedings 36th IEEE Conference on Decision and Control*, San Diego, CA, pp. 1044-1050, 1997.

[8] W. Lohmiller and J.-J.E. Slotine. "Applications of Contraction Analysis," *Proceedings IEEE International Conference on Control Applications*, Hartford, CT, pp. 699-704, 1997.

[9] W. Lohmiller and J.-J.E. Slotine. "Simple Observers for Hamiltonian Systems," *American Control Conference*, Albuquerque, NM, 1997.

[10] W. Lohmiller and J.-J.E. Slotine. "On Contraction Analysis for Nonlinear Systems," *Automatica*, vol. 34, pp. 683-696, 1998.

[11] D. Lovelock and H. Rund. *Tensors, Differential Forms, and Variational Principles*, Dover Publications, New York, 1989.

[12] A. H. Nayfeh and D. T. Mook. *Nonlinear Oscillations*, Wiley, New York, 1979.

[13] S. S. Oueini, A. H. Nayfeh and J. R. Pratt "A Nonlinear Vibration Absorber for Flexible Structures," *Nonlinear Dynamics*, vol. 15, pp. 259-282, 1998.

[14] K. L. Tuer, M. F. Golnaraghi and D. Wang. "Towards a Generalized Regulation Scheme for Oscillatory Systems via Coupling Effects," *IEEE Transactions on Automatic Control*, vol. 40, pp. 522-530, 1995.

Transformation to State Affine System and Observer Design

A. Glumineau and V. López-M.

Institut de Recherche en Cybernétique de NANTES, IRCyN, UMR 6597
1 rue de la Noë, B.P. 92101, 44321 Nantes cedex 3. FRANCE

1 Introduction

The observer design problem is completely solved for linear time invariant systems, whereas in the nonlinear case, there is no general theory. In order to tackle this problem, some methods have been employed: Lyapunov-like technique, linearizations, numerical differentiation, and geometric and algebraic methods (cf. [2, 10, 16, 18, 27, 6, 9, 21, 28, 31]). In order to combine the advantages and improve the shortcomings of two different approaches, structural and numerical differentiation have been sucessfully dealt with input time derivatives [25] and input and/or output time derivatives [21]. Table 1 summarizes the existing literature and shows some observer applications.

Table 1 **Linearization by input-output injection**

System	Approach	Applications
$\dot{\zeta} = A\zeta + \varphi(y, u)$	Geometric: [16, 22, 31] Algebraic: [9, 10, 19]	Motor: Shunt DC, Series DC: [5, 24], Flexible joint: [23].
$\dot{\zeta} = A\zeta + \varphi(y, u, \dot{u}, \cdots, u^{(w)})$	Geometric: [28, 15] Algebraic: [26, 25]	Biological systems:[28],
$\dot{\zeta} = A\zeta + \varphi(y, \cdots, y^{(s)}, u, \cdots, u^{(w)})$	[21]	Numerical differentiation:[6].

In order to extend the class of linearizable systems, some results about the transformation of nonlinear systems into state affine systems have been obtained. High-gain observers are useful for state affine systems as shown in [3, 12, 30] and the references therein. These observers are based on optimal Kalman's observer and used in physical processes, for instance chemical reactors, distilling columns and mechanical systems [8, 30, 1].

The following table summarizes the main contributions on the equivalence between a nonlinear system and a bilinear or state affine system, as well as some observer design applications.

Table 1.

System	Authors	Construct.	Applications
$\dot{\zeta} = A(u)\zeta + \varphi(u, y)$	[11]	No	-
	[14]	Yes	
$\dot{\zeta} = A(u, y)\zeta + \varphi(u, y)$	[1]	No	Synch. Generator [17]
			Inverse Pendulum [1]
	[20]	Yes	Chemical reactor [8]
			Distilling columns [30]

In the following our new results [20] are introduced.

One of the contributions of [20] is the definition of a first algorithm to compute the transformed system functions, from the I/O differential equation.

The chapter is organized as follows. Section 2 introduces some definitions and notation. Section 3 we state the problem of state affine transformation of nonlinear systems, and gives the aim of our approach we introduce by an example. We define an algorithm that permits to give a NSC in order to solve this problem. Section 4 obtains the synthesis observer for the state affine system founded in Section 3. This is achieved with a well defined coordinates transformation and a Kalman-like observer. Some conclusions are given in Section 5.

2   Definitions and Notation

Consider the nonlinear system:

$$\sum \left\{ \begin{array}{rcl} \dot{x} & = & f(x, u) \\ y & = & h(x) \end{array} \right. \tag{4.1}$$

with $x \in M$ where M is an open and dense subset of $\Re^n$, $u \in \Re^m$ and $y \in \Re$. The entries of $f(\cdot, \cdot)$ and $h(\cdot)$ are meromorphic functions of their arguments.

Let us define the state affine system, considered here

$$\sum_a \left\{ \begin{array}{l} \dot{z} = A(y(t), u(t))z + \varphi(y(t), u(t)) \\ y = Cz \end{array} \right. \tag{4.2}$$

where $z(t) \in \Re^n, y(t) \in \Re, u(t) \in \Re^m$.

When one measures $y(t)$, one can define $\vartheta := (y, u)$ as a new input and as recalled in [13] if it is regularly persistent [3], thus the system

$$\sum_o \left\{ \begin{array}{l} \dot{\hat{z}} = A(\vartheta) \cdot \hat{z} + \varphi(\vartheta) - S^{-1}C^T(C\hat{z} - y) \\ \dot{S} = -\theta S - A^T(\vartheta)S - SA(\vartheta) + C^TC \end{array} \right. \tag{4.3}$$

is a Kalman-like observer for $\sum_a$. Where $z(t) \in \mathfrak{R}^n, S(t) \in \mathfrak{R}^+$ is a symmetric positive definite matrix and $\theta > 0$. The norm of the estimation error converges locally exponentially to the origin. From now on, $\sum$ is supposed to be generically observable [25] and will be called *observable*.

3   Problem Statement

The goal is to find a state coordinates transformation $z = \Phi(x)$, such that system $\sum$ (4.1) is locally equivalent to system $\sum_a$ (4.2, in order to design the observer $\sum_o$ (4.3). The approach consists in checking that the I/O differential equation associated to the system $\sum$ has the same form than the $\sum_a$ one. The uniqueness of this equation for an observable system is shown in [29].

3.1   The Input-Output Differential Equation for State Affine Systems $\sum_a$

The I/O differential equation for $\sum_a$ verifies

$$
\begin{aligned}
P_a := y^{(n)} = {} & F_n(A_1, \cdots, A_{n-1}) + \\
& + F_{n-1}(A_1, \cdots, A_{n-1}, \varphi_1) + \\
& + A_1 F_{n-2}(A_2, \cdots, A_{n-1}, \varphi_2) + \cdots \\
& + A_1 A_2 \cdots A_{n-2} F_1(A_{n-1}, \varphi_{n-1}) + \\
& A_1 A_2 \cdots A_{n-1} F_0(\varphi_n),
\end{aligned}
\tag{4.4}
$$

where F_{n-j} $(0 \leq j \leq n)$, is the sum of all monomials ină

$$
y^{(n_1)q_1} \cdots y^{(n_p)q_p} \cdot u_1^{(k_1)^{l_1}} \cdots u_m^{(k_m)^{l_m}},
\tag{4.5}
$$

such that

$$
n_1 q_1 + \cdots + n_p q_p + k_1 l_1 + \cdots + k_m l_m = n - j.
$$

F_{n-j} is a function involving all monomials of "degree" $(n-j)$. For instance, F_n verifies

$$
F_n := Y
\begin{bmatrix}
\dot{f}_{11}(A_1, \cdots, A_{n-1}) \\
f_{21}^{(2)}(A_1, \cdots, A_{n-2}) + \delta_{21}(\cdot) \\
f_{31}^{(3)}(A_1, \cdots, A_{n-3}) + \delta_{31}(\cdot) \\
\cdots \\
f_{(n-2)1}^{(n-2)}(A_1, A_2) + \delta_{(n-2)1}(\cdot) \\
[\log A_1]^{(n-1)} + \delta_{(n-1)1}(\cdot)
\end{bmatrix}
\tag{4.6}
$$

where $Y := [y^{(n-1)} \cdots \dot{y}]$ and $\delta_{\beta 1}(\cdot)$ $(2 \le \beta \le n-1)$, involves all the functions not depending on I/O time derivatives of degree β. Whereas F_{n-j} verifies

$$
F_{n-j} \quad := \quad \varphi_j^{(n-j)} - \left[\varphi_j^{(n-j-1)} \varphi_j^{(n-j-2)} \cdots \varphi_j\right] \cdot
$$
$$
\begin{bmatrix}
f_{1j}(A_j, \cdots, A_{n-1}) \\
f_{2j}^{(2)}(A_j, \cdots, A_{n-2}) + \delta_{2j}(\cdot) \\
f_{3j}^{(3)}(A_j, \cdots, A_{n-3}) + \delta_{3j}(\cdot) \\
\cdots \\
[\log A_j]^{(n-j)} + \delta_{(n-j)j}(\cdot)
\end{bmatrix}
$$

for $1 \le j \le n-1$ and $F_0 := \varphi_n$.

Remark 4.1 *The cascade form of the f_{i1} functions in (4.6), is useful to compute A_i functions.*

Example 4.1

Let the 3-D nonlinear system $\sum_a$,

$$
\begin{aligned}
\dot{z}_1 &= A_1(u)z_2 + \varphi_1(y, u) \\
\dot{z}_2 &= A_2(y, u)z_3 + \varphi_2(y, u) \\
\dot{z}_3 &= \varphi_3(y, u) \\
y &= z_1
\end{aligned}
\tag{4.7}
$$

Its I/O differential equation (4.4) verifies

$$
\begin{aligned}
P_a \quad := \quad & y^{(3)} := F_3(A_1, A_2) + F_2(A_1, A_2, \varphi_1) \\
+ \quad & A_1 F_1(A_2, \varphi_2) + A_1 A_2 F_0(\varphi_3)
\end{aligned}
\tag{4.8}
$$

where, as defined in (4.6) and (3.1),

$$
\begin{aligned}
F_3 \quad &:= \quad Y \cdot f(A); \quad F_2 := \varphi_1^{(2)} - [\dot{\varphi}_1 \, \varphi_1] \cdot f(A) \\
f(A) \quad &:= \quad \left[\overline{\log(A_1^2 A_2)}, \; \overline{\log A_1} - \overline{\log A_1} \cdot \overline{\log A_1 A_2}\right]^T \\
F_1 \quad &:= \quad \dot{\varphi}_2 - \varphi_2 \overline{\log A_2} \\
F_0 \quad &:= \quad \varphi_3
\end{aligned}
\tag{4.9}
$$

where $Y := [y^{(2)} \, \dot{y}]$.

One of the contributions of [20] is the definition of a first algorithm to compute the A_i and φ_i functions from the I-O differential equation.

3.2   State Affine Transformation Algorithm

This algorithm proceeds in two steps. First, one derives all the A_i functions from F_n in (4.4). Then, one finds $n-1$ first order partial differential equations and gives a NSC for the existence of a solution. Secondly, substituting them in (4.4), one solves the following equation

$$
\begin{aligned}
y^{(n)} - F_n &= F_{n-1}(A_1, \cdots, A_{n-1}, \varphi_1) \\
&+ A_1 F_{n-2}(A_2, \cdots, A_{n-1}, \varphi_2) \\
&+ \cdots + A_1 \cdots A_{n-2} F_1(A_{n-1}, \varphi_{n-1}) \\
&+ A_1 \cdots A_{n-1} F_0(\varphi_n).
\end{aligned}
\tag{4.10}
$$

S.A.T. Algorithm.

Step 1. From I/O differential equation of $\sum (4.1)$ set $P_0 := y^{(n)}$.
For $k = 1$ to $n - 1$, $c_k := 1$ with $c_{n-1} := 0$ let:

$$
\omega_k := \frac{\partial^2 P_0}{\partial y^{(k)} \partial y^{(n-k)}} dy + \sum_{j=1}^{m} \frac{\partial^2 P_0}{\partial u_j^{(k)} \partial y^{(n-k)}} du_j
\tag{4.11}
$$

For $1 \le k \le n - 2$

- If $d\omega_k \wedge dy \ne 0$ or $d\omega_k \wedge du \ne 0$ then the problem has no solution.

- If $d\omega_k \wedge dy = 0$ and $d\omega_k \wedge du = 0$

then let the A_i functions be any solution of

$$
\omega_k = \frac{\partial^2 P_a}{\partial y^{(k)} \partial y^{(n-k)}} dy + \sum_{j=1}^{m} \frac{\partial^2 P_a}{\partial u_j^{(k)} \partial y^{(n-k)}} du_j
\tag{4.12}
$$

For $k = n - 1$

- If $d\omega_{n-1} \ne 0$, then the problem has no solution. The algorithm stops.

- If $d\omega_{n-1} = 0$, then let A_1 be any solution of

$$
\omega_{n-1} = \frac{1}{A_1} \sum_{j=1}^{m} \frac{\partial A_1}{\partial u_j} du_j,
\tag{4.13}
$$

where $P_a(A_1, \cdots, A_{n-1}, \varphi_1, \cdots \varphi_n)$ is the formal I/O equation (4.4) computed for the system $\sum_a$.

Step 2. Substitute all the A_i functions in (4.4).
For $r = 1$ to n let

$$
P_r := P_{r-1} - F_{n-r+1}
\tag{4.14}
$$

ă and let define $K_r := A_1 \cdot A_2 \cdots A_r$ with $A_n := 1$, and the differential form $\overline{\omega}_r$ as

$$
\overline{\omega}_r := \frac{1}{K_r} \left[\frac{\partial P_r}{\partial y^{(n-r)}} dy + \sum_{j=1}^{m} \frac{\partial P_r}{\partial u_j^{(n-r)}} du_j \right].
\tag{4.15}
$$

- If $d\bar{\omega}_r \wedge dy \neq 0$, $d\bar{\omega}_r \wedge du \neq 0$ then the problem has no solution.
- If $d\bar{\omega}_r \wedge dy = 0$, $d\bar{\omega}_r \wedge du = 0$ then φ_r is a solution of

$$A_r^{-1} \left[\frac{\partial \varphi_r}{\partial y} dy + \sum_{j=1}^m \frac{\partial \varphi_r}{\partial u_j} du_j - \frac{\varphi_r}{A_r} \cdot \right.$$
$$\left. \left(\frac{\partial A_r}{\partial y} dy + \sum_{j=1}^m \frac{\partial A_r}{\partial u_j} du_j \right) \right] = \bar{\omega}_r$$
$$\text{for } 1 \leq r \leq n-1,$$
$$A_1 \cdots A_{n-1}\varphi_n := P_n \quad \text{for } r = n.$$

$$\diamond$$

Theorem 4.1 : *The nonlinear system $\sum$ is locally equivalent to the system $\sum_a$ by a state coordinates transformation $z = \Phi(x)$ if and only if:*

$$d\omega_k \wedge dy = 0, \ d\omega_k \wedge du = 0 \tag{4.16}$$
$$\text{for } 1 \leq k \leq n-2$$
$$d\omega_{n-1} = 0 \tag{4.17}$$

$$d\bar{\omega}_r \wedge dy = 0, \ d\bar{\omega}_r \wedge du = 0 \tag{4.18}$$

where ω_k $(1 \leq k \leq n-1)$, and $\bar{\omega}_r$ $(1 \leq r \leq n)$ are defined by (4.11) and (4.15) respectively.

Thus, if $\sum$ is locally equivalent to $\sum_a$, then the state coordinates transformation $z = \Phi(x)$ is computed from $\sum_a$:

$$z_1 = y = h(x)$$
$$z_2 = \frac{1}{A_1(u)}(\dot{h} - \varphi_1(y, u))$$
$$z_3 = \frac{y^{(2)} - A_1\varphi_2 - \dot{\varphi}_1 - z_2\dot{A}_1}{A_1 \cdot A_2} \tag{4.19}$$
$$= \frac{h^{(2)} - A_1\varphi_2 - \dot{\varphi}_1 - \overline{\log A_1}(\dot{h} - \varphi_1)}{A_1 \cdot A_2}$$

$$\vdots$$
$$z_k = \frac{y^{(k-1)} - P_{k-1}}{\prod_{i=1}^{k-1} A_i} \tag{4.20}$$

where

$$P_k = \varphi_k \prod_{i=1}^{k-1} A_i + \dot{\overline{P}}_{k-1} + z_k \overline{\prod_{i=1}^{k-1} A_i} \tag{4.21}$$

for $1 \leq k \leq n$, $A_n := 0$ and $P_1 := \varphi_1$.
For the details of the proof, see [20]

4   Synthesis Observer for State Affine Systems

The observer $\sum_o$ of $\sum_a$, can be the realized solving

$$\dot{S} = -\theta S - A^T(\vartheta)S - SA(\vartheta) + C^T C \qquad (4.22)$$

The time varying solution is a definite positive matrice, and $\theta \in \Re^+$. In order to avoid the time varying computation, [4] propose a transformation, $\zeta = \Omega z$ where

$$\Omega := \text{diag}(\prod_{i=0}^{n-1} A_r), \quad \text{with} \ A_0 := 1. \qquad (4.23)$$

Applying it to system $\sum_a$, one gets

$$\begin{aligned} \dot{\zeta} &= \overline{A}\zeta + \Gamma(\vartheta, \zeta) \\ y &= C\zeta = \zeta_1 \end{aligned} \qquad (4.24)$$

where $\Gamma(\vartheta, \zeta) := \Omega\varphi(\vartheta) + \dot{\Omega}\Omega^{-1}\zeta$, $\overline{A} := \Omega A(\vartheta)\Omega^{-1}$ i.e.

$$\overline{A} = \begin{pmatrix} 0 & 1 & 0 & \cdots & 0 & 0 \\ 0 & 0 & 1 & \cdots & 0 & 0 \\ & & \vdots & \ddots & & \vdots \\ 0 & 0 & 0 & \cdots & 0 & 1 \\ 0 & 0 & 0 & \cdots & 0 & 0 \end{pmatrix}; \Omega\varphi(\vartheta) = \begin{pmatrix} \varphi_1 \\ A_1\varphi_2 \\ \vdots \\ \prod_{r=1}^{n-1} A_r\varphi_n \end{pmatrix};$$

$$\dot{\Omega}\Omega^{-1} = \text{Diag} \,(\log \prod_{r=1}^{\overset{i}{}} A_{r-1}), \ 1 \leq i \leq n.$$

where $A_0 := 1$.

Assume that $\|\Gamma(\vartheta, \zeta)\| \leq K\|\zeta\|_S$ with $\|\zeta\|_S := \zeta^T S\zeta$. Thus it is Lipchitzian with respect to ζ, uniformly w.r.t. ϑ with Lipschitz constant K. Then consider the system

$$\dot{\hat{\zeta}} = \overline{A}\hat{\zeta} + \Gamma(\vartheta, \hat{\zeta}) - S_\infty^{-1}C^T(C\hat{\zeta} - y) \qquad (4.25)$$

where S_∞ is the algebraic stationary solution of

$$0 = -\theta S_\infty - \overline{A}^T S_\infty - S_\infty \overline{A} + C^T C \qquad (4.26)$$

for θ large enough. Then, for inputs ϑ uniformly bounded by some $\vartheta \geq 0$: (4.25) is an observer for (4.24), i.e., for θ large enough

$$\|\hat{\zeta}(t) - \zeta(t)\| \leq K(\theta) \exp(-\theta t/3)\|\hat{\zeta}_0 - \zeta_0\| \qquad (4.27)$$

For the proof and the details cf. [7]. Thus from (4.23) this observer in the z coordinates have the following form

$$\dot{\hat{z}} = \Omega^{-1}\dot{\zeta} - \Omega^{-1}\dot{\Omega}\hat{z} \tag{4.28}$$

substituting (4.25) in (4.28) we finally have

$$\dot{\hat{z}} = A(\vartheta)\hat{z} + \varphi(\vartheta) - \Omega^{-1}S_\infty^{-1}C^T(C\hat{z} - y) \tag{4.29}$$

The transformation Ω and its inverse is always well defined since the observability assumption.

4.1   Physical Example

In order to illustrate the feasibility of S.A.T. Algorithm, Theorem 4.1 and the observer synthesis, we apply now the algorithm to a physical example.

The SISO system is a inverse pendulum, as considered in [1]. The dynamics of the system are

$$
\begin{aligned}
\dot{x}_1 &= x_2 \\
\dot{x}_2 &= \frac{\cos x_1\left(-al\sin x_1 x_2^2 + g\frac{\sin x_1}{\cos x_1} - \frac{a}{m}u\right)}{l(1 - a\cos^2 x_1)} \\
y &= x_1,
\end{aligned}
\tag{4.30}
$$

where x_1 denotes the angular displacement of the pendulum shaft from vertical position, l is its length, m the mass of the plumb bob, g the force of gravity and a the ratio $\frac{m}{m+M}$ with M the mass of the bearing trolley, u is the force applied to the car. Then, the system I/O differential equation is given by

$$y^{(2)} = \frac{-a\cos y\left(lm\sin y \cdot \dot{y}^2 + u - \frac{gm}{a}\frac{\sin y}{\cos y}\right)}{lm(1 - a\cos^2 y)} \tag{4.31}$$

and for $\sum_a$, one gets

$$P_a = y^{(2)} = F_2(A_1) + F_1(A_1, \varphi_1) + A_1 F_0(\varphi_2) \tag{4.32}$$

where

$$
\begin{aligned}
F_2 &:= \dot{y}\,\overline{\log A_1}; \; F_1 := \dot{\varphi}_1 - \overline{\log A_1}\varphi_1 \\
F_0 &:= \varphi_2
\end{aligned}
\tag{4.33}
$$

Applying the S.A.T Algorithm gives $A_1(y) = \dfrac{1}{(1 - a\cos^2 y)^{\frac{1}{2}}}$ and

$$\varphi_2 := \frac{P_2}{A_1} = \frac{au\cos y - gm\sin y}{lm(1 - a\cos^2 y)^{\frac{1}{2}}}. \tag{4.34}$$

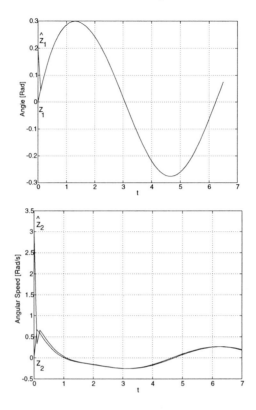

FIGURE 1. **Real and estimated state versus time (s).**

The system (4.30) is then locally equivalent to:

$$\dot{z} = \begin{pmatrix} 0 & A_1(y) \\ 0 & 0 \end{pmatrix} z + \begin{pmatrix} 0 \\ \varphi_2(y, u) \end{pmatrix} \qquad (4.35)$$
$$y = Cz = z_1$$

and from (4.20) the state coordinates transformation: $z_1 = h(x) = x_1$; $z_2 = x_2(1 - a\cos^2 x_1)^{\frac{1}{2}}$.

Thus the observer (4.29) has the following form

$$\dot{\hat{z}} = A(y)\hat{z} + \varphi(u, y) - \Omega^{-1}S_\infty^{-1}C^T(\hat{z}_1 - z_1) \qquad (4.36)$$

where S_∞ is the algebraic stationary solution of (4.25). Solving for S_∞ one gets $S_{11} = 1/\theta$, $S_{21} = -1/\theta^2$, $S_{22} = 2/\theta^3$. Then $S_\infty^{-1}C^T = (2\theta\ \theta^2)^T$ and $\Omega^{-1}S_\infty^{-1}C^T = (\ 2\theta,\ \ \theta^2/A_1(y)\ )^T$.

4.2   Simulation Results

We present the simulation results for a trajectory tracking $y_r := c_1\sin(c_2 t)$ where $c_1 = .3$, $c_2 = 1$, and $l = 0.36m$,ǎ $M = 2.4kg,$, $m = 0.23kg$, $\theta = 50$

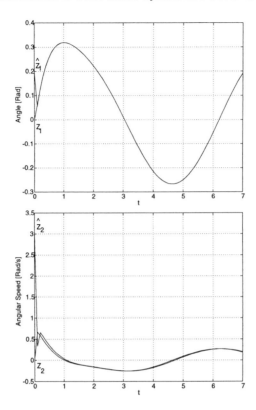

FIGURE 2. **Real and estimated state versus time (s).**

and $z_1(0) = z_2(0) := 0$, $\hat{z}_1(0) = 0.2$, $\hat{z}_2(0) = 3$.

In Fig. 1, we apply a tracking control with the real state and compare them with the estimated one. This error is practically neglectible. In Fig. 2, we measure only the output and apply the tracking control computed with the estimated states. As we have a pendulum set in our laboratory, next step of this work will be to check this observer in real conditions.

5   Conclusions

A NSC for the transformation to the state affine systems of MISO nonlinear systems has been obtained. The main result is stated in terms of the I/O differential equation and some one-forms obtained in a straightforward way and easy to check. With the help of a regular transformation, one avoids the time varying computation of the observer gain matrice an then a Kalman-like observer $\sum_o$ (4.3) for the transformed system $\sum_a$ (4.2) can be directly applied.

6 REFERENCES

[1] G. Besançon and G. Bornard. State equivalence based observer synthesis for nonlinear control systems. *Proc. IFAC 13th Triennial World Congress*, (San Francisco, USA, Vol. E, 287-292, 1996.

[2] J. Birk and M. Zeitz. Extended Luenberger observer for nonlinear multivariable systems, *Int. J. Control*, **47**, 1823-1836, 1988.

[3] G. Bornard, N. Couenne and F. Celle. Regularly persistent observers for bilinear systems Proc. in *29 I.C.N.S., New trends in nonlinear system theory*, (Vol. 122 Springer Verlag) June 1988.

[4] K. Busawon, M. Farza and H. Hammouri. A simple observer for a class of nonlinear systems, Personal Communication.

[5] J. N. Chiasson. 'Nonlinear differential-geometric techniques for control of a series DC motor', *IEEE Trans. Contr. Syst. Technology*, **2**, 35-42, 1994.

[6] S. Diop, J. W. Grizzle, P. E. Moraal and A. Stefanopoulou. Interpolation and numerical differentiation for observer design, *Proc. ACC 94, Evanston, Illinois*, 1329-1333, 1994.

[7] J. P. Gauthier, H. Hammouri and S. Othman. A simple observer for nonlinear systems applications to bioreactors. *IEEE TAC*, **37**(6), 885-880, 1992.

[8] D. Guillaume, P. Rouchon and J. Rudolph. Two simple observers for a class of polymerization reactors, 4th *ECC '97*, Bruxelles, 1997.

[9] A. Glumineau, C. H. Moog and F. Plestan. New algebro-geometric conditions for the linearization by input-output injection, *IEEE Trans. Autom. Control*, **41**, 598-603, 1996.

[10] H. Hammouri and J. P. Gauthier. Bilinearization up to output injection, *Syst. Contr. Letters*, **11**, 139-149, 1988.

[11] H. Hammouri and J. P. Gauthier. Global time varying linearization up to output injection, *SIAM J. Control Optim.*, **30** 1992, 1295-1310, 1992.

[12] H. Hammouri and M. de León. Observer synthesis for state-affine systems, *29 th. CDC*, Honolulu, Hawai. Dec. 5-7, 1990.

[13] H. Hammouri and M. de León. On systems equivalence and observer synthesis. *New Trends in Systems Theory*, 340-347, 1991.

[14] H. Hammouri and M. Kinnaert. A new procedure for time-varying linearization up to output injection, *Syst. Contr. Lett.*, **28**, 151-157, 1996.

[15] H. Keller. 'Nonlinear observer design by transformation into a generalized observer canonical form', *Int. J. Control,* **46**, 1915-1930, 1987.

[16] A. J. Krener and W. Respondek. Nonlinear observers with linearizable error dynamics, *SIAM J. Contr. Optim.,* **23**, 197-216, 1985.

[17] J. de León Morales and S. Acha Daza. Stabilization of a class of nonlinear systems, application to a synchronous generator, *Proc. of IASTE Int. Conf. High technology in the Power Industry, Orlando Fl., USA,* 23-26, 1998.

[18] J. Lévine and R. Marino, Nonlinear systems immersion, observers and finite dimensional filters, *Syst. Contr. Lett.,* **7** (1986), 133-142.

[19] V. López-M. and A. Glumineau. 'Further results on linearization of nonlinear systems by input output injection', *Proc. 36th. CDC IEEE,* (San Diego, USA), 1997.

[20] V. López-M. and J. de León Morales and A. Glumineau. Transformation of nonlinear systems into state affine control systems and observer synthesis, *IFAC CSSC,* Nantes, France, July 8-10, 1998.

[21] V. López-M., F. Plestan and A. Glumineau. Linearization by completely generalized input-output injection, *Proc. of 6th IEEE MCCA, Alghero,* Italy, 9-11, 1998.

[22] R. Marino and P. Tomei. 'Dynamic output feedback linearization and global stabilization', *Syst. Contr. Letters,* **17**, 115-121, 1991.

[23] F. Plestan and B. Cherki. 'An observer for a one flexible joint robot by an algebraic method', *Proc. IFAC Workshop on New Trends in Design of Control Systems NTDCS'94* (Smolenice, Slovakia), 41-46, 1994.

[24] F. Plestan and A. Glumineau. 'Linearization by generalized input-output injectionă for electrical motor observers', *Electrimacs 96* (Saint Nazaire, France), **2/3**, 569-574, 1996.

[25] F. Plestan and A. Glumineau. 'Linearization by generalized input-output injection', *Syst. Contr. Letters,* **31**, 115-128, 1997.

[26] T. Proychev and R. L. Mishkov. 'Transformation of nonlinear systems in observer canonical form with reduced dependency on derivatives of the input', *Automatica,* **29**, 495-498, 1993.

[27] J. Tsinias. Further results on the observer design problem, *Syst. Contr. Lett.* ,**14**, 411-418, 1990.

[28] D. Williamson. Observation of bilinear systems with application to biological systems, *Automatica,* **13**, 243-254, 1977.

[29] A. J. Van der Schaft. Representing a nonlinear state space system as a set of higher-order differential equations in the inputs and outputs, *Syst. Contr. Lett.*, **12**, 151-160, 1989.

[30] F. Viel. Stabilité des systèmes controlés par retour d'état estimé. Application aux réacteurs de polymérisation et aux colonnes à distiller, Thèse de Doctorat, Université Claude Bernard-Lyon 1, Lyon, 1994.

[31] X. H. Xia and W. B. Gao. Nonlinear observer design by observer error linearization, *SIAM J. Contr. Optim.*, **1**, 199-216, 1989.

On Existence of Extended Observers for Nonlinear Discrete-Time Systems

H. J. C. Huijberts

Department of Mathematics and Computing Science
Eindhoven University of Technology
P.O. Box 513, 5600 MB Eindhoven, The Netherlands

1 Introduction

In this chapter, we consider the design of observers for discrete-time non-linear systems by means of so called (extended) observer forms. Loosely speaking, a system in observer form is a linear observable (continuous-time or discrete-time) system that is interconnected with an output-dependent nonlinearity. Observers for this kind of systems may be built by building a classical linear Luenberger observer for the linear system, and adding the output-dependent nonlinearity to this observer. Thus, observer design for systems in observer form is relatively easy. By the same token, also observer design for systems that may be transformed into a system in observer form by means of a coordinate transformation and an output transformation is relatively easy.

Observer design for systems in observer form was first studied, in the continuous-time setting, in [10],[11] (see also [15]). In these papers, conditions were given under which a nonlinear continuous-time system may be transformed into a system in observer form by means of a coordinate transformation and an output transformation. Basically, these conditions were given in terms of the integrability of certain codistributions.

Later on, the observer design for discrete-time systems in observer form was studied (see [1],[12],[13] and the references therein), and conditions were given under which a nonlinear discrete-time system may be transformed into a system in observer form by means of a coordinate transformation and an output transformation. These conditions came down to the question whether certain functions could be factorized in a certain way. For single-output systems, conditions under which this factorization is indeed possible were given when only output transformations are allowed. (In fact, [13] also claims to give conditions for the multi-output case. However, these conditions seem to be incorrect.) One of the purposes of this chapter is to generalize the conditions given in [13] for single-output systems to the case where, besides a coordinate transformation, also an output transformation is allowed.

The conditions for the existence of an observer form for continuous-time systems and discrete-time systems given in [10],[11],[1],[12],[13] are quite restrictive. Therefore, generalizations have been considered, both in the continuous-time case and the discrete-time case. In the continuous-time case, so-called generalized observer forms were considered. These generalized observer forms consist of an observable linear system interconnected with a nonlinearity that depends on the output of the system and a finite number of its derivatives. In [6], differential geometric conditions were given under which a continuous-time system may be transformed into a generalized observer form by means of so called generalized coordinate transformations (i.e., transformations that, besides the state of the system, also depend on a finite number of time-derivatives of the outputs) and an output transformation. In the discrete-time context, the design of so called extended observers by using extended observer forms was studied in [7],[8]. Here, an extended observer is an observer that, besides the output of the system, also depends on a finite number of its past values, while a system in extended observer form consists of an observable linear system interconnected with a nonlinearity that depends on the output of the system and a finite number of past output values. In [7],[8], conditions were given under which a given single-output discrete-time system may be transformed into a system in extended observer form by means of an extended coordinate transformation (i.e., a coordinate transformation that depends on the state of the system and a finite number of past output values) and an output transformation. As in [1],[12], these conditions again boilt down to the question whether a given function may be factorized in a certain way. A corollary of the results obtained in [7],[8] is that when the number of past output values equals $n - 1$ (where n is the dimension of the state space of the system under consideration), an extended observer form always exists when the system under consideration is strongly observable (for the exact definition of strong observability, we refer to Section 3). In this chapter, differential geometric conditions for the existence of such a factorization for the cases that the number of past output values is smaller than $n - 1$ will be given.

It is to be noted that in principle the problem of observer design is a *global* problem. Therefore one would also like to obtain global conditions for existence of extended observer forms. However, if one knows that the system under consideration evolves on an invariant set, also existence conditions on this invariant set would suffice. In this chapter, all results obtained will be valid on open invariant sets on which some regularity assumptions hold. Of course, this also includes the case where one really would like to have global conditions.

The chapter is organized as follows. In the following section, an overview of results from the theory of differential forms that will be used in this chapter is given. In Section 3, we consider the existence of observer forms for single-output discrete-time systems. The results in this section reformulate and generalize the results in [13]. In Section 4, the existence of extended observer forms for single-output discrete-time systems is studied. Section 5 contains some conclusions.

2   Differential Forms

In this section we give an overview of results from the theory of differential forms that will be used in this chapter. For details, we refer to [2],[3],[4],[5],[14].

Let V be an r-dimensional vector space over $\mathbb{R}$. A k-form ω on V is a k-linear completely antisymmetric mapping $\omega : \underbrace{V \times \cdots \times V}_{k \text{ times}} \to \mathbb{R}$, i.e.,

$$(\forall \alpha, \beta \in \mathbb{R})(\forall v_1, v_1', v_2, \cdots, v_r \in V)(\omega(\alpha v_1 + \beta v_1', v_2, \cdots, v_k) =$$
$$\alpha\omega(v_1, \cdots, v_k) + \beta\omega(v_1', v_2, \cdots, v_k)) \tag{5.1}$$

$$(\forall v_1, \cdots, v_k \in V)(\omega(v_1, \cdots, v_k) =$$
$$-\omega(v_1, \cdots, v_{i-1}, v_{i+1}, v_i, v_{i+2}, \cdots, v_k)) \tag{5.2}$$

Note that a one-form on V is just an element of V^*, the dual of V. The space of all k-forms on V is denoted by $\Lambda^k(V^*)$. It is easily checked that the k-linearity and anti-symmetry of a k-form on V implies that all k-forms are zero for $k > r$. We define

$$\Lambda(V^*) := \Lambda^0(V^*) \oplus \Lambda^1(V^*) \oplus \cdots \oplus \Lambda^r(V^*) \tag{5.3}$$

where

$$\Lambda^0(V^*) := \mathbb{R}$$

We call $\Lambda(V^*)$ the *exterior algebra* over V^*. An element $\omega \in \Lambda(V^*)$ is called a *form* on V and may be written in a unique way as

$$\omega = \omega^0 + \omega^1 + \cdots + \omega^r \tag{5.4}$$

where $\omega^i \in \Lambda^i(V^*)$ $(i = 0, \cdots, r)$.

We next define a product on $\Lambda(V^*)$, the so called *wedge product* (or *exterior product*). This product will be denoted by "$\wedge$". First, let $\eta \in \Lambda^p(V^*)$,

$\omega \in \Lambda^q(V^*)$. Then $\eta \wedge \omega \in \Lambda^{p+q}(V^*)$ is defined by

$$(\eta \wedge \omega)(v_1, \cdots, v_{p+q}) =$$

$$\sum_{\pi}(\text{sign}(\pi))\eta(v_{\pi(1)}, \cdots, v_{\pi(p)})\omega(v_{\pi(p+1)}, \cdots, v_{\pi(p+q)}) \tag{5.5}$$

where the summation is over all possible permutations of $1, \cdots, p + q$, and $v_1, \cdots, v_{p+q} \in V$. For $\eta = \eta^0 + \cdots + \eta^r$, $\omega = \omega^0 + \cdots + \omega^r$, with $\eta^i, \omega^i \in \Lambda^i(V^*)$ $(i = 0, \cdots, r)$, we define

$$\eta \wedge \omega = \sum_{i,j=0}^{r} \eta^i \wedge \omega^j \tag{5.6}$$

Note that the wedge-product is associative and distributive, but not commutative. Instead, it satisfies

$$\eta \wedge \omega = (-1)^{pq}\omega \wedge \eta, \eta \in \Lambda^p(V^*), \omega \in \Lambda^q(V^*) \tag{5.7}$$

Let $v \in V$ be given. Then the *interior product* $v \lrcorner : \Lambda(V^*) \rightarrow \Lambda(V^*)$ is defined in the following way. First consider $\omega \in \Lambda^k(V^*)$. Then $v \lrcorner \omega \in \Lambda^{k-1}(V^*)$ is given by

$$(v \lrcorner \omega)(v_1, \cdots, v_{p-1}) = \omega(v, v_1, \cdots, v_{p-1}) \tag{5.8}$$

where $v_1, \cdots, v_{p-1} \in V$. If $\omega = \omega^0 + \cdots + \omega^r$, with $\omega^i \in \Lambda^i(V^*)$ $(i = 0, \cdots, r)$, then

$$(v \lrcorner \omega) = \sum_{i=0}^{r}(v \lrcorner \omega^i) \tag{5.9}$$

Next, consider an n-dimensional manifold M. Let $T_x^* M$ denote the cotangent space at $x \in M$ and let $T^* M$ denote the cotangent bundle of M. Since $T_x^* M$ is an n-dimensional vector space over $\mathbb{R}$, we may define $\Lambda^k(T_x^* M)$ for all $x \in M$, $k = 0, \cdots, n$, as

$$\Lambda(T_x^* M) := \sum_{k=0}^{n} \Lambda^k(T_x^* M)$$

We then define the bundles $\Lambda^k(T^* M)$, $\Lambda(T^* M)$ over M by

$$\Lambda^k(T^* M) := \bigcup_{x \in M} \Lambda^k(T_x^* M), (k = 0, \cdots, n) \tag{5.10}$$

$$\Lambda(T^* M) := \bigoplus_{k=0}^{n} \Lambda^k(T^* M) \tag{5.11}$$

A *differential form on* M is now defined to be a smooth section of the bundle $\Lambda(T^*M)$, while a *differential* k-*form on* M is defined to be a smooth section of the bundle $\Lambda^k(T^*M)$. So, roughly speaking, a differential $(k-)$form on M is a "prescription" that assigns a $(k-)$form ω_x on T_x^*M to every $x \in M$ in a smooth way. Note that by this definition a differential 0-form on M is just a smooth function on M. When no confusion arises, we will simply call a differential $(k-)$form on M a $(k-)$form on M in the sequel.

The *wedge product* of the forms η, ω on M is defined to be the form $(\eta \wedge \omega)$ satisfying

$$(\eta \wedge \omega)_x = \eta_x \wedge \omega_x, (\forall x \in M) \tag{5.12}$$

Let τ be a smooth vector field on M, and let ω be a form on M. Then the *interior product* $(\tau \lrcorner \omega)$ is defined to be the form satisfying

$$(\tau \lrcorner \omega)_x = \tau_x \lrcorner \omega_x, (\forall x \in M) \tag{5.13}$$

The *exterior differential operator* d maps a k-form ω into a $(k+1)$-form dω, called the *exterior derivative* of ω. The operator d is uniquely defined by the following properties:

1. d is linear:

$$(\forall \alpha, \beta \in I\!R)(\mathrm{d}(\alpha\eta + \beta\omega) = \alpha\mathrm{d}\eta + \beta\mathrm{d}\omega)$$

2. If η is a k-form, then

$$\mathrm{d}(\eta \wedge \omega) = \mathrm{d}\eta \wedge \omega + (-1)^k \eta \wedge \mathrm{d}\omega$$

3. $\mathrm{d}^2 = 0$.

4. If f is a 0-form, then df is the ordinary differential df of f.

5. d is local: if η and ω coincide on an open set U, then d$\eta = $ dω on U.

A k-form ω is called *closed* if d$\omega = 0$; it is called *exact* if there exists a $(k-1)$-form η such that $\omega = $ dη. Note that, since d$^2 = 0$, an exact k-form is closed. The converse does not need to hold globally. However, it does hold locally, as is reflected by the following theorem.

Theorem 5.1 (Poincaré Lemma) *If* M *is smoothly contractible to a point* $x_0 \in M$, *then every closed form* ω *on* M *is exact.*

Let τ be a smooth vector field on M. The *Lie-derivative* $\mathcal{L}_\tau$ maps a k-form ω into a k-form $\mathcal{L}_\tau\omega$. $\mathcal{L}_\tau$ is uniquely defined by the following properties:

1. If f is a 0-form on M, then

$$\mathcal{L}_\tau f = \tau \lrcorner df$$

2. $\mathcal{L}_\tau$ is a derivation:

$$\mathcal{L}_\tau(\eta \wedge \omega) = \eta \wedge \mathcal{L}_\tau \omega + \mathcal{L}_\tau \eta \wedge \omega$$

3. $\mathcal{L}_\tau$ commutes with d:

$$\mathcal{L}_\tau(\mathrm{d}\omega) = \mathrm{d}(\mathcal{L}_\tau \omega)$$

From the definitions of interior product, exterior derivative, and Lie-derivative one may derive the following identities that will be frequently used in the sequel (here σ, τ denote smooth vector fields on M, and ω denotes a k-form on M).

$$\mathcal{L}_\tau \omega = \mathrm{d}(\tau \lrcorner \omega) + \tau \lrcorner \mathrm{d}\omega \qquad (5.14)$$

$$[\sigma, \tau] \lrcorner \omega = \mathcal{L}_\sigma(\tau \lrcorner \omega) - \tau \lrcorner \mathcal{L}_\sigma \omega (\mathrm{Leibniz} - \mathrm{formula}) \qquad (5.15)$$

$$\mathcal{L}_{[\sigma,\tau]}\omega = \mathcal{L}_\sigma \mathcal{L}_\tau \omega - \mathcal{L}_\tau \mathcal{L}_\sigma \omega \qquad (5.16)$$

Denote by $\Omega^1(M)$ the set of all one-forms on M. Then $\Omega^1(M)$ has the structure of a finitely generated module over the ring of smooth functions on M. A finitely generated submodule of $\Omega^1(M)$ is called a *codistribution* on M. The minimal number of generators of a codistribution is called its *dimension*. A codistribution Ω on M is called *integrable* if it has a set of generators consisting of exact one-forms. The codistribution generated by a set of one-forms $\omega_1, \cdots, \omega_d$ is denoted by $\mathrm{span}\{\omega_1, \cdots, \omega_d\}$. For a one-form ω and a d-dimensional codistribution Ω, we will say that

$$\mathrm{d}\omega \equiv 0 \mathrm{mod}\Omega$$

if and only if

$$\mathrm{d}\omega \wedge \omega_1 \wedge \cdots \wedge \omega_d = 0 (\forall \omega_1, \cdots, \omega_d \in \Omega)$$

Theorem 5.2 (Frobenius Theorem) *Let M be smoothly contractible to a point $x_0 \in M$, and let Ω be a codistribution on M. Then the following statements are equivalent.*

(i) Ω is integrable.

(ii) For every $\omega \in \Omega$ we have that $\mathrm{d}\omega \equiv 0 \mathrm{mod}\Omega$.

(iii) Let $\{\omega_1, \cdots, \omega_d\}$ be a set of generators of Ω and let $\omega_{d+1}, \cdots, \omega_n$ be such that $\{\omega_1, \cdots, \omega_n\}$ generate $\Omega^1(M)$. Then Ω is integrable if and only if there exist smooth functions Γ_{ij}^k $(i, k = 1, \cdots, d; j = i + 1, \cdots, n)$ on M such that

$$\mathrm{d}\omega^k = \sum_{i=1}^{d} \sum_{j=i+1}^{n} \Gamma_{ij}^k \omega_i \wedge \omega_j (k = 1, \cdots, d) \qquad (5.17)$$

In what follows, we will also need the following result.

Theorem 5.3 (Cartan's Lemma) *If $\omega_1, \cdots, \omega_d$ are independent one-forms, and $\pi_1, \cdots, \pi_d$ are one-forms such that*

$$\pi_1 \wedge \omega_1 + \cdots + \pi_d \wedge \omega_d = 0$$

then

$$\pi_i \in \mathrm{span}\{\omega_1, \cdots, \omega_d\}$$

3 Observer Design using Observer Forms

We start our investigation of observer design for nonlinear discrete-time systems by considering a nonlinear discrete-time system $\tilde{\Sigma}$ of the form

$$\tilde{\Sigma} \begin{cases} z(k+1) &= Az(k) + \Phi(\tilde{y}(k)) \\ \tilde{y}(k) &= Cz(k) \end{cases} \tag{5.18}$$

with state $z \in \mathbb{R}^n$, output $\tilde{y} \in \mathbb{R}$, and where A, C are matrices of appropriate dimensions, the mapping $\Phi : \mathbb{R} \to \mathbb{R}^n$ is smooth, and the pair (C, A) is in Brunovsky form. Analogously to [10],[11], a system of this form is called a system in *observer form*. For systems in observer form, observer design is particularly simple. Namely, the fact that the pair (C, A) is in Brunovsky form, and thus in particular observable, implies that there exists a matrix K such that all eigenvalues of the matrix $A - KC$ are in the open unit disc. It is then straightforwardly checked that the following system is an observer for $\tilde{\Sigma}$:

$$\begin{cases} \hat{z}(k+1) &= A\hat{z}(k) + K(\tilde{y}(k) - \hat{\tilde{y}}(k)) + \Phi(\tilde{y}(k)) \\ \hat{\tilde{y}}(k) &= C\hat{z}(k) \end{cases} \tag{5.19}$$

We next consider a nonlinear discrete-time system Σ of the form

$$\Sigma \begin{cases} x(k+1) &= f(x(k)) \\ y(k) &= h(x(k)) \end{cases} \tag{5.20}$$

where $x \in \mathbb{R}^n$, $y \in \mathbb{R}$, and the mappings $f : \mathbb{R}^n \to \mathbb{R}^n$ and $h : \mathbb{R}^n \to \mathbb{R}$ are smooth. We will say that Σ can be put in observer form if there exist a diffeomorphism $P : \mathbb{R}^n \to \mathbb{R}^n$ of the state space and a diffeomorphism $p : \mathbb{R} \to \mathbb{R}$ of the output space such that in the new coordinates $z = P(x)$ and with the new output $\tilde{y} = p(y)$ the system Σ takes the form (5.18), where the pair (C, A) is in Brunovsky form. If Σ does admit an observer form, an observer for Σ may then be obtained by first building an observer (5.19) for the observer form (5.18) and then letting $\hat{x}(k) := P^{-1}(\hat{z}(k))$ be the estimate of $x(k)$.

We thus see that observer design for a discrete-time system Σ of the form (5.20) is relatively easy when Σ can be put in observer form. This raises the question under what conditions Σ can be put in observer form. To derive these conditions, we first introduce the so called *observable form* of Σ. Define the *observability map* $\psi : I\!\!R^n \to I\!\!R^n$ by

$$\psi(x) := \begin{pmatrix} h(x) \\ h \circ f(x) \\ \vdots \\ h \circ f^{n-1}(x) \end{pmatrix} \tag{5.21}$$

where $f^1 := f$, $f^k := f \circ f^{k-1}$. Assume that the origin is an equilibrium point of Σ (i.e., $f(0) = 0$) and that $h(0) = 0$. We then call Σ *strongly observable* on an open subset $U \subset I\!\!R^n$ containing the origin if ψ is a diffeomorphism on U. It follows that if Σ is strongly observable on U, then $s := \psi(x)$ forms a new set of local coordinates for Σ on U. In these new coordinates, Σ takes the form

$$\left\{ \begin{array}{rcl} s_1(k+1) &=& s_2(k) \\ &\vdots& \\ s_{n-1}(k+1) &=& s_n(k) \\ s_n(k+1) &=& f_s(s(k)) \\ y(k) &=& s_1(k) \end{array} \right. \tag{5.22}$$

where $f_s := h \circ f^{n-1} \circ \psi^{-1}$. The form (5.22) is referred to as the *observable form* of Σ.

We now have the following result.

Theorem 5.4 *Consider a nonlinear discrete-time system Σ of the form (5.20), and assume that $f(0) = 0$, $h(0) = 0$. Then Σ can be put in observer form on an open invariant subset U of (5.20) containing the origin if and only if*

(i) Σ is strongly observable on U.

(ii) There exist functions $p, \phi_1, \cdots, \phi_n : I\!\!R \to I\!\!R$, where p is a diffeomorphism on $h(U)$, such that on U the function f_s in (5.22) satisfies

$$p \circ f_s(s) = \phi_1(s_n) + \phi_2(s_{n-1}) + \cdots + \phi_n(s_1) \tag{5.23}$$

Proof. The proof for the case where only coordinate transformations are considered (or, in other words, the case where $p = \mathrm{id}_{I\!\!R}$) may be found in [13],[8],[9]. The case where also output transformations are considered is an (almost) immediate consequence of the case where $p = \mathrm{id}_{I\!\!R}$. ∎

If functions $p, \phi_1, \cdots, \phi_n : I\!\!R \to I\!\!R$ exist such that f_s satisfies (5.23), we define the following coordinate change:

$$z_i := p(s_i) - \sum_{j=1}^{i-1} \phi_j(s_{i-j})(i = 1, \cdots, n) \qquad (5.24)$$

In these new coordinates, we obtain the observer form for Σ:

$$\left\{ \begin{array}{rcl} z_1(k+1) & = & z_2(k) + \tilde{\phi}_1(\tilde{y}(k)) \\ & \vdots & \\ z_{n-1}(k) & = & z_n(k) + \tilde{\phi}_{n-1}(\tilde{y}(k)) \\ z_n(k) & = & \tilde{\phi}_n(\tilde{y}(k)) \\ \tilde{y}(k) & = & z_1(k) \end{array} \right. \qquad (5.25)$$

where $\tilde{\phi}_i := \phi_i \circ p^{-1}$.

From Theorem 5.4, it follows that the question whether or not Σ can be put in observer form may be reduced to the question whether or not the function f_s in (5.22) may be written in the special form (5.23). We will now derive conditions on f_s under which this is possible. To this end, we define one-forms $\omega_1, \cdots, \omega_n$ by

$$\omega_i := \sum_{j=1}^{i} \left(\frac{\partial f_s}{\partial s_j} \right) ds_j (i = 1, \cdots, n) \qquad (5.26)$$

For the case that $p = \mathrm{id}_{I\!\!R}$ we then have the following result.

Theorem 5.5 *Consider a discrete-time system Σ of the form (5.20) that is strongly observable on an open invariant subset $U \subset I\!\!R^n$ containing the origin. Assume further that U is smoothly contractible to the origin, and that the one-forms $\omega_1, \cdots, \omega_n$ in (5.26) generate a codistribution on U. Then Σ can be put in observer form with $p = \mathrm{id}_{I\!\!R}$ on U if and only if the one-forms $\omega_1, \cdots, \omega_n$ in (5.26) satisfy*

$$d\omega_i = 0(i = 1, \cdots, n) \qquad (5.27)$$

Proof. *(necessity)* Follows by direct verification.

(sufficiency) Assume that the one-forms ω_i in (5.26) satisfy (5.27). It then follows from (5.26,5.27) that

$$0 = d\omega_i = \cdots = \sum_{j=1}^{i} \sum_{k=i+1}^{n} \left(\frac{\partial^2 f_s}{\partial s_k \partial s_j} \right) ds_k \wedge ds_j (i = 1, \cdots, n) \qquad (5.28)$$

which is equivalent to

$$\left(\frac{\partial^2 f_s}{\partial s_k \partial s_j} \right) = 0(j, k = 1, \cdots, n; j \neq k) \qquad (5.29)$$

This implies that there exist functions $\phi_1, \cdots, \phi_n : \mathbb{R} \to \mathbb{R}$ such that f_s satisfies (5.23) with $p = \mathrm{id}_\mathbb{R}$. ■

We next give conditions under which f_s may be written in the form (5.23), where $p \neq \mathrm{id}_\mathbb{R}$. To do this, we let $\tau_1, \cdots, \tau_n$ be vector fields that are dual to the one-forms $\omega_1, \cdots, \omega_n$ in (5.26), i.e.,

$$\tau_i \lrcorner \omega_j = \delta_{ij}(i, j = 1, \cdots, n) \tag{5.30}$$

where δ_{ij} is the Kronecker delta.

Theorem 5.6 *Consider a discrete-time system Σ of the form (5.20) that is strongly observable on an open invariant subset $U \subset \mathbb{R}^n$ containing the origin. Assume further that U is smoothly contractible to the origin, and that the one-forms $\omega_1, \cdots, \omega_n$ in (5.26) generate a codistribution on U. Then the following statements are equivalent:*

(i) Σ can be put in observer form on U.

(ii) There exists a function $S : U \to \mathbb{R}$ such that

$$\mathrm{d}\omega_i = \mathrm{d}S \wedge \omega_i (i = 1, \cdots, n) \tag{5.31}$$

(iii) The one-forms $\omega_1, \cdots, \omega_n$ and the vector fields $\tau_1, \cdots, \tau_n$ satisfy

$$\mathrm{d}\omega_i \wedge \omega_j + \mathrm{d}\omega_j \wedge \omega_i = 0(i, j = 1, \cdots, n) \tag{5.32}$$

and

$$\mathcal{L}_{\tau_i}\left([\tau_i, \tau_j]\lrcorner\omega_i\right) = \mathcal{L}_{\tau_j}\left([\tau_j, \tau_i]\lrcorner\omega_j\right) \tag{5.33}$$

Proof. $(i) \Rightarrow (ii)$ Assume that Σ can be put in observer form on U. Define one-forms $\tilde{\omega}_i$ by

$$\tilde{\omega}_i := \sum_{j=1}^{i}\left(\frac{\partial(p \circ f_s)}{\partial s_j}\right)ds_j(i = 1, \cdots, n) \tag{5.34}$$

It then follows from Theorem 5.5 that we have that

$$\mathrm{d}\tilde{\omega}_i = 0(i = 1, \cdots, n) \tag{5.35}$$

Together with the fact that $\tilde{\omega}_i = (p' \circ f_s)\omega_i$, this gives that

$$0 = \mathrm{d}\tilde{\omega}_i = d\left(\frac{1}{p' \circ f_s}\right) \wedge \tilde{\omega}_i$$

$$(p' \circ f_s)d\left(\frac{1}{p' \circ f_s}\right) \wedge \omega_i = d(-\log|p' \circ f_s|) \wedge \omega_i \tag{5.36}$$

which establishes our claim for $S := -\log|p' \circ f_s|$.

$(ii) \Rightarrow (i)$ Assume that there exists a function $S : U \to \mathbb{R}$ such that (5.31) holds. Since $\omega_n = df_s$, (5.31) for $i = n$ gives that $dS \wedge df_s = 0$. By Cartan's Lemma, this gives that $dS \in \text{span}\{df_s\}$. Define $T := \exp(-S)$. We then also have that $dT \in \text{span}\{df_s\}$, which implies that there exists a function $\tilde{p} : \mathbb{R} \to \mathbb{R}$ such that $T = \tilde{p} \circ f_s$. Define $p : \mathbb{R} \to \mathbb{R}$ by $p := \int p(\tau) d\tau$, and define one-forms $\tilde{\omega}_1, \cdots, \tilde{\omega}_n$ as in (5.34). Note that we then have that $\tilde{\omega}_i = T\omega_i$ $(i = 1, \cdots, n)$, which implies that

$$d\tilde{\omega}_i = T d\omega_i + dT \wedge \omega_i = T(d\omega_i - dS \wedge \omega_i) = 0 (i = 1, \cdots, n) \qquad (5.37)$$

Together with Theorem 5.5 this establishes our claim.

$(ii) \Leftrightarrow (iii)$ We first show that (5.32) is equivalent to the existence of a *unique* one-form η such that

$$d\omega_i = \eta \wedge \omega_i (i = 1, \cdots, n) \qquad (5.38)$$

Note that if there exists a one-form η such that (5.38) holds, then (5.32) follows immediately. Conversely, assume that (5.32) holds. For $i = j$, (5.32) gives in particular that $d\omega_i \wedge \omega_i = 0$ $(i = 1, \cdots, n)$. It then follows from the Frobenius Theorem that there exist one-forms $\rho_1, \cdots, \rho_n$ such that

$$d\omega_i = \rho_i \wedge \omega_i (i = 1, \cdots, n) \qquad (5.39)$$

Writing $\rho_i = \sum_{k=1}^n \rho_{ik}\omega_k$ $(i = 1, \cdots, n)$, we then obtain from (5.32) that

$$0 = d\omega_i \wedge \omega_j + d\omega_j \wedge \omega_i = \cdots = \sum_{\substack{k=1 \\ k \neq i,j}}^n (\rho_{ik} - \rho_{jk})\omega_k \wedge \omega_i \wedge \omega_j \qquad (5.40)$$

which gives that

$$\rho_{ik} = \rho_{jk}(i, j, k = 1, \cdots, n; k \neq i, j) \qquad (5.41)$$

This immediately implies that indeed there exists a unique one-form η such that (5.38) holds. What remains to be done, is to show that there locally exists a function S such that $\eta = dS$ if and only if (5.33) holds. Writing $\eta = \sum_{j=1}^n \eta_j \omega_j$, we have that

$$d\eta = \sum_{j=1}^n (d\eta_j \wedge \omega_j + \eta_j d\omega_j) = \cdots = \sum_{j=1}^{n-1} \sum_{i=j+1}^n (\mathcal{L}_{\tau_i}\eta_j - \mathcal{L}_{\tau_j}\eta_i)\omega_i \wedge \omega_j$$

$$(5.42)$$

By the Poincaré Lemma, this gives that there exists a function $S : U \to \mathbb{R}$ such that $\eta = dS$ if and only if

$$\mathcal{L}_{\tau_i}\eta_j = \mathcal{L}_{\tau_j}\eta_i (i, j = 1, \cdots, n) \qquad (5.43)$$

Using (5.15),(5.14), it is straightforwardly checked that $\eta_j = [\tau_i, \tau_j] \lrcorner \omega_i$. Together with (5.43), this yields (5.33), which establishes our claim. ∎

4   Observer Design using Extended Observer Forms

In the previous section, we have seen that observer design for a nonlinear discrete-time system (5.20) is relatively easy when the system can be put in observer form (5.18). Unfortunately, however, the conditions for existence of an observer form given in Theorem 5.4 are quite restrictive. In this section, we will relax these conditions by considering so called *extended observers* and *extended observer forms*.

We will first explain what is meant by an extended observer. Consider a system Σ of the form (5.20), and let $N \in I\!N$ be given. We now assume that at every time instant $k \geq N$ we do not only know the output $y(k)$ at time k, but also the past outputs $y(k-1), \cdots, y(k-N)$. An *extended observer with buffer* N for Σ then is a dynamical system $\widehat{\Sigma}$ of the form

$$\widehat{\Sigma} : \widehat{x}(k+1) = \widehat{f}(\widehat{x}(k), y(k), \cdots, y(k-N)), k \geq N$$

where $\widehat{x} \in I\!R^n$ and the mapping $\widehat{f} : I\!R^{n+N+1} \to I\!R^n$ is smooth, with the property that $x(k) - \widehat{x}(k) \to 0$ $(k \to \infty)$, for all $x(0), \widehat{x}(N) \in I\!R^n$.

To study the design of extended observers for a discrete-time system Σ of the form (5.20), we first consider a system $\widetilde{\Sigma}^e$ of the form

$$\widetilde{\Sigma}^e \left\{ \begin{array}{rcl} z(k+1) & = & Az(k) + \Phi(\tilde{y}(k), \cdots, \tilde{y}(k-N)) \\ \tilde{y}(k) & = & Cz(k) \end{array} \right. \tag{5.44}$$

where the state $z \in I\!R^n$, the output $\tilde{y} \in I\!R$, A, C are matrices of appropriate dimensions, the mapping $\Phi : I\!R^{N+1} \to I\!R^n$ is smooth, and the pair (C, A) is in Brunovsky form. Note that for $N = 0$ the system $\widetilde{\Sigma}^e$ is identical to the system $\widetilde{\Sigma}$ in (5.18). Therefore, a system $\widetilde{\Sigma}^e$ of the form (5.44) will be referred to as a *system in extended observer form with buffer* N. As for a system in observer form, the design of an extended observer for a system in extended observer form is relatively easy. Namely, it is straightforwardly checked that the system

$$\left\{ \begin{array}{rcl} \widehat{z}(k+1) & = & A\widehat{z}(k) + K(\tilde{y}(k) - \widehat{\tilde{y}}(k)) + \Phi(\tilde{y}(k), \cdots, \tilde{y}(k-N)) \\ \widehat{\tilde{y}}(k) & = & C\widehat{z}(k) \end{array} \right.$$

$$\tag{5.45}$$

where the matrix K is such that all eigenvalues of $A - KC$ are in the open unit disc, is an extended observer for $\widetilde{\Sigma}^e$.

As in the previous section, we now consider the question under which conditions a given discrete-time system can be put in extended observer form for some $N \in I\!N$. The transformations we are going to use here, are more general than the ones in the previous section, in the sense that we also allow them to depend on the past output measurements $y(k-1), \cdots, y(k-N)$. More specifically, we will be looking at parametrized transformations $z = P(x, \xi_n, \cdots, \xi_N)$, where $z \in I\!R^n$, with the property that there exists a mapping $P^{-1}(\cdot, \xi_1, \cdots, \xi_N) : I\!\!R^n \to I\!\!R^n$ parametrized by $(\xi_1, \cdots, \xi_N)$ such that for all $(\xi_1, \cdots, \xi_N)$ we have that

$$P(P^{-1}(z, \xi_1, \cdots, \xi_N), \xi_1, \cdots, \xi_N) = z$$

A mapping having this property will be referred to as an *extended coordinate change*. We will then say that the system (5.20) *can be put in extended observer form with buffer* N if there exists an extended coordinate change $P(\cdot, \xi_1, \cdots, \xi_N) : I\!\!R^n \to I\!\!R^n$ parametrized by $(\xi_1, \cdots, \xi_N)$ and a diffeomorphism $p : I\!\!R \to I\!\!R$ of the output space such that the variable

$$z(k) := P(x(k), y(k-1), \cdots, y(k-N)) \tag{5.46}$$

satisfies (5.44), where $\tilde{y} := p(y)$, and the pair (C, A) is in Brunovsky form. As pointed out above, one may then build an extended observer (5.45) for $z(k)$ in (5.46). From this extended observer, one then obtains an estimate $\hat{x}(k)$ for $x(k)$ by inverting the extended transformation P:

$$\hat{x}(k) := P^{-1}(\hat{z}(k), y(k-1), \cdots, y(k-N)), k \geq N \tag{5.47}$$

The following generalization of Theorem 5.4 gives conditions under which a discrete-time system (5.20) can be put in extended observer form.

Theorem 5.7 *Consider a nonlinear discrete-time system Σ of the form (5.20), and assume that $f(0) = 0$, $h(0) = 0$. Let $N \in \{0, \cdots, n-1\}$ be given. Then Σ can be put in observer form with buffer N on an open invariant subset $U \subset I\!\!R^n$ containing the origin if and only if*

(i) Σ *is strongly observable on U.*

(ii) *There exist functions $\phi_1, \cdots, \phi_{n-N} : I\!\!R^{N+1} \to I\!\!R$ and a diffeomorphism p on $h(U)$, such that on U the function f_s in (5.22) satisfies*

$$p \circ f_s(s) = \sum_{i=1}^{n-N} \phi_{n+1-N-i}(s_{i+N}, \cdots, s_i) \tag{5.48}$$

Proof. The proof for the case where only coordinate transformations are considered (or, in other words, the case where $p = \mathrm{id}_{I\!\!R}$) may be found in [8]. The case where also output transformations are considered is an (almost) immediate consequence of the case where $p = \mathrm{id}_{I\!\!R}$. ∎

If there exist functions $p, \phi_1, \cdots, \phi_{n-N}$ such that f_s satisfies (5.48), one defines the following variables:

$$z_i(k) := s_i(k) -$$

$$\sum_{j=1}^{\alpha_i} \phi_j(s_{i-j}(k), \cdots, s_1(k), y(k-1), \cdots, y(k-N-1+i-j)) \tag{5.49}$$

$$(i = 1, \cdots, n)$$

where $\alpha_i := \min(i - 1, n - N)$. Further, define $\tilde{\phi}_1, \cdots, \tilde{\phi}_{n-N} : I\!R^{N+1} \to I\!R^{N+1}$ by

$$\tilde{\phi}_i(\xi_1, \cdots, \xi_{N+1}) := \phi_i(p^{-1}(\xi_1), \cdots, p^{-1}(\xi_{N+1}))$$

$$(i = 1, \cdots, n - N)$$

It may then be shown that in these variables we obtain the following extended observer form:

$$\left\{ \begin{array}{rcl} z_1(k+1) & = & z_2(k) + \tilde{\phi}_1(\tilde{y}(k), \cdots, \tilde{y}(k-N)) \\ & \vdots & \\ z_{n-N}(k+1) & = & z_{n-N+1}(k) + \tilde{\phi}_{n-N}(\tilde{y}(k), \cdots, \tilde{y}(k-N)) \\ z_{n-N+1}(k+1) & = & z_{n-N+2}(k) \\ & \vdots & \\ z_{n-1}(k+1) & = & z_n(k) \\ z_n(k+1) & = & 0 \\ \tilde{y}(k) & = & z_1(k) \end{array} \right. \tag{5.50}$$

Note that Theorem 5.7 generalizes Theorem 5.4. Further, from Theorem 5.7, we obtain the following result for $N = n - 1$ (see also [7],[8]).

Corollary 5.1 *Consider a nonlinear discrete-time system Σ of the form (5.20), and assume that $f(0) = 0$, $h(0) = 0$. Then Σ can be put in observer form with buffer N on an open invariant subset $U \subset I\!R^n$ containing the origin if and only if Σ is strongly observable on U.*

Thus, we see that every strongly observable system can be put in extended observer with buffer $N = n - 1$. From a practical point of view, e.g. when n is large, it may be desirable to reduce the size of the buffer. Therefore, we next investigate under which conditions an extended observer with buffer $N \in \{1, \cdots, n - 2\}$ exists. As in the previous section, these conditions are again given in terms of the one-forms ω_i in (5.26) and the vector fields τ_i in (5.30). Using the one-forms ω_i in (5.26), we define the following codistributions:

$$\Omega_i := \mathrm{span}\{\omega_k - \omega_{k-1} \mid k = i + 1, \cdots, \min(n, i + N)\}(i = 1, \cdots, n) \tag{5.51}$$

$$\tilde{\Omega}_i := \mathrm{span}\{\omega_i, \cdots, \omega_{i+N}\}(i = 1, \cdots, n - N - 1) \tag{5.52}$$

We first consider the case without output transformations, i.e., the case where in (5.48) we have that $p = \mathrm{id}_{I\!R}$. This result generalizes Theorem 5.5.

Theorem 5.8 *Consider a discrete-time system Σ of the form (5.20) that is strongly observable on an open invariant subset $U \subset I\!R^n$ containing the* *origin. Assume further that U is smoothly contractible to the origin, and that the one-forms $\omega_1, \cdots, \omega_n$ in (5.26) generate a codistribution on U. Let $N \in \{1, \cdots, n-2\}$ be given. Then for $p = \mathrm{id}_{\mathbb{R}}$, Σ can be put in extended observer form with buffer N on U if and only if the one-forms ω_i in (5.26) satisfy*

$$d\omega_i \equiv 0 \bmod \Omega_i (i = 1, \cdots, n) \tag{5.53}$$

Proof. *(necessity)* Follows by direct verification.

(sufficiency) Assume that the one-forms ω_i in (5.26) satisfy (5.53). Note that by the definition of the ω_i we have that

$$\Omega_i = \mathrm{span}\{ds_k \mid k = i+1, \cdots, \min(n, i+N)\}(i = 1, \cdots, n)$$

Defining $\alpha_i := \min(n, i+N)$ $(i = 1, \cdots, n)$, (5.53) then gives

$$0 = d\omega_i \wedge ds_{i+1} \wedge \cdots \wedge ds_{\alpha_i} = \cdots =$$

$$\sum_{j=1}^{i} \sum_{k=\alpha_i+1}^{n} \left(\frac{\partial^2 f_s}{\partial s_k \partial s_j} \right) ds_k \wedge ds_j \wedge ds_{i+1} \wedge \cdots \wedge ds_{\alpha_i} \tag{5.54}$$

$$(i = 1, \cdots, n)$$

which is equivalent to

$$\left(\frac{\partial^2 f}{\partial s_k s_j} \right) = 0 (j, k = 1, \cdots, n; |j - k| > n) \tag{5.55}$$

It is easily checked that this condition is equivalent to the existence of functions $\phi_1, \cdots, \phi_{n-N}$ such that f_s satisfies (5.48). ∎

For the case that in (5.48) we have that $p \neq \mathrm{id}_{\mathbb{R}}$, the following result holds.

Theorem 5.9 *Consider a discrete-time system Σ of the form (5.20) that is strongly observable on an open invariant subset $U \subset \mathbb{R}^n$ containing the origin. Assume further that U is smoothly contractible to the origin, and that the one-forms $\omega_1, \cdots, \omega_n$ in (5.26) generate a codistribution on U. Let $N \in \{1, \cdots, n-2\}$ be given. Then the following statements are equivalent:*

(i) Σ can be put in extended observer form with buffer N on U.

(ii) There exists a function $U : \mathbb{R}^n \to \mathbb{R}$ such that

$$d\omega_i - dS \wedge \omega_i \equiv 0 \bmod \Omega_i (i = 1, \cdots, n) \tag{5.56}$$

(iii) The one-forms $\omega_1, \cdots, \omega_n$ and the vector fields $\tau_1, \cdots, \tau_n$ satisfy

$$d\omega_i \equiv 0 \bmod \Omega_i + \mathrm{span}\{\omega_n\}(i = 1, \cdots, n-N-1) \tag{5.57}$$

$$dw_i \equiv 0 \bmod \tilde{\Omega}_i (i = 1, \cdots, n - N - 1) \tag{5.58}$$

$$dw_i \equiv 0 \bmod \Omega_i (i = n - N, \cdots, n - 1) \tag{5.59}$$

$$[\sigma_i, \tau_n] \lrcorner w_n = [\sigma_j, \tau_n] \lrcorner w_n (i, j = 1, \cdots, n - N - 1) \tag{5.60}$$

$$\mathcal{L}_{\tau_j}([\sigma_i, \tau_n] \lrcorner w_i) = 0 (i = 1, \cdots, n - N - 1; j = 1, \cdots, n - 1) \tag{5.61}$$

where the vector fields $\sigma_1, \cdots, \sigma_{n-N-1}$ are defined by

$$\sigma_i := \tau_i + \cdots + \tau_{i+N} (i = 1, \cdots, n - N - 1) \tag{5.62}$$

Proof. $(i) \Rightarrow (ii)$ Assume that there exist functions $p, \phi_1, \cdots, \phi_{n-N}$, such that p is a diffeomorphism on $h(U)$ and f_s satisfies (5.48). Define one-forms $\tilde{w}_1, \cdots, \tilde{w}_n$ by

$$\tilde{w}_i := \sum_{j=1}^{i} \left(\frac{\partial (p \circ f_s)}{\partial s_j} \right) ds_j (i = 1, \cdots, n) \tag{5.63}$$

Then it follows from Theorem 5.8 that

$$d\tilde{w}_i \equiv 0 \bmod \mathrm{span}\{\tilde{w}_k - \tilde{w}_{k-1} \mid k = i + 1, \cdots, \min(n, i + N)\}$$

$$(i = 1, \cdots, n) \tag{5.64}$$

Note that we have that $\tilde{w}_i = (p' \circ f_s) w_i$ $(i = 1, \cdots, n)$. Defining the function $S := -\log |p' \circ f_s|$, this gives

$$dw_i - dS \wedge w_i = d \left(\frac{1}{p' \circ f_s} \tilde{w}_i \right) + \left(\frac{1}{p' \circ f_s} d(p' \circ f_s) \right) \wedge \frac{1}{p' \circ f_s} \tilde{w}_i = \cdots =$$

$$\frac{1}{p' \circ f_s} d\tilde{w}_i (i = 1, \cdots, n) \tag{5.65}$$

Further, it follows that

$$\mathrm{span}\{\tilde{w}_k - \tilde{w}_{k-1} \mid k = i + 1, \cdots, \min(n, i + N)\} = \Omega_i (i = 1, \cdots, n) \tag{5.66}$$

Our claim is then established by combining (5.64),(5.65),(5.66).

$(ii) \Rightarrow (i)$ Assume that there exists a function $S : U \to \mathbb{R}$ such that (5.56) holds. Note that from (5.26) we have that $w_n = df_s$. Thus, (5.56) for $i = n$ gives that

$$0 = dw_n - dS \wedge w_n = -dS \wedge w_n \tag{5.67}$$

By Cartan's Lemma, this gives that $dS \in \mathrm{span}\{df_s\}$. Define $T := \exp(-S)$. Then we also have that $dT \in \mathrm{span}\{df_s\}$, and thus there exists a function $\tilde{p} : \mathbb{R} \to \mathbb{R}$ such that $T = \tilde{p} \circ f_s$. Define $p := \int \tilde{p}(\tau) d\tau$, and one-forms $\tilde{\omega}_1, \cdots, \tilde{\omega}_n$ as in (5.63). We then have that $\tilde{\omega}_i = T\omega_i$ $(i = 1, \cdots, n)$, and thus

$$d\tilde{\omega}_i = dT \wedge \omega_i + Td\omega_i = T(d\omega_i - dS \wedge \omega_i) \equiv 0 \mathrm{mod} \Omega_i (i = 1, \cdots, n) \tag{5.68}$$

Together with Theorem 5.8, this establishes our claim.

$(ii) \Leftrightarrow (iii)$ From the fact that $\omega_n = df_s$, it follows that the function S that needs to exist has to satisfy $dS \wedge \omega_n$. By Cartan's Lemma, this implies that there should exist a function α such that $dS = \alpha \omega_n$ and

$$d\alpha \wedge \omega_n = 0 \tag{5.69}$$

Thus, the existence of a function S such that (5.56) holds is equivalent to the existence of a function α satisfying (5.69) and

$$d\omega_i - \alpha \omega_n \wedge \omega_i \equiv 0 \mathrm{mod} \Omega_i (i = 1, \cdots, n-1) \tag{5.70}$$

We now have that a two-form ω^2 satisfies $\omega^2 \equiv 0 \mathrm{mod} \Omega$ for some codistribution Ω if and only if $X \lrcorner Y \lrcorner \omega^2$ for all $X, Y \in \Omega^\perp$. It is easily checked that we have

$$\Omega_i^\perp = \mathrm{span}\{\tau_1, \cdots, \tau_{i-1}, \tau_{i+N+1}, \cdots, \tau_n, \sigma_i\} \tag{5.71}$$
$$(i = 1, \cdots, n - N - 1)$$

and

$$\Omega_i^\perp = \mathrm{span}\{\tau_1, \cdots, \tau_{i-1}, \sigma_i\}(i = n - N, \cdots, n - 1) \tag{5.72}$$

where, analogously to (5.62), we have

$$\sigma_i := \tau_i + \cdots + \tau_n (i = n - N, \cdots, n - 1) \tag{5.73}$$

Now let $i \in \{1, \cdots, n - N - 1\}$ be given. We then have

$$0 = \tau_k \lrcorner \tau_\ell \lrcorner (d\omega_i - \alpha \omega_n \wedge \omega_i) = \tau_k \lrcorner \tau_\ell \lrcorner d\omega_i = [\tau_k, \tau_\ell] \lrcorner \omega_i \tag{5.74}$$
$$(k, \ell = 1, \cdots, i - 1, i + N + 1, \cdots, n)$$

$$0 = \tau_k \lrcorner \sigma_i \lrcorner (d\omega_i - \alpha \omega_n \wedge \omega_i) = \tau_k \lrcorner \sigma_i \lrcorner d\omega_i = [\tau_k, \sigma_i] \lrcorner \omega_i \tag{5.75}$$
$$(k = 1, \cdots, i - 1, i + N + 1, \cdots, n - 1)$$

and

$$0 = \sigma_i \lrcorner \tau_n \lrcorner (\mathrm{d}\omega_i - \alpha\omega_n \wedge \omega_i) = \sigma_i \lrcorner \tau_n \lrcorner \mathrm{d}\omega_i - \alpha = [\sigma_i, \tau_n] \lrcorner \omega_i - \alpha \tag{5.76}$$

Where in (5.74),(5.75),(5.76) the last equality follows by applying (5.15) and (5.14). Combining (5.74) and (5.75), we obtain (5.57) and (5.58). Further, (5.76) gives (5.60), while combining (5.69) and (5.76) we obtain (5.61). Next, let $i \in \{n - N, \cdots, n - 1\}$ be given. We then have

$$0 = \tau_k \lrcorner \tau_\ell \lrcorner (\mathrm{d}\omega_i - \alpha\omega_n \wedge \omega_i) = \tau_k \lrcorner \tau_\ell \lrcorner \mathrm{d}\omega_i = [\tau_k, \tau_\ell] \lrcorner \omega_i$$

$$(k, \ell = 1, \cdots, i - 1) \tag{5.77}$$

and

$$0 = \tau_k \lrcorner \sigma_i \lrcorner (\mathrm{d}\omega_i - \alpha\omega_n \wedge \omega_i) = \tau_k \lrcorner \sigma_i \lrcorner \mathrm{d}\omega_i = [\tau_k, \sigma_i] \lrcorner \omega_i$$

$$(k = 1, \cdots, i - 1) \tag{5.78}$$

Combining (5.77) and (5.78), we then obtain (5.59). ■

Remark 5.1 *Theorem 5.9 generalizes Theorem 5.6. From the first two items of both theorems this is seen immediately. If however, one considers the third item of both theorems the generalization is far from obvious at first sight. This is due to the fact that the equivalence (ii)$\Leftrightarrow$(iii) in Theorem 5.6 holds for* general *independent one-forms $\omega_1, \cdots, \omega_n$, while this equivalence in Theorem 5.9 only holds for one-forms $\omega_1, \cdots, \omega_n$ of the form (5.26).*

5   Conclusions

In this chapter, we have given conditions for the existence of extended observer forms and extended observers for single-output nonlinear discrete-time systems. All conditions are valid on an open invariant subset of the state space that is smoothly contractible to an equilibrium point of the system and on which some regularity assumptions are satisfied. This raises the question what can be said for the case that (some of) the regularity assumptions are not satisfied. This remains a topic for future research.

A further topic for future research would be the question when extended observer forms and extended observers for multi-output discrete-time systems exist. As also mentioned in the Introduction, it seems that the conditions given in [13] for the existence of an observer form when only coordinate transformations are allowed, seem to be incorrect. Preliminary investigations suggest that in fact the problem of coming up with correct conditions may be quite intractable. On the other hand however, it may be shown by using the same techniques as in [7],[8] that a strongly observable multi-output system may always be put in observer form with buffer $N = \kappa^* - 1$, where κ^* equals the maximal so called observability index of the system.

Acknowledgments

Part of this research was performed while the author was visiting the Laboratoire d'Automatique de Nantes, Nantes, France, supported by a grant from the Région Pays de la Loire.

6   REFERENCES

[1] M. Brodmann. *Beobachterentwurf für nichtlineare zeitdiskrete Systeme*, VDI Verlag, Düsseldorf, 1994.

[2] R. L. Bryant, S. S. Chern, R.B. Gardner, H.L. Goldschmidt and P.A. Griffiths, *Exterior differential systems*, Springer, New York, 1991.

[3] H. Cartan. *Formes différentielles*, Hermann, Paris, 1967.

[4] Y. Choquet-Bruhat and C. DeWitt-Morette (with M. Dillard-Bleick), *Anaysis, manifolds and physics*, Part I: Basics, North-Holland, Amsterdam, 1991.

[5] H. Flanders. *Differential forms with applications to the physical sciences*, Dover, New York, 1989.

[6] A. Glumineau, C.H. Moog and F. Plestan. *New algebro-geometric conditions for the linearization by input-output injection*, IEEE Trans. Automat. Control, **41**, pp. 598-603, 1996.

[7] H. J.C. Huijberts, T. Lilge and H. Nijmeijer. A control perspective on synchronization and the Takens-Aeyels-Sauer Reconstruction Theorem, *to appear in Phys. Rev. E*, 1999.

[8] H. J. C. Huijberts, T. Lilge and H. Nijmeijer. Synchronization and observers for nonlinear discrete time systems, *submitted to European Control Conference* 1999.

[9] H. J. C. Huijberts, H. Nijmeijer, and A.Yu. Pogromsky. Discrete-time observers and synchronization, in G. Chen (Ed.), *Controlling chaos and bifurcations in engineering systems*, CRC Press, Boca Raton, Florida, 1999.

[10] A. J. Krener and A. Isidori. Linearization by output injection and nonlinear observers, *Syst. Control Lett.*, **3**, pp. 47–52, 1983.

[11] A. J. Krener and W. Respondek. Nonlinear observers with linearizable error dynamics, *SIAM J. Control Optimiz.*, **23**, pp. 197–216, 1985.

[12] T. Lilge. On observer design for nonlinear discrete-time systems, *Eur. J. Control*, **4**, pp. 306-319, 1998.

[13] W. Lin and C.I. Byrnes. Remarks on linearization of discrete–time autonomous systems and nonlinear observer design, *Syst. Control Lett.*, **25**, pp. 31–40, 1995.

[14] M. Spivak. *A comprehensive introduction to differential geometry*, Volume I, Publish or Perish, Houston, 1979.

[15] X. Xia and W. Gao. Nonlinear observer design by observer canonical forms, *Int. J. Control*, **47**, pp. 1081-1100, 1988.

Stability Analysis and Observer Design for Nonlinear Diffusion Processes

Winfried Lohmiller and Jean-Jacques E. Slotine

Nonlinear Systems Laboratory
Massachusetts Institute of Technology
Cambridge, Massachusetts, 02139, USA

1   Introduction

The stability of a nonlinear reaction-diffusion process and its convergence rate can be determined very simply. This allows in turn the design of simple observers for such processes. The technique is based on an extension of the recently developed tools of contraction theory. Reaction-diffusion processes are pervasive in physics. Mathematical properties such as existence and smoothness of solutions are well understood for many such processes (Evans [3]). This paper shows that analyzing global stability − are initial conditions or temporary disturbances eventually "forgotten," and if so, how fast? − and determining convergence rates given boundary conditions is very simple for such processes, and shows how this result applies naturally to the design of observers. This is achieved by taking advantage of recent results on stability theory, referred to as contraction analysis (Lohmiller and Slotine [8]), and extending them to partial differential equations describing nonlinear reaction-diffusion processes. After a brief review of contraction analysis in Section 2, Section 3 analyzes the contraction properties of the Laplace operator with general boundary conditions. The result is applied to reaction-diffusion equations of the form

$$\frac{\partial \phi}{\partial t} \;=\; h(t)\,\nabla^2 \phi \;+\; g(t)\,\nabla \phi \;+\; f(\phi, \mathbf{x}, t) \tag{6.1}$$

whose stability and convergence rates are explicitly quantified, and then to exponentially convergent observer designs for these systems. Numerical aspects are discussed in Section 4, and extensions to other classes of distributed systems in Section 5.

2   Contraction Analysis

Differential approximation is the basis of all linearized stability analysis. What is new in contraction analysis is that differential stability analysis can be made *exact*, and in turn yield global results on the nonlinear system. We first summarize some basic results of (Lohmiller and Slotine [8]), to which the reader is referred for more details, and then discuss how these results apply naturally to the design of nonlinear observers.

2.1   Basic Tools

We consider general deterministic systems of the form

$$\dot{\Phi} = \mathbf{f}(\Phi, t) \tag{6.2}$$

where Φ is the $n \times 1$ state vector and $\mathbf{f}$ is an $n \times 1$ nonlinear vector field, All quantities are assumed to be real and smooth, so that we can write the exact differential relation

$$\delta\dot{\Phi} = \frac{\partial \mathbf{f}}{\partial \Phi}(\Phi, t)\, \delta\Phi \tag{6.3}$$

where $\delta\Phi$ is a virtual displacement − recall that a virtual displacement is an infinitesimal displacement *at fixed time*. Note that virtual displacements, pervasive in physics and in the calculus of variations, are also well-defined mathematical objects (Arnold [1] and Schwartz [10]). Consider now two neighboring trajectories in the flow field $\dot{\Phi} = \mathbf{f}(\Phi, t)$, and the virtual displacement $\delta\Phi$ between them (1). The squared distance between these two trajectories can be defined as $\delta\Phi^T \delta\Phi$, leading from (6.3) to the rate of change

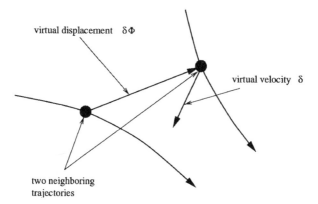

FIGURE 1. Virtual dynamics of two neighboring trajectories.

$$\frac{d}{dt}(\delta\Phi^T\delta\Phi) \;=\; 2\,\delta\Phi^T\delta\dot{\Phi} \;=\; 2\,\delta\Phi^T\frac{\partial\mathbf{f}}{\partial\Phi}\,\delta\Phi \;\leq\; 2\,\lambda_{max}\,\delta\Phi^T\delta\Phi \qquad (6.4)$$

where $\lambda_{max}(\Phi,t)$ denotes the largest eigenvalue of the *symmetric part* of the Jacobian $\frac{\partial\mathbf{f}}{\partial\Phi}$, i.e., the largest eigenvalue of $\frac{1}{2}(\frac{\partial\mathbf{f}}{\partial\Phi}+\frac{\partial\mathbf{f}}{\partial\Phi}^T)$. Assume now that $\lambda_{max}(\Phi,t)$ is uniformly strictly negative (i.e., $\exists\,\beta>0$, $\forall\Phi$, $\forall t\geq 0$, $\lambda_{max}(\Phi,t)\leq-\beta<0$.). Then, from (6.4) any infinitesimal length $\|\delta\Phi\|$ converges exponentially to zero. By path integration, this immediately implies that the length of any finite path converges exponentially to zero. Thus, as in stable linear time-invariant systems, the initial conditions are exponentially "forgotten." We can state the following definition and basic result (Lohmiller and Slotine [8])

Theorem 6.1 *The system* $\dot{\Phi}=\mathbf{f}(\Phi,t)$ *is said to be contracting if* $\frac{\partial\mathbf{f}}{\partial\Phi}$ *is uniformly negative definite. All system trajectories then converge exponentially to a single trajectory, with convergence rate* $|\lambda_{max}|$, *where* λ_{max} *is the largest eigenvalue of the symmetric part of* $\frac{\partial\mathbf{f}}{\partial\Phi}$.

The system is called semi-contracting if $\frac{\partial\mathbf{f}}{\partial\Phi}$ is only negative semi-definite, and indifferent if $\frac{\partial\mathbf{f}}{\partial\Phi}$ is skew-symmetric. More precise local versions of the above theorem can also be derived. In addition, departing further from classical Krasovskii-like results for autonomous systems (Krasosvkii [6], Hahn [5] and [7]), the approach can be vastly extended by allowing for a prior *differential* coordinate transformation, leading to a a necessary and sufficient condition for global exponential convergence (Lohmiller and Slotine [8]). Specifically, the line vector $\delta\Phi$ between two neighboring trajectories in Figure 1 can also be expressed using the differential coordinate transformation

$$\delta\Psi = \Theta\,\delta\Phi \qquad (6.5)$$

where $\Theta(\Phi,t)$ is a square matrix. This leads to a generalization of our earlier definition of squared length

$$\delta\Psi^T\delta\Psi = \delta\Phi^T\mathbf{M}\,\delta\Phi \qquad (6.6)$$

where $\mathbf{M}(\Phi,t)=\Theta^T\Theta$ represents a symmetric and continuously differentiable *metric* − formally, equation (6.6) defines a Riemann space (Lovelock and Rund [9]). Since (6.5) is in general not integrable, we cannot expect to find explicit new coordinates $\Psi(\Phi,t)$, but $\delta\Psi$ and $\delta\Psi^T\delta\Psi$ can always be defined. We require $\mathbf{M}$ to be uniformly positive definite, so that exponential convergence of $\delta\Phi$ to $\mathbf{0}$ also implies exponential convergence of $\delta\Psi$ to $\mathbf{0}$. Distance between two points P_1 and P_2 with respect to the metric $\mathbf{M}$ is defined as the shortest path length (i.e., the smallest path integral $\int_{P_1}^{P_2}\|\delta\Psi\|$) between these two points. Accordingly, a ball of center $\mathbf{c}$ and radius R is defined as the set of all points whose distance to $\mathbf{c}$ with respect to $\mathbf{M}$ is strictly less than R. Computing

$$\frac{d}{dt}\, \delta\Psi \;=\; \mathbf{F}\, \delta\Psi \qquad \text{where} \qquad \mathbf{F} = \left(\dot{\Theta} + \Theta\frac{\partial\mathbf{f}}{\partial\Phi}\right)\Theta^{-1} \qquad (6.7)$$

or alternatively

$$\frac{d}{dt}\left(\delta\Phi^T\mathbf{M}\,\delta\Phi\right) = \delta\Phi^T\left(\frac{\partial\mathbf{f}}{\partial\Phi}^T\mathbf{M} + \dot{\mathbf{M}} + \mathbf{M}\frac{\partial\mathbf{f}}{\partial\Phi}\right)\delta\Phi \qquad (6.8)$$

we can state the following definition and main result (Lohmiller and Slotine [8])

Definition 6.1 *Given the system equations $\dot{\Phi} = \mathbf{f}(\Phi, t)$, a region of the state space is called a contraction region with respect to a uniformly positive definite metric $\mathbf{M}(\Phi, t) = \Theta^T\Theta$ if $\mathbf{F}$ in (6.7) or equivalently $\frac{\partial\mathbf{f}}{\partial\Phi}^T\mathbf{M} + \mathbf{M}\frac{\partial\mathbf{f}}{\partial\Phi} + \dot{\mathbf{M}}$ is uniformly negative definite in that region.*

Regions where $\mathbf{F}$ is negative semi-definite are called semi-contracting, and regions where $\mathbf{F}$ is skew-symmetric are called indifferent.

Theorem 6.2 *Given the system equations $\dot{\Phi} = \mathbf{f}(\Phi, t)$, any trajectory, which starts in a ball of constant radius with respect to the metric $\mathbf{M}(\Phi, t)$, centered at a given trajectory and contained at all times in a contraction region with respect to $\mathbf{M}(\Phi, t)$, remains in that ball and converges exponentially to this trajectory. Furthermore global exponential convergence to the given trajectory is guaranteed if the whole state space is a contraction region with respect to the metric $\mathbf{M}(\Phi, t)$.*

It can be shown that the existence of a uniformly positive definite metric with respect to which the whole state space is a contraction region is actually a necessary condition for global exponential convergence. In the linear time-invariant case, a system is globally contracting if and only if it is strictly stable, with $\mathbf{F}$ simply being a normal Jordan form of the system and Θ the coordinate transformation to that form. Note that the metric is unchanged by an additional (perhaps time-varying or state-dependent) orthonormal transformation, i.e., by left-multiplying Θ by an orthonormal matrix.

2.2   Nonlinear Observer Design using Contraction Theory

By using a differential approach, contraction theory in a sense treats convergence analysis and limit behavior separately. Guaranteeing contraction means that after exponential transients the system's behavior will be independent of the initial conditions. In a control context, once contraction is guaranteed through feedback, specifying the final behavior reduces to the problem of shaping one particular solution, i.e. specifying an adequate

open-loop control input to be added to the feedback terms, a necessary step
of any control method. In an nonlinear observer context, contraction theory
is a rather natural tool, since once the observer contraction behavior has
been shown and quantified, one needs only verify that the observer equations contain the actual plant state as a particular solution to automatically
guarantee convergence to that state. If $\mathbf{y}(\Phi, t)$ is the available measurement
vector, and $\hat{\Phi}$ the estimated state, with $\hat{\mathbf{y}} = \mathbf{y}(\hat{\Phi}, t)$, this may be achieved
by simply copying the system dynamics (identity observer) and adding to
the right-hand side a term of the form $\mathbf{k}(\hat{\mathbf{y}}, t) - \mathbf{k}(\mathbf{y}, t)$, where the vector
field $\mathbf{k}$ is selected to guarantee or enhance the contraction behavior of the
observer. When the actual system is itself contracting, as will be the case
for most of the nonlinear diffusion processes considered in this paper, $\mathbf{k}$
needs only to be selected to enhance (speed up) the natural contraction
behavior of the system.

2.3 Weakly Contracting Systems

In this section we derive an exponential convergence condition for classes of
semi-contracting systems. The idea is simple: whereas in Section 2 we have
used only first time-derivatives of a virtual displacement to characterize a
flow field, we now perform a complete Taylor series expansion to analyze a
semi-contracting virtual dynamics. This section may be skipped in a first
reading, as it will only be needed for an extension in Section 5. Consider a
semi-contracting, analytic virtual dynamics in $\delta\Psi$

$$\frac{d}{dt}\delta\Psi = \mathbf{F}\,\delta\Psi$$

The corresponding virtual length dynamics is

$$\frac{d}{dt}\left(\delta\Psi^T\delta\Psi\right) = -2\delta\Psi^T\mathbf{F}_s\delta\Psi$$

with positive semi-definite $\mathbf{F}_s = -\frac{1}{2}\left(\mathbf{F} + \mathbf{F}^T\right)$. Factorizing $\mathbf{F}_s$ as $\sqrt{\mathbf{F}_s}^T\sqrt{\mathbf{F}_s}$,
say with a Cholesky factorization, allows one to compute the time-derivatives
of $\delta\Psi^T\delta\Psi$ as

$$\frac{d}{dt}\left(\delta\Psi^T\delta\Psi\right) = -2\,\delta\Psi^T\left(L^\circ\sqrt{\mathbf{F}_s}^T\,L^\circ\sqrt{\mathbf{F}_s}\right)\delta\Psi$$

$$\frac{d^2}{dt^2}\left(\delta\Psi^T\delta\Psi\right) = -2\,\delta\Psi^T\left(L^1\sqrt{\mathbf{F}_s}^T\,L^\circ\sqrt{\mathbf{F}_s} + L^\circ\sqrt{\mathbf{F}_s}^T\,L^1\sqrt{\mathbf{F}_s}\right)\delta\Psi$$

$$\frac{d^3}{dt^3}\left(\delta\Psi^T\delta\Psi\right) = -2\,\delta\Psi^T(L^2\sqrt{\mathbf{F}_s}^T\,L^\circ\sqrt{\mathbf{F}_s}$$
$$+ \ 2L^1\sqrt{\mathbf{F}_s}^T\,L^1\sqrt{\mathbf{F}_s} + L^\circ\sqrt{\mathbf{F}_s}^T\,L^2\sqrt{\mathbf{F}_s})\delta\Psi$$

$$\vdots$$

with the Lie derivatives $L^j \sqrt{\mathbf{F}}_s(\mathbf{x}, t)$ (Lovelock and Rund [9])

$$
\begin{aligned}
L^o \sqrt{\mathbf{F}}_s &= \sqrt{\mathbf{F}}_s \\
L^{j+1} \sqrt{\mathbf{F}}_s &= L^j \sqrt{\mathbf{F}}_s \mathbf{F} + \frac{d}{dt} L^j \sqrt{\mathbf{F}}_s \qquad \forall j \geq 0
\end{aligned}
$$

As a result $\delta \Psi^T \delta \Psi(t + T)$ along a trajectory $\mathbf{x}(t)$ can be written as the Taylor series expansion

$$
\begin{aligned}
\delta \Psi^T \delta \Psi(t + T) \;=\;& \delta \Psi^T \delta \Psi - \delta \Psi^T \left(L^o \sqrt{\mathbf{F}}_s^T \quad L^1 \sqrt{\mathbf{F}}_s^T \quad \cdots \right) \\
& \begin{pmatrix} T & \frac{T^2}{2!} & \frac{T^3}{3!} & \cdots \\ \frac{T^2}{2!} & \frac{2T^3}{3!} & \frac{3T^4}{4!} & \cdots \\ \frac{T^3}{3!} & \frac{3T^4}{4!} & \frac{4T^5}{5!} & \cdots \\ \vdots & \vdots & \vdots & \ddots \end{pmatrix} \begin{pmatrix} L^o \sqrt{\mathbf{F}}_s \\ L^1 \sqrt{\mathbf{F}}_s \\ \vdots \end{pmatrix} \delta \Psi \\
\leq \;& \delta \Psi^T \delta \Psi \\
& - \beta\, \delta \Psi^T (L^o \sqrt{\mathbf{F}}_s^T \quad L^1 \sqrt{\mathbf{F}}_s^T \; \cdots) \begin{pmatrix} L^o \sqrt{\mathbf{F}}_s \\ L^1 \sqrt{\mathbf{F}}_s \\ \vdots \end{pmatrix} \delta \Psi
\end{aligned}
$$

where all the terms on the right hand side are computed at time t. Note that for a given constant $T > 0$ the interior matrix above can be shown by complete induction to be uniformly positive definite, which implies a uniformly positive β. As a result we can conclude on exponential convergence of $\|\delta \Psi\|$ to zero for uniformly positive definite $\mathbf{C}^T \mathbf{C}$ with

$$
\mathbf{C} = \begin{pmatrix} L^o \sqrt{\mathbf{F}}_s \\ L^1 \sqrt{\mathbf{F}}_s \\ \vdots \end{pmatrix} \tag{6.9}
$$

Contrary to linear time-invariant systems the rank of $\mathbf{C}$ can be increased with additional Lie derivatives $L^j \sqrt{\mathbf{F}}_s$. However, once $\mathbf{C}^T \mathbf{C}$ is uniformly positive definite for some finite number of Lie derivatives the following ones do not influence the definiteness of $\mathbf{C}^T \mathbf{C}$ anymore. Consider similarly the corresponding semi-contracting metric dynamics

$$
\frac{d}{dt} \left(\delta \Phi^T \mathbf{M} \delta \Phi \right) = -2 \delta \Phi^T \mathbf{O} \delta \Phi \tag{6.10}
$$

with positive semi-definite $\mathbf{O} = \frac{\partial \mathbf{f}}{\partial \Phi}^T \mathbf{M} + \mathbf{M} \frac{\partial \mathbf{f}}{\partial \Phi} + \dot{\mathbf{M}}$. Factorizing $\mathbf{O}$ as $\sqrt{\mathbf{O}}^T \sqrt{\mathbf{O}}$, say with a Cholesky factorization, and a similar argumentation to the previous discussion allows to conclude on exponential convergence of $\|\delta\Phi\|_M = \sqrt{\delta\Phi^T M \delta\Phi}$ to zero for uniformly positive definite $\bar{C}^T\bar{C}$ with

$$
\bar{C} = \begin{pmatrix} L^o\sqrt{O} \\ L^1\sqrt{O} \\ \vdots \end{pmatrix}
\tag{6.11}
$$

and the Lie derivatives $L^j\sqrt{O}(\Phi,t)$

$$
\begin{aligned}
L^o\sqrt{O} &= \sqrt{O} \\
L^{j+1}\sqrt{O} &= L^j\sqrt{O}\,\frac{\partial f}{\partial\Phi} + \frac{d}{dt}L^j\sqrt{O} \qquad \forall j \geq 0
\end{aligned}
$$

This leads to the following definition and theorem

Definition 6.2 *Given the globally analytic system equations $\dot{\Phi} = f(\Phi, t)$, a semi-contraction region of the state space is called a weak-contraction region with respect to the metric $M(\Phi,t)$, if $C^T C$ in (6.9) or $\bar{C}^T\bar{C}$ in (6.11) is uniformly positive definite in that region.*

Theorem 6.3 *Given the system equations $\dot{\Phi} = f(\Phi, t)$, any trajectory, which starts in a ball of constant radius with respect to the metric $M(\Phi,t)$, centered at a given trajectory and contained at all times in a weak contraction region with respect to $M(\Phi,t)$, remains in that ball and converges exponentially to this trajectory. Furthermore global exponential convergence to the given trajectory is guaranteed if the whole state space is a contraction region with respect to the metric $M(\Phi,t)$.*

3   Nonlinear Diffusion Equations

We now extend and apply the above results to *partial* differential equations describing nonlinear reaction-diffusion processes. We shall use continuous state vectors Φ in Cartesian coordinates x, defined on a bounded m-dimensional region V. Bold characters will denote the continuous state-space quantities corresponding to pointwise terms. For instance, $\dot{\Phi}$ will denote the state-space vector of components $\frac{\partial\phi}{\partial t}$, and $\nabla^k\Phi$ the state-space vector of components $\nabla^k\phi$. Formally, Φ now lives on the Hilbert space $L^2(V)$, and differential length is defined by $\int_V \delta\phi^2 dV$, so that the derivation of Section 2.1 extends immediately. Furthermore, continuous state-space quantities can be computed as the limits of regularly discretized versions, as the discretization step tends to zero. For instance, on a one-dimensional continuum of length l

$$
\int_V \delta\phi^2 dV = \lim_{n\to+\infty} \frac{l^2}{n^2} \sum_{i=1}^{n} \delta\phi_i^T \delta\phi_i
$$

where the $\delta\phi_i$'s are the discretized values. This limiting process will be our main computation tool in assessing and quantifying the stability properties of the systems. Specifically, we first analyze the contraction properties of the ∇ and ∇^2 operators. We then analyze and quantify the convergence rate of the nonlinear reaction-diffusion equation (6.1).

3.1   Contraction Properties of Reaction-Diffusion Processes

One-Dimensional ∇ Operator

Consider first the one-dimensional ∇ operator on a one-dimensional continuum of length l, with given left and right boundary conditions $\phi_l(t)$ and $\phi_r(t)$. We can write $\nabla\Phi$ as the limit of an $n \times n$ discretization matrix as $n \to +\infty$,

$$
\nabla\Phi = \lim_{n\to+\infty} \frac{n+1}{2\,l}
\left(
\begin{pmatrix}
0 & 1 & 0 & 0 & \ddots \\
-1 & 0 & 1 & \ddots & \ddots \\
0 & -1 & \ddots & \ddots & 0 \\
0 & & \ddots & \ddots & 0 & 1 \\
\ddots & & \ddots & 0 & -1 & 0
\end{pmatrix}
\Phi
-
\begin{pmatrix}
\phi_l \\
0 \\
\vdots \\
\vdots \\
0
\end{pmatrix}
+
\begin{pmatrix}
0 \\
\vdots \\
0 \\
\phi_r
\end{pmatrix}
\right)
$$

where $\phi_l(t)$ and $\phi_r(t)$ are the given left and right boundary elements. The Jacobian

$$
\frac{\partial\nabla\Phi}{\partial\Phi} = \lim_{n\to+\infty} \frac{n+1}{2\,l}
\begin{pmatrix}
0 & 1 & 0 & 0 & \ddots \\
-1 & 0 & 1 & \ddots & \ddots \\
0 & -1 & \ddots & \ddots & 0 \\
0 & & \ddots & \ddots & 0 & 1 \\
\ddots & & \ddots & 0 & -1 & 0
\end{pmatrix}
\tag{6.12}
$$

is skew-symmetric, independently of $\phi_l(t)$ and $\phi_r(t)$. More generally, it is straightforward to show that any odd derivative $\nabla^k\Phi$, with k an odd positive integer, is skew-symmetric. One can easily show that the result is unchanged if $\nabla\phi_l(t)$ or $\nabla\phi_r(t)$ are given instead.

One-Dimensional Laplace Operator

Consider now the one-dimensional Laplace operator on the continuum, with given $\phi_l(t)$ and $\phi_r(t)$. We can write

$$
\nabla^2 \Phi = \lim_{n \to +\infty} \frac{(n+1)^2}{l^2} \left(\begin{pmatrix} -2 & 1 & 0 & 0 & \ddots \\ 1 & -2 & 1 & \ddots & \ddots \\ 0 & 1 & \ddots & \ddots & 0 \\ 0 & & \ddots & -2 & 1 \\ & \ddots & & 0 & 1 & -2 \end{pmatrix} \Phi + \begin{pmatrix} \phi_l \\ 0 \\ \vdots \\ 0 \end{pmatrix} + \begin{pmatrix} 0 \\ \vdots \\ 0 \\ \phi_r \end{pmatrix} \right)
$$

The corresponding Jacobian is

$$
\frac{\partial \nabla^2 \Phi}{\partial \Phi} = \lim_{n \to +\infty} \frac{(n+1)^2}{l^2} \begin{pmatrix} -2 & 1 & 0 & 0 & \ddots \\ 1 & -2 & 1 & \ddots & \ddots \\ 0 & 1 & \ddots & \ddots & 0 \\ 0 & & \ddots & -2 & 1 \\ & \ddots & & 0 & 1 & -2 \end{pmatrix} \tag{6.13}
$$

whose largest eigenvalue is shown in the Appendix A to be upper bounded by $-\frac{2\pi^2}{l^2}$. Thus, the one-dimensional Laplace operator $\nabla^2 \Phi$ with given boundary elements $\phi_l(t)$ and $\phi_r(t)$ is contracting. Consider instead the Laplace operator with given boundary elements $\phi_l(t)$ and $\nabla \phi_r(t)$ along a continuum of length l. By adding a mirror image of the system to the right we get a continuum of length $2l$ and given boundary elements $\phi_l(t)$ and $\phi_r(t)$. Thus the largest eigenvalue of the Jacobian is now $-\frac{\pi^2}{2l^2}$. If instead, $\nabla \phi_l(t)$ and $\nabla \phi_r(t)$ are given on the left and right boundary, then in (6.13) the upper left and lower right corners become -1. A similar derivation in the Appendix A shows that the resulting matrix is only negative semi-definite, i.e. the resulting Laplacian is only semi-contracting. Note that this can be expected physically, since the system might simply converge to a specific Φ with a constant error over the continuum.

Multi-dimensional Laplace Operator

Consider now an m-dimensional continuum, and the Laplacian

$$
\nabla^2 \Phi = \sum_{i=1}^{m} \frac{\partial^2 \Phi}{\partial x_i^2}
$$

with boundary condition $\phi_b(t)$, where the x_i are orthonormal Cartesian coordinates. Discretizing the region along any coordinate axis x_i, similarly to the previous section, the largest eigenvalue of $\frac{\partial^2 \Phi}{\partial x_i^2}$ is at most $-\frac{2\pi^2}{l_{i,max}^2}$, where $l_{i,max}$ is the diameter (maximal thickness) of the region in the direction x_i. Furthermore, since a discretization along any coordinate axis x_i can be transformed into a discretization along any other coordinate axis using an orthonormal coordinate transformation, the largest eigenvalue of the sum $\nabla^2 \Phi = \sum_{i=1}^{m} \frac{\partial^2 \Phi}{\partial x_i^2}$ is at most

$$\lambda_{\nabla^2} = -\sum_{i=1}^{m} \frac{2\pi^2}{l_{i,max}^2} \tag{6.14}$$

and thus the multi-dimensional Laplace operator $\nabla^2 \Phi$ with given boundary conditions $\phi_b(t)$ is contracting. If instead $\nabla\phi_b(t)$ is given on the whole boundary, then the Laplacian is only semi-contracting, with

$$\lambda_{\nabla^2} = 0 \tag{6.15}$$

Example 3.1: Consider the Laplace operator in spherical coordinates

$$\nabla^2 \Phi = \frac{\partial^2 \Phi}{\partial r^2} + \frac{2}{r}\frac{\partial \Phi}{\partial r} + \frac{1}{r^2}\frac{\partial^2 \Phi}{\partial \nu^2} + \frac{\cos \nu}{r^2 \sin \nu}\frac{\partial \Phi}{\partial \nu} + \frac{1}{r^2 \sin^2 \nu}\frac{\partial^2 \Phi}{\partial \theta^2}$$

with angles θ, ν, and radius r, and assume that $\phi_b(t)$ is given on the sphere $r = r_o$. The sphere's diameter is $2r_o$, so that the largest eigenvalue of this Laplacian is $\lambda_{\nabla^2} = -\frac{3}{2}\frac{\pi^2}{r_o^2}$. □

Reaction-Diffusion Equation

Collecting the above results and using Theorem 6.2 thus yields

Theorem 6.4 *Consider the reaction-diffusion equation*

$$\frac{\partial \phi}{\partial t} = h(t)\,\nabla^2\phi + g(t)\,\nabla\phi + f(\phi, \mathbf{x}, t) \tag{6.16}$$

with $h(t)$ a continuous and uniformly positive function of time ($\exists h_o > 0$, $\forall t \geq 0$, $h(t) \geq h_o$), and $g(t)$ an arbitrary continuous function of time. Assume that

$$h(t)\,\lambda_{\nabla^2} + \frac{\partial f}{\partial \phi}$$

is uniformly negative, where λ_{∇^2} is given by equation (6.14) or (6.15) depending on the boundary conditions. Then, all system trajectories converge exponentially to a single trajectory $\Phi_d(\mathbf{x}, t)$, with convergence rate $|h_o \lambda_{\nabla^2} + \frac{\partial f}{\partial \phi}|$. In the autonomous case ($f = f(\phi, \mathbf{x})$, $h(t) \equiv h_o$, $g(t) \equiv g_o$) and with constant boundary conditions, the system converges exponentially to a steady-state $\Phi_d(\mathbf{x})$, which is the unique solution of the generalized Poisson equation

$$0 = h_o\,\nabla^2\phi_d + g_o\,\nabla\phi_d + f(\phi_d, \mathbf{x})$$

The above development also implies that all the results on contracting systems in (Lohmiller and Slotine [8]) can be extended to contracting reaction-diffusion processes, with boundary conditions acting as additional inputs to the system. For instance, any autonomous contracting reaction-diffusion process, when subjected to boundary conditions periodic in time, will tend exponentially to a periodic solution of the same period. The convergence is robust to bounded or linearly increasing disturbances. (e.g., in the form of a discretization of the continuum for simulation purposes). Also note that the above result can be immediately extended to the case where the left-hand side of (6.16) is multiplied by a uniformly positive definite term $m(\mathbf{x})$, by using this term as the metric.

3.2 Observer Design for Nonlinear Diffusion Processes

When the nonlinear diffusion processes are naturally contracting, as described by the above theorem, observer designs beyond merely copying the system dynamics need only be selected to speed up the natural contraction behavior of the system, as discussed in Section 2.2. In addition, bounded measurement disturbances in a contracting obsever design do not affect the contraction behavior, though they may lead to a bounded offset in the limit behavior (Lohmiller and Slotine [8]). Finally, numerical implementations will discretize the continuum. The specific issues linked to this discretization will be discussed in Section 4. Let us now illustrate the design technique on a specific example, which also points out the possible role of boundary conditions as a design tool, specific to observers for distributed systems.

Example 3.2: Consider a wafer disk of radius r_o (Figure 2) subjected to continuous external light source, similarly e.g. to (Cho and Gyugyi [2]). The dynamic equations in radial coordinates θ and r are (Groeber et al. [4])

$$\dot{T} = h\nabla^2 T - f\left(T^4 - T_o^4\right)$$

with $T > 0$ the wafer temperature, $T_o = 500 + 200\sin\frac{\pi t}{20\,s}$ K the external temperature, $h = 1\,\mathrm{m}^2/\mathrm{s}$ a heat transfer constant, $f = 10^{-8}\mathrm{K}^{-3}\mathrm{s}^{-1}$ a radiation constant, $r_o = 20$ cm, and boundary conditions $\left(\frac{\partial T}{\partial r}\right)_{r_o} = 0$. According to Theorem 6.4, the system is naturally contracting with convergence rate $4fT^3$. This means that an open-loop identity observer

$$\dot{\hat{T}} = h\nabla^2 \hat{T} - f\left(\hat{T}^4 - T_o^4\right)$$

guarantees exponential convergence to T at the rate $4f\hat{T}^3$. The corresponding plant and observer response with initial conditions $T(r, t = 0) = 500 + 200\cos\frac{2\pi r}{r_0}$ K and $\hat{T}(r, t = 0) = 700$ K are illustrated in Figure 3. For computational simplicity the simulation exploits rotational symmetry in θ. Assume now that the actual temperature $T(r_o, t)$ is measured at the boundary $r = r_o$ of the disk. Using this measurement *as a boundary condition* $\hat{T}(r_o, t)$ increases

external radiation

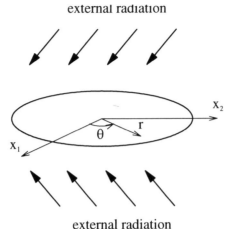

external radiation

FIGURE 2. Thermal processing of a wafer.

the convergence rate to $4f\hat{T}^3 + h\frac{\pi^2}{r_o^2}$, according to Theorem 6.4, while at the same time preserving consistency with the actual plant (i.e., keeping the actual T as a particular solution of the observer equations with their boundary condition). The corresponding observer response with the same initial conditions as above is shown in Figure 4.   □

4   Spatial Discretization and Numerical Implementation

This section briefly discusses some of the implications of the previous results for the numerical simulation of partial differential equations, which is of particular relevance in the context of observer implementation. Let us spatially discretize equation (6.16) by approximating the continuous state vector Ψ with $\Psi(\Phi, \mathbf{x})$, where Φ is a finite-dimensional state vector. This discretization leads to an error $\mathbf{e}$ in equation (6.16)

$$\dot{\Psi} = h\nabla^2\Psi + \mathbf{f}(\Psi, \mathbf{x}, t) + \mathbf{e} \tag{6.17}$$

We can minimize $\mathbf{e}$ by requiring $\frac{\partial\Psi}{\partial\Phi}^T\mathbf{e} = \mathbf{0}$ resulting in

$$\mathbf{M}\dot{\Phi} = \frac{\partial\Psi}{\partial\Phi}^T\left(h\nabla^2\Psi + \mathbf{f}\right)$$

with $\mathbf{M} = \frac{\partial\Psi}{\partial\Phi}^T\frac{\partial\Psi}{\partial\Phi}$. Taking the variation of (6.17) and $\delta(\frac{\partial\Psi}{\partial\Phi}^T\mathbf{e}) = \mathbf{0}$ lead to

$$\frac{1}{2}\frac{d}{dt}(\delta\Phi^T\mathbf{M}\delta\Phi) = \delta\Phi^T\frac{\partial\Psi}{\partial\Phi}^T\frac{\partial(h\nabla^2\Psi + \mathbf{f})}{\partial\Psi}\frac{\partial\Psi}{\partial\Phi}\delta\Phi - \delta\Phi\frac{\partial^2\Psi}{\partial\Phi^2}^T\mathbf{e}\,\delta\Phi$$

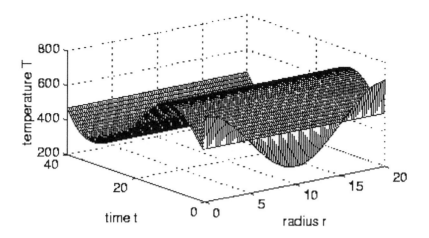

FIGURE 3. Wafer temperature as a function of time.

This dynamics can be simplified using the largest eigenvalue λ_{max} of $\frac{\partial \Psi}{\partial \Phi}^T \frac{\partial (h\nabla^2 \Psi + \mathbf{f})}{\partial \Psi} \frac{\partial \Psi}{\partial \Phi}$ with respect to $\mathbf{M}$, and the maximal principal curvature $|\kappa|_{max}$ of $\Psi(\Phi, \mathbf{x})$, obtained from $\det |\frac{\partial^2 \Psi}{\partial \Phi^2}^T \frac{\mathbf{e}}{\|\mathbf{e}\|} - \kappa \mathbf{M}| = 0$

$$\frac{1}{2}\frac{d}{dt}(\delta\Phi^T \mathbf{M} \delta\Phi) \leq (\lambda_{max} + \|\mathbf{e}\| \, |\kappa|_{max}) \, \delta\Phi^T \mathbf{M} \delta\Phi$$

If we assume Φ to be a minimal realization of Ψ, i.e. require $\mathbf{M}$ to be uniformly positive definite, then exponential convergence of $\delta\Phi^T \mathbf{M} \delta\Phi$ to zero implies exponential convergence of $\delta\Phi^T \delta\Phi$ to zero. The contraction behavior is hence unchanged in regions of small $\|\mathbf{e}\| \, |\kappa|_{max}$. This shows that we have to approximate regions with large $|\kappa|_{max}$ more precisely in order to preserve contraction.

5   Further Extensions

Besides actual reaction-diffusion or heat transfer processes, many other distributed physical processes can be written as reaction-diffusion equations using a nonlinear coordinate transformation (Evans [3]). This is e.g. the case of the Burgers equation, used to model turbulent flows and shock waves, by applying the so-called Hopf-Cole transformation. Reaction-diffusion equations are also used as computing paradigms in other fields, such as machine vision. In this section, we briefly discuss some further extensions of the analysis, from which exponentially convergent observers could be similarly designed.

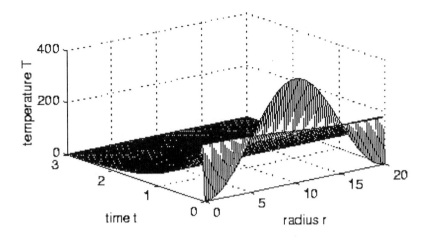

FIGURE 4. Estimation error of identity observer.

Nonlinear Functions of the Laplace Operator

Consider now the system

$$\frac{\partial \phi}{\partial t} = h(\nabla^2 \phi, \mathbf{x}, t)$$

with given boundary conditions $\phi_b(t)$, where h is continuous function of its argument, differentiable and uniformly *strictly monotonically increasing* with respect to its first argument $\nabla^2 \phi$. The corresponding function $\mathbf{h}(\nabla^2 \Phi, \mathbf{x}, t)$ is a continuous state-space vector of elements $h(\nabla^2 \phi, \mathbf{x}, t)$, whose Jacobian is

$$\frac{\partial \mathbf{h}}{\partial \Phi} = \frac{\partial \mathbf{h}}{\partial \nabla^2 \Phi} \frac{\partial \nabla^2 \Phi}{\partial \Phi}$$

where $\frac{\partial \mathbf{h}}{\partial \nabla^2 \Phi}$ is a diagonal matrix of elements $\frac{\partial h}{\partial \nabla^2 \phi}$. Using the symmetric positive definite Jacobian $-\frac{\partial \nabla^2 \Phi}{\partial \Phi}$ as the metric leads to the virtual length dynamics

$$\frac{d}{dt}\left(-\delta \Phi^T \frac{\partial \nabla^2 \Phi}{\partial \Phi} \delta \Phi\right) = -2 \, \delta \Phi^T \frac{\partial \nabla^2 \Phi}{\partial \Phi}^T \frac{\partial \mathbf{h}}{\partial \nabla^2 \Phi} \frac{\partial \nabla^2 \Phi}{\partial \Phi} \delta \Phi$$

The increase condition on h immediately implies that the system is contracting.

Wave Equation with Nonlinear Damping

Consider now the wave equation

$$m(\mathbf{x}) \frac{\partial^2 \phi}{\partial t^2} = \nabla^2 \phi - g(\frac{\partial \phi}{\partial t}, \mathbf{x}, t) - k(\mathbf{x})\phi$$

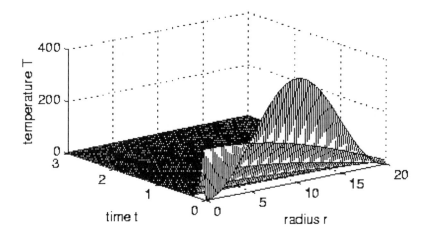

FIGURE 5. Estimation error with temperature measurement at r_o.

where all functions are assumed to be smooth, m and k are uniformly positive definite functions of $\mathbf{x}$. and g is *uniformly monotonically increasing* in $\frac{\partial \phi}{\partial t}$. The following virtual "energy" dynamics

$$\frac{d}{dt} \left(\delta \dot{\Phi}^T \text{diag}(\mathbf{m}) \, \delta \dot{\Phi} \, + \, \delta \Phi^T \left(\text{diag}(\mathbf{k}) - \frac{\partial \nabla^2 \Phi}{\partial \Phi} \right) \delta \Phi \right) = - \, 2 \, \delta \dot{\Phi}^T \frac{\partial \mathbf{g}}{\partial \dot{\Phi}} \delta \dot{\Phi} \tag{6.18}$$

guarantees semi-contraction behavior and hence bounded $\delta \Phi$ and $\delta \dot{\Phi}$. This implies that any time-derivative of (6.18) is bounded since it can always be expressed as functions of $\delta \Phi$ and $\delta \dot{\Phi}$. Barbalat's lemma then implies asymptotic convergence of $\delta \Phi$ and $\delta \dot{\Phi}$ to zero. Veryfing the condition in Theorem 6.3 even shows weak-contraction behavior i.e. that the system is globally exponentially convergent.

Chains of Contracting Processes

Chains of contracting processes are themselves contracting. More precisely, consider a smooth virtual dynamics of the form

$$\frac{d}{dt} \left(\begin{array}{c} \delta \Phi_1 \\ \delta \Phi_2 \end{array} \right) = \left(\begin{array}{cc} \frac{\partial \mathbf{f}_1}{\partial \Phi_1} & \mathbf{0} \\ \frac{\partial \mathbf{f}_2}{\partial \Phi_1} & \frac{\partial \mathbf{f}_2}{\partial \Phi_2} \end{array} \right) \left(\begin{array}{c} \delta \Phi_1 \\ \delta \Phi_2 \end{array} \right)$$

A uniformly negative definite $\frac{\partial \mathbf{f}_1}{\partial \Phi_1}$ implies exponential convergence of $\delta \Phi_1$ to zero. In turn, assuming that $\frac{\partial \mathbf{f}_2}{\partial \Phi_1}$ is bounded, $\frac{\partial \mathbf{f}_2}{\partial \Phi_1} \delta \Phi_1$ represents an exponentially decaying disturbance in the second equation, so that a uniformly negative definite $\frac{\partial \mathbf{f}_2}{\partial \Phi_2}$ implies exponential convergence of $\delta \Phi_2$ to zero. The result extends by recursion to higher-order chains.

Example 5.1: Consider the nuclear chain reaction

$$
\begin{aligned}
U_{92}^{238} + n_0^1 &\rightarrow U_{92}^{239} + \gamma \\
U_{92}^{239} &\rightarrow Np_{93}^{239} + e^- \\
Np_{93}^{239} &\rightarrow Pu_{94}^{239} + e^-
\end{aligned}
$$

whose dynamics is, with reaction constants k_i, diffusion constants h_i, and bounded uniformly positive neutron input $n_0^1(t)$

$$
\begin{aligned}
\dot{U}_{92}^{238} &= h_1 \nabla^2 U_{92}^{238} - k_1 n_0^1 U_{92}^{238} \\
\dot{U}_{92}^{239} &= h_2 \nabla^2 U_{92}^{239} + k_1 n_0^1 U_{92}^{238} - k_2 U_{92}^{239} \\
\dot{Np}_{93}^{239} &= h_3 \nabla^2 Np_{93}^{239} + k_2 U_{92}^{239} - k_3 Np_{93}^{239}
\end{aligned}
$$

Since the output of a contracting process is bounded for bounded input, using the above reasoning recursively shows that the chain reaction is globally exponentially convergent.   □

Nonlinear Schroedinger Equation

Finally, the nonlinear Schroedinger equation

$$
i\hbar \frac{\partial \phi}{\partial t} = -\frac{\hbar^2}{2m} \nabla^2 \phi + V(\phi, \mathbf{x}, t)
$$

can also be studied with only minor modifications, where $\hbar$ is the reduced Planck constant and m the mass. Defining squared length as $\delta \Phi^T \delta \Phi^*$ (where $*$ indicates complex conjugation), using

$$
i\hbar \delta \dot{\Phi} = -\frac{\hbar^2}{2m} \frac{\partial \nabla^2 \Phi}{\partial \Phi} \delta \Phi + \frac{\partial \mathbf{V}}{\partial \Phi} \delta \Phi
$$

leads to (assuming that $V(\phi, \mathbf{x}, t)$ is holomorphic in ϕ)

$$
\frac{d}{dt} \left(\delta \Phi^T \delta \Phi^* \right) = \delta \dot{\Phi}^T \delta \Phi^* + \delta \Phi^T \delta \dot{\Phi}^* = \frac{2}{\hbar} \delta \Phi^T \ Im \left(\frac{\partial \mathbf{V}}{\partial \Phi} \right) \delta \Phi^*
$$

In the linear case, the system is indifferent, a well-known result.

Acknowledgement

The authors are grateful to Christophe Bernard for his help in computing the contraction rates in Appendix A.

6   REFERENCES

[1] V. I. Arnold. *Mathematical Methods of Classical Mechanics*, Springer Verlag, 1978.

[2] Y. M. Cho and P. Gyugyi. Control of Rapid Thermal Processing: A System Theoretic Approach, *I.E.E.E. Transactions on control systems technology, 5(6)*, 1997.

[3] L. C. Evans. *Partial Differential Equations*, American Mathematical Society, 1998.

[4] H. Groeber, S. Erk and U. Grigull. *Die Grundgesetze der Waermeue-bertragung*, Springer Verlag, 1988.

[5] W. Hahn. *Stability of motion*, Springer Verlag, 1967.

[6] N. N. Krasovskii. *Problems of the Theory of Stability of Motion*, Mir, Moskow, 1959. English translation by Stanford University Press, 1963.

[7] P. Hartmann. *Ordinary differential equations*, second ed., Birkhauser, 1982.

[8] W. Lohmiller and J. J. E. Slotine. On Contraction Analysis for Nonlinear Systems, *Automatica, 34(6)*, 1998.

[9] D. Lovelockand H. Rund. Tensors, *differential forms, and variational principles*, Dover, 1989.

[10] L. Schwartz. *Analyse*, Hermann, Paris, 1993.

Appendix A: Computation of Contraction Rates

Let us compute an upper bound on the largest eigenvalue λ of the $n \times n$ matrix

$$-\frac{(n+1)^2}{l^2} \begin{pmatrix} 2 & -1 & 0 & 0 & \ddots \\ -1 & 2 & -1 & \ddots & \ddots \\ 0 & -1 & \ddots & \ddots & 0 \\ 0 & \ddots & \ddots & 2 & -1 \\ \ddots & \ddots & 0 & -1 & 2 \end{pmatrix} \tag{6.19}$$

as $n \to +\infty$. Since λ is the largest eigenvalue of (6.19), the $n \times n$ matrix

$$
\begin{pmatrix}
2 - 2\eta & -1 & 0 & 0 & \ddots \\
-1 & 2 - 2\eta & -1 & \ddots & \ddots \\
0 & -1 & \ddots & \ddots & 0 \\
0 & \ddots & \ddots & 2 - 2\eta & -1 \\
\ddots & \ddots & 0 & -1 & 2 - 2\eta
\end{pmatrix}
\tag{6.20}
$$

is positive semi-definite, where $\eta = -\dfrac{l^2}{2(n+1)^2}\lambda$. Now the principal minors Δ_k of (6.20), which must all be positive or zero, can be computed recursively

$$
\Delta_k = (2 - 2\eta)\Delta_{k-1} - \Delta_{k-2}
$$

with $\Delta_1 = 2 - 2\eta$ and $\Delta_2 = (2 - 2\eta)^2 - 1$. Noticing that $\Delta_1 \geq 0$ implies $\eta \leq 1$, this yields

$$
\Delta_k = \frac{\left(1 - \eta + \sqrt{(1-\eta)^2 - 1}\right)^{k+1} - \left(1 - \eta - \sqrt{(1-\eta)^2 - 1}\right)^{k+1}}{2\sqrt{(1-\eta)^2 - 1}}
$$

Thus, letting $\cos\alpha = 1 - \eta$ and $\sin\alpha = \sqrt{1 - (1-\eta)^2}$, with $0 \leq \alpha \leq \frac{\pi}{2}$, one can write

$$
\Delta_k = \frac{(\cos\alpha + i\sin\alpha)^{k+1} - (\cos\alpha - i\sin\alpha)^{k+1}}{2i\sin\alpha} = \frac{\sin(k+1)\alpha}{\sin\alpha}
$$

Thus, $\Delta_k \geq 0$ implies that $\alpha \leq \frac{\pi}{k+1}$, for any $k \geq 1$. This in turn implies that

$$
1 - \eta = \cos\alpha \geq 1 - \alpha^2 \geq 1 - \frac{\pi^2}{(k+1)^2}
$$

and thus that $\eta \leq \frac{\pi^2}{(n+1)^2}$. As $n \to +\infty$, the largest eigenvalue of (6.19) thus verifies

$$
\lambda_{\nabla^2} = -\lim_{n \to +\infty} \frac{2(n+1)^2}{l^2}\,\eta \leq -\frac{2\pi^2}{l^2}
$$

In the simpler case of boundary conditions in $\nabla\phi_l(t)$ and $\nabla\phi_r(t)$, the corresponding $n \times n$ matrix is

$$
\frac{(n+1)^2}{l^2}
\begin{pmatrix}
-1 & 1 & 0 & 0 & \ddots \\
1 & -2 & 1 & \ddots & \ddots \\
0 & 1 & \ddots & \ddots & 0 \\
0 & \ddots & \ddots & -2 & 1 \\
\ddots & \ddots & 0 & 1 & -1
\end{pmatrix}
$$

To show that this matrix is negative semi-definite as $n \to +\infty$ it suffices
to show that

$$
\begin{pmatrix}
1 & -1 & 0 & 0 & \ddots \\
-1 & 2 & -1 & \ddots & \ddots \\
0 & -1 & \ddots & \ddots & 0 \\
0 & \ddots & \ddots & 2 & -1 \\
\ddots & \ddots & 0 & -1 & 1
\end{pmatrix}
\tag{6.21}
$$

is positive semi-definite. Computing by induction the principal minors Δ_k
of the above matrix leads to $\Delta_k = 1$ for $1 \leq k \leq n-1$, and $\Delta_n = 0$, hence
the result. Note that, alternatively to this explicit derivation, standard
results on eigenvalues of the Laplacian operator may also be used, since
the associated Jacobian matrix is itself symmetric.

Nonlinear Passive Observer Design for Ships with Adaptive Wave Filtering

Jann Peter Strand[1] and Thor I. Fossen[2]

[1]ABB Industri AS
Oslo, Norway
[2]Department of Engineering Cybernetics
Norwegian University of Science and Technology
Trondheim, Norway

1   Introduction

Dynamic positioning (DP) systems have been commercially available for marine vessels since the 1960's. DP systems are often installed on advanced ships like cable-layers, ice-breakers semi-submersible rigs and offshore supply vessels, for maintaining the horizontal position and orientation by use of the thrusters. See the recent paper by Sørensen, Sagatun and Fossen [15] for an overview of DP systems and references to earlier work. More recently positioning mooring (PM) systems has been developed. This is a control system for thruster assistance of moored structures. Such systems are important for safe operation of floating oil production, storage and off-loading vessels (FPSO's) and semi-submersible rigs, which are moored to the seabed by an anchor system. Modeling and control system design of turret-moored ships is treated in Strand, Sørensen and Fossen [16].

Filtering and state estimation are important features of both DP and PM systems. In most cases, accurate measurements of the vessel velocities are not available. Hence, estimates of the velocities must be computed from noisy position and heading measurements through a state observer. Unfortunately, the position and heading measurements are corrupted with colored noise mainly caused by wind, waves and ocean currents. However, only the slowly-varying disturbances should be counteracted by the propulsion system whereas the oscillatory motion due to the waves (1st-order wave-induced disturbances) should *not* enter the feedback loop. This is done by using so-called wave filtering techniques, which separates the position and heading measurements into a low-frequency (LF) and a wave frequency (WF) position and heading part. The traditional Kalman filter-based estimators are linearized about a set of pre-defined constant yaw angles, typically 36 operating points in steps of 10 degrees, to cover the whole heading envelope between 0 and 360 degrees. When this estimator is used

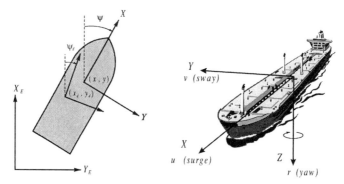

FIGURE 1. Reference frames (left) and definition of surge, sway and yaw modes of motion (right).

in conjunction with a linear (often linear quadratic) controller, there is no guarantee for global stability of the total system. In addition, linearization of the kinematic equations will naturally degrade the performance of the system.

A nonlinear observer that can replace the traditional Kalman filter based designs was proposed by Fossen and Strand [6] for free-floating ships. This observer includes wave filtering and bias state estimation and it is proven to be GES, through a passivation design. Compared to the Kalman filter, the number of tuning parameters is significantly reduced and the tuning parameters are coupled more directly to the physics of the system. By using a nonlinear formulation, the software algorithms in a practical implementation are simplified. The observer of Fossen and Strand [6] has been applied by Aarset, Strand and Fossen [1], Strand and Fossen [17] and Loria, Fossen and Panteley [9] in output feedback controller design.

Nonlinear observer designs for mechanical systems have been discussed by many. For instance, sliding observer designs for nonlinear systems is discussed by Slotine, Hendrick and Misawa [14]. Passivity and observer designs for mechanical systems are treated in Ortega, Loria, Nicklasson and Sira [13]. A similar problem to the ship positioning control is output feedback control of robot manipulators, see e.g. Berghuis and Nijmeijer [2], Canudas de Wit, Fixot and Astrom [3], Nicosia and Tomei [11], and Nicosia, Tornambe and Valigi [12].

In this paper nonlinear and adaptive observers for DP and PM systems are proposed. These are extensions of the nonlinear observer of Fossen and Strand [6]. The observers are proven to be passive and GES. An observer with adaptive wave filtering is derived, in order to adjust to slowly-varying sea states. The observers have been implemented and tested on a model ship and the results are reported. More details regarding nonlinear observer and controller designs for ships and experimental results can be found in Strand [18].

2 Modeling

2.1 Kinematics

Let the position (x, y) and the orientation (heading) ψ of the ship in the horizontal plane relative to an earth-fixed (EF) $X_E Y_E Z_E$-frame be represented by the vector $\eta = [x, y, \psi]^T$. An XYZ-frame is fixed to the vessel body, with the origin often located at the centre of gravity (CG), see Fig. 1. Body-fixed velocities are represented by the vector $\nu = [u, v, r]^T$ where u is the alongship velocity (surge), v is the athwartship velocity (sway) and r is the rotational velocity (yaw). A third frame, the so-called vessel parallel (VP) frame, specifies the desired position (x_d, y_d) and heading ψ_d of the ship relative to the $X_E Y_E Z_E$-frame, and is represented by the vector $\eta_d = [x_d, y_d, \psi_d]^T$. In the design of tracking control systems, the reference vector is a smooth time varying signal $\eta_d \in C^r$, where the corresponding reference velocities in the body-fixed frame is $\nu_d = [u_d, v_d, r_d]^T$. The reference trajectories are generated by a separate reference generator. The linear velocities of the ship and the reference model in the body-fixed and in earth-fixed frames are related by the transformations:

$$\dot{\eta} = J(\psi)\nu, \qquad \dot{\eta}_d = J(\psi_d)\nu_d, \tag{7.1}$$

where the rotation matrix in yaw $J(a) : \Re \rightarrow \Re^{3\times3}$ is defined as:

$$J(a) = \begin{bmatrix} \cos a & -\sin a & 0 \\ \sin a & \cos a & 0 \\ 0 & 0 & 1 \end{bmatrix}. \tag{7.2}$$

Note that $J^{-1}(a) = J^T(a)$.

2.2 Vessel Dynamics

In the mathematical modeling of ship dynamics, it is common to separate the model into a LF model and WF model. The WF motion of the ship is due to1st-order wave loads. The nonlinear LF equation of motion is driven by 2nd-order mean and slowly-varying wave, current and wind loads as well as thrust forces. In the case of moored ships, the restoring forces from the mooring system are treated as a function of the LF position and velocity of the ship. The total motion of the ship is given as the sum of the LF and the WF contributions, see Fig. 2.

Nonlinear Low-Frequency Model

Both DP and PM systems are typical *low-speed* applications. In DP operations the ship will either follow a pre-defined track or maintain a fixed position and heading in the horizontal plane. In tracking operations the

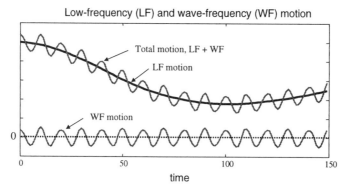

FIGURE 2. The total motion of a ship is modeled as a LF response with the WF response added as an output disturbance.

speed of the reference velocities are small. In PM systems the mooring system will naturally strongly limit the speed. With low speed we mean speed in the range of $0 - 3$ m/s. Under the assumption of low vessel speed, the following LF ship model is proposed (Fossen [5]):

$$M\dot{\nu} + D\nu + J^T(\psi)G\eta = \tau_{\text{thr}} + J^T(\psi)b. \tag{7.3}$$

Here $\tau_{\text{thr}} \in \Re^3$ is a control vector of forces and moment provided by the propulsion system (thrusters). $M \in \Re^{3\times3}$ is the inertia matrix including hydrodynamic added inertia, $D \in \Re^{3\times3}$ is a linear damping matrix, $G \in \Re^{3\times3}$ is a stiffness matrix, due to a mooring system and b is a bias term, accounting for unmodelled external forces and moment. A symmetrical, spread mooring system is assumed, where we for simplicity have placed the earth-fixed frame in the natural equilibrium point of the mooring system. Further, we assume that G is a constant, diagonal matrix. For more details on mooring systems, see Faltinsen [4], Strand $et\ al.$ [16] and Triantafyllou [19].

Based on the low speed assumption, the following statements are made:

P1 M is constant and positive definite (Newman [10]):

$$M = M^T > 0 \ \ \text{and} \ \ \dot{M} = 0. \tag{7.4}$$

P2 The linear damping matrix D is strictly positive:

$$D > 0. \tag{7.5}$$

Linear Wave Frequency Model

The WF motions are mainly generated by the 1st-order wave forces acting on the ship. Based on linear approximations of existing wave spectrum descriptions, see Fossen [5] for details, a linear WF model can be formulated as:

$$\dot{\xi} = A_w \xi + E_w w_w \qquad (7.6a)$$
$$\eta_w = C_w \xi \qquad (7.6b)$$

where p is the order of the WF model, $\xi \in \Re^{3 \cdot p}$, $w_w \in \Re^3$ is a zero-mean bounded disturbance vector and A_w, C_w and E_w are constant matrices of appropriate dimensions. The components of the WF motion is represented by the vector $\eta_w = [x_w, y_w, \psi_w]^T$. In one degree of freedom, this can for example be a 2nd-order damped oscillator ($p = 2$):

$$\eta_w^{\{i\}}(s) = \frac{\epsilon_{wi} s}{s^2 + 2\zeta_i \omega_{oi} s + \omega_{oi}^2} w_w^{\{i\}}, \quad (i = 1, 2, 3) \qquad (7.7)$$

where $(\cdot)^{\{i\}}$ denotes the *i-th* vector element. Here ζ_i is the relative damping ratio and ω_{oi} is the natural frequency, which is related to the *dominating wave frequency* of the incoming waves. From a practical point of view, these are slowly varying quantities, depending on the sea state. Typically, the periods of the dominating waves are in the range of 5 to 20 seconds in the North Sea. In the case of a 2nd-order WF model the matrices in (7.6a)--(7.6b) are:

$$A_w = \begin{bmatrix} 0 & I \\ -\Omega^2 & -2\Lambda\Omega \end{bmatrix}, \; C_w = \begin{bmatrix} 0 & I \end{bmatrix}, \; E_w = \begin{bmatrix} 0 \\ E_{w2} \end{bmatrix} \qquad (7.8)$$

where

$$\Lambda = \text{diag}\{\zeta_1, \zeta_2, \zeta_3\}, \qquad (7.9)$$
$$\Omega = \text{diag}\{\omega_{o1}, \omega_{o2}, \omega_{o3}\}, \qquad (7.10)$$
$$E_{w2} = \text{diag}\{\epsilon_{w1}, \epsilon_{w2}, \epsilon_{w3}\}. \qquad (7.11)$$

Bias Modeling

A frequently used bias model for marine control applications is:

$$\dot{b} = -T_b^{-1} b + E_b w_b \qquad (7.12)$$

where $b \in \Re^3$, $w_b \in \Re^3$ is a zero-mean bounded disturbance vector, $T_b \in \Re^{3 \times 3}$ is a diagonal matrix of bias time constants and E_b a diagonal matrix scaling the amplitude of w_b. The bias model accounts for slowly-varying forces and moment due to 2nd-order wave loads, ocean currents and wind.

In addition, a bias model will account for errors in modeling of the constant mooring loads, actuator thrust losses and other unmodelled slowly-varying dynamics.

Measurements

For conventional ships usually only position and heading measurements are available to the positioning system, whereas accurate velocity measurements are not available. Hence the measurement equation is written:

$$y = \eta + \eta_w + v_y \tag{7.13}$$

which consists of the LF and WF motions and measurement noise $v_y \in \Re^3$.

2.3 Total Ship Model

When designing the observer, the following assumptions are made in the Lyapunov analysis regarding the ship model:

A1 $J(\psi) \approx J(\psi + \psi_w) = J(\psi_y)$, where $\psi_y \triangleq \psi + \psi_w$ denotes the measured heading. This is a good assumption since the magnitude of the wave-induced yaw motion ψ_w will be less than 5 degrees in extreme weather situations and less than 1 degree during normal operation of the ship/rig.

A2 Position and heading sensor noise is omitted, $v_y = 0$, since this noise is negligeable compared to the wave-induced motion.

From Assumptions A1-A2 the total motion of moored and free-floating ships is represented by the following equations:

$$\dot{\xi} = A_w \xi + E_w w_w \tag{7.14a}$$
$$\dot{\eta} = J(\psi_y)\nu \tag{7.14b}$$
$$\dot{b} = -T_b^{-1}b + E_b w_b \tag{7.14c}$$
$$M\dot{\nu} = -D\nu - J^T(\psi_y)G\eta + J^T(\psi_y)b + \tau_{\text{thr}} \tag{7.14d}$$
$$y = \eta + \eta_w = \eta + C_w\xi. \tag{7.14e}$$

3 Non-Adaptive Observers

Two different non-adaptive observers will be derived in this section. The first one is similar to the observer of Fossen and Strand [6] for dynamically positioned (free-floating) ships, where here also the effect of a spread mooring system is taken into account. In the second design, the observer is augmented by a new filtered state of the innovation signals. This adds more flexibility to the observer design. By using feedback from the high-pass filtered innovation in the WF part of the observer there will be no steady-state offsets in the WF estimates. Moreover, by using the low-pass filtered innovation in the bias estimation, these estimates will be less noisy and can thus be used directly as a feedforward term in the control law. The adaptive observer proposed in Section 4 is based on the augmented observer. In the design we use an SPR-Lyapunov approach for obtaining passivity and stability of the observers. By including the synthetic wave model in the observer, wave-filtering is obtained, see Definition 1.

Definition 7.1 (Wave Filtering) *Wave filtering can be defined as the reconstruction of the LF motion components from noisy measurements of position and heading by means of an observer. In addition to this, noise-free estimates of the LF velocities should be produced. This is crucial in ship motion control systems since the WF part of the motion should **not** be compensated for by the positioning system. If the WF part of the motion enters the feedback loop, this will cause unnecessary tear and wear of the actuations and increase the fuel consumption.*

3.1 Observer in the EF frame

The observer in this section is similar to the observer in [6], except that the effect of a spread mooring system attached to the ship is included.

Observer Equations

A nonlinear observer copying the ship-mooring dynamics (7.14a)-(7.14e) is:

$$\dot{\hat{\xi}} = A_w\hat{\xi} + K_1\tilde{y} \tag{7.15a}$$

$$\dot{\hat{\eta}} = J(\psi_y)\hat{\nu} + K_2\tilde{y} \tag{7.15b}$$

$$\dot{\hat{b}} = -T_b^{-1}\hat{b} + K_3\tilde{y} \tag{7.15c}$$

$$M\dot{\hat{\nu}} = -D\hat{\nu} - J^T(\psi_y)G\hat{\eta} + J^T(\psi_y)\hat{b} + \tau_{\text{thr}} + J^T(\psi_y)K_4\tilde{y} \tag{7.15d}$$

$$\hat{y} = \hat{\eta} + C_w\hat{\xi} \tag{7.15e}$$

where $\tilde{y} = y - \hat{y}$ is the innovation vector and $K_1 \in \Re^{2 \cdot p \times 3}$, $K_2, K_3, K_4 \in \Re^{3 \times 3}$ are observer gain matrices to be determined later.

Observer Error Dynamics

The estimation errors are defined as $\tilde{\xi} = \xi - \hat{\xi}$, $\tilde{\eta} = \eta - \hat{\eta}$, $\tilde{b} = b - \hat{b}$ and $\tilde{\nu} = \nu - \hat{\nu}$. Hence, from (7.14a)-(7.14e) and (7.15a)-(7.15e) the observer error dynamics is:

$$\dot{\tilde{\xi}} = A_w\tilde{\xi} - K_1\tilde{y} + E_w w_w \tag{7.16a}$$

$$\dot{\tilde{\eta}} = J(\psi_y)\tilde{\nu} - K_2\tilde{y} \tag{7.16b}$$

$$\dot{\tilde{b}} = -T_b^{-1}\tilde{b} - K_3\tilde{y} + E_b w_b \tag{7.16c}$$

$$M\dot{\tilde{\nu}} = -D\tilde{\nu} - J^T(\psi_y)G\tilde{\eta} + J^T(\psi_y)\tilde{b} - J^T(\psi_y)K_4\tilde{y} \tag{7.16d}$$

$$\tilde{y} = \tilde{\eta} + C_w\tilde{\xi}. \tag{7.16e}$$

By defining a new output

$$\tilde{z}_o \triangleq K_4\tilde{y} + G\tilde{\eta} - \tilde{b} \triangleq C_o\tilde{x}_o \tag{7.17}$$

and the vectors

$$\tilde{x}_o \triangleq \begin{bmatrix} \tilde{\xi} \\ \tilde{\eta} \\ \tilde{b} \end{bmatrix}, \quad w \triangleq \begin{bmatrix} w_w \\ w_b \end{bmatrix}, \tag{7.18}$$

the error dynamics (7.16a)-(7.16d) can be written in compact form as:

$$M\dot{\tilde{\nu}} = -D\tilde{\nu} - J^T(\psi_y)C_o\tilde{x}_o \tag{7.19a}$$

$$\dot{\tilde{x}}_o = A_o\tilde{x}_o + B_oJ(\psi_y)\tilde{\nu} + E_o w \tag{7.19b}$$

where

$$A_o = \begin{bmatrix} A_w - K_1C_w & -K_1 & 0 \\ -K_2C_w & -K_2 & 0 \\ -K_3C_w & -K_3 & -T_b^{-1} \end{bmatrix},$$

$$C_o = \begin{bmatrix} K_4C_w & K_4 + G & -I \end{bmatrix},$$

$$B_o = \begin{bmatrix} 0 \\ I \\ 0 \end{bmatrix}, \quad E_o = \begin{bmatrix} E_w & 0 \\ 0 & 0 \\ 0 & E_b \end{bmatrix}.$$

Next the requirements on the observer gain matrices for stability and passivity of the observer error dynamics is provided.

Stability and Passivity

By rewriting the observer error dynamics as (7.19a)--(7.19b) stability of the observer is provided by a SPR-Lyapunov design. The error dynamics is shown in Figure 3 where two new error terms ε_z and ε_ν are defined as:

$$\varepsilon_z \triangleq -J^T(\psi_y)\tilde{z}_o, \quad \varepsilon_\nu \triangleq J(\psi_y)\tilde{\nu}. \tag{7.20}$$

Thus, the observer system consists of two linear blocks, interconnected through the bounded transformation matrix $J(\psi_y)$. Based on the physical properties of the ship dynamics, we can make the following statement:

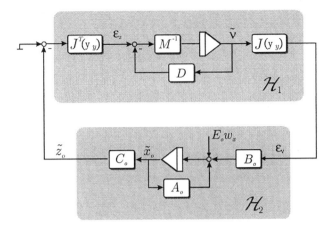

FIGURE 3. Block diagram of the observer error dynamics.

Proposition 7.1 *The mapping $\varepsilon_z \mapsto \tilde{\nu}$ is state strictly passive and the block $\mathcal{H}_1$ in Fig. 3 is strictly passive.*
Proof. *Let*

$$S_1 = \frac{1}{2}\tilde{\nu}^T M \tilde{\nu} \tag{7.21}$$

be a positive definite storage function. From (7.19a) we have:

$$\dot{S}_1 = -\frac{1}{2}\tilde{\nu}^T (D + D^T)\tilde{\nu} + \tilde{\nu}^T \varepsilon_z \tag{7.22}$$

$$\Downarrow$$

$$\tilde{\nu}^T \varepsilon_z \geq \dot{S}_1 + \beta \tilde{\nu}^T \tilde{\nu} \tag{7.23}$$

where $\beta = \frac{1}{2}\lambda_{\min}(D + D^T) > 0$ and $\lambda_{\min}(.)$ denotes the minimum eigenvalue. Thus, (7.23) proves that $\varepsilon_z \mapsto \tilde{\nu}$ is state strictly passive [8]. Moreover, since this mapping is strictly passive, post-multiplication with the bounded transformation matrix $J(\psi_y)$ and pre-multiplication by it's transpose will not affect the passivity properties. Hence the block $\mathcal{H}_1$ is strictly passive. □

Passivity and stability of the total system will be provided if the observer gain matrices $K_1, ..., K_4$ can be chosen such that the mapping $\varepsilon_\nu \mapsto \tilde{z}_o$ is passive. This is obtained if the matrices A_o, B_o, C_o in (7.19a)--(7.19b) satisfies the KYP Lemma which is stated as below:

Lemma 7.1 (Kalman-Yakubovich-Popov) *Let $\mathcal{Z}(s) = \bar{C}(sI - \mathcal{A})^{-1}\mathcal{B}$ be a $n \times n$ transfer function matrix, where $\mathcal{A}$ is Hurwitz, $(\mathcal{A}, \mathcal{B})$ is controllable, and $(\mathcal{A}, \bar{C})$ is observable. Then, $\mathcal{Z}(s)$ is strictly positive real (SPR) if and* *only if there exist positive definite matrices* $\mathcal{P} = \mathcal{P}^T$ *and* $\mathcal{Q} = \mathcal{Q}^T$ *such that:*

$$\mathcal{P}\mathcal{A} + \mathcal{A}^T\mathcal{P} = -\mathcal{Q}, \qquad \mathcal{B}^T\mathcal{P} = \bar{\mathcal{C}}. \tag{7.24}$$

Proof: See e.g. Khalil [8]. □

Given a set of observer gains $K_1, ..., K_4$ the existence of the system to satisfy the KYP Lemma can be checked numerically by using the *Frequency Theorem*, originally formulated by Yakubovich [20], explicitly contained in Gelig, Leonov and Yakubovich [7]:

Theorem 7.1 (Frequency Theorem) *Consider the system*

$$\dot{x} = \mathcal{A}x + \mathcal{B}u \tag{7.25a}$$
$$y = \mathcal{C}^T x \tag{7.25b}$$

where $x \in \Re^n$, $u \in \Re^m$, $y \in \Re^m$ *and* $\mathcal{A}, \mathcal{B}, \mathcal{C}$ *are real matrices of appropriate dimensions. Suppose the pair* $(\mathcal{A},\mathcal{B})$ *is stabilizable and* $\det(j\omega I_n - \mathcal{A}) \neq 0$ $\forall \omega \in \Re^1$. *There exists a* $\mathcal{P} = \mathcal{P}^T > 0$ *with*

$$\mathcal{P} + \mathcal{A}^T\mathcal{P} < 0, \qquad \mathcal{P}\mathcal{B} + \mathcal{C} = 0 \tag{7.26}$$

if the following conditions

$$\mathrm{Re}\left(\mathcal{C}^T \left(j\omega I_n - \mathcal{A}\right)^{-1}\mathcal{B}\right) < 0, \quad \forall \omega \in \Re^1 \tag{7.27}$$

$$\lim_{\omega \to \infty} \omega^2 \mathrm{Re}\left(\mathcal{C}^T \left(j\omega I_n - \mathcal{A}\right)^{-1}\mathcal{B}\right) < 0 \tag{7.28}$$

hold. □

If the Frequency Theorem is satisfied for $\mathcal{A} = A_o$, $\mathcal{B} = B_o$, $\mathcal{C}^T = -C_o$, the mapping $\varepsilon_\nu \mapsto \tilde{z}_o$ (block $\mathcal{H}_2$ in Fig. 3) is SPR and the observer error dynamics system is passive and GES as stated in the following:

Theorem 7.2 (Passive Observer) *The nonlinear observer (7.15a)–(7.15d) is passive.*
Proof. *Since it is established that* $\mathcal{H}_1$ *is strictly passive and* $\mathcal{H}_2$ *is SPR, the nonlinear observer is passive.* □

Theorem 7.3 (ISS and GES Observer) *The observer (7.15a)-(7.15d) with disturbance* w *is input-to-state stable (ISS). In addition, the observer error dynamics is rendered GES if we disregard the zero-mean disturbance,* $w \equiv 0$.
Proof. *Consider the following Lyapunov function candidate:*

$$V_o = \tilde{\nu}M\tilde{\nu} + \tilde{x}_o^T P_o \tilde{x}_o$$

Time differentiation of V_o along the trajectories of $\tilde{\nu}$ and $\tilde{x}_o$ yields:

$$\dot{V}_o = -\tilde{\nu}^T(D + D^T)\tilde{\nu} + \tilde{x}_o^T\left(P_oA_o + A_o^TP_o\right)\tilde{x}_o + 2\tilde{\nu}^TJ^T(\psi_y)B_o^TP_o\tilde{x}_o$$
$$- 2\tilde{\nu}^TJ^T(\psi_y)C_o\tilde{x}_o + 2\tilde{x}_o^TP_oE_ow. \tag{7.29}$$

If the KYP Lemma is satisfied for the mapping $\varepsilon_\nu \mapsto \tilde{z}_o$, with $P_oA_o + A_o^TP_o = -Q_o$ and $B_o^TP_o = C_o$, $\dot{V}_o$ can be written as:

$$\dot{V}_o = -\tilde{\nu}^T(D + D^T)\tilde{\nu} - \tilde{x}_o^TQ_o\tilde{x}_o + 2\tilde{x}_o^TP_oE_ow. \tag{7.30}$$

From (7.30) it is seen that

$$\dot{V}_o < 0, \quad \|\tilde{x}_o\| > 2\left\|Q_o^{-1}E_ow\right\| \tag{7.31}$$

which shows that the observer is ISS. Moreover, in the disturbance-free case, $w \equiv 0$, the equilibrium point of the error dynamics is GES. □

Regarding the choice of observer gain matrices, the tuning procedure can be similar as for the observer for free-floating ships in Fossen and Strand [6]. Pole placement techniques can also be applied.

3.2 Augmented Observer

The proposed observer in Section 3.1 can be further refined by augmenting a new state. The augmented design provides more flexibility and it is the basis for the adaptive observer in Section 4. We start by adding a new state, x_f, in the observer, which is the low-pass filtered innovation $\tilde{y}$:

$$\dot{x}_f = -T_f^{-1}x_f + \tilde{y} = -T_f^{-1}x_f + \tilde{\eta} + C_w\tilde{\xi} \tag{7.32}$$

where $x_f \in \Re^3$ and $T_f =\mathrm{diag}\{T_{f1}, T_{f2}, T_{f3}\}$ contains positive filter constants. High-pass filtered innovation signals can be derived from x_f by:

$$\tilde{y}_f = -T_f^{-1}x_f + \tilde{y} = -T_f^{-1}x_f + \tilde{\eta} + C_w\tilde{\xi} \tag{7.33}$$

Thus, both the low-pass and high-pass filtered innovation is available for feedback. Moreover,

$$\left.\begin{array}{l} x_f^{\{i\}}(s) = \frac{1}{1+T_{fi}s}\tilde{y}^{\{i\}}(s) \\ \tilde{y}_f^{\{i\}}(s) = \frac{T_{fi}s}{1+T_{fi}s}\tilde{y}^{\{i\}}(s) \end{array}\right\}, \quad (i = 1, 2, 3) \tag{7.34}$$

The cut-off frequency in the filters should be below the frequencies of the dominating waves in the WF model (7.6a)--(7.6b).

Augmented Observer Equations

The augmented observer is formulated as:

$$\dot{\hat{\xi}} = A_w\hat{\xi} + K_{1h}\tilde{y}_f \tag{7.35a}$$

$$\dot{\hat{\eta}} = J(\psi_y)\hat{\nu} + K_2\tilde{y} + K_{21}x_f + K_{2h}\tilde{y}_f \tag{7.35b}$$

$$\dot{\hat{b}} = -T_b^{-1}\hat{b} + K_3\tilde{y} + K_{31}x_f \tag{7.35c}$$

$$M\dot{\hat{\nu}} = -D\hat{\nu} - J^T(\psi_y)G\hat{\eta} + J^T(\psi_y)\hat{b} + \tau_{\text{thr}}$$
$$+ J^T(\psi_y)(K_4\tilde{y} + K_{41}x_f + K_{4h}\tilde{y}_f) \tag{7.35d}$$

$$\hat{y} = \hat{\eta} + C_w\hat{\xi} \tag{7.35e}$$

where x_f is the low-pass filtered innovation vector and $\tilde{y}_f$ is the high-pass filtered innovation given by (7.32) and (7.33), respectively. Here $K_{1h} \in \Re^{6x3}$ and $K_{21}, K_{2h}, K_{31}, K_{41}, K_{4h} \in \Re^{3x3}$ are new observer gain matrices.

Augmented Observer Error Dynamics

The augmented observer error dynamics can be written compactly as:

$$M\dot{\tilde{\nu}} = -D\tilde{\nu} - J^T(\psi_y)C_a\tilde{x}_a \tag{7.36a}$$

$$\dot{\tilde{x}}_a = A_a\tilde{x}_a + B_aJ(\psi_y)\tilde{\nu} + E_aw \tag{7.36b}$$

where

$$\tilde{x}_a \triangleq \begin{bmatrix} \tilde{\xi}^T & \tilde{\eta}^T & x_f^T & \tilde{b}^T \end{bmatrix}^T, \tag{7.37}$$

$$\tilde{z}_a \triangleq K_4\tilde{y} + K_{41}x_f + K_{4h}\tilde{y}_f + G\tilde{\eta} - \tilde{b} \triangleq C_a\tilde{x}_a, \tag{7.38}$$

and

$$A_a = \begin{bmatrix} A_w - K_{1h}C_w & -K_{1h} & K_{1h}T_f^{-1} & 0 \\ -(K_2 + K_{2h})C_w & -(K_2 + K_{2h}) & K_{2f}T_f^{-1} - K_{21} & 0 \\ C_w & I & -T_f^{-1} & 0 \\ -K_3C_w & -K_3 & -K_{31} & -T_b^{-1} \end{bmatrix}$$

$$B_a = \begin{bmatrix} 0 \\ I \\ 0 \\ 0 \end{bmatrix}, \qquad E_a = \begin{bmatrix} E_w & 0 \\ 0 & 0 \\ 0 & 0 \\ 0 & E_b \end{bmatrix}$$

$$C_a = \begin{bmatrix} (K_4 + K_{4h})C_w & (K_4 + K_{4h}) + G & -K_{4h}T_f^{-1} + K_{41} & -I \end{bmatrix}$$

The signals $\tilde{y}_f$ and x_f are extracted from $\tilde{x}_a$ by $\tilde{y}_f = C_h\tilde{x}_a$ and $x_f = C_l\tilde{x}_a$ where

$$C_h = \begin{bmatrix} C_w & I & -T_f^{-1} & 0 \end{bmatrix}, \quad C_l = \begin{bmatrix} 0 & 0 & I & 0 \end{bmatrix}. \tag{7.39}$$

Passivity and Stability

Let

$$V_a = \tilde{\nu}^T M \tilde{\nu} + \tilde{x}_a^T P_a \tilde{x}_a \qquad (7.40)$$

be a Lyapunov function for the observer error dynamics (7.36a)--(7.36b). As before, the cross-terms of $\tilde{\nu}$ and $\tilde{x}_a$ in the expression for $\dot{V}_a$ are cancelled by using an SPR-Lyapunov design, where it is required that:

$$A_a^T P_a + P_a A_a = -Q_a, \qquad B_a^T P_a = C_a \qquad (7.41)$$

The existence of a $P_a = P_a^T > 0$, satisfying the KYP Lemma for the augmented observer error dynamics, can be tested numerically for a fixed set of observer gains by using the Frequency Theorem with $\mathcal{P} = P_a$, $\mathcal{A} = A_a$, $\mathcal{B} = B_a$, $\mathcal{C}^T = -C_a$ and $u = \varepsilon_\nu$. If so, the passivity and stability properties are similar as for the observer in Section 3.1 where:

$$\dot{V}_a = -\tilde{\nu}^T (D + D^T) \tilde{\nu} - x_a^T Q_a x_a + 2\tilde{x}_a^T P_a E_a w \qquad (7.42)$$

and

$$\dot{V}_a < 0, \qquad \|\tilde{x}_a\| > 2 \|Q_a^{-1} E_a w\| \qquad (7.43)$$

4 Adaptive Observer

In this section we treat the problem when the parameters of A_w in the WF model (7.6a) are *not* known. The parameters vary with the different sea-states in which the ship is operating. Gain-scheduling techniques, using off-line frequency trackers and external sensors such as wind velocity and wave radars can be used to adjust to the WF model to varying sea states. However, this can be circumvented by using an adaptive observer design. Since the wave models are decoupled, Λ and Ω in A_w are diagonal matrices, and we have:

$$A_w(\theta_w) = \begin{bmatrix} 0 & I \\ -\Omega^2 & -2\Lambda\Omega \end{bmatrix} \triangleq \begin{bmatrix} 0 & I \\ -\mathrm{diag}(\theta_{w1}) & -\mathrm{diag}(\theta_{w2}) \end{bmatrix} \qquad (7.44)$$

where $\theta_w = [\theta_{w1}^T, \theta_{w2}^T]$, $\theta_{w1}, \theta_{w2} \in \Re^3$ contains the unknown wave model parameters to be estimated. We start with the following assumption:

A3 **(Constant environmental parameters).** It is assumed that the unknown parameters Ω and Λ in the WF model (7.7) are slowly varying and within the range of

$$\left. \begin{array}{c} 0 < \omega_{o,\min} < \omega_{oi} < \omega_{o,\max} \\ 0 < \zeta_{\min} < \zeta_i < \zeta_{\max} \end{array} \right\}, \quad i = 1, 2, 3 \qquad (7.45)$$

such that A_w is Hurwitz. Hence, the unknown wave model parameters are treated as constants in the analysis, such that

$$\dot{\theta}_w = 0 \qquad (7.46)$$

4.1 Adaptive Observer Equations

The adaptive version of the observer is equal to the augmented observer (7.35a)-(7.35e), except from the WF part which is now using the estimated WF parameters, $\hat{\theta}_w$, such that:

$$\dot{\hat{\xi}} = A_w(\hat{\theta}_w)\hat{\xi} + K_{1h}\tilde{y}_f. \tag{7.47a}$$

The adaptive update law for $\hat{\theta}_w$ remains to be decided.

4.2 Adaptive Observer Error Dynamics

The adaptive WF observer error dynamics is:

$$\dot{\tilde{\xi}} = A_w\xi - A_w(\hat{\theta}_w)\hat{\xi} - K_{1h}\tilde{y}_f + E_w w_w \tag{7.48}$$

By adding and subtracting $A_w\hat{\xi}$, defining

$$B_w\Upsilon_w^T(\hat{\xi})\tilde{\theta}_w \triangleq (A_w - A_w(\hat{\theta}_w))\hat{\xi} \tag{7.49}$$

where $\tilde{\theta}_w = \hat{\theta}_w - \theta_w$ denotes the estimation error,

$$\Upsilon_w^T(\hat{\xi}) \triangleq \left[\ \text{diag}(\hat{\xi}_1)\quad \text{diag}(\hat{\xi}_2)\ \right], \tag{7.50}$$

$$B_w \triangleq \left[\ 0\quad I\ \right]^T, \tag{7.51}$$

where $\hat{\xi} = [\hat{\xi}_1^T, \hat{\xi}_2^T]^T$, $\hat{\xi}_1, \hat{\xi}_2 \in \Re^3$, and by using (7.33) then (7.48) can be rewritten as:

$$\dot{\tilde{\xi}} = (A_w - K_{1f}C_w)\tilde{\xi} - K_{1f}\tilde{\eta} + B_w\Upsilon_w^T(\hat{\xi})\tilde{\theta}_w$$
$$+ K_{1f}T_f^{-1}x_f + E_w w_w \tag{7.52}$$

The observer error dynamics can be written compactly as:

$$M\dot{\tilde{\nu}} = -D\tilde{\nu} - J^T(\psi_y)C_a\tilde{x}_a \tag{7.53a}$$

$$\dot{\tilde{x}}_a = A_a\tilde{x}_a + B_aJ(\psi_y)\tilde{\nu} + H_a\Upsilon_w^T(\hat{\xi})\tilde{\theta}_w + E_a w \tag{7.53b}$$

where

$$H_a = \left[\ B_w^T\quad 0\quad 0\quad 0\ \right]^T \tag{7.54}$$

4.3 Stability and Passivity

Let

$$V_{ad} = \tilde{\nu}^T M\tilde{\nu} + \tilde{x}_a^T P_a\tilde{x}_a + \tilde{\theta}_w^T\Gamma_w^{-1}\tilde{\theta}_w \tag{7.55}$$

be a Lyapunov function for the observer error dynamics (7.53a)--(7.53b), where $\Gamma_w = \Gamma_w^T > 0 \in \Re^{6 \times 6}$ is an adaptive update gain matrix. Thus:

$$
\begin{aligned}
\dot{V}_{ad} = &-\tilde{\nu}^T (D + D^T)\tilde{\nu} - 2\tilde{\nu}^T J^T(\psi_y) C_a \tilde{x}_a + \tilde{x}_a^T (A_a^T P_a + P_a A_a)\tilde{x}_a \\
&+ 2\tilde{\nu}^T J^T(\psi_y) B_a^T P_a \tilde{x}_a + 2\tilde{\theta}_w^T \Upsilon_w(\hat{\xi}) H_a^T P_a \tilde{x}_a \\
&+ 2\tilde{x}_a^T P_a E_a w + 2\tilde{\theta}^T \Gamma^{-1} \dot{\tilde{\theta}}_w
\end{aligned}
\tag{7.56}
$$

In the adaptive case, we want the WF adaptive law to be updated by the high-pass filtered innovations signals. Hence, it is required that:

$$
A_a^T P_a + P_a A_a = -Q_a \tag{7.57a}
$$
$$
B_a^T P_a = C_a \tag{7.57b}
$$
$$
H_a^T P_a = C_h \tag{7.57c}
$$

The existence of a $P_a = P_a^T > 0$ that satisfies (7.57a)–(7.57c) can be tested numerically by using the Frequency Theorem with $\mathcal{P} = P_a$, $\mathcal{A} = A_a$, $u = [\varepsilon_\nu^T, \tilde{\theta}_w^T]^T$ and

$$
\mathcal{B} = \begin{bmatrix} B_a & H_a \end{bmatrix}, \quad \mathcal{C}^T = -\begin{bmatrix} C_a \\ C_h \end{bmatrix}. \tag{7.58}
$$

Since there exists a solution, the expression for $\dot{V}_{ad}$ becomes:

$$
\begin{aligned}
\dot{V}_{ad} = &-\tilde{x}_a^T Q_a \tilde{x}_a - \tilde{\nu}^T (D + D^T)\tilde{\nu} + 2\tilde{x}_a^T P_a E_a w \\
&+ 2\tilde{\theta}_w^T \Upsilon_w(\hat{\xi}) C_h \tilde{x}_a + 2\tilde{\theta}_w^T \Gamma_w^{-1} \dot{\tilde{\theta}}_w
\end{aligned}
\tag{7.59}
$$

which suggests that the adaptive update law should be:

$$
\dot{\hat{\theta}}_w \overset{\dot{\theta}_w=0}{=} \dot{\tilde{\theta}}_w = -\Gamma_w \Upsilon_w(\hat{\xi}) C_h \tilde{x}_a = -\Gamma_w \Upsilon_w(\hat{\xi})\tilde{y}_f \tag{7.60}
$$

Thus,

$$
\dot{V}_{ad} = -\tilde{x}_a^T Q_a \tilde{x}_a - \tilde{\nu}^T (D + D^T)\tilde{\nu} + 2\tilde{x}_a^T P_a E_a w \tag{7.61}
$$

and error dynamics of the adaptive observer is:

$$
M\dot{\tilde{\nu}} = -D\tilde{\nu} - J^T(\psi_y) C_a \tilde{x}_a \tag{7.62a}
$$
$$
\dot{\tilde{x}}_a = A_a \tilde{x}_a + B_a J(\psi_y)\tilde{\nu} + H_a \Upsilon_w^T(\hat{\xi})\tilde{\theta}_w + E_a w \tag{7.62b}
$$
$$
\dot{\tilde{\theta}} = -\Gamma_w \Upsilon_w(\hat{\xi}) C_h \tilde{x}_a \tag{7.62c}
$$

From (7.61) we see that the system is ISS. If the ship is exposed to WF motions, the system will be PE and $\hat{\theta}_w \to \theta_w$.

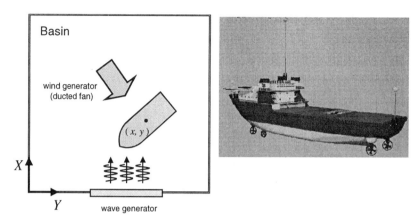

FIGURE 4. Left: Experimental setup. Right: Picture of Cybership I.

5   Experimental Results

Both the augmented and the adaptive observer have been implemented and tested at the Guidance, Navigation and Control (GNC) Laboratory, Department of Engineering Cybernetics at the Norwegian University of Science and Technology (NTNU). We have used Cybership I, a model ship of scale 1:70, see Fig. 4.

A nonlinear PID controller is used for maintaining the ship at the desired position (x_d, y_d) and heading ψ_d. The PID control law is written:

$$\tau_c = -MJ^T(\psi_y - \psi_d)\left(K_p e + K_d \dot{e} + K_i \int_0^t e\, d\delta\right) \tag{7.63}$$

where e is the estimated LF position deviation, defined as:

$$e \triangleq J^T(\psi_d)(\hat{\eta} - \eta_d) \tag{7.64}$$

An illustration of the experimental setup is given in Fig. 4. The experimental results are scaled to full scale by requiring that the *Froude number*

$$Fn = \frac{U}{\sqrt{Lg}} = \text{constant}, \tag{7.65}$$

where U is the vessel speed, L is the length of the ship and g is the acceleration of gravity. The scaling factors are given in Table 7.1 where m is the mass and the subscripts m and s denote the model and the full-scale ship, respectively. The length of the model ship is $L_m = 1.19$ m and the mass is $m_m = 17.6$ kg. A full scale ship similar to Cybership I has typically a length of 70-90 meters and mass of 4000-5000 tones. In the scaling we used $L_s = 70L_m$ meters and $m_s = 4500$ tones.

position:	L_s/L_m
linear velocity:	$\sqrt{L_s/L_m}$
angular velocity:	$\sqrt{L_m/L_s}$
linear acceleration:	1
angular acceleration:	L_m/L_s
force:	m_s/m_m
moment:	$\frac{m_s L_s}{m_m L_m}$
time:	$\sqrt{L_s/L_m}$

TABLE 7.1. Scaling factors used in the experiments (Bis scaling).

The experiment was performed in three phases:

- *Phase I (No waves)*. Initially the ship is maintaining the desired position and heading with no environmental loads acting on the ship (calm water). When the data acquisition starts, a wind fan is switched on. There is no adaptive wave filtering and the observer is identical to the augmented design in Section 3.2. The effect of the wind loads are reflected in the bias estimates in Fig. 6.

- *Phase II (Waves, adaptive wave filter off)*. After 1700 seconds the wave generator is started. In this phase we can see the performance of the observer without adaptive wave filter. In the wave model we are assuming that the dominating wave period is 9.2 seconds and the relative damping is 0.1, see Fig. 5.

- *Phase III (Waves, adaptive wave filter on)*. After 2800 seconds the adaptive wave filter is activated. The estimates of dominating wave period and relative damping are plotted in Fig. 5 for surge, sway and yaw.

A spectrum analysis of the position and heading measurements shows that the estimated wave periods converge to the true values, that is wave periods of approximately 7.8 seconds and relative damping ratios of 0.07, see Fig. 5. In Fig. 6 the measured position deviation and heading are plotted together with the corresponding LF estimates. The effect of the adaptive wave filtering is clearly seen in Fig. 8, where the innovation signals are significantly reduced in Phase III, when the adaptation is active and the wave model parameters start converging to their true values. The effect of *bad* wave filtering is reflected by noisy control action by the thrusters in phase II, see Fig. 8. A zoom-in of the heading measurement together with the LF estimate is given is Fig. 7 both for phase II and III. Here we see that the LF estimates have a significant WF contribution when the adaptive wave filter

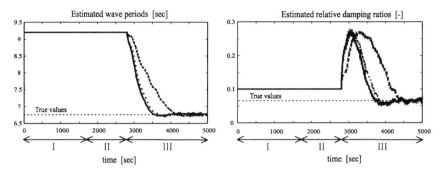

FIGURE 5. Estimated wave periods (left) and estimated relative damping ratios (right) for surge (solid), sway (dashed) and yaw (dashed). The adaptive wave-filter is activated after 2800 seconds.

is off. This is the reason for the noisy control action in phase II. The other zoom-in shows excellent LF estimation when the adaptive wave filter is active and the wave model parameters have converged to their true values. Hence, it can be concluded that adaptive wave filtering yields a significant improvement in performance compared to a filter with fixed parameters and varying sea states.

6   Conclusions

In this paper we have derived a nonlinear passive observer for moored and free-floating ships. By adding low- and high-pass filtered innovation signals in the design, we have additional flexibility in the design. This augmented design is extended to a new observer with adaptive wave filtering. Experiments with a model ship shows that the adaptive observer will significantly improve the performance of the ship positioning system. This results in reduced magnitude of the observer innovation, better filtering properties and reduced control action by the propulsion system.

ACKNOWNLEDGEMENTS

The authors are grateful to Dr. Anton Shiriaev, Dept. of Eng. Cybernetics, NTNU, for valuable discussions on passivity and frequency domain methods. This work is sponsored by the Research Council of Norway and ABB Industri AS.

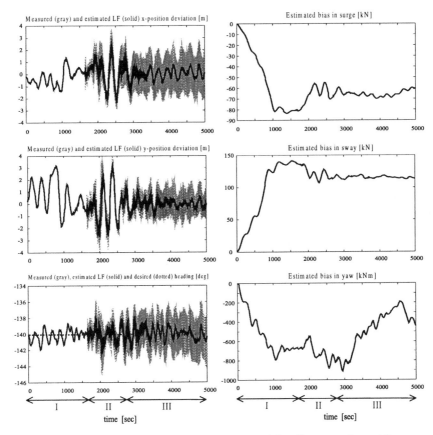

FIGURE 6. Left column: Measured position and heading together with corresponding LF estimates. Right column: Estimated bias in surge, sway and yaw.

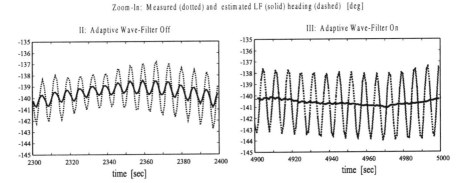

FIGURE 7. Zoom-in of measured and estimated LF heading. Left: Observer without adaptive wave-filtering. Right: Observer with adaptive wave-filtering.

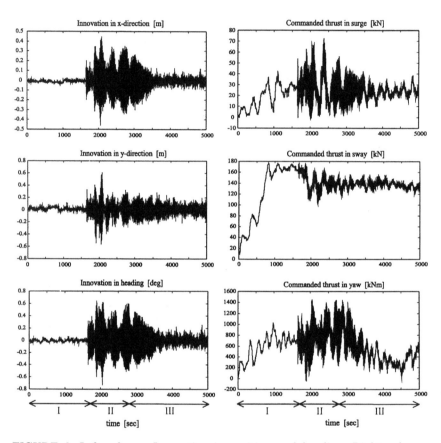

FIGURE 8. Left column: Innovation in position and heading. Right column: Commanded thrust in surge, sway and yaw.

7 REFERENCES

[1] M. F. Aarset, J. P. Strand and T. I. Fossen. Nonlinear Vectorial Observer Backstepping With Integral Action and Wave Filtering for Ships. Proc. of the IFAC Conference on Control Applications in Marine Systems (CAMS'98), Fukuoka, Japan, 1998, pp. 83-89, 1998.

[2] H. Berghuis and H. Nijmeijer. A Passivity Approach to Controller-Observer Design for Robots. IEEE Trans. on Robotics and Automation, **RA-8**(6):740-754, 1993.

[3] C. Canudas de Wit, N. Fixot and K. J. Astrom. Trajectory Tracking in Robot Manipulators via Nonlinear Estimated State Feedback. IEEE Trans. on Robotics and Automation, **RA-8**:138-144, 1992.

[4] O. M. Faltinsen. *Sea Loads on Ships and Offshore Structures*. Cambridge University Press, 1990.

[5] T. I. Fossen. *Guidance and Control of Ocean Vehicles*. John Wiley & Sons Ltd, 1994.

[6] T. I. Fossen and J. P. Strand. Passive Nonlinear Observer Design for Ships Using Lyapunov Methods: Full-Scale Experiments With a Supply Vessel. *Automatica*. **AUT-35**(1)3–16, 1999.

[7] A. H. Gelig, G. A. Leonov and V. A. Yakubovich. *Stability of Nonlinear Systems with Unique Stationary Points*. Nauka. Moscow, 1978.

[8] H. K. Khalil. *Nonlinear Systems*. Prentice Hall Inc., 1996.

[9] A. Loria, T. I. Fossen and E. Panteley. A Cascaded Approach to a Separation Principle for Dynamic Ship Positioning. IEEE Trans. on Control Systems Technology (*submitted*).

[10] J. N. Newman. *Marine Hydrodynamics*. MIT Press. Cambridge, Massachusetts, 1977.

[11] S. Nicosia and P. Tomei. Robot Control by Using only Joint Position Measurements. IEEE Trans. Automat. Contr., **AC-35**:1058-1061, 1990.

[12] S. Nicosia, A. Tornambe and P. Valigi. Experimental Results in State Estimation of Industrial Robots. Proc. Conf. Decision and Control, Honolulu, HI, Dec. 1990, pp. 360-365, 1990.

[13] R. Ortega, A. Loria, P. Nicklasson and H. Sira. *Passivity-Based Control of Euler-Lagrange systems*. Springer-Verlag, 1998.

[14] J. J.-E. Slotine, J. K. Hedrick and E. A. Misawa. Sliding Observers for Nonlinear Systems. ASME J. Dynam. Syst., Measurement, Control, Vol. 109, pp. 245-252, 1987.

[15] A. J. Sørensen, S. I. Sagatun and T. I. Fossen. Design of a Dynamic Positioning System Using Model-Based Control. Journal of Control Engineering Practice. **CEP-4**(3):359–368, 1996.

[16] J. P. Strand, A. J. Sørensen and T. I. Fossen. Design of Automatic Thruster Assisted Position Mooring Systems for Ships. *Modeling, Identification and Control,* **MIC-19**(2):61-75, 1998.

[17] J. P. Strand and Thor I. Fossen. Nonlinear Output Feedback and Locally Optimal Control of Dynamically Positioned Ships: Experimental Results. Proc. of the IFAC Conf. on Control Applications in Marine Systems (CAMS'98), Fukuoka, Japan, pp. 89-95, 1998.

[18] J. P. Strand. *Nonlinear Position Control Systems Design for Marine Vessels.* Doctoral Dissertation. Department of Engineering Cybernetics, NTNU, Trondheim, Norway, 1999.

[19] M. S. Triantafyllou. Cable Mechanics for Moored Floating Systems. BOSS'94 pp.67-77, Boston, MA, 1994.

[20] V. A. Yakubovich. A Frequency Theorem in Control Theory. *Siberian Mathematical Journal* **SMJ-14**(2):384–420, 1973.

Nonlinear Observer Design for Integration of DGPS and INS

Bjørnar Vik, Anton Shiriaev and Thor I. Fossen

Department of Engineering Cybernetics
Norwegian University of Science and Technology
N-7034 Trondheim, Norway

1 Introduction

This section contains the notation used in this chapter and the main motivation for the work with references to previous research.

1.1 Nomenclature

In the rest of this chapter, we will use the notation:

$$\omega_{ab}^c \in \Re^3 \quad : \quad \text{Angular velocity of frame } a \text{ relative to frame } b, \text{ decomposed in frame } c.$$

$\mathbf{v}^a \in \Re^3$: Linear velocity decomposed in frame a.

$\mathbf{S}(\omega_{ab}^c) \in \Re^{3\times 3}$: 3×3 skew-symmetric cross product matrix.

$\mathbf{R}_a^b \in SO(3)$: Rotation matrix from the b-frame to the a-frame.

The navigation frames used in the following are:

i-frame : The inertial frame located at the center of the Earth (ECI-frame).

e-frame : Earth-Centered, Earth-Fixed (ECEF) frame.

l-frame : Local geographic, North East Down (NED) frame.

b-frame : Body-frame.

The origins of the i- and e-frames are assumed to coincide. These frames are, however, rotated relative to each other with the Earth rotation rate. The l-frame origin is located in the center of the navigation system. The body frame (b-frame) is assumed to be aligned with the vehicle's *roll, pitch,* and *yaw* axes with the origin located in the vehicle's rotation point. In this chapter, it is assumed that the platform containing the inertial measurement unit (sensors) is in the b-frame. It is also assumed that there is no lever arm between the GPS and the inertial measurements. The position and velocity variables are defined according to:

$$\omega_{lb}^b \quad = \quad [p, q, r]^T \qquad \text{: Angular velocity of the body frame}$$
relative to the local frame.

$$\mathbf{p}^e \quad = \quad [x, y, z]^T \qquad \text{: Distance from center of Earth to the}$$
navigation system in Earth coordinates.

$$\mathbf{v}^e \quad = \quad [v_x, v_y, v_z]^T \qquad \text{: Velocity of vehicle in Earth coordinates.}$$

$$\mathbf{v}^l \quad = \quad [v_N, v_E, v_D]^T \qquad \text{: North, East and Down velocities.}$$

$$\rho \quad = \quad [\rho_1, ..., \rho_n]^T \qquad \text{: Vector of } n \text{ pseudorange observations.}$$

$$\delta \quad = \quad [\delta_1, ..., \delta_n]^T \qquad \text{: Vector of } n \text{ deltarange observations.}$$

$$\mathbf{q} \quad = \quad [q_1, q_2, q_3, q_4]^T \qquad \text{: Euler parameters (unit quaternions).}$$

1.2   Motivation

The development of GPS technology the last two decades has had a great impact on navigation. For local area operations, it is now possible to achieve meter accuracy by using *Differential GPS* (DGPS), and sub-decimeter accuracy by using *Carrier Differential GPS* (CDGPS). The development of *wide area augmentation systems* (WAAS) is expected to give meter accuracy across entire continents. However, the integrity and reliability of GPS are not on a sufficient level. Interference on data links, and several types of disturbance causing loss of lock on satellites, are common problems for many GPS applications. The latter can be a problem at latitudes between 60 and 70 degrees North, where the satellite coverage is low to begin with. In addition to the time spent reaquiring satellite signals, the ambiguity integers must also be recalculated after each loss of lock in applications where CDGPS is used.

A strapdown *Inertial Navigation System* (INS) is basically a cluster of accelerometeres and gyros known as an *Inertial Measurement Unit* (IMU), and a computer. The INS computer calculates position, velocity, heading and attitude using the strapdown equations. The INS is self-contained, and does not depend on an external signal to function properly. However, the long term drift resulting from integrating small measurement errors in the gyros and accelerometers is a major problem, and only very expensive units can be expected to have sufficient accuracy for a long period of time without re-initialization. The complementary properties of GPS and INS, suggest that integration of GPS with low-cost INS will give an affordable, highly reliable, and accurate solution.

There are several levels of integration possible, see Maybeck [4]. The most common approaches shown in Figures 1 and 2, both subtract the INS position measurement from that of the GPS system, and feed this error signal into a Kalman filter (KF). The KF model is usually a Markov process or integrated white noise for each of the error sources in the GPS and INS systems. The estimated INS errors are subsequently either subtracted from the INS measurement to give the output signal (*open-loop*), or fed back to the INS strapdown computation processor (*closed-loop*). When the position measurements from GPS are lost, the KF is not updated, and the INS has

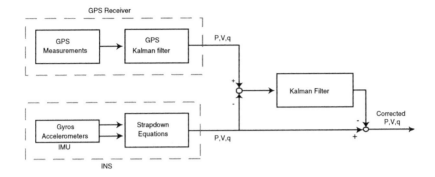

FIGURE 1. Open loop integration of INS and GPS.

to coast through the outage using the last error update from the filter. This is often referred to as uncoupled integration in the open-loop case, and loose integration in the closed-loop case. The filter can be designed with a slow update rate due to the slow dynamics of the error model. GPS receivers have an internal KF that computes position, velocity and time (PVT) solutions, and the outputs are correlated typically with 1–10 seconds correlation time. In order to avoid problems with the second KF, this filter needs to be updated sufficiently slower. In addition, the timing and lags between the two filters can be difficult to predict and control, see Tazartes and Mark [9].

A different approach to integration, often referred to as tight integration, is to use the raw accelerometer, gyro, pseudorange, and deltarange measurements from the sensors. One main advantage with this approach is that even in the case where less than the necessary amount of satellites are available for a PVT solution, the remaining pseudoranges and deltaranges provide information that will help the INS coast through the outage. Also, the accuracy of the solution is known to be better when using uncorrelated measurements in the estimator. The cost is increased complexity and bandwidth because the filter dynamics includes the vehicle dynamics. However, with the increased computing power available in the future, this is going to be less of a problem. When the nonlinear strapdown equations enter into the integration filter, a linear KF should not be used. In this chapter, we therefore propose nonlinear observers for tight integration that estimate the accelerometer and gyro biases, scale factor errors and misalignments using GPS measurements.

The outline of the chapter is as follows: In Section 2 some GPS basics are reviewed, and in Section 3 the INS strapdown equations are given. The stability and convergence properties of the proposed observers are analyzed in Section 4, and Section 5 contains a case study. Our conclusions are given

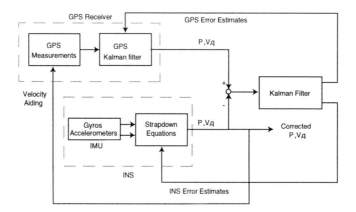

FIGURE 2. Closed loop integration of INS and GPS.

in Section 6.

2   Review of GPS Fundamentals

The raw GPS measurement is called *pseudorange*, and it is the distance from the receiver antenna (here assumed to be located at the origin of the body frame) to the satellite antenna. There are two basic ways of measuring the pseudorange, *code* and *phase* measurements. Phase measurements are the most accurate, but also the least reliable measurements.

To convert pseudoranges into user position, four measurements are needed. In addition to the three unknown position variables, the clock bias is unknown. With DGPS, the satellite clock errors cancel out when the differential corrections are applied, and only the receiver clock bias will be unknown. The clock bias multiplied with the speed of light gives the position error, which can be very large. Given the cost of very precise atomic clocks, it is a better solution to include the clock bias among the unknowns using four satellites instead of three. Given n satellites, the basic pseudorange measurement equation is:

$$\rho^i = \sqrt{(\mathbf{x}_s^i - \mathbf{x})^T(\mathbf{x}_s^i - \mathbf{x})} + c\tau^* + \varepsilon_\rho^i \qquad (8.1)$$

where $i \in [1, \cdots, n]$, and $\mathbf{x}_s^i \in \Re^3$ is the position of satellite i. This position is transmitted with the navigation message. c is the speed of light, τ^* is the clock bias and ε_ρ^i is the composite of other smaller pseudorange errors and noise. Note that τ^* is common to all n equations. The position is usually found by linearizing (8.1), that is:

$$\Delta \rho^i = -\lambda^i \Delta \mathbf{x} + \tau + \varepsilon_\rho^i \qquad (8.2)$$

where

$$\lambda^i = \frac{(\mathbf{x}_s^i - \mathbf{x}_0)^T}{\sqrt{(\mathbf{x}_s^i - \mathbf{x}_0)^T (\mathbf{x}_s^i - \mathbf{x}_0)}} \qquad (8.3)$$

and $\tau = c\tau^*$, $\mathbf{x}_0$ is an estimate of $\mathbf{x}$, and $\Delta \mathbf{x}$ is the position error. If we stack n measurements together in a vector, we obtain:

$$\Delta \rho = \Lambda_* \Delta \mathbf{x}_* \qquad (8.4)$$

where $\Delta \mathbf{x}_* = \left[\Delta \mathbf{x}^T, \ \Delta \tau\right]^T \in \Re^4$, $\Delta \rho \in \Re^n$ and

$$\Lambda_* = \begin{bmatrix} -\lambda^{1x} & -\lambda^{1y} & -\lambda^{1z} & 1 \\ -\lambda^{2x} & -\lambda^{2y} & -\lambda^{2z} & 1 \\ \vdots & \vdots & \vdots & \vdots \\ -\lambda^{nx} & -\lambda^{ny} & -\lambda^{nz} & 1 \end{bmatrix} \in \Re^{nx4} \qquad (8.5)$$

$\Delta \mathbf{x}_*$ can now be found from

$$\Delta \mathbf{x}_* = \Lambda_*^\dagger \Delta \rho \qquad (8.6)$$

Notice that, the pseudo-inverse $\Lambda_*^\dagger = \left[\Lambda_*^T \Lambda_*\right]^{-1} \Lambda_*^T$ of Λ_* always exists. Hence, by using a KF for instance, $\mathbf{x}$ and τ can be computed from a minimum of four pseudorange measurements. The distance from the user to a satellite in view is approximately $20,000$ km, and Λ_* is therefore not very sensitive to position errors since λ^i is the *line-of-sight* (LOS) unit vector from the user to satellite i. Thus, with one exception which will be addressed later, we can consider Λ_* constant throughout this chapter, even though it is a function of both satellite and user positions.

When calculating position only, and not clock bias, we will use the following LOS matrix

$$\Lambda = \begin{bmatrix} -\lambda^{1x} & -\lambda^{1y} & -\lambda^{1z} \\ -\lambda^{2x} & -\lambda^{2y} & -\lambda^{2z} \\ \vdots & \vdots & \vdots \\ -\lambda^{nx} & -\lambda^{ny} & -\lambda^{nz} \end{bmatrix} \in \Re^{nx3} \qquad (8.7)$$

The pseudo-inverse $\Lambda^\dagger$ of Λ will always exist.

The deltarange equation is written:

$$\delta^i = \lambda^i (\mathbf{v}_s^i - \mathbf{v}) + cf^* + \varepsilon_\delta^i \qquad (8.8)$$

where $\mathbf{v}_s^i$ is the velocity of satellite i, f^* is the clock drift (or frequency bias), and ε_δ^i is the composite of other smaller deltarange errors and noise. $\mathbf{v}$ and $f = cf^*$ can be calculated equivalently to (8.6) from

$$\Delta \mathbf{v}_* = \Lambda_*^\dagger \Delta \delta \qquad (8.9)$$

where $\Delta \mathbf{v}_* = [\Delta \mathbf{v}^T, \ \Delta f]^T$, or directly from

$$\mathbf{v}_* = \Lambda_*^\dagger (\delta - \delta_{\text{sat}}) \qquad (8.10)$$

where $\mathbf{v}_* = [\mathbf{v}^T, f]^T$, and

$$\delta_{\text{sat}} = [\lambda^1 \mathbf{v}_s^1, \cdots, \lambda^n \mathbf{v}_s^n]^T \qquad (8.11)$$

is the known satellite velocity component of the deltarange vector. The main focus of this chapter is the INS dynamics, and for simplicity it is assumed that $\varepsilon_\rho^i = 0$ and $\varepsilon_\delta^i = 0$. The clock bias and clock drift are the only GPS errors that will be estimated together with the INS errors. More details about GPS navigation processing can be found in Parkinson and Spilker [5], Chapter 9.

3   Strapdown Equations

PVA measurements from an INS are derived from acceleration measurements along the three orthogonal axes in the body (vessel-fixed) frame, and from angular velocity (gyroscope) measurements of the rotation of the inertial frame relative to the body frame. For applications on the Earth, the inertial frame can be well approximated by the Earth-Centered-Inertial (ECI) frame. The GPS measurements are usually given in the ECEF frame, while the raw INS measurements are given in the body frame. The best choice of reference frame for integration will depend on the application and the priorities of the system integrator.

3.1   Local Frame Representation

The large majority of strapdown implementations are mechanized in the local frame. The basic velocity equation decomposed in the local frame can be found in standard textbooks, e.g. Britting [1], and it is written:

$$\dot{\mathbf{v}}^l = \mathbf{R}_b^l \mathbf{a}^b + \bar{\mathbf{g}}^l - \left[2\mathbf{S}(\omega_{ie}^l) + \mathbf{S}(\omega_{el}^l) \right] \mathbf{v}^l \qquad (8.12)$$

$$\dot{\mathbf{p}}^e = \mathbf{R}_l^e \mathbf{v}^l \qquad (8.13)$$

where $\mathbf{a}^b \in \Re^3$ is the measured acceleration, and $\bar{\mathbf{g}}^l = \mathbf{g}^l - \mathbf{S}(\omega_{ie}^l)\mathbf{S}(\omega_{ie}^l)\mathbf{p}^l$ is the *plumb-bob gravity* where

$$\bar{\mathbf{g}}^l = [0, \ 0, \ g_D]^T \qquad (8.14)$$

$\mathbf{p}^e \in \Re^3$ is the position vector from the center of the Earth to the body frame decomposed in the Earth-fixed frame. ω_{ie}^l is the rotation rate of the Earth decomposed in the local frame, and ω_{el}^l is the rotation rate of the local frame relative to the Earth-fixed frame (due to vehicle movement over the Earth surface) decomposed in the local frame. ω_{ie}^l and ω_{el}^l are in general slow compared to the angular velocity ω_{lb}^b of the body frame relative to the NED frame. They can therefore, even though they are functions of the linear velocity $\mathbf{v}^l$, be assumed known. Finally, g_D is the acceleration of gravity, which can be estimated from the following equation:

$$g_D = G_1(1 - 2h/r_0 + 2e\sin^2\mu) - G_2(1 - 3\sin^2\mu)$$
$$- G_3(1 - 3\sin^2\mu + h/r_0)\cos^2\mu \qquad (8.15)$$

where G_1, G_2, G_3, r_0 and e are positive Earth model constants, h is altitude and μ is geodetic latitude. From (8.15) we see that the acceleration of gravity decreases as the height increases. The differential equation for the height is:

$$\ddot{h} = -g_D - a_z \qquad (8.16)$$

Thus, the INS stand-alone vertical channel is unstable since this is a pure double integrator. Integration with GPS height measurements can be used to stabilize the vertical channel. However, in the event of a GPS dropout, the vertical channel of the INS would go unstable. Therefore, a barometer is usually employed in the integration filter for applications in the air, while for marine applications a height constraint can be used.

3.2   Earth Frame Representation

It has been suggested, Wei and Schwartz [12], that an Earth frame mechanization can offer some advantages over the local level frame. An increase in computational speed was found, while the accuracy was found to be comparable to that of a local level implementation. On the other hand, as seen from (8.14), the gravity equations are rather simple and directed only along the vertical axis in the local level frame. In the Earth frame, however, difficulties arise in representing the Earth's gravitational field precisely in the navigation computer, see Titterton and Weston [10]. Gravity will have a component along all three axes, and all three axes need to be stabilized. The use of barometer data to stabilize all three axes would require more investigation into the practicability and detail of implementation, see e.g. Tang and Howell [8]. In this chapter, we will not deal with this problem, because GPS measurements are assumed to be present at all times, but we present an observer mechanized in the e-frame as an alternative to the system integrator. We will therefore consider the velocity equation decomposed in the Earth frame, that is:

$$\dot{\mathbf{v}}^e = \mathbf{R}_l^e \mathbf{R}_b^l \mathbf{a}^b + \mathbf{R}_l^e \bar{\mathbf{g}}^l - 2\mathbf{S}(\omega_{ie}^e)\mathbf{v}^e \qquad (8.17)$$
$$\dot{\mathbf{p}}^e = \mathbf{v}^e \qquad (8.18)$$

3.3 Angular Velocity Equations

The rotation matrix $\mathbf{R}_b^l$, which gives us *roll, pitch,* and *yaw* angles, can be calculated from

$$\dot{\mathbf{R}}_b^l = \mathbf{R}_b^l \mathbf{S}(\omega_{lb}^b) \qquad (8.19)$$

The gyros measure $\omega_{ib}^b = \omega_{il}^b + \omega_{lb}^b$ which leads to

$$\dot{\mathbf{R}}_b^l = \mathbf{R}_b^l \mathbf{S}(\omega_{ib}^b - \omega_{il}^b)$$
$$= \mathbf{R}_b^l \mathbf{S}(\omega_{ib}^b) - \mathbf{S}(\omega_{il}^l)\mathbf{R}_b^l \qquad (8.20)$$

where $\omega_{il}^l = \omega_{ie}^l + \omega_{el}^l$. Another representation of attitude is Euler parameters (*unit quaternions*). The set $\mathcal{H}$ of *unit quaternions* is defined as

$$\mathcal{H} = \left\{ \mathbf{q} \mid \mathbf{q}^T\mathbf{q} = 1, \mathbf{q} = [\eta, \ \varepsilon^T]^T, \eta \in \Re, \varepsilon \in \Re^3 \right\} \qquad (8.21)$$

The most common error representation used for *unit quaternions* is $\tilde{\mathbf{q}} = \bar{\mathbf{q}} \otimes \mathbf{q} \in \mathcal{H}$, where the product $\otimes$ is defined as

$$\tilde{\mathbf{q}} = \bar{\mathbf{q}} \otimes \mathbf{q} = \begin{pmatrix} \hat{\eta} \\ -\hat{\varepsilon} \end{pmatrix} \otimes \begin{pmatrix} \eta \\ \varepsilon \end{pmatrix} \triangleq \begin{bmatrix} \hat{\eta} & \hat{\varepsilon}^T \\ -\hat{\varepsilon} & \hat{\eta} - \mathbf{S}(\hat{\varepsilon}) \end{bmatrix} \begin{bmatrix} \eta \\ \varepsilon \end{bmatrix} \qquad (8.22)$$

This is equivalent to $\tilde{\mathbf{R}} = \hat{\mathbf{R}}^T\mathbf{R}$. When the error is zero, $\tilde{\mathbf{R}} = \mathbf{I}$ and $\tilde{\mathbf{q}} = [\pm 1, 0, 0, 0]^T$.

The dynamics associated with the *unit quaternions* takes the form:

$$\dot{\mathbf{q}} = \begin{bmatrix} \dot{\eta} \\ \dot{\varepsilon} \end{bmatrix} = \frac{1}{2} \begin{bmatrix} -\varepsilon^T \\ [\eta\mathbf{I} + \mathbf{S}(\varepsilon)] \end{bmatrix} \omega_{ib}^b - \frac{1}{2} \begin{bmatrix} -\varepsilon^T \\ [\eta\mathbf{I} - \mathbf{S}(\varepsilon)] \end{bmatrix} \omega_{il}^l \qquad (8.23)$$

Finally, we note that the relation between the rotation matrix $\mathbf{R}_b^l$ and $\mathbf{q}$ is

$$\mathbf{R}_b^l(\mathbf{q}) = \mathbf{I}_{3x3} + 2\eta\mathbf{S}(\varepsilon) + 2\mathbf{S}^2(\varepsilon) \qquad (8.24)$$

4 Nonlinear Observer Design

The main task of the observers is to estimate the most important errors of accelerometers and gyros. In addition, filtering of GPS and INS positions,

velocities, and angular velocity measurements are achieved. The most dominating errors are usually biases, scale-factor errors and misalignment errors.
The biases and scale-factor errors are usually precompensated through cal-
ibration and thermometer feedback, and the misalignment errors (which
include non-orthogonality of the sensor axes) are made as small as possible during mounting of the IMU. However, there are still residual errors
that need to be estimated. In the following, the errors will be modeled in
a standard way as:

$$\omega_{ib}^b = [\mathbf{I} + \Delta(\kappa, \alpha)\omega_{\text{imu}}] + \mathbf{b}_1 \tag{8.25}$$

$$\mathbf{a}^b = [\mathbf{I} + \Delta(\epsilon, \beta)\mathbf{a}_{\text{imu}}] + \mathbf{b}_2 \tag{8.26}$$

where

$$\Delta = \Delta(\sigma, \phi) = \begin{bmatrix} \sigma_x & \phi_{xy} & \phi_{xz} \\ \phi_{yx} & \sigma_y & \phi_{yz} \\ \phi_{zx} & \phi_{zy} & \sigma_z \end{bmatrix} \tag{8.27}$$

Here $\sigma = [\sigma_x, \sigma_y, \sigma_z]^T$ are three scale factor errors and

$$\phi = [\phi_{xy}, \phi_{xz}, \phi_{yx}, \phi_{yz}, \phi_{zx}, \phi_{zy}]^T$$

are six small misalignment angles. $\mathbf{b}_1 \in \Re^3$ and $\mathbf{b}_2 \in \Re^3$ represent gyro
and accelerometer biases, respectively. Notice that all error signals are decomposed in the body frame. The magnitude of these errors are in general
directly proportional to the price of the IMU.

The effect of white noise on stability will not be considered in this chapter. However, all results can easily be extended to show that the solutions
will converge to a ball around the origin when white noise is considered.
Without loss of generality, we will therefore restrict our analyses to the
case of zero noise.

4.1 Angular Velocity Observer

When bias, scale-factor and misalignment errors are included, (8.23) is
written:

$$\begin{bmatrix} \dot{\eta} \\ \dot{\varepsilon} \end{bmatrix} = \frac{1}{2} \begin{bmatrix} -\varepsilon^T \\ [\eta\mathbf{I} + \mathbf{S}(\varepsilon)] \end{bmatrix} [(\mathbf{I} + \Delta_1)\omega_{\text{imu}} + \mathbf{b}_1]$$

$$- \frac{1}{2} \begin{bmatrix} -\varepsilon^T \\ [\eta\mathbf{I} - \mathbf{S}(\varepsilon)] \end{bmatrix} \omega_{il}^l \tag{8.28}$$

where $\Delta_1 = \Delta(\kappa, \alpha)$. The gyro error models are assumed to be described
by the 1st-order models:

$$\dot{\mathbf{b}}_1 = -\mathbf{T}_1^{-1}\mathbf{b}_1 + \mathbf{w}_1 \tag{8.29}$$

$$\dot{\kappa} = -\mathbf{T}_2^{-1}\kappa + \mathbf{w}_2 \tag{8.30}$$

$$\dot{\alpha} = -\mathbf{T}_3^{-1}\alpha + \mathbf{w}_3 \tag{8.31}$$

driven by bounded signals $\mathbf{w}_i \in \Re^3$ $(i = 1, 2, 3)$. The matrices $\mathbf{T}_i > 0$ $(i = 1, 2, 3)$ are diagonal matrices of time constants, $\kappa = [\kappa_x, \kappa_y, \kappa_z]^T$ are three gyro scale factor errors, and $\alpha = [\alpha_{xy}, \alpha_{xz}, \alpha_{yx}, \alpha_{yz}, \alpha_{zx}, \alpha_{zy}]^T$ are six small gyro misalignment angles. $\mathbf{b}_1 = -\mathbf{b}_{\text{gyro}}$ represents the biases of the gyros. The nonlinear observer of Salcudean [6] is extended to include bias and error update laws according to:

$$
\begin{bmatrix} \dot{\hat{\eta}} \\ \dot{\hat{\varepsilon}} \end{bmatrix} = \frac{1}{2} \begin{bmatrix} -\hat{\varepsilon}^T \\ [\hat{\eta}\mathbf{I} + \mathbf{S}(\hat{\varepsilon})] \end{bmatrix} \left[(\mathbf{I} + \hat{\Delta}_1)\omega_{\text{imu}} + \hat{\mathbf{b}}_1 + \mathbf{K}_1 \tilde{\varepsilon} \operatorname{sgn}(\tilde{\eta}) \right]
$$
$$
- \frac{1}{2} \begin{bmatrix} -\hat{\varepsilon}^T \\ [\hat{\eta}\mathbf{I} - \mathbf{S}(\hat{\varepsilon})] \end{bmatrix} \omega_{il}^l \tag{8.32}
$$

$$
\dot{\hat{\mathbf{b}}}_1 = -\mathbf{T}_1^{-1}\hat{\mathbf{b}}_1 + \frac{1}{2}\tilde{\varepsilon} \operatorname{sgn}(\tilde{\eta}) \tag{8.33}
$$

$$
\dot{\hat{\kappa}} = -\mathbf{T}_2^{-1}\hat{\kappa} + \frac{1}{2}\operatorname{diag}(\tilde{\varepsilon}) \omega_{\text{imu}} \operatorname{sgn}(\tilde{\eta}) \tag{8.34}
$$

$$
\dot{\hat{\alpha}} = -\mathbf{T}_3^{-1}\hat{\alpha} + \frac{1}{2}\Gamma(\tilde{\varepsilon}) \omega_{\text{imu}} \operatorname{sgn}(\tilde{\eta}) \tag{8.35}
$$

where

$$
\Gamma(\tilde{\varepsilon}) = \begin{bmatrix} 0 & \tilde{\varepsilon}_1 & 0 \\ 0 & 0 & \tilde{\varepsilon}_1 \\ \tilde{\varepsilon}_2 & 0 & 0 \\ 0 & 0 & \tilde{\varepsilon}_2 \\ \tilde{\varepsilon}_3 & 0 & 0 \\ 0 & \tilde{\varepsilon}_3 & 0 \end{bmatrix} \tag{8.36}
$$

The error model is found by combining (8.29)–(8.35):

$$
\begin{bmatrix} \dot{\tilde{\eta}} \\ \dot{\tilde{\varepsilon}} \end{bmatrix} = \frac{1}{2} \begin{bmatrix} -\tilde{\varepsilon}^T \\ [\tilde{\eta}\mathbf{I} + \mathbf{S}(\tilde{\varepsilon})] \end{bmatrix} \left[\tilde{\Delta}_1\omega_{\text{imu}} + \tilde{\mathbf{b}}_1 - \mathbf{K}_1 \tilde{\varepsilon} \operatorname{sgn}(\tilde{\eta}) \right] \tag{8.37}
$$

$$
\dot{\tilde{\mathbf{b}}}_1 = -\mathbf{T}_1^{-1}\tilde{\mathbf{b}}_1 - \frac{1}{2}\tilde{\varepsilon} \operatorname{sgn}(\tilde{\eta}) \tag{8.38}
$$

$$
\dot{\tilde{\kappa}} = -\mathbf{T}_2^{-1}\tilde{\kappa} - \frac{1}{2}\operatorname{diag}(\tilde{\varepsilon}) \omega_{\text{imu}} \operatorname{sgn}(\tilde{\eta}) \tag{8.39}
$$

$$
\dot{\tilde{\alpha}} = -\mathbf{T}_3^{-1}\tilde{\alpha} - \frac{1}{2}\Gamma(\tilde{\varepsilon}) \omega_{\text{imu}} \operatorname{sgn}(\tilde{\eta}) \tag{8.40}
$$

Note that the equilibrium points $(\tilde{\eta}, \tilde{\varepsilon}, \tilde{\mathbf{b}}_1, \tilde{\kappa}, \tilde{\alpha}) = (\pm 1, 0, 0, 0, 0)$.

Theorem 8.1 (Exponentially Stable Angular Velocity Observer)
The equilibrium points $(\pm 1, 0, 0, 0, 0)$ of the error model (8.37)–(8.40) are exponentially stable.

Proof. Consider the following Lyapunov function candidate:

$$
V = \frac{1}{2}\tilde{\mathbf{b}}_1^T\tilde{\mathbf{b}}_1 + \frac{1}{2}\tilde{\kappa}^T\tilde{\kappa} + \frac{1}{2}\tilde{\alpha}^T\tilde{\alpha} + \frac{1}{2} \begin{cases} (\tilde{\eta} - 1)^2 + \tilde{\varepsilon}^T\tilde{\varepsilon} & \text{if } \tilde{\eta} \geq 0 \\ (\tilde{\eta} + 1)^2 + \tilde{\varepsilon}^T\tilde{\varepsilon} & \text{if } \tilde{\eta} < 0 \end{cases} \tag{8.41}
$$

The time derivative along the trajectory of (8.37)–(8.40) is:

$$\dot{V} = \tilde{\mathbf{b}}_1^T \dot{\tilde{\mathbf{b}}}_1 + \tilde{\kappa}^T \dot{\tilde{\kappa}} + \tilde{\alpha}^T \dot{\tilde{\alpha}} + \begin{cases} -\dot{\tilde{\eta}} & \text{if } \tilde{\eta} \geq 0 \\ \dot{\tilde{\eta}} & \text{if } \tilde{\eta} < 0 \end{cases}$$

$$= -\tilde{\mathbf{b}}_1^T \mathbf{T}_1^{-1} \tilde{\mathbf{b}}_1 - \frac{1}{2} \tilde{\mathbf{b}}_1^T \tilde{\varepsilon} \, \text{sgn}(\tilde{\eta})$$

$$- \tilde{\kappa}^T \mathbf{T}_2^{-1} \tilde{\kappa} - \frac{1}{2} \tilde{\kappa}^T \text{diag}(\tilde{\varepsilon}) \, \omega_{\text{imu}} \, \text{sgn}(\tilde{\eta})$$

$$- \tilde{\alpha}^T \mathbf{T}_3^{-1} \tilde{\alpha} - \frac{1}{2} \tilde{\alpha}^T \Gamma(\tilde{\varepsilon}) \, \omega_{\text{imu}} \, \text{sgn}(\tilde{\eta})$$

$$+ \frac{1}{2} \tilde{\varepsilon}^T \left[\tilde{\Delta}_1 \omega_{\text{imu}} + \tilde{\mathbf{b}}_1 - \mathbf{K}_1 \tilde{\varepsilon} \, \text{sgn}(\tilde{\eta}) \right] \text{sgn}(\tilde{\eta})$$

$$= -\tilde{\mathbf{b}}_1^T \mathbf{T}_1^{-1} \tilde{\mathbf{b}}_1 - \tilde{\kappa}^T \mathbf{T}_2^{-1} \tilde{\kappa} - \tilde{\alpha}^T \mathbf{T}_3^{-1} \tilde{\alpha} - \frac{1}{2} \tilde{\varepsilon}^T \mathbf{K}_1 \tilde{\varepsilon}$$

$$< 0, \quad \forall \, \tilde{\mathbf{b}}_1, \tilde{\kappa}, \tilde{\alpha}, \tilde{\varepsilon} \neq \mathbf{0} \tag{8.42}$$

where we have used:

$$\tilde{\eta} \dot{\tilde{\eta}} + \tilde{\varepsilon}^T \dot{\tilde{\varepsilon}} \equiv 0 \tag{8.43}$$

$$\tilde{\varepsilon}^T \tilde{\Delta}_1 \omega_{\text{imu}} = \tilde{\kappa}^T \, \text{diag}(\tilde{\varepsilon}) \omega_{\text{imu}} + \tilde{\alpha}^T \Gamma(\tilde{\varepsilon}) \, \omega_{\text{imu}} \tag{8.44}$$

From the relation $\tilde{\eta}^2 + \tilde{\varepsilon}^T \tilde{\varepsilon} \equiv 1$ it is seen that as $\tilde{\varepsilon} \to \mathbf{0}$ exponentially, $|\tilde{\eta}| \to 1$ exponentially. This completes the proof. ■

4.2   Velocity and Position Observers

In this section, we will show how tightly integrated observers for position and velocity can be designed. The observers will be updated with raw GPS measurements, in order to take advantage of GPS measurements when there is less than four satellites available. With less than four satellites available, these observers will still perform better than the stand-alone INS. See e.g. Lewantowicz [3] for a study on graceful degradation of an integrated solution with less than four satellites in view. The stability proofs below will of course *not* be valid in this case. Note that the raw GPS pseudorange and deltarange measurements are scalars, and as such not decomposed in a reference frame. However, the LOS unit vectors that give the direction from the GPS receiver antenna to the satellite antennas, are easiest to calculate in the Earth frame. The main reason for this is that the satellite coordinates in the navigation message are given in the Earth frame. This means that we need to transform positions and velocities to the Earth frame before the transformation to pseudoranges and deltaranges can be done.

We have chosen to design three different observers which are mechanized in the local frame, Earth frame, and LOS-subspace of the Earth frame, respectively.

Local Frame Formulation

The update mechanism of the observers takes the following form:

$$
\begin{bmatrix} \tilde{\mathbf{p}}^e \\ \tilde{\tau} \end{bmatrix} = (\Lambda^e)_*^\dagger \left(\rho_{\text{gps}} - \sqrt{(\mathbf{p}_s^e - \hat{\mathbf{p}}^e)^T(\mathbf{p}_s^e - \hat{\mathbf{p}}^e)} - c\hat{\tau} \right)
$$

$$
= \begin{bmatrix} (\Lambda^e)^\dagger \\ (\lambda^e)_*^\dagger \end{bmatrix} \left(\rho_{\text{gps}} - \sqrt{(\mathbf{p}_s^e - \hat{\mathbf{p}}^e)^T(\mathbf{p}_s^e - \hat{\mathbf{p}}^e)} - c\hat{\tau} \right) \qquad (8.45)
$$

for position, while velocity is given by

$$
\begin{bmatrix} \tilde{\mathbf{v}}^l \\ \tilde{f} \end{bmatrix} = \begin{bmatrix} (\mathbf{R}_l^e)^T(\Lambda^e)^\dagger \\ (\lambda^e)_*^\dagger \end{bmatrix} \left(\delta_{\text{gps}} - \delta_{\text{sat}} - \Lambda^e \mathbf{R}_l^e \hat{\mathbf{v}}^l - c\hat{f} \right) \qquad (8.46)
$$

where $\tilde{\mathbf{p}}^e$ and $\tilde{\mathbf{v}}^l$ are the position and velocity estimation errors, ρ_{gps} and δ_{gps} are the pseudorange and deltarange vectors measured by GPS, δ_{sat} is given by (8.11), Λ_*^e and Λ^e are defined in (8.5) and (8.7), and $(\lambda^e)_*^\dagger \in \Re^4$ is the last row of $(\Lambda^e)_*^\dagger$. The superscript e is used to emphasize that the LOS vectors are decomposed in the Earth frame. $\hat{\mathbf{p}}^e, \hat{\mathbf{v}}^l, \hat{\tau}$ and $\hat{f}$ are estimated Earth frame position, local frame velocity, receiver clock bias, and frequency bias.

The l-frame velocity and position update equations (8.12)–(8.13) with bias, scale-factor and misalignment errors included are written:

$$
\dot{\mathbf{v}}^l = \mathbf{R}_b^l \left[(\mathbf{I} + \Delta_2)\mathbf{a}_{\text{imu}} + \mathbf{b}_2 \right] + \bar{\mathbf{g}}^l - \left[2\mathbf{S}(\omega_{ie}^l) + \mathbf{S}(\omega_{el}^l) \right] \mathbf{v}^l \qquad (8.47)
$$

$$
\dot{\mathbf{p}}^e = \mathbf{R}_l^e \mathbf{v}^l \qquad (8.48)
$$

$$
\dot{\mathbf{b}}_2 = -\mathbf{T}_4^{-1}\mathbf{b}_2 + \mathbf{w}_4 \qquad (8.49)
$$

$$
\dot{\epsilon} = -\mathbf{T}_5^{-1}\epsilon + \mathbf{w}_5 \qquad (8.50)
$$

$$
\dot{\beta} = -\mathbf{T}_6^{-1}\beta + \mathbf{w}_6 \qquad (8.51)
$$

$$
\dot{f} = -t_7^{-1}f + w_7 \qquad (8.52)
$$

$$
\dot{\tau} = f + w_8 \qquad (8.53)
$$

where $\Delta_2 = \Delta(\epsilon, \beta), \epsilon = [\epsilon_x, \epsilon_y, \epsilon_z]^T$ are three accelerometer scale factor errors, and $\beta = [\beta_{xy}, \beta_{xz}, \beta_{yx}, \beta_{yz}, \beta_{zx}, \beta_{zy}]^T$ are six small accelerometer misalignment errors. $\mathbf{b}_2 = -\mathbf{b}_{\text{acc}}$ represents the biases of the accelerometers, τ and f are the scalar GPS clock and frequency biases, and $\mathbf{w}_i \in \Re^3$ ($i = 4, 5, 6$) and $w_7, w_8 \in \Re$ are white noise. $\mathbf{T}_i$ ($i = 4, 5, 6$) are large positive definite time constant matrices, and $t_7 > 0$ is a large time constant, all assumed to be known. The following observer is proposed for (8.47)–(8.53):

$$\dot{\hat{\mathbf{v}}}^l = \hat{\mathbf{R}}_b^l \left[(\mathbf{I} + \hat{\Delta}_2)\mathbf{a}_{\mathrm{imu}} + \hat{\mathbf{b}}_2 \right] + \hat{\mathbf{g}}^l - \left[2\mathbf{S}(\omega_{ie}^l) + \mathbf{S}(\omega_{el}^l) \right] \hat{\mathbf{v}}^l$$
$$+ \mathbf{K}_2 \tilde{\mathbf{v}}^l + (\mathbf{R}_l^e)^T \tilde{\mathbf{p}}^e \tag{8.54}$$

$$\dot{\hat{\mathbf{p}}}^e = \mathbf{R}_l^e \hat{\mathbf{v}}^l + \mathbf{K}_3 \tilde{\mathbf{p}}^e \tag{8.55}$$

$$\dot{\hat{\mathbf{b}}}_2 = -\mathbf{T}_4^{-1}\hat{\mathbf{b}}_2 + (\hat{\mathbf{R}}_b^l)^T \tilde{\mathbf{v}}^l \tag{8.56}$$

$$\dot{\hat{\epsilon}} = -\mathbf{T}_5^{-1}\hat{\epsilon} + \mathrm{diag}(\mathbf{a}_{\mathrm{imu}})(\hat{\mathbf{R}}_b^l)^T \tilde{\mathbf{v}}^l \tag{8.57}$$

$$\dot{\hat{\beta}} = -\mathbf{T}_6^{-1}\hat{\beta} + \Upsilon^T(\mathbf{a}_{\mathrm{imu}})(\hat{\mathbf{R}}_b^l)^{\mathbf{T}} \tilde{\mathbf{v}}^l \tag{8.58}$$

$$\dot{\hat{f}} = -t_7^{-1}\hat{f} + k_4 \tilde{f} + \tilde{\tau} \tag{8.59}$$

$$\dot{\hat{\tau}} = \hat{f} + k_5 \tilde{\tau} \tag{8.60}$$

where Υ is defined in (8.92), $\mathbf{K}_i = \mathbf{K}_i^T > 0$, $(i = 2,3)$ are two 3×3 gain matrices, and k_4 and k_5 are scalar gains. The error dynamics takes the form:

$$\dot{\tilde{\mathbf{v}}}^l = \hat{\mathbf{R}}_b^l \left[\tilde{\Delta}_2 \mathbf{a}_{\mathrm{imu}} + \tilde{\mathbf{b}}_2 \right] + \mathbf{E} \left[(\mathbf{I} + \Delta_2)\mathbf{a}_{\mathrm{imu}} + \mathbf{b}_2 \right]$$
$$- \left[2\mathbf{S}(\omega_{ie}^l) + \mathbf{S}(\omega_{el}^l) \right] \tilde{\mathbf{v}}^l - \mathbf{K}_2 \tilde{\mathbf{v}}^l - (\mathbf{R}_l^e)^T \tilde{\mathbf{p}}^e \tag{8.61}$$

$$\dot{\tilde{\mathbf{p}}}^e = \mathbf{R}_l^e \tilde{\mathbf{v}}^l - \mathbf{K}_3 \tilde{\mathbf{p}}^e \tag{8.62}$$

$$\dot{\tilde{\mathbf{b}}}_2 = -\mathbf{T}_4^{-1}\tilde{\mathbf{b}}_2 - (\hat{\mathbf{R}}_b^l)^T \tilde{\mathbf{v}}^l \tag{8.63}$$

$$\dot{\tilde{\epsilon}} = -\mathbf{T}_5^{-1}\tilde{\epsilon} - \mathrm{diag}(\mathbf{a}_{\mathrm{imu}})(\hat{\mathbf{R}}_b^l)^T \tilde{\mathbf{v}}^l \tag{8.64}$$

$$\dot{\tilde{\beta}} = -\mathbf{T}_6^{-1}\tilde{\beta} - \Upsilon^T(\mathbf{a}_{\mathrm{imu}})(\hat{\mathbf{R}}_b^l)^{\mathbf{T}} \tilde{\mathbf{v}}^l \tag{8.65}$$

$$\dot{\tilde{f}} = -t_7^{-1}\tilde{f} - k_4 \tilde{f} - \tilde{\tau} \tag{8.66}$$

$$\dot{\tilde{\tau}} = \tilde{f} - k_5 \tilde{\tau} \tag{8.67}$$

where $\mathbf{E} = \mathbf{R}_b^l - \hat{\mathbf{R}}_b^l$. Now, (8.61)–(8.67) can be written:

$$\dot{\mathbf{x}} = \mathbf{f}(\mathbf{x}, t) + \mathbf{g}(t) \tag{8.68}$$

where $\mathbf{x} = \left[\tilde{\mathbf{v}}^l, \tilde{\mathbf{p}}^e, \tilde{\mathbf{b}}_2, \tilde{\epsilon}, \tilde{\beta}, \tilde{f}, \tilde{\tau} \right]^T$,

$$\mathbf{f}(\mathbf{x}, t) = \begin{bmatrix} \hat{\mathbf{R}}_b^l \left[\tilde{\Delta}_2 \mathbf{a}_{\mathrm{imu}} + \tilde{\mathbf{b}}_2 \right] - \left[2\mathbf{S}(\omega_{ie}^l) + \mathbf{S}(\omega_{el}^l) \right] \tilde{\mathbf{v}}^l - \mathbf{K}_2 \tilde{\mathbf{v}}^l - (\mathbf{R}_l^e)^T \tilde{\mathbf{p}}^e \\ \mathbf{R}_l^e \tilde{\mathbf{v}}^l - \mathbf{K}_3 \tilde{\mathbf{p}}^e \\ -\mathbf{T}_4^{-1}\tilde{\mathbf{b}}_2 - (\hat{\mathbf{R}}_b^l)^T \tilde{\mathbf{v}}^l \\ -\mathbf{T}_5^{-1}\tilde{\epsilon} - \mathrm{diag}(\mathbf{a}_{\mathrm{imu}})(\hat{\mathbf{R}}_b^l)^T \tilde{\mathbf{v}}^l \\ -\mathbf{T}_6^{-1}\tilde{\beta} - \Upsilon^T(\mathbf{a}_{\mathrm{imu}})(\hat{\mathbf{R}}_b^l)^{\mathbf{T}} \tilde{\mathbf{v}}^l \\ -t_7^{-1}\tilde{f} - k_4 \tilde{f} - \tilde{\tau} \\ \tilde{f} - k_5 \tilde{\tau} \end{bmatrix}$$
$$\tag{8.69}$$

and

$$\mathbf{g}(t) = [\mathbf{E}\,[(\mathbf{I} + \Delta_2)\mathbf{a}_{\mathrm{imu}} + \mathbf{b}_2]\ \ 0\ 0\ 0\ 0\ 0\ 0]^T \qquad (8.70)$$

By recognizing that the rotation matrix $\mathbf{R}_b^l$, as seen from (8.24), is a non-linear function of the quaternion $\mathbf{q}$, and using the angular velocity observer above, it can be seen that the error matrix $\mathbf{E} = \mathbf{R}_b^l(\mathbf{q}) - \mathbf{R}_b^l(\hat{\mathbf{q}}) \rightarrow \mathbf{0}$ as $t \rightarrow \infty$. Thus, it is established that $\mathbf{g}(t) \rightarrow \mathbf{0}$ as $t \rightarrow \infty$ since $(\mathbf{I} + \Delta_2)\mathbf{a}_{\mathrm{imu}} + \mathbf{b}_2 = \mathbf{a}$, the true acceleration of the vehicle, obviously is bounded. The following two lemmas will be useful for analyzing the stability of (8.68).

Lemma 8.1 (Yakubovich, 1964) *Suppose that $\exists$ a function $V(\mathbf{x})$ satisfying the conditions:*

(I) $V(\mathbf{x})$ is Lipschitz in every bounded region.

(II) $V(\mathbf{x}) \rightarrow \infty$ as $\mathbf{x} \rightarrow \infty$.

(III) $\exists\,\xi > 0$ and a continuos function $\alpha(\mathbf{x}) > 0$, well defined when $|\mathbf{x}| > \xi$ such that for any solution $\mathbf{x}(t)$ of the system

$$\dot{\mathbf{x}} = \mathbf{f}(\mathbf{x}, t) \qquad (8.71)$$

when $|\mathbf{x}| > \xi$ the relation

$$\frac{d}{dt} V(\mathbf{x}(t)) \leq -\alpha(\mathbf{x}(t)) \qquad (8.72)$$

hold almost everywhere.

Choose the number $\eta > 0$ such that

$$F = E\,\{V(\mathbf{x}) \leq \eta\} \supset E\,\{|\mathbf{x}| \leq \xi\} \qquad (8.73)$$

Then:

a) F is an invariant set of (8.71).

b) For any solution $\mathbf{x}(t)$ of the system (8.71) $\exists T \geq 0$ where $\mathbf{x}(T) \in F$.

Proof. See Yakubovich [11]. ∎

Remark 8.1 *Lemma 8.1 remains true if the relation (8.72) becomes valid for any solution after some time $T^* > 0$, which may depend on initial conditions, provided that this solution of (8.71) is bounded on $[0, T^*]$.*

Lemma 8.2 (Asymptotic Stability of Forced Systems) *Suppose that for the system* $\dot{\mathbf{x}} = \mathbf{f}(\mathbf{x}, t)$, *there exist a scalar C^1-smooth function $V(\mathbf{x})$ and a continuos function $\alpha(\mathbf{x}) > 0$, such that*

(I) *There exist class $\mathcal{K}$ functions κ_1, κ_2 such that $\forall \mathbf{x}$*

$$\kappa_1(|\mathbf{x}|) \le V(\mathbf{x}) \le \kappa_2(|\mathbf{x}|) \tag{8.74}$$

(II) *Along any solution $\mathbf{x}(t)$ of the system $\dot{\mathbf{x}} = \mathbf{f}(\mathbf{x}, t)$ the following relation holds*

$$\frac{d}{dt} V(\mathbf{x}(t)) \le -\alpha(\mathbf{x}(t)) \tag{8.75}$$

(III)

$$\limsup_{|\mathbf{x}| \to \infty} \frac{\left| \frac{\partial V(\mathbf{x})}{\partial \mathbf{x}} \right|}{\alpha(\mathbf{x})} = 0 \tag{8.76}$$

Then,

a) *Any solution $\mathbf{x}(t)$ of the system*

$$\dot{\mathbf{x}} = \mathbf{f}(\mathbf{x}, t) + \mathbf{g}(t) \tag{8.77}$$

where $\mathbf{g}(t) \to \mathbf{0}$ as $t \to \infty$, tends to zero as $t \to \infty$.

b) $\forall \varepsilon > 0$ *there exist a $\delta > 0$ and a $T^{**} = T^{**}(\mathbf{g}(t), \varepsilon) \ge 0$ such that if $|\mathbf{x}_0| < \delta$, then $\forall t \ge T^{**} : |\mathbf{x}(t, \mathbf{x}_0)| < \varepsilon$.*

Proof. Given $\mathbf{g}(t)$. Suppose that $|\mathbf{x}(t)|$ does not tend to zero as $t \to \infty$. Then there are two possible cases: $\limsup_{t \to \infty} |\mathbf{x}(t)| \to \infty$ or $|\mathbf{x}(t)| \le R$.

1) Let $\limsup_{t \to \infty} |\mathbf{x}(t)| \to \infty$. Choose any $\varepsilon > 0$. Then $\exists T \ge 0$ such that $\forall t \ge T : |\mathbf{g}(t)| < \varepsilon$. Due to the assumptions, the derivative of $V(\mathbf{x})$ along any solution of (8.77) satisfies the inequality

$$\frac{d}{dt} V(\mathbf{x}(t)) \le -\alpha(\mathbf{x}(t)) + \frac{\partial V(\mathbf{x})}{\partial \mathbf{x}} \mathbf{g}(t) \tag{8.78}$$

Due to the assumption (8.76), there exist a $N > 0$ such that

$$\left| \frac{\partial V(\mathbf{x})}{\partial \mathbf{x}} \right| < \frac{1}{2\varepsilon} \alpha(\mathbf{x}) \ \forall x \in \Re^n \backslash B(0, N) \tag{8.79}$$

Therefore,

$$\frac{d}{dt}V(\mathbf{x}(t)) < -\alpha(\mathbf{x}(t)) + \left|\frac{\partial V(\mathbf{x})}{\partial \mathbf{x}}\right|\varepsilon$$

$$< -\alpha(\mathbf{x}(t))(1 - \frac{1}{2}) = -\frac{1}{2}\alpha(\mathbf{x}(t)) \; \forall x \in \Re^n \backslash B(0, N) \tag{8.80}$$

Due to Lemma 8.1 and (8.74), $\mathbf{x}(t)$ is bounded. The contradiction proves that $|\mathbf{x}(t)| \leq R$.

2) Let $|\mathbf{x}(t)|$ be bounded, and assume that $\mathbf{x} \nrightarrow 0$ as $t \rightarrow \infty$.

Take any small $\delta > 0$. If $\mathbf{x} \nrightarrow 0$ as $t \rightarrow \infty$, $\exists \{T_n\}_{n=1}^{\infty}$, $T_n \rightarrow +\infty$ such that

$$|\mathbf{x}(T_n)| > \delta > 0, :: \forall n \geq 1 \tag{8.81}$$

Due to (8.74) there exist a $\delta_1 > 0$ and a $\delta_2 > 0$ such that

$$E\{|\mathbf{x}| \leq \delta_2\} \subset E\{V(\mathbf{x}) \leq \delta_1\} \subset E\{|\mathbf{x}| \leq \delta\} \tag{8.82}$$

Choose

$$\varepsilon = \frac{1}{2}\left[\max_{\delta_2 < |\mathbf{x}(t)| < R}\left|\frac{\partial V(\mathbf{x})}{\partial \mathbf{x}}\right|\right]^{-1}\left[\min_{\delta_2 < |\mathbf{x}(t)| < R}\alpha(\mathbf{x})\right]$$

and a $T^* = T^*(\varepsilon) \geq 0$ such that $\forall t \geq T^* : |g(t)| < \varepsilon$. Again using (8.78), we have that $\forall t \geq T^*$, when $\delta_2 < |\mathbf{x}(t)| < R$

$$\frac{d}{dt}V(\mathbf{x}(t)) \leq -\alpha(\mathbf{x}(t)) + \frac{\partial V(\mathbf{x})}{\partial \mathbf{x}}g(t)$$

$$< -\alpha(\mathbf{x}(t)) + \left|\frac{\partial V(\mathbf{x})}{\partial \mathbf{x}}\right|\varepsilon$$

$$\leq -\frac{1}{2}\alpha(\mathbf{x}(t)) \tag{8.83}$$

Taking advantage of Lemma 8.1 and Remark 8.1, we conclude that there exist $T^{**} > T^*$ such that the solution $\mathbf{x}(t)$ will belong to

$$E\{V(\mathbf{x}) \leq \delta_1\} \tag{8.84}$$

This is a contradiction to (8.81)–(8.82), and thus $|\mathbf{x}(t)| \rightarrow 0$ as $t \rightarrow \infty$. This completes the proof of part a).

3) To prove b), it is necessary to point out that due to Step 1, if the initial condition $\mathbf{x}_0$ belongs to $E\{|\mathbf{x}_0| \leq \delta\}$ then there exist a R_δ such that

$$\mathbf{x}(t, \mathbf{x}_0) \in E\{|\mathbf{x}| \leq R_\delta\} \; \forall \mathbf{x}_0 \in E\{|\mathbf{x}_0| \leq \delta\}, \; \forall t \geq 0 \tag{8.85}$$

Choose any small $\varepsilon > 0$. Due to (8.74) there exists a $\delta_1 > 0$, and a $\delta_2 > 0$ such that

$$E\{|\mathbf{x}| \leq \delta_2\} \subset E\{V(\mathbf{x}) \leq \delta_1\} \subset E\{|\mathbf{x}| \leq \varepsilon\} \tag{8.86}$$

Fix any $\delta > 0$. Then following Step 2, there exists a $T^* : \forall t \geq T^*$, $\delta_2 < |\mathbf{x}(t)| < R_\delta$ where the relation

$$\frac{d}{dt}V(\mathbf{x}(t)) \leq -\frac{1}{2}\alpha(\mathbf{x}(t)) \tag{8.87}$$

is valid. Then,

$$V(\mathbf{x}(t)) = V(\mathbf{x}(T^*)) + \int_{T^*}^{t} \dot{V}(\mathbf{x}(\tau))d\tau$$

$$\leq V(\mathbf{x}(T^*)) - \frac{1}{2}(t - T^*)\alpha_0 \tag{8.88}$$

where $\alpha_0 = \min_{\delta_2 < |\mathbf{x}(t)| < R_\delta} \alpha(\mathbf{x})$.

The inequality (8.88) implies that any solution $\mathbf{x}(t)$ starting with $E\{|\mathbf{x}| \leq R_\delta\}$ at time $t = T^*$ should reach the set $E\{|\mathbf{x}| \leq \delta_2\}$ at least at time

$$T^{**} = T^* + 2 \max_{|\mathbf{x}| \leq R_\delta} V(\mathbf{x})\alpha_0^{-1} \tag{8.89}$$

Otherwise, it will contradict with the positive definiteness of V. Due to Lemma 8.1, the set $E\{V(\mathbf{x}) \leq \delta_1\}$ is invariant, and due to the inclusion (8.86) $\forall t \geq T^{**}$

$$\mathbf{x}(t) \in E\{|\mathbf{x}| \leq \varepsilon\}$$

This completes the proof of part b)

∎

Remark 8.2 (Quasi-Equi Asymptotic Stability (QEAS)) *The results of Lemma 8.2 guarantees that the equilibrium point $\mathbf{x} = \mathbf{0}$ of the forced system $\dot{\mathbf{x}} = \mathbf{f}(\mathbf{x},t) + \mathbf{g}(t)$ is quasi-equi asymptotically stable in the large, as defined by Lakshmikantham et al. [2].*

The next theorem considers the convergence and stability properties of the system (8.68):

Theorem 8.2 (QEAS l-Frame Observer) *For any solution of the l-frame observer error equation (8.68), $|\mathbf{x}(t)| \to \mathbf{0}$ as $t \to \infty$ and QEAS follows from Remark 8.2.*

Proof. Consider the unforced system $\dot{\mathbf{x}} = \mathbf{f}(\mathbf{x}, t)$, where $\mathbf{f}(\mathbf{x}, t)$ is given by (8.69). Let V be a Lyapunov Function Candidate for $\dot{\mathbf{x}} = \mathbf{f}(\mathbf{x}, t)$:

$$V = \frac{1}{2}(\tilde{\mathbf{v}}^l)^T \tilde{\mathbf{v}}^l + \frac{1}{2}(\tilde{\mathbf{p}}^e)^T \tilde{\mathbf{p}}^e + \frac{1}{2}\tilde{\mathbf{b}}_2^{\ T}\tilde{\mathbf{b}}_2$$
$$+ \frac{1}{2}\tilde{\epsilon}^T\tilde{\epsilon} + \frac{1}{2}\tilde{\beta}^T\tilde{\beta} + \frac{1}{2}\tilde{f}^2 + \frac{1}{2}\tilde{\tau}^2 \tag{8.90}$$

It is easily seen that the conditions on V in Lemma 8.2 are satisfied. The derivative of V along the system trajectories is:

$$\dot{V} = (\tilde{\mathbf{v}}^l)^T\dot{\tilde{\mathbf{v}}}^l + (\tilde{\mathbf{p}}^e)^T\dot{\tilde{\mathbf{p}}}^e + \tilde{\mathbf{b}}_2^{\ T}\dot{\tilde{\mathbf{b}}}_2 + \tilde{\epsilon}^T\dot{\tilde{\epsilon}} + \tilde{\beta}^T\dot{\tilde{\beta}} + \tilde{f}\dot{\tilde{f}} + \tilde{\tau}\dot{\tilde{\tau}}$$
$$(\tilde{\mathbf{v}}^l)^T \left[\hat{\mathbf{R}}_b^l \left(\tilde{\Delta}_2 \mathbf{a}_{\text{imu}} + \tilde{\mathbf{b}}_2 \right) - (2\mathbf{S}(\omega_{ie}^l) + \mathbf{S}(\omega_{el}^l)) \tilde{\mathbf{v}}^l - \mathbf{K}_2 \tilde{\mathbf{v}}^l \right]$$
$$- (\tilde{\mathbf{v}}^l)^T(\mathbf{R}_l^e)^T\tilde{\mathbf{p}}^e + (\tilde{\mathbf{p}}^e)^T \left[\mathbf{R}_l^e \tilde{\mathbf{v}}^l - \mathbf{K}_3 \tilde{\mathbf{p}}^e \right]$$
$$- \tilde{\mathbf{b}}_2^T \left[\mathbf{T}_4^{-1}\tilde{\mathbf{b}}_2 + (\hat{\mathbf{R}}_b^l)^T\tilde{\mathbf{v}}^l \right] - \tilde{\epsilon}^T \text{diag}(\mathbf{a}_{\text{imu}})(\hat{\mathbf{R}}_b^l)^T\tilde{\mathbf{v}}^l$$
$$- \tilde{\beta}^T\Upsilon^T(\mathbf{a}_{\text{imu}})(\hat{\mathbf{R}}_b^l)^T\tilde{\mathbf{v}}^l - \tilde{\epsilon}^T\mathbf{T}_5^{-1}\tilde{\epsilon} - \tilde{\beta}^T\mathbf{T}_6^{-1}\tilde{\beta}$$
$$- (t_7^{-1} + k_4)\tilde{f} - \tilde{f}\tilde{\tau} - k_5\tilde{\tau}^2 + \tilde{\tau}\tilde{f}$$
$$= -(\tilde{\mathbf{v}}^l)^T\mathbf{K}_2\tilde{\mathbf{v}}^l - (\tilde{\mathbf{p}}^e)^T\mathbf{K}_3\tilde{\mathbf{p}}^e - \tilde{\mathbf{b}}_2^{\ T}\mathbf{T}_4^{-1}\tilde{\mathbf{b}}_2$$
$$- \tilde{\epsilon}^T\mathbf{T}_5^{-1}\tilde{\epsilon} - \tilde{\beta}^T\mathbf{T}_6^{-1}\tilde{\beta} - (t_7^{-1} + k_4)\tilde{f}^2 - k_5\tilde{\tau}^2$$
$$= -\alpha(\mathbf{x}) < 0 \ \forall \mathbf{x} \neq \mathbf{0} \tag{8.91}$$

where Υ have been designed such that

$$\tilde{\Delta}_2 \mathbf{a}_{\text{imu}} - \Upsilon(\mathbf{a}_{\text{imu}})\tilde{\beta} - \text{diag}(\mathbf{a}_{\text{imu}})\tilde{\epsilon} = \mathbf{0} \tag{8.92}$$

Thus, Condition II in Lemma 8.2 is satisfied. In addition it is seen that

$$\limsup_{\mathbf{x}\to\infty} \frac{\left|\frac{\partial V}{\partial \mathbf{x}}\right|}{\alpha(\mathbf{x})} = 0 \tag{8.93}$$

Hence, Conditions I–III of Lemma 8.2 are satisfied, and it follows that $|\mathbf{x}(t)| \to \mathbf{0}$ as $t \to \infty$ and therefore the system is QEAS. ∎

Earth Frame Formulation

The update equations for position and clock bias are the same as for the local frame observer. For velocity and frequency biases, we now have the update equations:

$$\begin{bmatrix} \tilde{\mathbf{v}}^e \\ \tilde{f} \end{bmatrix} = \begin{bmatrix} (\Lambda^e)^\dagger \\ (\lambda^e)_*^\dagger \end{bmatrix} \left(\delta_{\text{gps}} - \delta_{\text{sat}} - \Lambda^e \hat{\mathbf{v}}^e - c\hat{f} \right) \tag{8.94}$$

The Earth frame velocity and position update equations (8.17)–(8.18) with bias, scale-factor and misalignment errors included are written:

$$\dot{\mathbf{v}}^e = \mathbf{R}_l^e \mathbf{R}_b^l \left[(\mathbf{I} + \Delta_2)\mathbf{a}_{\mathrm{imu}} + \mathbf{b}_2 \right] + \mathbf{R}_l^e \bar{\mathbf{g}}^l - 2\mathbf{S}(\omega_{ie}^e)\mathbf{v}^e \tag{8.95}$$

$$\dot{\mathbf{p}}^e = \mathbf{v}^e \tag{8.96}$$

The proposed observer for (8.95)–(8.96) is:

$$\dot{\hat{\mathbf{v}}}^e = \mathbf{R}_l^e \hat{\mathbf{R}}_b^l \left[(\mathbf{I} + \hat{\Delta}_2)\mathbf{a}_{\mathrm{imu}} + \hat{\mathbf{b}}_2 \right] + \mathbf{R}_l^e \bar{\mathbf{g}}^l - 2\mathbf{S}(\omega_{ie}^e)\hat{\mathbf{v}}^e$$

$$\qquad + \mathbf{K}_2 \tilde{\mathbf{v}}^e + \tilde{\mathbf{p}}^e \tag{8.97}$$

$$\dot{\hat{\mathbf{p}}}^e = \hat{\mathbf{v}}^e + \mathbf{K}_3 \tilde{\mathbf{p}}^e \tag{8.98}$$

$$\dot{\hat{\mathbf{b}}}_2 = -\mathbf{T}_4^{-1}\hat{\mathbf{b}}_2 + (\hat{\mathbf{R}}_b^l)^T (\mathbf{R}_l^e)^T \tilde{\mathbf{v}}^e \tag{8.99}$$

$$\dot{\hat{\epsilon}} = -\mathbf{T}_5^{-1}\hat{\epsilon} + \mathrm{diag}(\mathbf{a}_{\mathrm{imu}})(\hat{\mathbf{R}}_b^l)^T (\mathbf{R}_l^e)^T \tilde{\mathbf{v}}^e \tag{8.100}$$

$$\dot{\hat{\beta}} = -\mathbf{T}_6^{-1}\hat{\beta} + \Upsilon^T(\mathbf{a}_{\mathrm{imu}})(\hat{\mathbf{R}}_b^l)^T (\mathbf{R}_l^e)^T \tilde{\mathbf{v}}^e \tag{8.101}$$

$$\dot{\hat{f}} = -t_7^{-1}\hat{f} + k_4 \tilde{f} + \tilde{\tau} \tag{8.102}$$

$$\dot{\hat{\tau}} = \hat{f} + k_5 \tilde{\tau} \tag{8.103}$$

and the error dynamics is:

$$\dot{\tilde{\mathbf{v}}}^e = \mathbf{R}_l^e \hat{\mathbf{R}}_b^l \left[\tilde{\Delta}_2 \mathbf{a}_{\mathrm{imu}} + \tilde{\mathbf{b}}_2 \right] + \mathbf{R}_l^e \mathbf{E} \left[(\mathbf{I} + \Delta_2)\mathbf{a}_{\mathrm{imu}} + \mathbf{b}_2 \right]$$

$$\qquad - 2\mathbf{S}(\omega_{ie}^e)\tilde{\mathbf{v}}^e - \mathbf{K}_2 \tilde{\mathbf{v}}^e - \tilde{\mathbf{p}}^e \tag{8.104}$$

$$\dot{\tilde{\mathbf{p}}}^e = \tilde{\mathbf{v}}^e - \mathbf{K}_3 \tilde{\mathbf{p}}^e \tag{8.105}$$

$$\dot{\tilde{\mathbf{b}}}_2 = -\mathbf{T}_4^{-1}\tilde{\mathbf{b}}_2 - (\hat{\mathbf{R}}_b^l)^T (\mathbf{R}_l^e)^T \tilde{\mathbf{v}}^e \tag{8.106}$$

$$\dot{\tilde{\epsilon}} = -\mathbf{T}_5^{-1}\tilde{\epsilon} - \mathrm{diag}(\mathbf{a}_{\mathrm{imu}})(\hat{\mathbf{R}}_b^l)^T (\mathbf{R}_l^e)^T \tilde{\mathbf{v}}^e \tag{8.107}$$

$$\dot{\tilde{\beta}} = -\mathbf{T}_6^{-1}\tilde{\beta} - \Upsilon^T(\mathbf{a}_{\mathrm{imu}})(\hat{\mathbf{R}}_b^l)^T (\mathbf{R}_l^e)^T \tilde{\mathbf{v}}^e \tag{8.108}$$

$$\dot{\tilde{f}} = -t_7^{-1}\tilde{f} - k_4 \tilde{f} - \tilde{\tau} \tag{8.109}$$

$$\dot{\tilde{\tau}} = \tilde{f} - k_5 \tilde{\tau} \tag{8.110}$$

Corollary 8.1 (QEAS e-Frame Observer) *Any solution of the e-frame observer error equations (8.104)–(8.110) will tend to zero as time goes to infinity and QEAS follows from Remark 8.2.*

Proof. The proof is analogous to that of Theorem 8.2. We write the system (8.104)–(8.110) on the form $\dot{\mathbf{x}} = \mathbf{f}(\mathbf{x}, t) + \mathbf{g}(t)$. The following Lyapunov Function Candidate for the unforced system $\dot{\mathbf{x}} = \mathbf{f}(\mathbf{x}, t)$ satisfies the conditions on V in Lemma 8.2:

$$V = \frac{1}{2}(\tilde{\mathbf{v}}^e)^T \tilde{\mathbf{v}}^e + \frac{1}{2}(\tilde{\mathbf{p}}^e)^T \tilde{\mathbf{p}}^e + \frac{1}{2}\tilde{\mathbf{b}}_2^T \tilde{\mathbf{b}}_2$$

$$\qquad + \frac{1}{2}\tilde{\epsilon}^T \tilde{\epsilon} + \frac{1}{2}\tilde{\beta}^T \tilde{\beta} + \frac{1}{2}\tilde{f}^2 + \frac{1}{2}\tilde{\tau}^2 \tag{8.111}$$

The derivative along the system trajectories is

$$\dot{V} = -(\tilde{\mathbf{v}}^e)^T \mathbf{K}_2 \tilde{\mathbf{v}}^e - (\tilde{\mathbf{p}}^e)^T \mathbf{K}_3 \tilde{\mathbf{p}}^e - \tilde{\mathbf{b}}_2{}^T \mathbf{T}_4^{-1} \tilde{\mathbf{b}}_2$$
$$- \tilde{\epsilon}^T \mathbf{T}_5^{-1} \tilde{\epsilon} - \tilde{\beta}^T \mathbf{T}_6^{-1} \tilde{\beta} - (t_7^{-1} + k_4) \tilde{f}^2 - k_5 \tilde{\tau}^2$$
$$= -\alpha(\mathbf{x}) < 0, \quad \forall \mathbf{x} \neq \mathbf{0} \tag{8.112}$$

Further arguments are the same as in Theorem 8.2, and it can be concluded that $|\mathbf{x}(t)| \to \mathbf{0}$ as $t \to \infty$. ∎

LOS Formulation

In this section, we will show how the velocity equation can be transformed to pseudorange and deltarange differential equations. This makes it possible to use raw GPS measurements directly to update the observer velocity equation. The outputs of the observer are pseudoranges and deltaranges. The vehicle position and velocities can be calculated from (8.6) and (8.9), or from other algorithms that doesn't require linearization. The equations (8.95)–(8.96) can be transformed to satellite LOS space by defining $\dot{\rho} = \delta + \delta_{\text{sat}} = \Lambda^e \mathbf{v}^e + \delta_{\text{sat}}$, and further that $\dot{\delta} = \dot{\Lambda} \mathbf{v}^e + \Lambda^e \dot{\mathbf{v}}^e$. Note that the dynamics of the δ in this case only is due to the dynamics measured by the IMU. This gives the system:

$$\dot{\delta} = \Lambda^e \mathbf{R}_l^e \mathbf{R}_b^l [(\mathbf{I} + \Delta_2) \mathbf{a}_{\text{imu}} + \mathbf{b}_2] + \Lambda^e \mathbf{R}_l^e \bar{\mathbf{g}}^l$$
$$- (2\Lambda^e \mathbf{S}(\omega_{ei}^e) - \dot{\Lambda}^e)(\Lambda^e)^\dagger \delta \tag{8.113}$$
$$\dot{\rho} = \delta + \delta_{\text{sat}} \tag{8.114}$$

where $\delta \in \Re^n$, and $\rho \in \Re^n$.

Remark 8.3 $|\lambda^i| = 1$ $(i = 1, ..., n)$. *Hence, $\dot{\Lambda}^e$ will mainly be caused by the angular motion due to the movement of the satellites. It is therefore a very small term, but it will have significant effect when integrated over a long period of time. $\dot{\Lambda}^e$ can, however, be assumed to be known.*

Remark 8.4 *Note that when the number of satellites the receiver is locked onto goes below four, the matrix Λ^e should keep the LOS vector(s) of the last satellite(s) that fell out until there is four satellites in view again. If this is not done, position and velocity solutions cannot be computed. Satellite orbits can be predicted accurately for hours after the last satellite orbit data download.*

The update part is now simply

$$\begin{bmatrix} \tilde{\rho} \\ \tilde{\tau} \end{bmatrix} = \begin{bmatrix} \mathbf{I} \\ \lambda_*^i \end{bmatrix} (\rho_{\text{gps}} - \hat{\rho} - \hat{\tau}) \tag{8.115}$$

$$\begin{bmatrix} \tilde{\delta} \\ \tilde{f} \end{bmatrix} = \begin{bmatrix} \mathbf{I} \\ (\lambda^e)_*^\dagger \end{bmatrix} \left(\delta_{\text{gps}} - \delta_{\text{sat}} - \hat{\delta} - \hat{f} \right) \tag{8.116}$$

The following observer is proposed for (8.113)-(8.114):

$$\dot{\hat{\delta}} = \Lambda^e \mathbf{R}_l^e \hat{\mathbf{R}}_b^l \left[(\mathbf{I} + \hat{\Delta}_2)\mathbf{a}_{\mathrm{imu}} + \hat{\mathbf{b}}_2 \right] + \Lambda^e \mathbf{R}_l^e \hat{\bar{\mathbf{g}}}^l - (2\Lambda^e \mathbf{S}(\omega_{ie}^e) - \dot{\Lambda}^e)(\Lambda^e)^\dagger \hat{\delta}$$
$$+ \mathbf{K}_2 \tilde{\delta} + \tilde{\rho} \tag{8.117}$$

$$\dot{\hat{\rho}} = \hat{\delta} + \delta_{\mathrm{sat}} + \mathbf{K}_3 \tilde{\rho} \tag{8.118}$$

$$\dot{\hat{\mathbf{b}}}_2 = -\mathbf{T}_4^{-1}\hat{\mathbf{b}}_2 + (\hat{\mathbf{R}}_b^l)^T(\mathbf{R}_l^e)^T(\Lambda^e)^T\tilde{\delta} \tag{8.119}$$

$$\dot{\hat{\epsilon}} = -\mathbf{T}_5^{-1}\hat{\epsilon} + \mathrm{diag}(\mathbf{a}_{\mathrm{imu}})(\hat{\mathbf{R}}_b^l)^T(\mathbf{R}_l^e)^T(\Lambda^e)^T\tilde{\delta} \tag{8.120}$$

$$\dot{\hat{\beta}} = -\mathbf{T}_6^{-1}\hat{\beta} + \Upsilon^T(\mathbf{a}_{\mathrm{imu}})(\hat{\mathbf{R}}_b^l)^T(\mathbf{R}_l^e)^T(\Lambda^e)^T\tilde{\delta} \tag{8.121}$$

$$\dot{\hat{f}} = -t_7^{-1}\hat{f} + k_4\tilde{f} + \tilde{\tau} \tag{8.122}$$

$$\dot{\hat{\tau}} = \hat{f} + k_5\tilde{\tau} \tag{8.123}$$

The closed-loop system is:

$$\dot{\tilde{\delta}} = \Lambda^e \mathbf{R}_l^e \hat{\mathbf{R}}_b^l \left[\tilde{\Delta}_2 \mathbf{a}_{\mathrm{imu}} + \tilde{\mathbf{b}}_2 \right] + \Lambda^e \mathbf{R}_l^e \mathbf{E} \left[(\mathbf{I} + \Delta_2)\mathbf{a}_{\mathrm{imu}} + \mathbf{b}_2 \right]$$
$$- (2\Lambda^e \mathbf{S}(\omega_{ie}^e) - \dot{\Lambda}^e)(\Lambda^e)^\dagger \tilde{\delta} - \mathbf{K}_2 \tilde{\delta} - \tilde{\rho} \tag{8.124}$$

$$\dot{\tilde{\rho}} = \tilde{\delta} - \mathbf{K}_3 \tilde{\rho} \tag{8.125}$$

$$\dot{\tilde{\mathbf{b}}}_2 = -\mathbf{T}_4^{-1}\tilde{\mathbf{b}}_2 - (\hat{\mathbf{R}}_b^l)^T(\mathbf{R}_l^e)^T(\Lambda^e)^T\tilde{\delta} \tag{8.126}$$

$$\dot{\tilde{\epsilon}} = -\mathbf{T}_5^{-1}\tilde{\epsilon} - \mathrm{diag}(\mathbf{a}_{\mathrm{imu}})(\hat{\mathbf{R}}_b^l)^T(\mathbf{R}_l^e)^T(\Lambda^e)^T\tilde{\delta} \tag{8.127}$$

$$\dot{\tilde{\beta}} = -\mathbf{T}_6^{-1}\tilde{\beta} - \Upsilon^T(\mathbf{a}_{\mathrm{imu}})(\hat{\mathbf{R}}_b^l)^T(\mathbf{R}_l^e)^T(\Lambda^e)^T\tilde{\delta} \tag{8.128}$$

$$\dot{\tilde{f}} = -t_7^{-1}\tilde{f} - k_4\tilde{f} - \tilde{\tau} \tag{8.129}$$

$$\dot{\tilde{\tau}} = \tilde{f} - k_5\tilde{\tau} \tag{8.130}$$

Corollary 8.2 (QEAS Observer for LOS Formulation) *Any solution of the observer error equations (8.124)–(8.130) will tend to zero as time goes to infinity, and QEAS follows from Remark 8.2.*

Proof. The proof is also here analogous to that of Theorem 8.2. We write the system (8.124)–(8.130) on the form $\dot{\mathbf{x}} = \mathbf{f}(\mathbf{x}, t) + \mathbf{g}(t)$. The following Lyapunov Function Candidate for the unforced system $\dot{\mathbf{x}} = \mathbf{f}(\mathbf{x}, t)$ satisfies the conditions on V in Lemma 8.2:

$$V = \frac{1}{2}\tilde{\delta}^T\tilde{\delta} + \frac{1}{2}\tilde{\rho}^T\tilde{\rho} + \frac{1}{2}\tilde{\mathbf{b}}_2^T\tilde{\mathbf{b}}_2$$
$$+ \frac{1}{2}\tilde{\epsilon}^T\tilde{\epsilon} + \frac{1}{2}\tilde{\beta}^T\tilde{\beta} + \frac{1}{2}\tilde{f}^2 + \frac{1}{2}\tilde{\tau}^2 \tag{8.131}$$

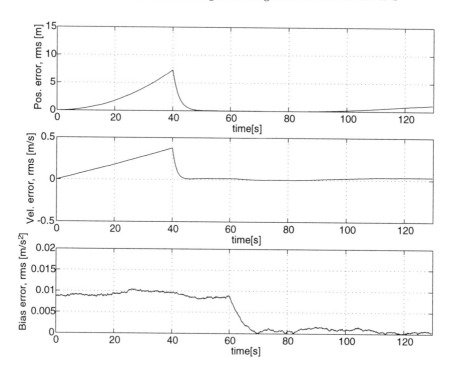

FIGURE 3. The figure shows the estimated position, velocity and acceleration bias errors.

Differentiating V along the system trajectories gives

$$\dot{V} = -\tilde{\delta}^T \left[\mathbf{K}_2 + (2\Lambda^e \mathbf{S}(\omega_{ie}^e) - \dot{\Lambda}^e)(\Lambda^e)^\dagger \right] \tilde{\delta} - \tilde{\rho}^T \mathbf{K}_3 \tilde{\rho}$$
$$- \tilde{\mathbf{b}}_2 \mathbf{T}_4^{-1} \tilde{\mathbf{b}}_2 - \tilde{\epsilon}^T \mathbf{T}_5^{-1} \tilde{\epsilon} - \tilde{\beta}^T \mathbf{T}_6^{-1} \tilde{\beta} - (t_7^{-1} + k_4)\tilde{f}^2 - k_5 \tilde{\tau}^2$$
$$= -\alpha(\mathbf{x}) \tag{8.132}$$

If the gain matrices are chosen according to:

$$\mathbf{K}_2 - (2\Lambda^e \mathbf{S}(\omega_{ei}^e) - \dot{\Lambda}^e)(\Lambda^e)^\dagger > 0 \tag{8.133}$$
$$\mathbf{K}_3 > 0 \tag{8.134}$$
$$k_4, k_5 > 0 \tag{8.135}$$

this results in:

$$\dot{V} = -\alpha(\mathbf{x}) < 0, \quad \forall \, \mathbf{x} \neq \mathbf{0} \tag{8.136}$$

Further arguments are the same as in Theorem 8.2, and it can be concluded that $|\mathbf{x}(t)| \to \mathbf{0}$ as $t \to \infty$. ∎

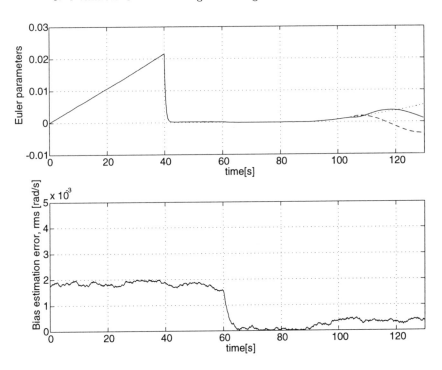

FIGURE 4. The top figure shows $\tilde{\varepsilon}$. The maximum error corresponds to about 2 (deg) heading error. The lower figure shows the error of the estimated gyro bias.

5   Case Study

The l-frame observer have been simulated on a simple ship trajectory with rather slow dynamics. The scale factor observability and misalignments are best under highly dynamic conditions, while biases are observable also at low speed maneuvers. The accelerometer bias was set to 0.5 mg, and the gyro bias was set to 0.06 deg/s, for all three axes. Thus, the sensors are at the lower end of the price range. Accelerometer and gyro measurements were available at 50 Hz, while GPS position, velocity, attitude and heading measurements were available at 5 Hz.

The following scenario is simulated: At time $t = 0$ each bias is set, and there is no update from GPS. At $t = 40$ s, the GPS reference measurement is enabled, while at $t = 60$ s, bias estimation is started. At $t = 80$ s, the GPS reference falls out again and the bias estimator is stopped. From Fig. 3 we can see that the position, velocity, and bias errors tend to zero, and that the position and velocity errors of the free running INS diverge slower after the biases are estimated. Fig. 4 shows a similar result for the angular velocity observer. This is one of the main reasons for estimating biases, because it enables the system to handle longer periods of GPS absence.

6 Conclusions and Future Work

In this chapter, nonlinear observers for integration of GPS and INS position velocity and attitude have been presented. The angular velocity observer was proven to be exponentially stable, while the position and velocity observer was proven to be Quasi-Equi Asymptotically Stable in the large. This guarantees that all estimation errors tends to zero. The position and velocity observer have been designed for three different coordinate frames, making it possible to use the observer in different system architectures. Simulations have shown that the observers perform well. Future work will include more extensive comparative simulations as well as experimental validation.

Acknowledgements

This work was sponsored by Navia Maritime, Division Seatex and the Norwegian Research Council. The authors are grateful to Dr. John-Morten Godhavn at Navia Maritime for his comments on nonlinear observer design for GPS/INS navigation systems.

The first author is also grateful to Halgeir Ludvigsen at the Department of Engineering Cybernetics, NTNU for commenting on the nonlinear stability analyses.

7 REFERENCES

[1] K. R. Britting. *Inertial Navigation Systems Analysis.* Wiley Interscience, 1971.

[2] V. Lakshmikantham, V. M. Matrosov and S. Sivasundaram. *Vector Lyapunov Functions and Stability Analysis of Nonlinear Systems,* Kluwer Academic Publishers, 1991.

[3] Z. H. Lewantowicz and D. W. Keen. Graceful Degradation of GPS/INS Performance With Fewer Than Four Satellites. *Proceedings of the Institute of Navigation, National Technical Meeting,* 1991.

[4] P. S. Maybeck. Stochastic Models, Estimation and Control, *Vol. 1. Academic Press, New York,* 1979.

[5] B. W. Parkinson and J. J. Spilker (Eds.). *Global Positioning System: Theory and Applications, Volume I.* American Institute of Aeronautics and Astronautics, Inc., Washington DC, USA, 1995.

[6] S. Salcudean. A Globally Convergent Angular velocity Observer for Rigid Body Motion. *IEEE Transaction on Automatic Control,* **TAC-36**:1493–1497, 1991

[7] P. G. Savage. Strapdown Inertial Navigation, *Lecture Notes. Strapdown Associates Inc., Minnetonka, MN, USA*, 1990.

[8] W. Tang and G. Howell. Integrated GPS/INS Kalman Filter Implementation Issues. *Proceedings of the ION GPS-93, pp. 217–224, Salt Lake City, Utah, USA*, 1993.

[9] D. A. Tazartes and J. G. Mark. Integration of GPS Receivers into Existing Inertial Navigation Systems. *Journal of the Institute of Navigation*, **JIN-35**:105–119, 1988.

[10] D. H. Titterton and J. L. Weston. *Strapdown Inertial Navigation Technology*. IEE, London, UK, 1997.

[11] V. A. Yakubovich. Method of Matrix Inequalities in the Theory of Stability of Nonlinear Control Systems. Part I: Absolute Stability of Forced Oscillations. *Journal of Automation & Remote Control*, **JARC-7**:1017–1029 (English translation pp. 905–917), 1964.

[12] M. Wei and K. P. Schwartz. A Strapdown Inertial Algorithm Using an Earth-Fixed Cartesian Frame. *Journal of the Institute of Navigation*, **JIN-37**:153–167, 1990.

Variants of Nonlinear Normal Form Observer Design

J. Schaffner[1] and M. Zeitz[2]

[1]Institute for Systems, Informatics and Safety,
European Commission Joint Research Centre, Ispra, Italy
[2]Institut für Systemdynamik und Regelungstechnik,
University of Stuttgart, Germany

1 Introduction

Normal form coordinates are a very appropriate basis for analysis and design of nonlinear systems. But the transformation of physical coordinates of a nonlinear system into a normal form representation is requiring very restrictive assumptions [10, 24]. Therefore, various modifications of nonlinear normal form approaches have been developed in order to enable an extended application of the normal form coordinates. In this paper, three variants of the nonlinear normal form observer design are presented in the development of which the authors have been involved.

The normal form observer design is in some sense dual to the exact input-state linearization (confer [8, 1, 9, 7, 21] or see text books [6, 12]). In the next section, this design method will be recapitulated for nonlinear single output systems Σ without input

$$\Sigma: \quad \dot{x} = f(x), \quad x(0) = x_0, \quad y = h(x), \quad x \in M \subset {I\!\!R}^n, \quad y \in Y \subset {I\!\!R}^1 \quad (9.1)$$

where f, $h \in C^\infty$, $f(0) = 0$, $h(0) = 0$. The observability properties of (9.1) required for the considered observer design variants are related to the observation space $\mathcal{O}$ or the observability map

$$M \to {I\!\!R}^n: \quad q(x) = \left[h(x), L_f h(x), \dots, L_f^{n-1} h(x) \right]^T \quad (9.2)$$

and the observation codistribution $d\mathcal{O}$ or the observability matrices

$$dq(x) = \begin{bmatrix} dh(x) \\ L_f dh(x) \\ \vdots \\ L_f^{n-1} dh(x) \end{bmatrix}, \quad dq(0) = \begin{bmatrix} c^T \\ c^T A \\ \vdots \\ c^T A^{n-1} \end{bmatrix} \quad \begin{aligned} &\text{with} \\ & A = \frac{\partial f}{\partial x}(0), \\ & c^T = \frac{\partial h}{\partial x}(0). \end{aligned} \quad (9.3)$$

The injectivity of the map $q(x)$ and the regularity of the matrices $dq(x)$ and $dq(0)$ are used to define the observability of system (9.1) and of its Taylor linearization (A, c^T), respectively.

2   Normal Form Observer

The design of a nonlinear observer becomes very simple, if the system (9.1) can be represented by means of the state transformation

$$x^* = \Phi(x) \tag{9.4}$$

in the observer normal form coordinates [6, 12]:

$$\Sigma^* : \quad \dot{x}^* = A^* x^* + \alpha(y), \quad x^*(0) = \Phi(x_0) = x_0^*, \quad y = c^{*T} x^* = x_1^*. \tag{9.5}$$

This form is characterized by the matrices (A^*, c^{*T}) in dual Brunovsky form and by the output injection $\alpha(y)$. The coordinates x^* allow the design of a nonlinear observer with an exactly linear error dynamics. This property is related to the output dependent vector field $\alpha(y)$, which is reproduced exactly in the observer equations

$$\hat{\Sigma}^* : \quad \dot{\hat{x}}^* = A^* \hat{x}^* + \alpha(y) + l^* \cdot (y - c^{*T} \hat{x}^*), \quad \hat{x}^*(0) = \hat{x}_0^*, \quad \hat{x} = \Phi^{-1}(\hat{x}^*)$$
$$\text{with} \quad l^* : \operatorname{Re}\{\lambda_i(A^* - l^* c^{*T})\} < 0, \quad i = 1, \ldots, n . \tag{9.6}$$

The normal form observer $\hat{\Sigma}^*$ comprises a dynamical part and the algebraic transformation Φ^{-1} from coordinates $\hat{x}^*$ to original coordinates $\hat{x}$. The design of the constant gain vector l^* is based on the linear observer error dynamics $\tilde{\Sigma}^* : \quad \dot{\tilde{x}}^* = (A^* - l^* c^{*T}) \tilde{x}^*, \quad \tilde{x}^*(0) = \hat{x}_0^* - x_0^*$, such that $\tilde{\Sigma}^*$ possesses eigenvalues with negative real part.

The normal form observer design is applicable, if the transformation (9.4) does exist and can be determined from the first order partial differential equations (pdes)

$$\frac{\partial \Phi}{\partial x} \left[ad_{-f}^{n-1} \tau(x), \ldots, ad_{-f}^{1} \tau(x), \tau(x) \right] = I_n \quad \text{with} \tag{9.7}$$
$$\tau(x) = dq^{-1}(x) [0, \ldots, 0, 1]^T, \quad I_n - (n \times n) \text{ unity matrix.} \tag{9.8}$$

The n^2 pdes (9.7) for the components $\Phi_i(x), i = 1, \ldots, n$ have state dependent coefficients, which are formulated by means of $\tau(x)$, i.e. the last column of the inverse $dq^{-1}(x)$ of the observability matrix (9.3). The following two conditions for the existence of the diffeomorphism $\Phi(x)$ are corresponding to the existence of the vector $\tau(x)$ in (9.8) and to the integrability of the pdes (9.7) [6, 12]:

Theorem 9.1 (Existence of Normal Form Observer) *The state transformation (9.4) of system (9.1) into the observer normal form (9.5) exists and can be calculated from (9.7), (9.8), if and only if*

$$\operatorname{rank} dq(x) = n , \tag{9.9}$$
$$\left[ad_f^i \tau(x), ad_f^j \tau(x) \right] = 0 , \quad i, j = 0, 1, \ldots, n - 1 . \tag{9.10}$$

The rank condition (9.9) is sufficient for the local observability of the system (9.1), which is satisfied in most cases. The commutativity condition (9.10) is yet very restrictive, especially for higher order systems. Moreover, an analytical solution of the pdes (9.7) for $n > 2$ is often rather difficult to find. However, the elegance of the normal form observer design and of the exact linearization of the observer error dynamics, respectively, have caused a lot of investigations to loosen these restrictions.

In the next sections, three variants of the nonlinear normal form observer design are presented, which extend its application to a wider class of nonlinear systems. The *continuous (normal form) observer* [22, 25] can be applied if the observability rank condition (9.9) is not fulfilled, but the system has an injective observability map (9.2). If the commutatitvity condition (9.10) is violated, it is worth considering the design of an *extended Luenberger observer* [1, 23, 2, 3]. This approach is similar to the extended Kalman filter and requires only the observability rank condition (9.9). The third variant concerns the *block-triangular (normal form) observer* [16, 17, 18, 19], where the presence of multiple outputs is used to decompose the original system in subsystems that have to be connected in a cascade. This structure allows a decentral (normal form) observer design.

3   Continuous Observer

Most state estimation methods for smooth nonlinear systems (9.1) are smooth and require that the Taylor linearization $(\mathbf{A}, \mathbf{c}^T)$ of the system is observable (detectable) [20], i.e.:

$$\operatorname{rank} d\mathbf{q}(\mathbf{0}) = n. \tag{9.11}$$

Under the assumption that the observability rank condition (9.9) of Theorem 9.1 is fulfilled, normal form observers (9.6) are even generically smooth. If condition (9.9) is violated, a continuous normal form observer may still be possible. The potential of continuous observers was pointed out by Krener in [11], where the following example has been explicitly constructed to show the non-existence of a smooth observer and the intuitive set-up of a continuous observer:

$$\Sigma: \quad \dot{x} = \lambda x, \ y = x^3; \quad \hat{\Sigma}: \quad \dot{\hat{x}} = \lambda \hat{x} + l \cdot (\sqrt[3]{y} - \hat{x}) \ \text{ with } l > \lambda. \tag{9.12}$$

The first order system Σ has no smooth observer, because the Taylor linearization $(A = \lambda, c = 0)$ is not observable, i.e. $\operatorname{rank} dq(0) = 0$. But Σ is in fact observable, also in the origin, because of the injectivity of the observability map $q(x) = x^3$, $q^{-1}(y) = \sqrt[3]{y}$. The constructed observer $\hat{\Sigma}$ has an asymptotically stable error dynamics $\tilde{\Sigma}: \dot{\tilde{x}} = (\lambda - l)\tilde{x}$.

The first systematic design method of continuous observers for a class of nonlinear systems was developed in [22]. The systematics includes the continuous normal form observer as a special case [25]. The design method of continuous observers is based on the smoothness properties of the observability map $q \in C^\infty$ and its inverse $q^{-1} \in C^0$ as in Example (9.12). In [22], such a map is named as a *semi-diffeomorphism*:

Definition 9.1 (Semi-Diffeomorphism) *A smooth map* $\Phi : \mathbb{R}^n \to \mathbb{R}^n$ *with a continuous inverse* $\Phi^{-1} : \mathbb{R}^n \to \mathbb{R}^n$ *is called a* semi-diffeomorphism.

It is easy to see that the scalar map $\Phi(x) = x^p$ ($p > 1$, odd) is a semi-diffeomorphism, but not a diffeomorphism.

If Φ is a semi-diffeomorphism, denoted $\boldsymbol{\xi} = \Phi(\boldsymbol{x})$, then one can formally transform system (9.1) into

$$\dot{\boldsymbol{\xi}} = \frac{\partial \Phi}{\partial \boldsymbol{x}} \boldsymbol{f} \circ \Phi^{-1}(\boldsymbol{\xi}) = \bar{\boldsymbol{f}}(\boldsymbol{\xi}), \ \boldsymbol{\xi}(0) = \Phi(\boldsymbol{x}_0) = \boldsymbol{\xi}_0, \ y = h \circ \Phi^{-1}(\boldsymbol{\xi}) = \bar{h}(\boldsymbol{\xi})$$
(9.13)

where $\bar{\boldsymbol{f}}$ and $\bar{h}$ are smooth or continuous. The uniqueness of the solution $\boldsymbol{\xi}(t; \boldsymbol{\xi}_0)$ requires that the components $\bar{f}_i(\boldsymbol{\xi})$, $i = 1, \ldots, n$ are Lipschitz continuous. Under this condition, the trajectories of the two system representations (9.1) and (9.13) are equivalent. This has led in [22] to

Definition 9.2 (Trajectory-Equivalence) *By means of the semi-diffeomorphism* Φ *the system (9.1) is trajectory-equivalent to the transformed system (9.13), if the latter has the property of uniqueness of solutions:*

$$\boldsymbol{x}(t; \boldsymbol{x}_0) = \Phi^{-1}(\boldsymbol{\xi}(t; \Phi(\boldsymbol{x}_0))) \ \text{and} \ \boldsymbol{\xi}(t; \boldsymbol{\xi}_0) = \Phi(\boldsymbol{x}(t; \Phi^{-1}(\boldsymbol{\xi}_0))) \quad (9.14)$$

Trajectory-equivalence of the two system representations can be used for the design of an observer in order to study its convergence in the most appropriate coordinates. This is the key idea of the proposed design method of continuous observers [22, 25]. The design coordinates for the continuous observer are defined by the semi-diffeomorphic observability map (9.2), by which the system (9.1) is transformed into observability normal form [10, 24, 6, 12]:

$$\bar{\Sigma}: \ \dot{\boldsymbol{\xi}} = \bar{\boldsymbol{f}}(\boldsymbol{\xi}) = \begin{bmatrix} \xi_2 \\ \vdots \\ \xi_n \\ \varphi(\boldsymbol{\xi}) \end{bmatrix}, \ \boldsymbol{\xi}(0) = q(\boldsymbol{x}_0) = \boldsymbol{\xi}_0, \ y = \bar{h}(\boldsymbol{\xi}) = \xi_1 \quad (9.15)$$

with (i) $\varphi(\boldsymbol{\xi}) = L_f^n h \circ q^{-1}(\boldsymbol{\xi}) \in C^i$, $i \geq 0$ (Lipschitz–C^0),

(ii) $\boldsymbol{\xi}(t; \boldsymbol{\xi}_0) = q(\boldsymbol{x}(t; q^{-1}(\boldsymbol{\xi}_0)))$, $\boldsymbol{x}(t; \boldsymbol{x}_0) = q^{-1}(\boldsymbol{\xi}(t; q(\boldsymbol{x}_0)))$
(trajectory-equivalence of Σ and $\bar{\Sigma}$),

(iii) $\bar{q}(\boldsymbol{\xi}) = \boldsymbol{\xi}$, $d\bar{q} = \boldsymbol{I}_n$ ($\bar{\Sigma}$ – structurally observable).

The properties (i)–(iii) permit the design of a continuous observer with the following structure

$$\hat{\Sigma}: \quad \dot{\hat{\xi}} = \bar{f}(\hat{\xi}) + \bar{l}(\hat{\xi}, y) \text{ with } \bar{l}(\xi, y) = 0, \quad \hat{\xi}(0) = \hat{\xi}_0, \quad \hat{x} = q^{-1}(\hat{\xi}). \quad (9.16)$$

The continuous observer $\hat{\Sigma}$ comprises a dynamical part and the algebraic transformation q^{-1} from coordinates $\hat{\xi}$ to original coordinates $\hat{x}$. At least the algebraic part is C^0-continuous. Therefore, no differential equation exists for the estimation $\hat{x}(t)$. Due to the observability property (iii) of $\bar{\Sigma}$, the design of the injected correction $\bar{l}(\hat{\xi}, y)$ is possible with an arbitrary method, where only the particular form of $\bar{f}(\xi)$ or $\varphi(\xi)$, respectively, must be considered. The desired convergence $\hat{\xi}(t) \to \xi(t)$ can be obtained locally by a linear weighting of the difference between the output and its estimation: $\bar{l}(\hat{\xi}, y) = l \cdot (y - \hat{\xi}_1)$, $l \in \mathbb{R}^n$. Summing up, it can be stated that:

Theorem 9.2 (Existence of Continuous Observer) *A continuous observer $\hat{\Sigma}$ (9.16) exists, if the system (9.1) possesses a semi-diffeomorphic observability map $q \in C^\infty$, $q^{-1} \in C^0$ and a Lipschitz continuous observability normal form $\bar{\Sigma}$ (9.15).*

The design of a continuous observer is requiring two steps. At first the observability map (9.2) and the observability normal form (9.15) are determined. Then the convergence of the dynamical part of the observer must be guaranteed by a suitable choice of the correction vector $\bar{l}(\hat{\xi}, y)$.

If these design steps are applied on Krener's example (9.12), a different continuous observer is obtained

$$\hat{\Sigma}: \quad \dot{\hat{\xi}} = 3\lambda\hat{\xi} + \bar{l} \cdot (y - \hat{\xi}) \text{ with } \bar{l} > 3\lambda, \quad \hat{x} = (\hat{\xi})^{1/3} \quad (9.17)$$

with a dynamical part in the coordinates of the observability normal form $\bar{\Sigma}: \dot{\xi} = 3\lambda\xi$, $y = \xi$. In this special case, $\bar{\Sigma}$ is linear which enables a simple design of the observer gain $\bar{l}$.

Krener's example (9.12) can also be used to illustrate a first application of the proposed design of continuous observers. For the parameter $\lambda = 0$, system (9.12) presents a so-called *critical observation problem*, which is in a loose sense dual to the *critical stabilization problem* [4, 11]:

Definition 9.3 (Critical Observation Problem) *The critical observation or observer design problem is characterized by non-observable (non-detectable) modes of a marginally stable Taylor linearization $(\mathbf{A}, \mathbf{c}^T)$ having eigenvalues with zero real part:*

$$\mathrm{Re}\{\lambda_i(\mathbf{A})\} = \mathrm{Re}\{\lambda_i(\mathbf{A} - \mathbf{l}\mathbf{c}^T)\} = 0, \quad i \in \{1, \dots, n\} \ \forall \mathbf{l} \in \mathbb{R}^n. \quad (9.18)$$

Of course, both continuous observers $\hat{\Sigma}$ in (9.12) and in (9.17) also represent a solution of the critical observation problem.

The second example of a critical observation problem (9.18)

$$\Sigma: \quad \dot{x} = \begin{bmatrix} x_2^p \\ x_1 x_2 \end{bmatrix} \quad (p > 1, \text{ odd}), \quad y = x_1 \tag{9.19}$$

is leading to the design of a continuous normal form observer of second order. In the Taylor linearization ($\mathbf{A} = \mathbf{0}, c^T = [1,0]$), the mode $x_2(t)$ is marginally stable and non-observable, i.e. represents the critical mode. By means of the semi-diffeomorphic observability map $\boldsymbol{\xi} = q(\boldsymbol{x}) = [x_1, x_2^p]^T$, $\boldsymbol{x} = q^{-1}(\boldsymbol{\xi}) = [\xi_1, \xi_2^{1/p}]^T$, the observability normal form (9.15) is obtained

$$\bar{\Sigma}: \quad \dot{\boldsymbol{\xi}} = \begin{bmatrix} \xi_2 \\ p\xi_1\xi_2 \end{bmatrix}, \quad y = \xi_1 \tag{9.20}$$

which is smooth. Moreover, the representation $\bar{\Sigma}$ fulfills both conditions of Theorem 9.1 for the transformation $\boldsymbol{x}^* = \bar{\boldsymbol{\Phi}}(\boldsymbol{\xi})$ into the observer normal form (9.5):

$$d\bar{q}(\boldsymbol{\xi}) = \boldsymbol{I}_2 \quad \rightarrow \quad \boldsymbol{\tau}(\boldsymbol{\xi}) = [0, 1]^T \quad \rightarrow \quad [\boldsymbol{\tau}(\boldsymbol{\xi}), ad_{\bar{f}}\boldsymbol{\tau}(\boldsymbol{\xi})] = \mathbf{0}.$$

A solution of the four pdes (9.7) is $\boldsymbol{x}^* = [\xi_1, \xi_2 - p\xi_1^2/2]^T$, by means of which the observer normal form $\Sigma^*: \quad \dot{\boldsymbol{x}}^* = [x_2^* + py^2/2, 0]^T, \quad y = x_1^*$ can be derived from (9.20). These coordinates allow an exactly linear design of the dynamical part of the continuous normal form observer with constant gains $l_{1,2}^* > 0$

$$\hat{\Sigma}^*: \dot{\hat{\boldsymbol{x}}}^* = \begin{bmatrix} \hat{x}_2^* + py^2/2 + l_1^* \cdot (y - \hat{x}_1^*) \\ l_2^* \cdot (y - \hat{x}_1^*) \end{bmatrix}, \quad \hat{\boldsymbol{x}} = \begin{bmatrix} \hat{x}_1^* \\ (\hat{x}_2^* + p\hat{x}_1^{*2}/2)^{1/p} \end{bmatrix}. \tag{9.21}$$

The algebraic part comprises the transformation $\hat{\boldsymbol{x}} = q^{-1} \circ \bar{\boldsymbol{\Phi}}^{-1}(\hat{\boldsymbol{x}}^*)$ from $\hat{\boldsymbol{x}}^*$ via observability coordinates $\hat{\boldsymbol{\xi}}$ to original coordinates $\hat{\boldsymbol{x}}$ and is continuous due to $q^{-1} \in C^0$. This continuous normal form observer design explained for Example (9.19) means an extension of the smooth normal form observer (9.6), since the observability rank condition (9.9) is no longer required in original coordinates:

Theorem 9.3 (Existence of Continuous Normal Form Observer)
A continuous normal form observer (9.16) with an algebraic part $\hat{\boldsymbol{x}} = q^{-1} \circ \bar{\boldsymbol{\Phi}}^{-1}(\hat{\boldsymbol{x}}^*)$ exists, if and only if the observability normal form (9.15) fulfills the commutativity condition (9.10).

The design of a continuous normal form observer is rather similar to the two-step transformation into the observer normal form Σ^* via the observability normal form $\bar{\Sigma}$ as proposed by Keller [7] and by Phelps [13]. Yet in those contributions, the observability rank condition (9.9) is required in order to obtain a diffeomorphic transformation $\boldsymbol{x}^* = \bar{\boldsymbol{\Phi}} \circ q(\boldsymbol{x})$. Thus, the smooth normal form observer (9.6) can be seen as a special case of the continuous one. Of course, the continuous (normal form) observer design can be applied also to systems with multiple outputs and with smooth inputs [25].

4 Extended Luenberger Observer

The rather restrictive commutativity condition (9.10) and the analytical solution of the pdes (9.7) in course of the normal form observer design are bypassed by using the extended Luenberger observer [1, 23, 2, 3]. This observer has a classical structure consisting of a simulation part and an injected difference of the measured and the estimated output weighted by an $\hat{x}$–dependent gain vector $l(\hat{x})$

$$\hat{\Sigma} : \ \dot{\hat{x}} = f(\hat{x}) + l(\hat{x}) \cdot [y - h(\hat{x})], \quad \hat{x}(0) = \hat{x}_0 \quad \text{with} \tag{9.22}$$

$$l(\hat{x}) = \{ [p_0 ad_{-f}^0 + \ldots + p_{n-1} ad_{-f}^{n-1} + ad_{-f}^n] \circ \tau(\hat{x}) \} / \beta(\hat{x}), \tag{9.23}$$

$$\tau(\hat{x}) = dq^{-1}(\hat{x}) [0, \ldots, 0, \beta(\hat{x})]^T. \tag{9.24}$$

The nonlinear Ackermann formula (9.23) for the gain vector $l(\hat{x})$, which contains the dynamical parameters p_i, $i = 0, \ldots, n-1$ and the functional degree of freedom $\beta(\hat{x}) \neq 0$, is derived from the observer error dynamics

$$\tilde{\Sigma} : \quad \dot{\tilde{x}} = f(\hat{x}) - f(x) + l(\hat{x}) \cdot [h(x) - h(\hat{x})], \quad \tilde{x}(0) = \hat{x}_0 - x_0. \tag{9.25}$$

In a first step, $\tilde{\Sigma} = \hat{\Sigma} - \Sigma$ is brought into the observer normal form representation $\tilde{\Sigma}^* = \hat{\Sigma}^* - \Sigma^*$. For this operation, the state transformation $x^* = \Phi(x)$ defined by (9.7), (9.24) and the output transformation

$$y^* = \gamma(y) = c^{*T} x^* = x_1^* \tag{9.26}$$

are applied to (9.1) and (9.22). The inverse output transformation $\gamma^{-1}(x_1^*)$ is related to the functional degree of freedom in (9.23), (9.24) by

$$\beta(\hat{x}) = \frac{d\gamma^{-1}}{d\hat{x}_1^*} \circ \Phi_1(\hat{x}). \tag{9.27}$$

In the second step to derive the formula (9.23), the observer error dynamics transformed into normal form coordinates

$$\tilde{\Sigma}^* : \ \dot{\tilde{x}}^* = A^* \tilde{x}^* + \alpha \circ \gamma^{-1}(\hat{x}_1^*) - \alpha \circ \gamma^{-1}(x_1^*) + l^*(\hat{x}^*) \cdot [\gamma^{-1}(x_1^*) - \gamma^{-1}(\hat{x}_1^*)],$$
$$\tilde{x}^*(0) = \hat{x}_0^* - x_0^* \quad \text{with} \tag{9.28}$$

$$l^*(\hat{x}^*) = \left(\frac{\partial \Phi}{\partial \hat{x}} \right) l \circ \Phi(\hat{x}^*), \quad \left(\frac{\partial \Phi}{\partial \hat{x}} \right)^{-1} \frac{d\alpha}{dy} \beta(\hat{x}) = ad_{-f}^n \tau(\hat{x})$$

is linearized along the estimated trajectory $\hat{x}_1^*(t) = \Phi_1(\hat{x}(t))$:

$$\tilde{\Sigma}^* : \quad \dot{\tilde{x}}^* = (\mathbf{A}^* - \mathbf{p}\mathbf{c}^{*T})\tilde{x}^* + O^2(|\tilde{x}^*|) \quad \text{with} \tag{9.29}$$
$$\det\left(\lambda \mathbf{I}_n - \mathbf{A}^* + \mathbf{p}\mathbf{c}^{*T}\right) = p_0 + p_1\lambda + \ldots + p_{n-1}\lambda^{n-1} + \lambda^n .$$

For this, the equations (9.7), (9.23), (9.24) and (9.27) have been inserted into (9.28). By choosing the dynamical parameters $\mathbf{p} = [p_{n-1}, \ldots, p_0]^T$ in (9.29) appropriately, the eigenvalues of the matrix $(\mathbf{A}^* - \mathbf{p}\mathbf{c}^{*T})$ can be assigned in a desired manner.

The development of the design formula of the extended Luenberger observer (9.22)–(9.24) is leading to the following:

Definition 9.4 (Extended Taylor Linearization) *The extended Luenberger observer (9.22)–(9.24) has an extended Taylor linearization (9.29) of the error dynamics (9.28) in observer normal form coordinates, which can be stabilized by an eigenvalue assignment.*

Theorem 9.4 (Existence of Extended Luenberger Observer) *The extended Luenberger observer (9.22)–(9.24) exists, if the system (9.1) fulfills the observability rank condition (9.9), and if the observer error $\tilde{x}^*(t)$ is sufficiently small ($|\tilde{x}^*| \ll 1$) allowing an extended Taylor linearization (9.29) of the error dynamics (9.28).*

Remark 9.1 (Choice of $\beta(\hat{x})$) *By a suitable choice of the function $\beta(\hat{x})$, the calculation (9.23) of the observer gain vector $l(\hat{x})$ can be simplified such that the recursive application of the ad–operator becomes as simple as possible. A computer–aided calculation of $l(\hat{x})$ can be carried out easily using a symbolic programming language [3], e.g. MATHEMATICA or MAPLE.*

The design of the extended Luenberger observer for systems with inputs and multiple outputs is shown in [2, 3]. Compared to single output systems, multiple output variables y_i, $i = 1, \ldots, m$, offer more degrees of freedom $\beta_i(\hat{x}), i = 1, \ldots, m$. This fact can be used to simplify the calculation of the Ackermann formula (9.23) even more.

The design of an extended Luenberger observer (9.22)–(9.24) is demonstrated for the second order polynomial system

$$\Sigma : \quad \dot{x} = \begin{bmatrix} ax_1 - bx_1x_2 \\ cx_1x_2 - dx_2 \end{bmatrix} \quad (x_{1,2} > 0), \quad y = x_2 . \tag{9.30}$$

For this system, the observability matrix (9.3)

$$d\mathbf{q}(x) = \begin{bmatrix} 0 & 1 \\ cx_2 & cx_1 - d \end{bmatrix}$$

has full rank for $cx_2 \neq 0$. According to (9.24), the vector

$$\boldsymbol{\tau}(\hat{x}) = [\,\beta(\hat{x})/(c\hat{x}_2)\,,\, 0\,]^T$$

is calculated. It is simplified by choosing the function $\beta(\hat{\boldsymbol{x}}) = c\hat{x}_2 \neq 0$. Now the Ackermann formula (9.23)

$$l(\hat{\boldsymbol{x}}) = \left\{ \left[p_0 ad^0_{-f} + p_1 ad^1_{-f} + ad^2_{-f} \right] \circ \begin{bmatrix} 1 \\ 0 \end{bmatrix} \right\} / \beta(\hat{\boldsymbol{x}}) = \tag{9.31}$$

$$= \left\{ p_0 \begin{bmatrix} 1 \\ 0 \end{bmatrix} + p_1 \begin{bmatrix} a - b\hat{x}_2 \\ c\hat{x}_2 \end{bmatrix} + \begin{bmatrix} (a - b\hat{x}_2)^2 - bd\hat{x}_2 \\ c\hat{x}_2(a - b\hat{x}_2) \end{bmatrix} \right\} / (c\hat{x}_2)$$

and the corresponding observer equations (9.22) are containing less analytical operations (by a factor of 5) than without a functional degree of freedom $\beta(\hat{\boldsymbol{x}})$ or without an output transformation (9.26), respectively.

Moreover, the output transformation $\gamma(y) = \frac{1}{c} \ln y$, which is calculated from (9.27) with $\beta(\boldsymbol{x}) = cx_2$, allows the transformation $\boldsymbol{x}^* = \boldsymbol{\Phi}(\boldsymbol{x})$ of system (9.30) into the observer normal form (9.5), since the commutativity condition (9.10) is fulfilled by $\boldsymbol{\tau} = [1, 0]^T$ and $ad_f \boldsymbol{\tau} = [(bx_2 - a), -cx_2]^T$. Hence, a normal form observer (9.6) does exist for Example (9.30):

$$\hat{\Sigma}^* : \quad \dot{\hat{\boldsymbol{x}}}^* = \begin{bmatrix} \hat{x}_2^* + \frac{1}{c}(a \ln y - by - d) & + l_1^* \cdot \left(\frac{1}{c} \ln y - \hat{x}_1^* \right) \\ \frac{d}{c}(a - b_2 y) & + l_2^* \cdot \left(\frac{1}{c} \ln y - \hat{x}_1^* \right) \end{bmatrix}, \tag{9.32}$$

$$\hat{\boldsymbol{x}} = \begin{bmatrix} a\hat{x}_1^* - \frac{b}{c} \exp(c\hat{x}_1^*) + \hat{x}_2^* \\ \exp(c\hat{x}_1^*) \end{bmatrix}.$$

The constant observer gains $l_{1,2}^*$ can be specified by eigenvalue assignment for the linear error dynamics in normal form coordinates

$$\tilde{\Sigma}^* : \quad \dot{\tilde{\boldsymbol{x}}}^* = \begin{bmatrix} -l_1^* & 1 \\ -l_2^* & 0 \end{bmatrix} \tilde{\boldsymbol{x}}^* .$$

For the gains $l_1^* = p_1$ and $l_2^* = p_0$, $\tilde{\Sigma}^*$ is identical with the Taylor linearization of the error dynamics of the extended Luenberger observer with the gain vector (9.31).

Finally, a relation between the extended Luenberger observer and the normal form observer design for linear time-variant systems will be established [3]. Therefore, at first the observer error dynamics (9.25) is linearized. Then the extended Taylor linearization is transformed into normal form coordinates. This means a reverse order of the two operations compared with the derivation of the extended Luenberger observer gain formula (9.23).

Under the assumption that the observer error $\tilde{\boldsymbol{x}}(t)$ is sufficiently small ($|\tilde{\boldsymbol{x}}| \ll 1$), an extended Taylor linearization of observer error dynamics (9.25) along the trajectory $\hat{\boldsymbol{x}}(t)$ yields

$$\tilde{\Sigma} : \quad \dot{\tilde{\boldsymbol{x}}} = [\mathbf{A}(t) - l(t)\boldsymbol{c}^T(t)]\tilde{\boldsymbol{x}} + O^2(|\tilde{\boldsymbol{x}}(t)|) \quad \text{with} \tag{9.33}$$

$$\mathbf{A}(t) = \frac{\partial \boldsymbol{f}}{\partial \boldsymbol{x}}(\hat{\boldsymbol{x}}(t)), \quad \boldsymbol{c}^T(t) = \frac{\partial h}{\partial \boldsymbol{x}}(\hat{\boldsymbol{x}}(t)), \quad l(t) \hat{=} l(\hat{\boldsymbol{x}}(t)).$$

In order to design the time-variant gain vector $l(t)$, the linear system $(\mathbf{A}(t), c^T(t))$ is transformed by $x^* = \Phi(t)x$ into the time-variant observer normal form

$$\Sigma^* : \dot{x}^* = \mathbf{A}^* x^* + \alpha(t)y, \; x^*(0) = \Phi(0)x_0 = x_0^*, \; y = \beta(t)c^{*T}x^* = \beta(t)x_1^*. \tag{9.34}$$

This form is characterized by the constant matrices $(\mathbf{A}^*, c^{*T})$ in dual Brunovsky form, the time-dependent output injection $\alpha(t)y$, and the time-variant degree of freedom $\beta(t) \neq 0$. Moreover, (9.34) is consistent with the nonlinear observer normal form (9.5), if there, additionally, the output transformation (9.26) is taken into account. The defining equations for the time-variant transformation matrix $\Phi(t)$ are [15]:

$$\Phi(t)\left[N_A^{n-1}\tau(t), \dots, N_A^1\tau(t), \tau(t)\right] = \mathbf{I}_n \quad \text{with} \tag{9.35}$$

$$\tau(t) = dq^{-1}(t)\left[0, \dots, 0, \beta(t)\right]^T, \quad N_A\tau = -\dot{\tau} + \mathbf{A}(t)\tau, \tag{9.36}$$

$$dq(t) = \begin{bmatrix} c^T(t) \\ M_A c^T(t) \\ \vdots \\ M_A^{n-1}c^T(t) \end{bmatrix}, \quad M_A c^T = \dot{c}^T + c^T \mathbf{A}(t). \tag{9.37}$$

These linear equations are corresponding to the nonlinear ones (9.7), (9.8) and (9.3), because the linear time-variant differential operators M_A and N_A can be represented as Lie derivatives L_f and ad_f for the vector field $f(x) = \mathbf{A}(t)x$ applied to $dh(x(t)) \hat{=} c^T(t)$ and $\tau(x(t)) \hat{=} \tau(t)$:

$$M_A c^T(t) \hat{=} L_{Ax}dh(x), \quad N_A\tau(t) \hat{=} ad_{Ax}\tau(x).$$

Unlike in Theorem 9.1 for the existence of the nonlinear observer normal form (9.5), only the condition rank $dq(t) = n$ or the observability of single-output systems $(\mathbf{A}(t), c^T(t))$, respectively, is required for the determination of the transformation matrix $\Phi(t)$.

As a result of the extended Taylor linearization and of the linear time-variant normal form representation of (9.25), the error dynamics (9.29) of the extended Luenberger observer is obtained, if the time-variant Ackermann formula [3, 15]

$$l(t) = \{[p_0 N_A^0 + \dots + p_{n-1}N_A^{n-1} + N_A^n] \circ \tau(t)\}/\beta(t) \tag{9.38}$$

is inserted into (9.33). This formula is consistent with the nonlinear one (9.23), when the correspondence between $N_A\tau(t)$ and $ad_{Ax}\tau(x)$ is considered.

The calculation of the gain vector $l(t)$ as well as of the transformation matrix $\Phi(t)$ are essentially simplified by an appropriate choice of the function $\beta(t) \neq 0$ in (9.34) and (9.36), which is similar to the nonlinear case.

The time-variant Ackermann formula (9.38) has received a new application in course of the flatness-based tracking control, where a nonlinear observer with time-variant gains has to be designed based on the extended Taylor linearization along a reference trajectory [14, 15].

5 Block-Triangular Observer

Block-triangular observers [16, 17, 18, 19] are designed in a decentral approach, i.e. they are designed by considering subsystems. The decentral method uses degrees of freedom, which are generally available in the observer design for systems with *multiple outputs*

$$\Sigma: \quad \dot{x} = f(x), \quad y = h(x), \quad x \in M \subset \mathbb{R}^n, \quad y \in Y \subset \mathbb{R}^p, \quad p > 1. \quad (9.39)$$

A survey of the literature on decentral observer design can be found in [19].

The first step of the decentral approach is to decompose the system (9.39) into single output observable subsystems

$$\Sigma_i: \quad \dot{x}_i = f_i(x), \quad y_i = h_i(x), \quad x_i \in \mathbb{R}^{n_i}, \quad y_i \in \mathbb{R}, \quad \sum_{i=1}^{p} n_i = n. \quad (9.40)$$

For the decomposition, the multiple output observability map

$$q(x) = \Big[h_1(x), L_f h_1(x), \dots, L_f^{n_1-1} h_1(x), \dots, \quad (9.41)$$

$$\dots, h_p(x), L_f h_p(x), \dots, L_f^{n_p-1} h_p(x) \Big]^T \quad \text{with} \quad \sum_{i=1}^{p} n_i = n$$

is used where the parameters n_i indicate possible subsystem dimensions. To ensure an observer design by subsystems, the single output subsystems (9.40) must be observable by their related output y_i alone.

However, in (9.41) there is a condition only for the sum of the subsystem dimensions n_i. This fact is related to the general ambiguity, how a system can be observed by means of multiple outputs. Because of these degrees of freedom, there are usually several possibilities to decompose a system. Moreover, the decompositions may not only differ in the dimensions n_i of the subsystems but also in the type of the connections: Subsystems

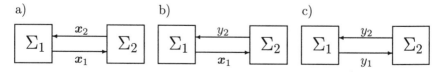

FIGURE 1. Possible connections of two subsystems Σ_1 and Σ_2 [16, 17, 18, 19].

(9.40) can be connected mutually by their states (see Figure 1a), or they can be connected in a *cascade* or *block-triangular form* (see Figure 1b), respectively. A mutual connection by the outputs (Figure 1c) is a special case of a block-triangular form.

Definition 9.5 (Block-Triangular Form) *A multiple output system representation*

$$\Sigma^{\triangle} : \dot{z} = f^{\triangle}(z, y), \qquad\qquad y = h^{\triangle}(z),$$

$$\text{with } z = \left[z_1^T, \ldots, z_p^T\right]^T, z_i = [z_{i1}, \ldots, z_{in_i}]^T, i = 1, \ldots, p \quad \text{and}$$

$$
\begin{aligned}
\Sigma_1^{\triangle} : \dot{z}_1 &= f_1^{\triangle}(z_1, y_2, \ldots, y_p), & y_1 &= h_1^{\triangle}(z_1) \\
\Sigma_2^{\triangle} : \dot{z}_2 &= f_2^{\triangle}(z_1, z_2, y_3, \ldots, y_p), & y_2 &= h_2^{\triangle}(z_2) & (9.42) \\
&\vdots & &\vdots \\
\Sigma_i^{\triangle} : \dot{z}_i &= f_i^{\triangle}(z_1, \ldots, z_i, y_{i+1}, \ldots, y_p), & y_i &= h_i^{\triangle}(z_i) \\
&\vdots & &\vdots \\
\Sigma_p^{\triangle} : \dot{z}_p &= f_p^{\triangle}(z_1, \ldots, z_p), & y_p &= h_p^{\triangle}(z_p)
\end{aligned}
$$

is called a block-triangular form. *The subsystems $\Sigma_i^{\triangle}$ of this form are assumed to be locally observable by their related outputs y_i.*

Remark 9.2 (Output Transformation) *A further degree of freedom to produce a block-triangular form (9.42) is an additional output transformation $y^{\triangle} = \gamma(y)$, by which a triangular state connection of the output equations can be considered [16, 17, 18, 19].*

Definition 9.6 (Block-Triangular Observer) *The set of subsystem observers $\hat{\Sigma}_i^{\triangle}$, $i = 1, \ldots, p$ which in a decentral approach are designed for a system in block-triangular form (9.42) is called a* block-triangular observer.

The block-triangular decomposition is very suitable for a decentral observer design. This fact will be shown below for the special case of the *block-triangular normal form observer*, which was introduced by Rudolph and Zeitz in [16].

Definition 9.7 (Block-Triangular Observer Normal Form) *A multiple output system representation with subsystems*

$$\Sigma_i^{*\triangle}: \quad \dot{z}_i^* = A_i^* z_i^* + \alpha_i^*(z_{j<i}^*, y_i, y_{j>i}), \quad y_i = c_i^{*T} z_i^* = z_{i1}^*, \quad i, j = 1, \ldots, p \tag{9.43}$$

with (A_i^*, c_i^{*T}) in dual Brunovsky form is called a block-triangular observer normal form.

The block-triangular observer normal form permits the design of an approximate normal form observer with constant gain vectors $l_i^{*\triangle}$

$$\hat{\Sigma}_i^{*\triangle}: \quad \dot{\hat{z}}_i^* = A_i^* \hat{z}_i^* + \alpha_i^{\triangle}(\hat{z}_{j<i}^*, y_i, y_{j>i}) + l_i^{*\triangle} \cdot (y_i - c_i^{*T} \hat{z}_i^*). \tag{9.44}$$

The error $\tilde{z}_i^* = \hat{z}_i^* - z_i^*$ of this subsystem observer has an approximately linear dynamics

$$\tilde{\Sigma}_i^{*\triangle}: \quad \dot{\tilde{z}}_i^* = (\mathbf{A}_i^* - \boldsymbol{l}_i^{*\triangle} \boldsymbol{c}_i^{*T}) \tilde{z}_i^* +$$
$$\boldsymbol{\alpha}_i^\triangle(\boldsymbol{z}_{j<i}^* + \tilde{\boldsymbol{z}}_{j<i}^*, y_i, y_{j>i}) - \boldsymbol{\alpha}_i^\triangle(\boldsymbol{z}_{j<i}^*, y_i, y_{j>i}),$$

if it is guaranteed that the errors $\tilde{z}_{j<i}^*$ of the preceding subsystem observers $\hat{\Sigma}_{j<i}^{*\triangle}$ converge sufficiently fast, i.e. $\tilde{z}_{j<i}^*(t) \to 0$. Therefore, block-triangular normal form observers $\hat{\Sigma}_i^{*\triangle}$ are designed successively; the design must ensure the convergence property by a suitable eigenvalue assignment for $\tilde{\Sigma}_i^{*\triangle}$. Because of the block-triangular structure, in a subsystem $\Sigma_i^\triangle$ the influence of unknown variables $z_{j>i}$ (or $z_{j>i}^*$, respectively) of subsequent subsystems is excluded. This fact enables the decentral observer design.

Since the decentral approach does not permit exactly linear subsystem error dynamics, in general, the conventional multiple output normal form observer, which is designed in a central approach, will be preferred. The structure of the multiple output observer normal form

$$\Sigma^*: \quad \dot{x}^* = \mathbf{A}^* x^* + \boldsymbol{\alpha}(y), \quad y = \mathbf{C}^* x^* = [x_{11}^*, \dots, x_{p1}^*]^T \text{ with} \quad (9.45)$$

$$\mathbf{A}^* = \begin{bmatrix} \mathbf{A}_1^* & \cdots & 0 \\ \vdots & \ddots & \vdots \\ 0 & \cdots & \mathbf{A}_p^* \end{bmatrix}, \quad \mathbf{C}^* = \begin{bmatrix} \boldsymbol{c}_1^{*T} & \cdots & 0 \\ \vdots & \ddots & \vdots \\ 0 & \cdots & \boldsymbol{c}_p^{*T} \end{bmatrix}$$

and the $(n_i \times n_i)$-matrices $\mathbf{A}_i^*$ and $(1 \times n_i)$-vectors $\boldsymbol{c}_i^{*T}$ in dual Brunovsky form is determined by the observability indices n_i, confer e.g. [10]. However, the form (9.45) may as well be interpreted as a special case of a block-triangular form (9.42) with a cascade of subsystems connected only by the outputs (see Figure 1c). To obtain (9.45) from (9.39), a state transformation $x^* = \boldsymbol{\Phi}(x)$ (9.4) must exist and can be calculated [9, 6, 12] from the pdes

$$\frac{\partial \boldsymbol{\Phi}}{\partial x} \left[ad_{-f}^{n_i - r} \boldsymbol{\tau}_i(x) \mid i = 1, \dots, p; \ r = 1, \dots, n_i \right] = \mathbf{I}_n \text{ with} \quad (9.46)$$

$$\boldsymbol{\tau}_i(x) \text{ from } L_{\boldsymbol{\tau}_i} L_f^l h_j(x) = \begin{cases} 0, & 0 \le l < n_j - 1, \\ \delta_{ij}, & l = n_j - 1; i, j = 1, \dots, p. \end{cases} \quad (9.47)$$

Theorem 9.5 (Existence of Multi Output Observer Normal Form)
The state transformation (9.4) of a system (9.39) into a multiple output observer normal form (9.45) does exist, if and only if

$$\text{rank } dq(x) = n, \quad (9.48)$$
$$\left[ad_f^l \boldsymbol{\tau}_i(x), \ ad_f^r \boldsymbol{\tau}_j(x) \right] = \mathbf{0} \quad (9.49)$$
$$\text{for } i, j = 1, \dots, p; \ l = 0, \dots, n_i - 1; \ r = 0, \dots, n_j - 1.$$

If however the conditions (9.48) and/or (9.49) of the central approach are violated, it is worth checking whether a decentral approach will lead further.

Yet, a multiple output system (9.39) seldom has a block-triangular observer normal form (9.43) in its original coordinates. A concept for the corresponding state transformation was introduced by Rudolph and Zeitz [16], which was further investigated in [17]. The starting point of this transformation is the system (9.39) in a multiple output observability normal form [10, 24, 6, 12]:

$$
\bar{\Sigma}_i : \ \dot{\boldsymbol{\xi}}_i = \bar{\boldsymbol{f}}_i(\boldsymbol{\xi}) =
\begin{bmatrix}
\xi_{i2} \\
\vdots \\
\xi_{in_i} \\
\varphi_i(\boldsymbol{\xi}_1, \dots, \boldsymbol{\xi}_p)
\end{bmatrix}, \ \ y_i = \bar{h}_i(\boldsymbol{\xi}_i) = \xi_{i1}, \ \ i = 1, \dots, p
$$

$$
\text{with} \ \ \boldsymbol{\xi} = \left[\boldsymbol{\xi}_1^T, \dots, \boldsymbol{\xi}_p^T \right]^T, \ \ \boldsymbol{\xi}_i = [\xi_{i1}, \dots, \xi_{in_i}]^T, \tag{9.50}
$$

$$
\varphi_i(\boldsymbol{\xi}_1, \dots, \boldsymbol{\xi}_p) = L_f^{n_i} h_i \circ \boldsymbol{q}^{-1}(\boldsymbol{\xi}) \ .
$$

It is determined by using the multiple output observability map $\boldsymbol{\xi} = \boldsymbol{q}(\boldsymbol{x})$ (9.41), which for observable systems is diffeomorphic or injective. The further transformation from (9.50) into (9.43) is obtained in two steps that have to be performed successively for each subsystem [16]:

(A) A state transformation $\boldsymbol{z} = \boldsymbol{\Psi}(\boldsymbol{\xi})$ from observability normal form (9.50) into block-triangular form (9.42).

(B) A state transformation $\boldsymbol{z}^* = \boldsymbol{\Phi}(\boldsymbol{z})$ from block-triangular form (9.42) into block-triangular observer normal form (9.43).

Hence, the total state transformation from original to block-triangular observer normal form coordinates (and vice versa) consists of three parts:

$$
\boldsymbol{z}^* = \boldsymbol{\Phi} \circ \boldsymbol{\Psi} \circ \boldsymbol{q}(\boldsymbol{x}) , \tag{9.51}
$$

$$
\boldsymbol{x} = \boldsymbol{q}^{-1} \circ \boldsymbol{\Psi}^{-1} \circ \boldsymbol{\Phi}^{-1}(\boldsymbol{z}^*) . \tag{9.52}
$$

Remark 9.3 (Structure of Block-Triangular Observer) *In the new coordinates $\boldsymbol{z}^*$ a block-triangular normal form observer can be designed; the resulting estimates $\hat{\boldsymbol{z}}^*$ have to be transformed back to original coordinate estimates $\hat{\boldsymbol{x}}$ by means of the inverse transformation (9.52). Thus, the observer comprises a dynamical part (9.44) and an algebraic part (9.52) for the estimates. If the observability map $\boldsymbol{q}(\boldsymbol{x})$ (9.41) is a semi-diffeomorphism (Definition 9.1), the block-triangular observer is continuous (Theorem 9.2).*

The idea how to transform a system from observability normal form (9.50) to a block-triangular form (9.42) is shown below:

(A)
$$\bar{\Sigma}_i : \quad \dot{\boldsymbol{\xi}}_i = \bar{\boldsymbol{f}}_i(\boldsymbol{\xi}_{j<i}, \boldsymbol{\xi}_i, y_{j>i}, \dot{y}_{j>i}, \dots, \overset{(\alpha_{ij})}{y}_{j>i}), \quad y_i = \xi_{i1},$$

$$\downarrow \quad \boldsymbol{z}_i = \boldsymbol{\Psi}_i(\boldsymbol{\xi}_i, \dots, \boldsymbol{\xi}_p), \quad \alpha_{ij} \le n_j - 1, \quad i, j = 1, \dots, p,$$

$$\Sigma_i^\triangle : \quad \dot{\boldsymbol{z}}_i = \boldsymbol{f}_i^\triangle(\boldsymbol{z}_{j<i}, \boldsymbol{z}_i, y_{j>i}), \quad y_i = h_i^\triangle(\boldsymbol{z}_i) = z_{i1}.$$

Subsystems $\bar{\Sigma}_i$ of an observability normal form (9.50) are usually mutually connected (see Figure 1a). Since the observability coordinates $\boldsymbol{\xi}_i$ of a subsystem $\bar{\Sigma}_i$ can be interpreted as the output y_i and its time-derivatives $\dfrac{d^r y_i}{dt^r}$, $r = 1, \dots, n_i - 1$ (see (9.41) and (9.50)), the desired transformation $\boldsymbol{\Psi}_i$ has just to eliminate the output derivatives of subsequent subsystems $\bar{\Sigma}_{j>i}$, i.e. $\dfrac{d^r y_j}{dt^r}$, $r = 1, \dots, n_j - 1$ for $j > i$. Thus, a cascade connection of subsystems is ensured. The influence of subsequent outputs $y_{j>i}$ in $\bar{\Sigma}_i$ is no problem in a decentral observer design, for they are measured and, hence, known variables.

In order to remove output derivatives $\dfrac{d^r y_{j>i}}{dt^r}$, $r = 1, \dots, n_j - 1$ in a subsystem $\bar{\Sigma}_i$, the outputs $y_{j>i}$ are interpreted as inputs of the subsystem. Now, the results of Delaleau and Respondek [5] for the elimination of input derivatives can be applied as published in [16].

To calculate the transformation $\boldsymbol{\Psi}_i(\boldsymbol{\xi}_i, \dots, \boldsymbol{\xi}_p)$, $i = 1, \dots, p-1$, (the last subsystem needs not to be transformed, see the structure of $\Sigma_p^\triangle$ in (9.42)) the extended subsystem state

$$\boldsymbol{\xi}_{i,e} = \left[\boldsymbol{\xi}_i^T, y_{j>i}, \dots, \overset{(\alpha_{ij})}{y}_{j>i} \right]^T \tag{9.53}$$

and the extended vector field

$$\boldsymbol{F}_i = \bar{\boldsymbol{f}}_i + \sum_{j=i+1}^{p} \sum_{l=1}^{\alpha_{ij}+1} \overset{(l)}{y}_j \frac{\partial}{\partial \overset{(l-1)}{y}_j} \tag{9.54}$$

$$\text{with} \quad \bar{\boldsymbol{f}}_i = \sum_{r=1}^{n_i-1} \xi_{ir+1} \frac{\partial}{\partial \xi_{ir}} + \varphi_i(\boldsymbol{\xi}) \frac{\partial}{\partial \xi_{in_i}}$$

are introduced. In the following theorem, which is taken from [16], sufficient conditions are given for the existence of the state transformation

$$\boldsymbol{z}_i = \boldsymbol{\Psi}_i(\boldsymbol{\xi}_{i,e}), \quad i = 1, \dots, p - 1. \tag{9.55}$$

Theorem 9.6 (Existence of Block-Triangular Form) *A single output subsystem $\bar{\Sigma}_i$, $i = 1, \dots, p-1$ of an observability normal form (9.50) can locally be transformed into a subsystem $\Sigma_i^\triangle$ of a block-triangular form (9.42)* *by a state transformation* $z_i = \mathbf{\Psi}_i(\boldsymbol{\xi}_{i,e})$, *if*

$$\left[ad_{F_i}^q \frac{\partial}{\partial \, y_j^{(\alpha_{ij})}}, \; ad_{F_i}^r \frac{\partial}{\partial \, y_l^{(\alpha_{il})}}\right] = \mathbf{0} \quad where \quad \begin{cases} 1 \le i < j, l \le p \\ 0 \le q \le \alpha_{ij} \\ 0 \le r \le \alpha_{il} \end{cases} \tag{9.56}$$

The transformation $\mathbf{\Psi}_i(\boldsymbol{\xi}_{i,e})$ is a solution of a system of $n_i \cdot \sum\limits_{j=i+1}^{p} (\alpha_{ij}+1)$ first order pdes, denoted in vector form

$$\frac{\partial \mathbf{\Psi}_i}{\partial \boldsymbol{\xi}_{i,e}} ad_{-F_i}^q \frac{\partial}{\partial \, y_j^{(\alpha_{ij})}} = \mathbf{0}, \quad q = 0, \dots \alpha_{ij}, \quad 1 \le i < j \le p . \tag{9.57}$$

Theorem 9.7 (Necessity and Sufficiency of Theorem 9.6) *The conditions (9.56) of Theorem 9.6 for the existence of a state transformation are necessary and sufficient for a subsystem decomposition of orders $n_i \le n_j$, $1 \le i < j \le p$. For subsystems of orders $n_i > n_j$, $1 \le i < j \le p$, the conditions (9.56) are only sufficient.*

In [16, 17, 19], the two theorems are proved and discussed.

After the state transformations into a block-triangular form (9.42), the generated subsystems are sometimes in block-triangular observer normal form (9.43). If not, state transformations $z_i^* = \mathbf{\Phi}_i(z)$ into (9.43) have to be tried [16]:

$$\Sigma_i^{\triangle}: \quad \dot{z}_i = f_i^{\triangle}(z_{j<i}, z_i, y_{j>i}), \quad y_i = h_i^{\triangle}(z_i) = z_{i1},$$

(B) $\quad\downarrow \quad z_i^* = \mathbf{\Phi}_i(z_i, \dots, z_p), \quad i, j = 1, \dots, p-1,$

$$\Sigma_i^{*\triangle}: \quad \dot{z}_i^* = \mathbf{A}_i^* z_i^* + \boldsymbol{\alpha}_i^{\triangle}(z_{j<i}^*, y_i, y_{j>i}), \quad y_i = z_{i1}^* .$$

Theorem 9.8 (Existence of Block-Triangular Observer Normal Form) *A subsystem $\Sigma_i^{\triangle}$ in a block-triangular form (9.42) is transformable into a block-triangular observer normal form subsystem $\Sigma_i^{*\triangle}$ (9.43) if*

(i) $\left[ad_{f_i^{\triangle}}^r \boldsymbol{\tau}_i, \; ad_{f_i^{\triangle}}^s \boldsymbol{\tau}_i\right] = \mathbf{0}, \quad 0 \le r, s \le n_i - 1 \quad with \; \boldsymbol{\tau}_i \; from$ (9.58)

$\quad L_{\tau_i} L_{f_i^{\triangle}}^k h_i^{\triangle}(z_i) = 0, \; k = 0, \dots, n_i - 2, \quad L_{\tau_i} L_{f_i^{\triangle}}^{n_i-1} h_i^{\triangle}(z_i) = 1,$ (9.59)

(ii) $\left[ad_{f_i^{\triangle}}^r \boldsymbol{\tau}_i, \; \frac{\partial}{\partial y_j}\right] = \mathbf{0}, \quad 0 \le r \le n_i - 1, \quad j = i+1, \dots, p .$ (9.60)

Condition (i) of this theorem is identical to condition (9.10) for the state transformation of a single output system (9.1) into an observer normal form (9.5), whereby the other states $z_j \neq z_i$ are considered as parameters. That transformation is now applied on a single output subsystem of (9.42) and must produce a very special subsystem right hand side. However, at the same time, it may not destroy the triangular connection of subsystems, which was generated by the first transformation step $z_i = \Psi_i(\xi_{i,e})$. This requirement is ensured by condition (ii).

If the conditions of Theorem 9.8 are not fulfilled, a different observer method has to be used for the respective subsystem. Moreover, one can still go back to the original system (9.39), select a new set of subsystems, and check whether a block-triangular observer normal form may be generated now. This flexibility in using every possible degree of freedom is the advantage of the decentral approach.

The two transformation steps from observability normal form (9.50) via block-triangular form (9.42) to the block-triangular observer normal form (9.43) are illustrated in the following example:

$$\Sigma: \quad \dot{\boldsymbol{x}} = \begin{bmatrix} x_2 \\ x_2 + x_1 x_4 + x_2 x_3 \\ x_4 \\ (x_2)^2 \end{bmatrix}, \quad \boldsymbol{y} = [x_1, \ x_3]^T . \tag{9.61}$$

For this system, no state transformation into a multiple output observer normal form (9.45) exists. Since (9.61) is already given in observability coordinates, i.e. $\boldsymbol{\xi} = \boldsymbol{q}(\boldsymbol{x}) = \boldsymbol{x}$, in a decentral approach two observable subsystems can be chosen easily by grouping the state variables into corresponding subsystem state vectors:

$$\bar{\Sigma}_1 : \boldsymbol{\xi}_1 = [\xi_{11}, \xi_{12}]^T = [x_1, x_2]^T , \quad \bar{\Sigma}_2 : \boldsymbol{\xi}_2 = [\xi_{21}, \xi_{22}]^T = [x_3, x_4]^T .$$

These subsystems are mutually connected like in Figure 1a. To bring the system into a block-triangular form (9.42), the influence of $x_4 = \xi_{21} = \dot{y}_2$ in $\bar{\Sigma}_1$ has to be eliminated by a state transformation of the first subsystem, $z_1 = \Psi_1(\xi_{1,e})$. For this, the extended subsystem state $\xi_{1,e} = [\xi_{11}, \xi_{12}, y_2, \dot{y}_2]^T$ and the extended vector field $\boldsymbol{F}_1 = \bar{\boldsymbol{f}}_1(\boldsymbol{\xi}_1, y_2, \dot{y}_2) + \dot{y}_2 \dfrac{\partial}{\partial y_2} + \ddot{y}_2 \dfrac{\partial}{\partial \dot{y}_2}$ have to be introduced. The conditions (9.56) of Theorem 9.6 reduce to the expression $\left[ad_{F_1} \dfrac{\partial}{\partial \dot{y}_2}, \dfrac{\partial}{\partial \dot{y}_2} \right] \overset{!}{=} \boldsymbol{0}$, which is fulfilled by the two vector fields $\dfrac{\partial}{\partial \dot{y}_2}$ and $ad_{F_1} \dfrac{\partial}{\partial \dot{y}_2} = -\xi_{11} \dfrac{\partial}{\partial \xi_{12}} - \dfrac{\partial}{\partial y_2}$. Hence, the state transformation does exist. From the first order partial differential equations $\dfrac{\partial \Psi_1}{\partial \dot{y}_2} = \boldsymbol{0}$ and $\xi_{11} \dfrac{\partial \Psi_1}{\partial \xi_{12}} + \dfrac{\partial \Psi_1}{\partial y_2} = \boldsymbol{0}$, the state transformation $z_1 = [\xi_{11}, \ \xi_{12} - \xi_{11}\xi_{21}]^T$ is obtained. Since the second subsystem $\bar{\Sigma}_2$ is already in a block-triangular form $\Sigma_2^\triangle$, no transformation is necessary, i.e. $z_2 = [\xi_{21}, \xi_{22}]^T$.

Hence, Example (9.61) has the block-triangular form

$$
\Sigma^\triangle : \quad \dot{z} =
\begin{bmatrix}
z_{12} + y_1 y_2 \\
z_{12} + y_1 y_2 \\
z_{22} \\
(z_{12} + y_1 y_2)^2
\end{bmatrix}, \quad
y = [z_{11}, \ z_{21}]^T . \tag{9.62}
$$

Because subsystem $\Sigma_1^\triangle$ of this representation is still not in a block-triangular normal form, the additional state transformation $z_1^* = \Phi_1(z)$ has to be performed. Using the vector fields $\tau_1 = [0, \ 1]^T$ and $ad_{f_1^\triangle} \tau_1 = [-1, \ -1]^T$, the conditions (i) and (ii) of Theorem 9.8 are fulfilled; and the state transformation $z_1^* = [z_{11}, \ z_{12} - z_{11}]^T$ is determined. For the second subsystem $\Sigma_2^\triangle$ again no transformation is needed, i.e. $z_2^* = [z_{21}, \ z_{22}]^T$.

The final block-triangular observer normal form of Example (9.61) is calculated as

$$
\Sigma^{*\triangle} : \quad \dot{z}^* =
\begin{bmatrix}
z_{12}^* + y_1(1 + y_2) \\
0 \\
z_{22}^* \\
[z_{12}^* + y_1(1 + y_2)]^2
\end{bmatrix}, \quad
y = [z_{11}^*, \ z_{21}^*]^T . \tag{9.63}
$$

Now for the first subsystem $\Sigma_1^{*\triangle}$ a single output normal form observer with constant gains l_{1i}^*, $i = 1, 2$ can be designed:

$$
\hat{\Sigma}_1^{*\triangle} : \quad \dot{\hat{z}}_1^* =
\begin{bmatrix}
\hat{z}_{12}^* + y_1(1 + y_2) + & l_{11}^* \cdot (y_1 - \hat{z}_{11}^*) \\
& l_{12}^* \cdot (y_1 - \hat{z}_{11}^*)
\end{bmatrix}. \tag{9.64}
$$

Ensuring that this observer is converging sufficiently fast by assigning the eigenvalues of the error dynamics properly, the variable z_{12}^* is approximately known by its estimate $\hat{z}_{12}^*$. Thus, for the second subsystem $\Sigma_2^{*\triangle}$ a single output normal form observer with constant gains l_{2i}^*, $i = 1, 2$ is possible as well:

$$
\hat{\Sigma}_2^{*\triangle} : \quad \dot{\hat{z}}_2^* =
\begin{bmatrix}
\hat{z}_{22}^* & + & l_{21}^* \cdot (y_2 - \hat{z}_{21}^*) \\
[\hat{z}_{12}^* + y_1(1 + y_2)]^2 & + & l_{22}^* \cdot (y_2 - \hat{z}_{21}^*)
\end{bmatrix}. \tag{9.65}
$$

The estimates in original coordinates are calculated in the algebraic part of the observer (Remark 9.3):

$$
\hat{x} = q^{-1} \circ \Psi^{-1} \circ \Phi^{-1}(\hat{z}^*) = [\hat{z}_{11}^*, \ \hat{z}_{11}^* + \hat{z}_{12}^* + \hat{z}_{11}^* \hat{z}_{21}^*, \ \hat{z}_{21}^*, \ \hat{z}_{22}^*]^T . \tag{9.66}
$$

For slightly modified equations of Example (9.61), i.e. for $f_2(x) = x_1 x_4 + x_2 x_3$, the block-triangular representation is even identical with the block-triangular observer normal form. For $f_2(x) = x_1 x_4$, the system (9.61) can be transformed to a block-triangular form (9.42), but not to a block-triangular observer normal form (9.43), since condition (ii) of Theorem 9.8 is violated. In such a case, an additional output transformation may be considered [17], or a different appropriate observer design method for the subsystem has to be chosen [3].

6   Conclusions

The study of the normal form observer design and its variants is confirming the experience that the observer design is far more complicated to solve than the feedback or controller design of nonlinear systems. The presented normal form observer design variants illustrate that a progress in the solution of the nonlinear observer design problem is possible in various directions: via the application of a semi-diffeomorphic state transformation derived from the observability properties in case of the *continuous observer;* through an extended Taylor linearization and the additional output transformation used by the *extended Luenberger observer;* and by a subtle utilization of degrees of freedoms of multiple outputs for a subsystem design of the *block–triangular observer.*

ACKNOWLEDGEMENT: As it can be concluded from the list of references, the presented nonlinear observer design methods are originating from cooperations with D. Bestle, J. Birk, A. J. Krener, R. Rothfuß, J. Rudolph, and X. Xia. The work was supported by Deutsche Forschungsgemeinschaft.

7   REFERENCES

[1] D. Bestle and M. Zeitz. Canonical form observer design for non-linear time-variable systems. *Int. Journal of Control* **38**, 419–431, 1983.

[2] J. Birk and M. Zeitz. Extended Luenberger observer for nonlinear multivariable systems. *Int. Journal of Control* **47**, 1823–1836, 1988.

[3] J. Birk. Rechnergestützte Analyse und Lösung nichtlinearer Beobachtungsaufgaben. VDI-Fortschritt-Berichte Nr. 8/294, *VDI-Verlag*, Düsseldorf, 1992.

[4] R. W. Brocket. Asymptotic stability and feedback stabilization. In: *R. W. Brocket, R. S. Millmann and H. J. Sussmann (eds.)*: Differential Geometric Control Theory. *Birkhäuser*, Bosten, 181–191, 1983.

[5] E. Delaleau and W. Respondek. Lowering the orders of derivatives of controls in generalized state space systems. *Journal of Mathematical Systems, Estimation, and Control* **5** , 1–27, 1995.

[6] A. Isidori. *Nonlinear Control Systems (3rd edition).* Springer-Verlag, London 1995.

[7] H. Keller. Non-linear observer design by transformation into a generalized observer canonical form. *Int. Journal of Control* **46**, 1915–1930, 1987.

[8] A. J. Krener and A. Isidori. Linearization by output injection and nonlinear observers. *Systems and Control Letters* **3**, 47–52, 1983.

[9] A. J. Krener and W. Respondek. Nonlinear observers with linearizable error dynamics. *SIAM J. on Control and Optim* **23**, 197–216, 1985.

[10] A. J. Krener. Normal forms for linear and nonlinear systems. In: *M. Luksic, C. Martin and W. Shadwick (eds.)*: Differential Geometry: The Interface between Pure and Applied Mathematics. *American Mathematical Society*, Providence, 157–189, 1987.

[11] A. J. Krener. Nonlinear stabilizability and detectability. In: *U. Helmke, R. Mennicken and J. Saurer (eds.)*: Systems and Networks: Mathematical Theory and Applications, *Vol. I, Akademie Verlag*, Berlin, 231–250, 1994.

[12] H. Nijmeijer and A. J. Van der Schaft. *Nonlinear Dynamical Control Systems*. Springer-Verlag, Berlin 1990.

[13] A. R. Phelps. On constructing nonlinear observers. *SIAM J. on Control and Optim.* **29**, 516–534, 1991.

[14] R. Rothfuß, J. Rudolph and M. Zeitz. Flatness-based control of a nonlinear chemical reactor model. *Automatica* **32**, 1433–1439, 1996.

[15] R. Rothfuß. Anwendung der flachheitsbasierten Analyse und Regelung nichtlinearer Mehrgrößensysteme. *VDI-Fortschritt-Berichte Nr. 8/664, VDI-Verlag*, Düsseldorf 1997.

[16] J. Rudolph and M. Zeitz. A block triangular nonlinear observer normal form. *Systems and Control Letters* **23**, 1–8, 1994.

[17] J. Schaffner. Zum Beobachterentwurf für nichtlineare Systeme mit mehreren Meßgrößen. *VDI-Fortschritt-Berichte Nr. 8/620, VDI-Verlag, Düsseldorf*, 1997.

[18] J. Schaffner and M. Zeitz. Decentralized block triangular observer design for nonlinear systems. *Proc. European Control Conference ECC'97*, Brussels, 1997.

[19] J. Schaffner and M. Zeitz. Decentral nonlinear observer design using a block-triangular form. *To appear in Int. J. of Systems Science*, 1999.

[20] X. H. Xia and W. Gao. On exponential observers for nonlinear systems. *Systems and Control Letters* **11**, 319–325, 1988.

[21] X. H. Xia and W. Gao. Nonlinear observer design by observer error linearization. *SIAM J. on Control and Optim.* **27**, 199–216, 1989.

[22] X. H. Xia and M. Zeitz. On nonlinear continuous observers. *Int. Journal of Control* **66**, 943–954, 1997.

[23] M. Zeitz. The extended Luenberger observer for nonlinear systems. *Systems and Control Letters* **9**, 149–156, 1987.

[24] M. Zeitz. Canonical normal forms for nonlinear systems. In: *A. Isidori, (ed.): Nonlinear Control Systems Design – Selected Papers from IFAC-Symposium, Capri/Italy 1989, Pergamon, Oxford*, 33–38, 1989.

[25] M. Zeitz. Nichtlineare stetige Beobachter. *ZAMM – Zeitschrift für Angewandte Mathematik und Mechanik* **78**, S1137–S1140, 1998.

Part II

Output Feedback Control Design

In this part emphasis is placed on the derivation and design of observers used in closed loop with controllers. The idea is that a state feedback controller is available that achieves stability or asymptotic tracking but, since the full state is not available for feedback, an observer must reconstruct the unmeasured states. In the context of linear systems, the celebrated "separation principle" guarantees that the controller and observer design can be performed independently, so that the combined closed-loop system yields the desired stable behavior. For nonlinear systems the situation is essentially different since without extra assumptions the "separation principle" will not hold, and in fact one can easily develop examples in which a separate design with a stable observer and a stable controller does not produce a stable observer/controller when combined. In the context of output feedback controllers this research has developed into two directions. On one hand, one tries to identify classes of nonlinear systems for which the "separation principle" does hold, that is a so-called "nonlinear separation principle". On the other hand, one seeks a specific observer or a class of observers, that in closed loop with a given controller yields a stable system. It is, however, interesting to note that the dual approach where one seeks a specific controller along with a given observer, has not yet been systematically studied. The reader can find contributions discussing both approaches in this part of the book.

Separation Results for Semiglobal Stabilization of Nonlinear Systems via Measurement Feedback

Stefano Battilotti

Dipartimento di Informatica e Sistemistica
Università di Roma "La Sapienza"
Via Eudossiana 18–00184, Italy

1 Introduction

Recently, the problem of *semiglobally* (rather than globally) stabilizing a nonlinear system via output feedback has gained a renewed interest ([4], [5], [16], [18], [17], [14], [15], [8]). Here, *semiglobally* means that one is requiring that the region of attraction of the equilibrium point contains at least an *a priori* given compact set of the state space. The earlier works of Esfandiari and Khalil ([4], [5]) have shown that fully feedback linearizable nonlinear systems, which are generally not *globally* stabilizable via dynamic output feedback, are instead *semiglobally* stabilizable. The basic ingredients are *input saturations* and *high–gain observers*: large values of the observer gain guarantee that the error between the state and its estimate, generated by the observer itself, goes to zero "sufficiently fast", while input saturations rule out destabilizing effects such as "peaking" ([12]). The combination of these two successful ingredients, together with the key concept of "complete uniform observability", has led to the unifying work of Teel and Praly ([14]): for the class of systems

$$
\begin{aligned}
\dot{x} &= f(x, u, d(t)) \\
y &= C(x, d(t))
\end{aligned}
\tag{1.1}
$$

where $d(t)$ is some exogenous disturbance, it is proved that semiglobal stabilization via *state–feedback* plus *complete uniform observability* imply semiglobal stabilization via *dynamic output feedback*. Complete uniform observability implies that a state feedback $u(x)$, which semiglobally stabilizes (1.1), can be written as a known function Ψ of y, u and their higher order derivatives (but *independent* of $d(t)$).

Besides, a *high–gain* observer is designed to reconstruct the higher order derivatives of y. A key feature of this design procedure is the possibility of taking y as a state, i.e. the dynamical behavior of y and their higher order derivatives can be modeled through

$$\dot{y} = y^{(1)}$$
$$\vdots$$
$$y^{(n_y+1)} = C_{n_y+1}(x(t), u(t), \ldots, u^{(m_u)}(t), d(t)) \qquad (1.2)$$

This indeed allow to act as if the output were not corrupted by uncertainties.

An important problem has been left open so far: what can we do either when (1.2) is not available or the observer gain cannot be pushed arbitrarily large? Say, consider the uncertain system

$$\dot{x}(t) = Ax(t) + B_2 u(t) + B_1 \widetilde{\Phi}(u(t), x(t), t)$$
$$y(t) = C_2 x(t) + C_1(u(t), x(t), t)\widetilde{\Phi}(u(t), x(t), t) \qquad (1.3)$$

where $\widetilde{\Phi}(u(t), x(t), t)$ is some exogenous disturbance (vector), of which nothing but some *bounds* are known. Note that (1.3) is made up of a *nominal linear* system, perturbed by nonlinear terms and uncertainties. Under which conditions (1.3) is semiglobally stabilizable via *measurement* feedback, i.e. when only the measure of y is available? In this paper, we give a general theorem on the *regional* stabilization of (1.3) via *measurement* feedback (Theorem 1.1), which stands as a generalization of a previous result on quadratic stabilization of linear uncertain systems ([6]). Our approach is completely different from the ones pursued in the literature and it is strongly based on $\mathcal{H}_\infty$ control tools. Moreover, we recover into a general framework most of the existing results on the semiglobal stabilization via *output feedback* (see Section 3) and generalize them into the more appealing perspective of stabilization via *measurement feedback*. In particular, in Section 3.2.1, we will discuss the case $C_1(u(t), x(t), t) = 0$ ("uncorrupted outputs") and see how high gain observers arise in this case; in Sections 3.2.2 and 3.2.3, we discuss the case of input saturations and output saturations, which can be seen as particular nonlinearities perturbing a linear system. Finally, using Theorem 1.1, in Section 3.3, we give some general tools for semiglobally stabilizing via *arbitrarily bounded measurement feedback* a significant class of nonlinear uncertain systems (Theorem 1.2), which include at least feedforward systems and homogenous systems ([18], [13]). Some key features of our design are given by allowing block–state equations and uncertainties in the outputs: moreover, our design procedures end up with *linear* controllers and *quadratic* Lyapunov functions. As a particularization of our results, we obtain that systems

$$
\begin{aligned}
\dot{z}_1 &= z_2 + p_{11}(z, u, t) \\
\dot{z}_2 &= z_3 + p_{21}(z, u, t) \\
\vdots &= \vdots \\
\dot{z}_n &= u + p_{n1}(z, u, t) \\
y_j &= z_j + p_{j2}(z, u, t), \qquad j = 1, \ldots, n
\end{aligned} \tag{1.4}
$$

with $z_j, y, u \in \mathbb{R}$ and $z = col(z_1, \ldots, z_n)$, are always semiglobally stabilizable via arbitrarily bounded measurement feedback as long as $p_{j1}(z, u, t)$ and $p_{j2}(z, u, t)$, $j = 1, \ldots, n$, are *higher order* in $(z_{j+1}, \ldots, z_n, u)$, uniformly with respect to t and $z_1, \ldots, z_j$ (see [18] for state feedback), while systems in the form

$$
\begin{aligned}
\dot{z} &= Az + Bu + \widetilde{p}_1(z, u)u \tag{1.5} \\
y &= z + \widetilde{p}_2(z, u)u \tag{1.6}
\end{aligned}
$$

with $\widetilde{p}_j(z, 0) = 0$, $j = 1, 2$, for all t and z, (A, B) in Brunowski form and $z \in \mathbb{R}^n$, $y \in \mathbb{R}^n$ and $u \in \mathbb{R}$, are semiglobally stabilizable via arbitrarily bounded measurement feedback as long as $\widetilde{p}_1(z, u)u$ and $\widetilde{p}_2(z, u)u$ are *of at least order one and zero*, respectively, with respect to the "generalized" dilation $\delta_l(z, u) = (l^{1-n}z_1, \ldots, l^{-1}z_{n-1}, z_n, lu)$ and uniformly w.r.t. t (see [13] for definitions and related results for state feedback).

2 Notations

- By $\lambda_{max}\{S\}$ ($\lambda_{min}\{S\}$) we denote the maximum (minimum) eigenvalue of a given matrix S;

- if $\|v\|$ denotes the 2–norm of any given vector v, by $\|A\|$ we denote the induced 2–norm of any given matrix A and we have $\|A\| = \sqrt{\lambda_{max}\{A^T A\}}$; by $\|v\|_A$ we denote the A–norm of v, i.e. $\|v\|_A = \sqrt{v^T A v}$;

- by $\mathcal{SP}^n$ we denote the set of $n \times n$ positive definite symmetric matrices;

- for any vector–valued function $\eta : \mathbb{R}^s \rightarrow \mathbb{R}^r$, we denote by η_i its i–th component; for any matrix M we denote by M_i its i–th row.

3 Regional Stabilization via Measurement Feedback

3.1 Tools

Let us consider the system

$$
\begin{aligned}
\dot{x}(t) &= Ax(t) + B_2 u(t) + B_1 \widetilde{\Phi}(u(t), x(t), t) \\
y(t) &= C_2 x(t) + C_1(u(t), x(t), t)\widetilde{\Phi}(u(t), x(t), t)
\end{aligned}
\tag{1.7}
$$

with $t \in I\!\!R$ a.e., state vector $x(t) \in I\!\!R^n$, input vector $u(t) \in I\!\!R^m$, $\widetilde{\Phi}(u(t), x(t), t) \in I\!\!R^k$, A, B_j, $j = 1, 2$, and C_2 (resp. $C_1(\cdot, \cdot, \cdot)$) *known* matrices (resp. matrix–valued function) with appropriate dimensions.

The structure of (1.7) is characterized by a *nominal linear* system perturbed by nonlinear (unknown) terms. The vector $\widetilde{\Phi}(\cdot, \cdot, \cdot)$ captures both the uncertainties and exogenous disturbances, which perturb the nominal system. We say that $\widetilde{\Phi} : I\!\!R^m \times I\!\!R^n \times I\!\!R \to I\!\!R^k$ is an *admissible* uncertainty if

- it is *continuous* w.r.t. the first two arguments and *Lebesgue measurable* w.r.t. the third argument

- for each $t \in I\!\!R$, $x \in I\!\!R^n$ and $u \in I\!\!R^m$

$$
\widetilde{\Phi}(u, x, t) \in \mathcal{U}(u, x) \overset{\text{def}}{=} \{v \in I\!\!R^k : \|v\| \le s(u, x)\}
\tag{1.8}
$$

  for some known continuous positive semidefinite function $s : I\!\!R^m \times I\!\!R^n \to I\!\!R$.

The class of *admissible* feedback laws we consider is characterized as follows

$$
\begin{aligned}
u &= k(y, \sigma) \\
\dot{\sigma} &= v(y, \sigma), \qquad \sigma \in I\!\!R^q
\end{aligned}
\tag{1.9}
$$

with continuous $k : I\!\!R^p \times I\!\!R^q \to I\!\!R^m$ and $v : I\!\!R^p \times I\!\!R^q \to I\!\!R^q$, vanishing at the origin.

In this section, we are interested in the *regional* stabilization of (1.7) under the constraint (1.8). Let $\mathcal{D}$ be the set of admissible uncertainties $\widetilde{\Phi} : I\!\!R^m \times I\!\!R^n \times I\!\!R \to I\!\!R^k$ and let $(x(t, x_0, \sigma_0), \sigma(t, x_0, \sigma_0))$ denote the trajectories of (1.7)–(1.9), at time t with initial condition (x_0, σ_0). We say that the system (1.7) is *uniformly locally asymptotically stabilizable via measurement-feedback (ULASMF) with region of attraction containing* $\Omega^e \subset I\!\!R^n \times I\!\!R^q$ if there exists an admissible control law (1.9) such that along the trajectories of (1.7)–(1.9)

1. (**Uniform stability**) $\forall \epsilon \in I\!\!R^+$ there exists $\delta_\epsilon \in I\!\!R^+$ such that

$$\left\| \begin{pmatrix} x(t, x_0, \sigma_0) \\ \sigma(t, x_0, \sigma_0) \end{pmatrix} \right\| \leq \epsilon \qquad (1.10)$$

for all $(x_0, \sigma_0) \in \left\{ v \in \Omega^e : \|v\| \leq \delta_\epsilon \right\}$, $t \in [0, +\infty)$ a.e. and $\widetilde{\Phi} \in \mathcal{D}$;

2. (**Uniform boundedness plus attraction**) there exists $M \in I\!\!R^+$ such that

$$\left\| \begin{pmatrix} x(t, x_0, \sigma_0) \\ \sigma(t, x_0, \sigma_0) \end{pmatrix} \right\| < M \qquad (1.11)$$

for all $(x_0, \sigma_0) \in \Omega^e$, $t \in [0, +\infty)$ a.e. and $\widetilde{\Phi} \in \mathcal{D}$, and $\forall \epsilon \in I\!\!R^+$ there exists $T_\epsilon \in I\!\!R^+$ such that

$$\left\| \begin{pmatrix} x(t, x_0, \sigma_0) \\ \sigma(t, x_0, \sigma_0) \end{pmatrix} \right\| < \epsilon \qquad (1.12)$$

for all $(x_0, \sigma_0) \in \Omega^e$, $t \in [T_\epsilon, +\infty)$ a.e. and $\widetilde{\Phi} \in \mathcal{D}$.

If, in addition, Ω^e contains an *a priori* given compact set $\mathcal{S} \times \mathcal{W} \subset I\!\!R^n \times I\!\!R^q$ we will say that (1.7) is *uniformly semiglobally asymptotically stabilizable*. Throughout the paper, we will consider compact sets containing the origin of the state space.

Assume that $s^2(u, x)$ can be taken some *quadratic* function of u and x, whenever $\|u\| \leq \Delta(c)$ and x lives in

$$\Omega(c) \stackrel{\text{def}}{=} \{ v \in I\!\!R^n : \|v\|_{P_{SF}(c)}^2 \leq c^2 \}$$

for some positive number c and where $P_{SF}(c)$ is a positive definite symmetric solution of the Riccati equation, associated to the state–feedback $\mathcal{H}_\infty$ linear control problem (see [3]), with $\Phi(u, x, t)$ as exogenous disturbance and $s(u, x)$ as penalty variable. Assume also that one is capable to find a positive definite symmetric solution $P_{OI}(c)$ of the *dual* Riccati equation, associated to the output injection $\mathcal{H}_\infty$ linear problem (see [3]), with $\widetilde{\Phi}$ as exogenous disturbance and $s(u, x)$ as penalty variable. If $P_{SF}(c)$ is sufficiently "small" than $P_{OI}(c)$, then (1.7) is ULASMF with a region of attraction $\Omega^e(c)$, which in general may suffer from an intrinsically nonlinear phenomenon, pointed out in [7], and commonly referred to as *vanishing region of attraction*. This negative phenomenon can be counteracted if an additional *nonlinear* inequality, involving $P_{SF}(c)$ and $P_{OI}(c)$, is satisfied. This inequality requires the knowledge of two continuous functions $\eta(\cdot)$ and $\delta(\cdot)$: the first one is instrumental in constructing a stabilizing controller (1.9), since $k(y, \sigma)$ is taken as the composition of $\eta(\cdot)$ with a *linear* controller $v = F_2(c)\sigma$, where $F_2(c)$ comes from solving the state–feedback

$\mathcal{H}_\infty$ linear control problem, with $\widetilde{\Phi}(u, x, t)$ as exogenous disturbance and $s(u, x)$ as penalty variable; the second one pops up in the candidate Lyapunov function of the closed–loop system (1.7)–(1.9) and, if suitably designed, prevents the phenomenon of vanishing region of attraction so that any given *a priori* compact set $\mathcal{S} \times \mathcal{W}$ can be included in the region of attraction of the closed–loop system. Different choices of these functions $\eta(\cdot)$ and $\delta(\cdot)$ allow to recover into a general framework most of the existing results on the semiglobal stabilization via *output feedback* of (1.7) (see next sections) and to put them into the more general perspective of stabilization via *measurement feedback*.

The above discussion leads directly to our first theorem. For simplicity of computations and resolving formulas, we will assume that $B_1 C_1^T(u, x, t) = 0$ for all $u \in \mathbb{R}^m$, $x \in \mathbb{R}^n$ and $t \in \mathbb{R}$.

Theorem 1.1 *Assume $B_1 C_1^T(u, x, t) = 0$ for all $u \in \mathbb{R}^m$, $x \in \mathbb{R}^n$ and $t \in \mathbb{R}$. Moreover, assume that for each $c \in \mathbb{R}^+$*

- **(state feedback)** *there exist $P_{SF}(c), Q_{SF}(c) \in \mathcal{SP}^n$, such that*

    - *for some $\gamma(c) \in \mathbb{R}^+$ and $\Delta(c) \in (0, \infty]$ and for some known matrices $E_1(c)$ and $E_2(c)$ one has*

$$\|\widetilde{\Phi}(u, x, t)\| \leq s(u, x) \overset{\text{def}}{=} \frac{1}{\gamma^2(c)} \left[\|E_1(c)x\|^2 + \|E_2(c)u\|^2 \right]$$
$$(1.13)$$

    for all $t \in \mathbb{R}$, $u \in \mathbb{R}^m$ such that $\|u\| \leq \Delta(c)$ and

$$x \in \Omega(c) \overset{\text{def}}{=} \{v \in \mathbb{R}^n : \|v\|^2_{P_{SF}(c)} \leq c^2\}$$

    - *$R_1(c) \overset{\text{def}}{=} E_2^T(c)E_2(c) \in \mathcal{SP}^m$;*
    - *the following Riccati equation is satisfied*

$$\mathcal{H}_{SF}(c) \overset{\text{def}}{=} A^T P_{SF}(c) + P_{SF}(c)A$$
$$+ \quad P_{SF}(c)(\frac{1}{\gamma^2(c)}B_1 B_1^T - B_2 R_1^{-1}(c)B_2^T)P_{SF}(c)$$
$$+ \quad E_1^T(c)E_1(c) = -Q_{SF}(c) \qquad (1.14)$$

- **(output injection)** *there exist $R_2(c) \in \mathcal{SP}^p$ and $P_{OI}(c), Q_{OI}(c) \in \mathcal{SP}^n$, such that*

    - *for all $u \in \mathbb{R}^m$ such that $\|u\| \leq \Delta(c)$, $x \in \Omega(c)$ and $t \in \mathbb{R}$*

$$R_2(c) \geq C_1(u, x, t)C_1^T(u, x, t) \qquad (1.15)$$

- *the following Riccati inequality is satisfied*

$$\mathcal{H}_{OI}(c) \stackrel{\text{def}}{=} A^T P_{OI}(c) + P_{OI}(c)A + P_{OI}(c)B_1 B_1^T P_{OI}(c)$$
$$+ \frac{E_1^T(c)E_1(c)}{\gamma^2(c)} - C_2^T R_2^{-1}(c)C_2 \leq -Q_{OI}(c) \quad (1.16)$$

- **(relative speed)**

  - $Q_m(c) \stackrel{\text{def}}{=} \gamma^2(c)Q_{OI}(c) - Q_{SF}(c) > 0$
  - $P_m(c) \stackrel{\text{def}}{=} \gamma^2(c)P_{OI}(c) - P_{SF}(c) > 0$

- **(nonlinear coupling)** *there exist $a \in (0,1)$ and C^0 functions $\delta :$* $\mathbb{R} \to (0, 1-a]$ *and* $\eta : \mathbb{R}^m \to \mathbb{R}^m$ *such that if*

$$F_2(c) \stackrel{\text{def}}{=} -R_1^{-1}(c)B_2^T P_{SF}(c) \qquad\qquad (1.17)$$

$$\varphi(s) \stackrel{\text{def}}{=} as + \int_0^s \delta(\vartheta)d\vartheta \qquad\qquad (1.18)$$

one has

$$\eta_i(F_2(c)x) = F_{2i}(c)x, \qquad i = 1,\dots,m \qquad (1.19)$$

and

$$\frac{\|\eta(F_2(c)(x-e)) - F_2(c)x\|_{R_1(c)}^2 - \|x\|_{Q_{SF}(c)}^2}{\|F_2(c)e\|_{R_1(c)}^2 + \|e\|_{Q_m(c)}^2} < a + \delta(\|e\|_{P_m(c)}^2)$$
$$(1.20)$$

for all $x \in \Omega(c)$ and $e \in \mathbb{R}^n$ such that $0 < \varphi(\|e\|_{P_m(c)}) \leq c^2$.

Under the above assumptions, (1.7) is ULASMF with region of attraction containing

$$\Omega^e(c) \stackrel{\text{def}}{=} \{(v_1, v_2) \in \mathbb{R}^n \times \mathbb{R}^n : \|v_1\|_{P_{SF}(c)}^2 + \varphi(\|v_1 - v_2\|_{P_m(c)}^2) \leq c^2\}$$
$$(1.21)$$

whenever $\|\eta(F_2(c)\sigma(t))\| \leq \Delta(c)$ for all $t \geq 0$. An admissible stabilizing controller is given by

$$u = \eta(F_2(c)\sigma)$$
$$\dot{\sigma} = H(c)\sigma + B_2\eta(F_2(c)\sigma) + G(c)y \qquad (1.22)$$

with

$$H(c) = A + \frac{1}{\gamma^2(c)}B_1 B_1^T P_{SF}(c) - G(c)C_2$$
$$G(c) = \gamma^2(c)P_m^{-1}(c)C_2^T R_2^{-1}(c) \qquad\qquad (1.23)$$

Remark 1.1 The value $a = 0$ can be allowed in Theorem 1.1 as long as $\int_0^s \delta(\vartheta)d\vartheta \in \mathcal{K}_\infty$.

Remark 1.2 The design parameter $\Delta(c)$ allows to take into account input saturations, see Section 3.2.2.

Remark 1.3 Theorem 1.1 states that *regional* stabilization can be achieved if the following can be achieved: a *semiglobal state feedback* stabilization problem, with controller $u = F_2(c)x$ and region of attraction being the points inside some level set $\mathcal{L}$ of a Lyapunov function V_{SF}, an *output injection* stabilization problem, assuming the state x remain inside $\mathcal{L}$, a *linear* coupling condition which guarantees the observer error converge to zero "sufficiently" faster than the state x does whenever (1.7) is plugged with $u = F_2(c)x$ and a *nonlinear* coupling condition which determines the width of the region of attraction. If $C_1(u, x, t)$ is constant and (1.13) holds for all x, u and t, Theorem 1.1 recovers a well known result on *quadratic* stabilization of linear uncertain systems ([6]), since in this case (1.20) is satisfied with $a = 0$, $\delta(s) = 1$ and $\eta(s) = s$ for all s. Moreover, the region of attraction is *all of* $\mathbb{R}^n \times \mathbb{R}^q$. For a general system (1.7), this simple choice of $\delta(s)$ and $\eta(s)$, although still satisfying (1.20), does not prevent the phenomenon of *vanishing region of attraction* to appear (see [7]). The flexibility in the choice of the functions $\delta(s)$ and $\eta(s)$ is crucial in recovering most of the existing results on robust *semiglobal* stabilization, as it will be shown in the next sections.

Remark 1.4 From the proof of Theorem 1.1 it follows that the conclusions of Theorem 1.1 hold if the inequality in (1.14) is replaced by inequality and (1.16) together with the relative speed constraint are replaced by the existence of matrices $P_m(c), Q_m(c) \in \mathcal{SP}^n$ such that

$$
\left[A + \frac{1}{\gamma^2(c)}B_1 B_1^T P_{SF}(c) - G(c)C_2\right]^T P_m(c)
$$
$$
+ \quad P_m(c)\left[A + \frac{1}{\gamma^2(c)}B_1 B_1^T P_{SF}(c) - G(c)C_2\right] + F_2^T(c)R_1(c)F_2(c)
$$
$$
+ \quad \frac{1}{\gamma^2(c)}P_m(c)(B_1 B_1^T + G(c)R_2(c)G^T(c))P_m(c) \leq -Q_m(c) \quad (1.24)
$$

with $G(c)$ as in (1.23).

3.2 Applications

3.2.1 Uncorrupted Outputs Revisited

This section is devoted to recover in the framework of measurement feedback the case in which $C_1(u, x, t) = 0$ for all $u \in \mathbb{R}^m$, $x \in \mathbb{R}^n$ and $t \in \mathbb{R}$ ([11], [4], [5], [8], [14], [15]). If this is the case, for each $c \in \mathbb{R}^+$ the matrix $R_2(c)$ (see (1.15)) can be taken *any* positive definite matrix. If (A, B_1, C_2) is (possibly after a change of coordinates) *left invertible with no zero dynamics* ([9]), the output injection, relative speed and nonlinear coupling constraints can be met as follows.

Assume that for each $c \in \mathbb{R}^+$ there exist $P_{SF}(c), Q_{SF}(c) \in \mathcal{SP}^n$ satisfying (1.14) for *all* $\gamma(c) \in \mathbb{R}^+$ (i.e. the gain between w and $z = s(u, x)$ can be taken arbitrarily small). Without loss of generality, we can assume that $E_1(\cdot)$ and $\gamma(\cdot)$ have been chosen so that $\gamma(c) < \|E_1(c)\|$ for all $c \in \mathbb{R}^+$. Moreover, for each given compact set S assume the existence of $c \in \mathbb{R}^+$ such that $S \subset \Omega(\frac{c}{\sqrt{2}})$.

Under the above assumptions, one can find continuous functions $Q_{OI}^{(\cdot)}(c)$, $P_{OI}^{(\cdot)}(c)$ and $R_2^{(\cdot)}(c) : (\frac{\|E_1(c)\|^2}{\gamma^2(c)}, +\infty) \to \mathcal{SP}^n$ and $\epsilon_{OI}^{(\cdot)} : (\frac{\|E_1(c)\|^2}{\gamma^2(c)}, +\infty) \to \mathbb{R}^+$ with the following properties:

1. the following Riccati equation is satisfied

$$\mathcal{H}_{OI}^{(l_{OI})}(c) \overset{\text{def}}{=} A^T P_{OI}^{(l_{OI})}(c) + P_{OI}^{(l_{OI})}(c)A + P_{OI}^{(l_{OI})}(c)B_1 B_1^T P_{OI}^{(l_{OI})}(c)$$
$$+ \frac{E_1^T(c)E_1(c)}{\gamma^2(c)} - C_2^T(R_2^{(l_{OI})}(c))^{-1}C_2 \le -Q_{OI}^{(l_{OI})}(c) \qquad (1.25)$$

for all $l_{OI} \in (\frac{\|E_1(c)\|^2}{\gamma^2(c)}, +\infty)$

2.

$$Q_{OI}^{(l_{OI})}(c) = \epsilon_{OI}^{(l_{OI})}(c) P_{OI}^{(l_{OI})}(c) \qquad (1.26)$$

$$\lim_{l_{OI} \to +\infty} \frac{\|P_{OI}^{(l_{OI})}(c)\|}{l_{OI}^k} = 0, \qquad \text{for some } k > 1 \qquad (1.27)$$

$$\lim_{l_{OI} \to +\infty} \lambda_{min}\{P_{OI}^{(l_{OI})}(c)\} = +\infty \qquad (1.28)$$

$$\lim_{l_{OI} \to +\infty} \frac{\epsilon_{OI}^{(l_{OI})}(c)}{l_{OI}} = +\infty \qquad (1.29)$$

$$\lim_{l_{OI} \to +\infty} \|R_2^{(l_{OI})}(c)\| = 0 \qquad (1.30)$$

Using (1.26), (1.28) and (1.29), choose $l_{OI,1}(c) \in (\frac{\|E_1(c)\|^2}{\gamma^2(c)}, +\infty)$ such that for all $l_{OI} \in [l_{OI,1}(c), +\infty)$

1. $P_m^{(l_{OI})}(c) \overset{\text{def}}{=} \gamma^2(c)P_{OI}^{(l_{OI})}(c) - P_{SF}(c) > 0$

2. $Q_m^{(l_{OI})}(c) \overset{\text{def}}{=} \gamma^2(c)Q_{OI}^{(l_{OI})}(c) - Q_{SF}(c) > 0$

and pick $L(c) \in \mathbb{R}^+$ such that, if

$$u_{max,i} = \max_{x \in \Omega(c)} \{|F_{2i}(c)x|\}$$

$$\eta_i(s_i) = s_i \min\{1, \frac{u_{max,i}}{|s_i|}\}$$

$$\eta(s_1, \ldots, s_m) = col(\eta_1(s_1), \ldots, \eta_m(s_m))$$

$$\delta(s) = \begin{cases} \frac{1}{l_{OI}(1+s)} & \text{if } s \geq 0 \\ \frac{1}{l_{OI}} & \text{if } s < 0 \end{cases} \tag{1.31}$$

then (1.20) holds for all $l_{OI} \in [l_{OI,1}(c), +\infty)$, $x \in \Omega(c)$ and $e \in \mathbb{R}^n$ such that $\|e\|^2_{P_m^{(l_{OI})}(c)} \geq L(c)$.

Note that with our choices the region of attraction of the closed–loop system (1.7)–(1.22) is given by

$$\Omega^e(c) \overset{\text{def}}{=} \{(v_1, v_2) \in \mathbb{R}^n \times \mathbb{R}^n : \|v_1\|^2_{P_{SF}(c)}$$

$$+ \frac{1}{l_{OI}} \ln(1 + \|v_1 - v_2\|^2_{P_m^{(l_{OI})}(c)}) \leq c^2\} \tag{1.32}$$

Since

$$\lim_{s \to +\infty} \frac{\ln s^k}{s} = 0, \qquad \forall k \geq 1$$

and from (1.27), it follows that, for each pair of compact sets $\mathcal{S}, \mathcal{W} \subset \mathbb{R}^n$,

$$\lim_{l_{OI} \to +\infty} \sup_{x \in \mathcal{S}; \sigma \in \mathcal{W}} \left\{ \frac{1}{l_{OI}} \ln(1 + \|x - \sigma\|^2_{P_m^{(l_{OI})}(c)}) \right\} = 0$$

We conclude that for each pair of compact sets $\mathcal{S}, \mathcal{W} \subset \mathbb{R}^n$ one can pick $l_{OI} \in (\frac{\|E_1(c)\|^2}{\gamma^2(c)}, +\infty)$ sufficiently *large* in such a way that $\Omega^e(c)$ contains $\mathcal{S} \times \mathcal{W}$. For the class of systems (1.7), by using the above arguments one can also recover the fact that *semiglobal stabilizability via state feedback (in the sense of (1.14)) plus complete uniform observability implies semiglobal stabilization via output feedback*. Indeed, under these assumptions, the output can be taken as a *state* and, thus, one can assume $C_1(u, x, t) = 0$ for all x, u and t. However, since throughout this paper we consider a nominal systems which is *linear*, Teel and Praly's separation result remains still more general than ours as long as $C_1(u, x, t) = 0$ for all x, u and t and the dynamical model (1.2) is available.

3.2.2 Input Saturations

Let us consider the system

$$
\begin{aligned}
\dot{x}(t) &= Ax(t) + B_2 u(t) + B_1 \widetilde{\Phi}(u(t), x(t), t) \\
y(t) &= C_2 x(t) + C_1(u(t), t) \widetilde{\Phi}(u(t), x(t), t)
\end{aligned}
\tag{1.33}
$$

with $B_1 C_1^T(u, t) = 0$ for all $u \in \mathbb{R}^m$ and $t \in \mathbb{R}$ (the case $B_1 C_1^T(u, t) \neq 0$ for some u and t can be studied as well). We will make the following assumptions

(H1) *the pairs (A, B_1) and (A, B_2) are stabilizable and the eigenvalues of A have nonpositive real part*

(H2) *the pair (C_2, A) is detectable*

(H3) *there exist continuous functions $Q_{SF}^{(\cdot)}, P_{SF}^{(\cdot)} : (0, 1] \to \mathcal{SP}^n$ and $\gamma, \Delta \in \mathbb{R}^+$ such that*

1. $\|\widetilde{\Phi}(u, x, t)\|^2 \le \frac{1}{\gamma^2} \|E_2 u\|^2$ *for some $E_2 \in \mathbb{R}^{n \times m}$, for all $t \in \mathbb{R}$, $x \in \mathbb{R}^n$ and $u \in \mathbb{R}^m$ such that $\|u\| \le \Delta$*

2. $R_1 \stackrel{\text{def}}{=} E_2^T E_2$ *is invertible and there exists $R_2 \in \mathcal{SP}^p$ such that*

$$
R_2 \ge C_1(u, t) C_1^T(u, t)
$$

   for all $u \in \mathbb{R}^m$ such that $\|u\| \le \Delta$ and for all $t \in \mathbb{R}$;

3. *the following Riccati equation is satisfied*

$$
\begin{aligned}
\mathcal{H}_{SF}^{(l_{SF})} \stackrel{\text{def}}{=} \; & A^T P_{SF}^{(l_{SF})} + P_{SF}^{(l_{SF})} A \\
& + P_{SF}^{(l_{SF})} \left(\frac{B_1 B_1^T}{\gamma^2} - B_2 R_1^{-1} B_2^T \right) P_{SF}^{(l_{SF})} \\
= \; & -Q_{SF}^{(l_{SF})}
\end{aligned}
\tag{1.34}
$$

   for all $l_{SF} \in (0, 1]$

4. $\lim\limits_{l_{SF} \to 0} \|Q_{SF}^{(l_{SF})}\| = 0$ *and* $\lim\limits_{l_{SF} \to 0} \|P_{SF}^{(l_{SF})}\| = 0$

The interest in the class of systems (1.33) relies on the possibility of taking into account input saturations. As an example, consider

$$
\begin{aligned}
\dot{x}(t) &= Ax(t) + B\sigma_1(u(t)) \\
\widetilde{y}(t) &= C_2 x(t) + C\sigma_2(u(t))
\end{aligned}
\tag{1.35}
$$

where σ_1, σ_2 are locally Lipschitz continuous, uniformly with respect to t, and such that

$$
C(\sigma_2(u, t) - u) = C_1(u, t)(\sigma_1(u, t) - u) + \sigma_3(u, t)
$$

for some continuous $C_1(\cdot, \cdot)$ and $\sigma_3(\cdot, \cdot)$, with $\sigma_3(0, t) = 0$ for all $t \in I\!\!R$. Clearly, (1.35) can be rewritten as (1.33) with $\widetilde{\Phi}(u(t), x(t), t) = \sigma_1(u(t), t) - u(t)$, $B_1 = B_2 = B$ and $y(t) = \widetilde{y}(t) - Cu(t) - \sigma_3(u(t), t)$. In (1.35) the term $\sigma_2(u)$ may also capture any (unknown) erro affecting $C_2 x$, due to torque disturbances, etc. etc.

Assumptions (H1)–(H3) are exactly the same invoked in [16], Lemma 3.1 (see also ([8]). By (H2), there exist $Q_{OI}, P_{OI} \in \mathcal{SP}^n$ such that the following Riccati inequality is satisfied

$$\mathcal{H}_{OI} \overset{\text{def}}{=} A^T P_{OI} + P_{OI} A + P_{OI} B_1 B_1^T P_{OI} - C_2^T R_2^{-1} C_2 \leq -Q_{OI} \quad (1.36)$$

Let $l_{SF}^* \in I\!\!R^+$ be such that $P_m^{(l_{SF})} \overset{\text{def}}{=} \gamma^2 P_{OI} - P_{SF}^{(l_{SF})} > 0$ for all $l_{SF} \in (0, l_{SF}^*]$. This choice is always feasible by H3.4.

Fix compact sets $\mathcal{S}, \mathcal{W} \subset I\!\!R^n$ and let $c \in I\!\!R^+$ be such that

$$c^2 \geq \sup_{x \in \mathcal{S}; \sigma \in \mathcal{W}; l_{SF} \in (0, l_{SF}^*]} \{ \|x\|^2_{P_{SF}^{(l_{SF})}} + \|x - \sigma\|^2_{P_m^{(l_{SF})}} \}$$

Define $a = 0$, $\eta(s) = s$ and $\delta(s) = 1$. Pick $l_{SF} \in (0, l_{SF}^*]$ such that

1. if $F_2^{(l_{SF})} \overset{\text{def}}{=} -R_1^{-1} B_2^T P_{SF}^{(l_{SF})}$, one has $\|F_2^{(l_{SF})} \sigma\| \leq \Delta$ for all σ such that $(x, \sigma) \in \Omega^e(c)$

2. $Q_m^{(l_{SF})} \overset{\text{def}}{=} \gamma^2 Q_{OI} - Q_{SF}^{(l_{SF})} > 0$

Property 1 above can be satisfied by H3.4 and since

$$\|F_2 \sigma\| \leq (k_1 \| \sqrt{P_{SF}^{(l_{SF})}} \| + k_2 \|P_{SF}^{(l_{SF})}\|) c$$

for some $k_1, k_2 \in I\!\!R^+$ and for all σ such that

$$(x, \sigma) \in \Omega^e \overset{\text{def}}{=} \left\{ (v_1, v_2) \in I\!\!R^n \times I\!\!R^n : \|v_1\|^2_{P_{SF}^{(l_{SF})}} + \|v_1 - v_2\|^2_{P_m^{(l_{SF})}} \leq c^2 \right\}$$

With our choices, (1.20) is satisfied for all $x \in \Omega(c)$ and $e \in I\!\!R^n \setminus \{0\}$. By applying Theorem 1.1, one conclude that the region of attraction Ω^e contains at least $\mathcal{S} \times \mathcal{W}$.

3.2.3 Output Saturations

Let us consider the system

$$\begin{aligned} \dot{x}(t) &= Ax(t) + B_2 u(t) \\ y(t) &= C_2 x(t) + C_1(u(t), x(t), t)\widetilde{\Phi}(u(t), x(t), t) \end{aligned} \quad (1.37)$$

with $x \in I\!\!R^n$, satisfying the following assumptions:

(H1) (A, B_2, C_2) *is in prime canonical form ([9])*

(H2) *for any C^0 function $c^{(\cdot)} : (0,1] \to I\!\!R^+$ such that $\lim_{l \to 0} c^{(l)} = +\infty$ there exist C^0 functions $\gamma^{(\cdot)}, R_2^{(\cdot)} : (0,1] \to I\!\!R^+$ such that*

$$\|\widetilde{\Phi}(u,x,t)\|^2 \quad \leq \quad \frac{\|C_2 x\|^2}{(\gamma^{(l)})^2} \tag{1.38}$$

$$\|C_1(u,x,t)C_1^T(u,x,t)\| \quad \leq \quad R_2^{(l)} \tag{1.39}$$

$$M \leq (\gamma^{(l)})^2 (R_2^{(l)})^{-1} - 1 \quad > \quad 0 \tag{1.40}$$

for some $M \in I\!\!R^+$, for all $l \in (0,1]$, $t \in I\!\!R$, $u \in I\!\!R^m$ and $x \in I\!\!R^n$ such that $\|C_2 x\| \leq c^{(l)}$ and, in addition,

$$\lim_{l \to 0} (c^{(l)})^{2n-1-2h} ((\gamma^{(l)})^2 (R_2^{(l)})^{-1} - 1)^{\frac{h}{2n-1}} \quad = \quad +\infty \tag{1.41}$$

where $h = n-1, n-2, \ldots, 0$.

Note that if $y = \text{sat}(C_2 x)$, where sat $(s) = s \min\{1, \frac{1}{|s|}\}$, one can take $\widetilde{\Phi}(u,x,t) = \text{sat}(C_2 x) - C_2 x$, $C_1(u,x,t) = 1$, $R_2^{(l)} = 1$ and $\gamma^{(l)} = \frac{c^{(l)}}{c^{(l)}-1}$, with $\lim_{l \to 0} \gamma^{(l)} = 1$ and $\|\widetilde{\Phi}(u,x,t)\|^2 \leq \frac{\|C_2 x\|^2}{(\gamma^{(l)})^2}$ for all $x \in I\!\!R^n$ such that $\|C_2 x\| \leq c^{(l)}$. Moreover, $(\gamma^{(l)})^2 (R_2^{(l)})^{-1} - 1 = \frac{2c^{(l)}-1}{(c^{(l)}-1)^2}$ so that (1.41) is satisfied (w.l.o.g. we can assume that $c^{(l)} > \frac{1}{2}$ for all $l \in (0,1]$).

From (H2) it follows that, whatever the C^0 function $R_1^{(\cdot)} : (0,1] \to I\!\!R^+$ is, one has

$$\|\widetilde{\Phi}(u,x,t)\|^2 \quad \leq \quad \frac{\|C_2 x\|^2 + \|u\|^2_{R_1^{(l)}}}{(\gamma^{(l)})^2} \tag{1.42}$$

for all $t \in I\!\!R$, $l \in (0,1]$, $u \in I\!\!R^m$ and $x \in I\!\!R^n$ such that $\|C_2 x\| \leq c^{(l)}$. Let $R_1^{(\cdot)}$ be such that $\lim_{l \to 0} R_1^{(l)} = 0$. It is easy to see that there exist $l^* \in (0,1]$ and C^0 functions $P_{SF}^{(\cdot)}, Q_{SF}^{(\cdot)}, P_m^{(\cdot)}, Q_m^{(\cdot)} : (0,1] \to \mathcal{SP}^n$ such that

$$A^T P_{SF}^{(l)} + P_{SF}^{(l)} A - P_{SF}^{(l)} B_2 (R_1^{(l)})^{-1} B_2^T P_{SF}^{(l)} + C_2^T C_2 = -Q_{SF}^{(l)} \tag{1.43}$$

$$A^T P_m^{(l)} + P_m^{(l)} A + P_{SF}^{(l)} B_2 (R_1^{(l)})^{-1} B_2^T P_{SF}^{(l)} - (\gamma^{(l)})^2 R_2^{-1} C_2^T C_2 \leq -Q_m^{(l)} \tag{1.44}$$

for all $l \in (0, l^*]$. Indeed, define $h_1 = \frac{1}{4n}$ and $h_2 = \frac{1}{2n}$. Let $\overline{P}_{SF,0}$ be the unique (stabilizing) positive definite symmetric solution of

$$\overline{P}_{SF,0} A + A^T \overline{P}_{SF,0} - \overline{P}_{SF,0} B_2 B_2^T \overline{P}_{SF,0} + C_2^T C_2 = 0 \tag{1.45}$$

For each fixed C^0 function $\overline{Q}_{SF}^{(\cdot)} : (0,1] \to \mathcal{SP}^n$ such that $\lim_{l \to 0} \overline{Q}_{SF}^{(l)} = 0$ pick C^0 function $\overline{P}_{SF}^{(\cdot)} : (0,1] \to \mathcal{SP}^n$ such that

$$\overline{P}_{SF}^{(l)} A + A^T \overline{P}_{SF}^{(l)} - \overline{P}_{SF} B_2 B_2^T \overline{P}_{SF}^{(l)} + C_2^T C_2 = -\overline{Q}_{SF}^{(l)} \tag{1.46}$$

for all $l \in (0,1]$ and $\lim_{l \to 0} \overline{P}_{SF}^{(l)} = \overline{P}_{SF,0}$. Define

$$P_{SF}^{(l)} = Z^{(l)} \overline{P}_{SF}^{(l)} Z^{(l)} \tag{1.47}$$

$$Q_{SF}^{(l)} = Z^{(l)} \overline{Q}_{SF}^{(l)} Z^{(l)} (R_1^{(l)})^{-h_2} \tag{1.48}$$

where $Z^{(l)} = \text{diag}\{(R_1^{(l)})^{h_1}, (R_1^{(l)})^{h_1+h_2}, \ldots, (R_1^{(l)})^{h_1+h_2(n-1)}\}$.

Let $c^{(l)} = \epsilon(R_1^{(l)})^{-h_1}$, with $\epsilon \in \mathbb{R}^+$ (independent of l but dependent on the compact set $\mathcal{S} \times \mathcal{W}$ to be included in the region of attraction) to be specified later. Moreover, define $k^{(l)} = ((\gamma^{(l)})^2 (R_2^{(l)})^{-1} - 1)^{\frac{1}{2n-1}}$ and pick

$$P_0^{(l)} = W^{(l)} \overline{P}_0 W^{(l)} \tag{1.49}$$

where $W^{(l)} = \text{diag}\{(k^{(l)})^{n-1}, \ldots, 1\}$, $\overline{P}_0$ is any positive definite symmetric solution of

$$\overline{P}_0 A + A^T \overline{P}_0 - C_2^T C_2 = -\overline{Q}_0$$

for $\overline{Q}_0 \in \mathcal{SP}^n$. Using (1.41)–(1.41) and $2h_1 = h_2$, pick $\overline{Q}_{SF}^{(l)}$ such that for some $l^* \in (0,1]$

$$Q_{SF}^{(l)} \leq k^{(l)} \frac{W^{(l)} \overline{Q}_0 W^{(l)}}{2} \tag{1.50}$$

$$P_{SF}^{(l)} \leq P_0^{(l)} \tag{1.51}$$

for all $l \in (0, l^*]$. Define

$$P_m^{(l)} = P_0^{(l)} - P_{SF}^{(l)} \tag{1.52}$$

$$Q_m^{(l)} = k^{(l)} \frac{W^{(l)} \overline{Q}_0 W^{(l)}}{2} \tag{1.53}$$

Finally, fix compact sets $\mathcal{S}, \mathcal{W} \subset \mathbb{R}^n$ and pick $\epsilon \in \mathbb{R}^+$ such that

$$\epsilon \geq \sup_{x \in \mathcal{S}; \sigma \in \mathcal{W}; l \in (0,1]} \left\{ \|x\|_{P_{SF}^{(l)}}^2 + \|x - \sigma\|_{P_m^{(l)}}^2 \right\} \tag{1.54}$$

(note that $\epsilon < +\infty$ by (1.40)). This concludes the proof of (1.43) and (1.44).

By Remark 1.4 and (1.54), it follows that, under (H1) and (H2), (1.37) is semiglobally stabilizable by arbitrarily bounded measurement feedback.

3.3 Semiglobal Stabilization of Uncertain Nonlinear Systems

In this section, using Theorem 1.1 and the arguments of Section 3.2.2, we give some general tools for achieving *large regions of attraction* for uncertain system using *arbitrarily bounded measurement feedback*. A basic feature of our design is that we allow for *block state equations* and *uncertainties in the outputs* and we end up with *linear controllers* and *quadratic Lyapunov functions*. We will recover and generalize some recent result on the semiglobal stabilization via state feedback of

$$
\begin{aligned}
\dot{z}_1(t) &= A_1 z_1(t) + B_1 \widetilde{\Psi}_1(z(t), \widetilde{z}_{n+1}(t), t) \\
\widetilde{y}_1(t) &= C_{12} z_1(t) + \widetilde{C}_{11} \widetilde{\Psi}_1(z(t), \widetilde{z}_{n+1}(t), t) \\
&\;\;\vdots \;\; = \;\; \vdots \\
\dot{z}_n(t) &= A_n z_n(t) + B_n \widetilde{\Psi}_n(z(t), \widetilde{z}_{n+1}(t), t) \\
\widetilde{y}_n(t) &= C_{n2} z_n(t) + \widetilde{C}_{n1} \widetilde{\Psi}_n(z(t), \widetilde{z}_{n+1}(t), t)
\end{aligned}
\tag{1.55}
$$

with $z_j \in \mathbb{R}^{r_j}$, $z = col(z_1 \cdots z_n)^T$, $\widetilde{z}_{n+1} \in \mathbb{R}^m$ the input vector, $\widetilde{y}_j \in \mathbb{R}^{p_j}$ the output vector, A_j, B_j, $\widetilde{C}_{j1}$ and C_{j2} matrices of suitable dimensions, $\widetilde{\Psi}_j(z, \widetilde{z}, t) \in \mathbb{R}^{k_j}$ are admissible uncertainties and $\sum_{j=1}^{n} r_j = r$. We will assume that $\widetilde{C}_{j1} B_{j1}^T = 0$ for all $j = 1, \ldots, n$ and rank $(B_j) = r_j$ for all $j = 2, \ldots, n$ (these assumptions can be relaxed and are motivated by simpler calculations and formulas).

We propose the following family of measurement feedback controllers. Let

$$
\begin{aligned}
\widetilde{z}_{n+1} &= F_{n2}^{(l)} \sigma_n \\
\dot{\sigma}_j &= \Pi_j^{(l)} \sigma_j + B_{j2} \widetilde{z}_{j+1} + G_j^{(l)} y_j, \quad \sigma_j \in \mathbb{R}^{r_j}, \; j = 1 \ldots n
\end{aligned}
\tag{1.56}
$$

where

$$
\begin{aligned}
y_1 &= \widetilde{y}_1 - \widetilde{C}_{11} K_1 \widetilde{z}_2 \\
y_j &= \widetilde{y}_j - C_{j2} \widetilde{z}_j - \widetilde{C}_{j1} K_j \widetilde{z}_{j+1}, \qquad 2 \leq j \leq n
\end{aligned}
\tag{1.57}
$$

and

$$
\begin{aligned}
\widetilde{z}_{j+1} &= F_{j2}^{(l)} \sigma_j \\
F_{j2}^{(l)} &= -R_{j1}^{-1} B_{j2}^T P_{SF,j}^{(l)} \\
\Pi_j^{(l)} &= A_j + \frac{1}{\gamma_j^2} B_{j1} B_{j1}^T P_{SF,j}^{(l)} - G_j^{(l)} C_{j2} \\
G_j^{(l)} &= \gamma_j^2 (P_{mj}^{(l)})^{-1} C_{j2}^T (R_{j2}^{(l)})^{-1}
\end{aligned}
\tag{1.58}
$$

with $l \in (0, 1]$, $B_{j1} = B_j$, $\gamma_j \in \mathbb{R}^+$, $P_{SF,j}^{(\cdot)}$, $P_{mj}^{(\cdot)}$ and $R_{j2}^{(\cdot)}$ are C^0 functions with domain $(0, 1]$ and codomain $\mathcal{SP}^{r_j}$ and $\mathcal{SP}^{p_j}$, respectively, and B_{j2} and R_{j1} are matrices of appropriate dimensions.

In order to understand how to choose the parameters characterizing the above family of controllers, we need some more definitions. For, define

$$\zeta_j = \begin{cases} z_1 & \text{if } j = 1 \\ z_j - \tilde{z}_j & \text{if } 2 \leq j \leq n \\ 0 & \text{if } j = n+1 \end{cases} \tag{1.59}$$

$$Z_j = col(z_j, \ldots, z_n) \tag{1.60}$$

$$\tilde{Z}_j = col(\tilde{z}_j, \ldots, \tilde{z}_{n+1}) \tag{1.61}$$

$$P_{mj}^{(l)} \stackrel{def}{=} \gamma_j^2 P_{OI,j}^{(l)} - P_{SF,j}^{(l)} \tag{1.62}$$

$$V_{SF,j}(v_1) \stackrel{def}{=} \|v_1\|_{P_{SF,j}^{(l)}}^2, \quad j = 1, \ldots, n \tag{1.63}$$

$$W_j(v_1, v_2) \stackrel{def}{=} \|v_1\|_{P_{SF,j}^{(l)}}^2 + \|v_1 - v_2\|_{P_{mj}^{(l)}}^2 \tag{1.64}$$

If no ambiguity arises, we will use alternatively the notations $\tilde{Z}_2$ and $\tilde{z}$. Given $h \in I\!\!R^+$, compact sets $\tilde{S}_j, \tilde{W}_j$, $j = 1, \ldots, n$, and C^0 matrix–valued functions $N_{ji}(\cdot, \cdot)$, $1 \leq j \leq n$, $j+1 \leq i \leq n$, and $\tilde{N}_{ji}(\cdot, \cdot)$, $2 \leq j \leq n$, $j \leq i \leq n$, of appropriate dimensions define the following C^0 functions

$$c_n^{(l)} \stackrel{def}{=} 2^{n-1} \sup_{v_1 \in \tilde{S}_n; v_2 \in \tilde{W}_n} \left\{ W_n(v_1, v_2) \right\} \tag{1.65}$$

$$\tilde{\delta}_n^{(l)} \stackrel{def}{=} \|R_{n1}^{-1} B_{n2}^T \sqrt{P_{SF,n}^{(l)}}\| \left(1 + \sqrt{\frac{1}{h}}\right) \sqrt{c_n^{(l)}} \tag{1.66}$$

$$\tilde{\delta}_{n-1}^{(l)} \stackrel{def}{=} \max_{V_{SF,n}(\zeta_n) \leq c_n^{(l)}} \left\{ \|\zeta_n\| \right\} \tag{1.67}$$

$$k_n^{(l)} \stackrel{def}{=} 4 \cdot \sup_{0 < V_{SF,n}(\zeta_n) \leq c_n^{(l)}; \|\tilde{z}_{j+1}\| \leq \tilde{\delta}_j^{(l)}, j=n-1,n} \left\{ \frac{\|\Gamma_{n-1,n}(Z_n, \tilde{Z}_n)\zeta_n\|^2}{\|\zeta_n\|_{Q_{SF,n}^{(l)}}^2} \right\} \tag{1.68}$$

$$\vdots \quad \stackrel{def}{=} \quad \vdots$$

$$c_i^{(l)} \stackrel{def}{=} c_{i+1}^{(l)} k_{i+1}^{(l)} \tag{1.69}$$

$$\tilde{\delta}_{i-1}^{(l)} \stackrel{def}{=} \max_{V_{SF,i}(\zeta_i) \leq c_i^{(l)}} \left\{ \|\zeta_i\| \right\} \tag{1.70}$$

$$k_i^{(l)} \stackrel{def}{=} 2^{n-i+2} \cdot \sup_{0 < V_{SF,s}(\zeta_s) \leq c_s^{(l)}, s=i,\ldots,n; \|\tilde{z}_{j+1}\| \leq \tilde{\delta}_j^{(l)}, j=i-1,\ldots,n}$$
$$\left\{ \frac{\|\Gamma_{i-1,j}(Z_i, \tilde{Z}_i)\zeta_j\|^2}{\|\zeta_j\|_{Q_{SF,j}^{(l)}}^2}, j = i, \ldots, n \right\} \tag{1.71}$$

$$\vdots \quad \stackrel{def}{=} \quad \vdots$$

$$c_1^{(l)} \stackrel{def}{=} c_2^{(l)} k_2^{(l)} \tag{1.72}$$

with Γ_{ji} as follows:

$$\Gamma_{1i}^T(Z_2, \widetilde{Z}_2)\Gamma_{1i}(Z_2, \widetilde{Z}_2) \;=\; N_{1i}^T(Z_2, \widetilde{Z}_2)N_{1i}(Z_2, \widetilde{Z}_2) \qquad (1.73)$$

for $i = 2, \ldots, n$ and

$$\Gamma_{jj}^T(Z_j, \widetilde{Z}_j)\Gamma_{ji}(Z_j, \widetilde{Z}_j) \;=\; \frac{4\gamma_j^2}{\gamma_{j-1}^2} \widetilde{N}_{jj}^T(Z_j, \widetilde{Z}_j)\widetilde{N}_{jj}(Z_j, \widetilde{Z}_j) \qquad (1.74)$$

$$\Gamma_{ji}^T(Z_{j+1}, \widetilde{Z}_{j+1})\Gamma_{ji}(Z_{j+1}, \widetilde{Z}_{j+1}) \;=\; \frac{4\gamma_j^2}{\gamma_{j-1}^2} \widetilde{N}_{ji}^T(Z_{j+1}, \widetilde{Z}_{j+1})\widetilde{N}_{ji}(Z_{j+1}, \widetilde{Z}_{j+1})$$
$$+\; 2N_{ji}^T(Z_{j+1}, \widetilde{Z}_{j+1})N_{ji}(Z_{j+1}, \widetilde{Z}_{j+1})$$
$$(1.75)$$

for all $j = 2, \ldots, n$ and $i = j + 1, \ldots, n$. Finally, let

$$D_j^{(l)} \;=\; B_{j+1,1}^+ \Big[A_{j+1}R_{j1}^{-1}B_{j2}^T P_{SF,j}^{(l)} + R_{j1}^{-1}B_{j2}^T \Big(A_j^T P_{SF,j}^{(s)}$$
$$+\; Q_{SF,j}^{(l)} + P_{SF,j}^{(s)}G_j^{(l)}C_{j2} \Big) \Big]$$
$$S_j^{(l)} \;=\; B_{j+1,1}^+ \Big[A_{j+1}R_{j1}^{-1}B_{j2}^T P_{SF,j}^{(l)} + R_{j1}^{-1}B_{j2}^T \Big(A_j^T P_{SF,j}^{(s)} + Q_{SF,j}^{(l)} \Big) \Big]$$
$$(1.76)$$

The main result of this section is the following.

Theorem 1.2 *Let $\mathcal{S}_j, \mathcal{W}_j, \widetilde{\mathcal{S}}_j, \widetilde{\mathcal{W}}_j \subset \mathbb{R}^{r_j}$, $j = 1, \ldots, n$, be given compact sets with $\mathcal{S}_j \subset \widetilde{\mathcal{S}}_j$ and $\mathcal{W}_j = \widetilde{\mathcal{W}}_j$. If there exist $\gamma_j \in \mathbb{R}^+$, $R_{j1} \in \mathcal{SP}^{r_j+1}$, $K_j \in \mathbb{R}^{k_j \times r_j+1}$, C^0 functions $P_{SF,j}^{(\cdot)}, Q_{SF,j}^{(\cdot)}, P_{OI,j}^{(\cdot)}, Q_{OI,j}^{(\cdot)} : (0,1] \to \mathcal{SP}^{r_j}$, $R_{j2}^{(l)} \in \mathcal{SP}^{p_j}$, $j = 1, \ldots, n$ and matrix–valued C^0 functions $E_{ji}(\cdot, \cdot)$, $1 \leq j \leq n$, $j + 1 \leq i \leq n+1$, $\widetilde{E}_{ji}(\cdot, \cdot)$, $2 \leq j \leq n$, $j \leq i \leq n+1$, $N_{ji}(\cdot, \cdot)$, $1 \leq j \leq n$, $j + 1 \leq i \leq n$, $\widetilde{N}_{ji}(\cdot, \cdot)$, $2 \leq j \leq n$, $j \leq i \leq n$, of appropriate dimensions, such that*

(A1) *for all $l \in (0,1]$ and $j = 1, \ldots, n$*

$$A_j^T P_{SF,j}^{(l)} + P_{SF,j}^{(l)}A_j + P_{SF,j}^{(l)}\Big(\frac{B_{j1}B_{j1}^T}{\gamma_j^2} - B_{j2}R_{j1}^{-1}B_{j2}^T \Big)P_{SF,j}^{(l)} = -Q_{SF,j}^{(l)}$$

$$A_j^T P_{OI,j}^{(l)} + P_{OI,j}^{(l)}A_j + P_{OI,j}^{(l)}B_{j1}B_{j1}^T P_{OI,j}^{(l)} - C_{j2}^T(R_{j2}^{(l)})^{-1}C_{j2} \leq -Q_{OI,j}^{(l)}$$
$$(1.77)$$

with $B_{j1} = B_j$, $B_{j2} = B_jK_j$ and $R_{j2}^{(l)} \geq \widetilde{C}_{j1}\widetilde{C}_{j1}^T$;

(A2) *there exists $l_1^* \in (0,1]$ such that for all $l \in (0,l_1^*]$ and $j = 1, \ldots, n$*

$$\gamma_j^2 P_{OI,j}^{(l)} - P_{SF,j}^{(l)} \;\geq\; hP_{SF,j}^{(l)} \qquad (1.78)$$

$$\gamma_j^2 Q_{OI,j}^{(l)} - Q_{SF,j}^{(l)} \;\geq\; hQ_{SF,j}^{(l)} \qquad (1.79)$$

for some $h \in \mathbb{R}^+$ and, in addition,

$$\lim_{l \to 0} \|P_{SF,j}^{(l)}\| = 0 \tag{1.80}$$

$$\lim_{l \to 0} \|P_{OI,j}^{(l)}\| < +\infty \tag{1.81}$$

for $j = 1, \ldots, n$;

(A3) *for all $l \in (0,1]$*

$$\begin{aligned}
\|\widetilde{\Psi}_j(z, \widetilde{z}_{n+1}, t) - K_j \widetilde{z}_{j+1}\|^2 \;\leq\; & \frac{1}{\gamma_j^2}\Big[\sum_{j+1 \leq i \leq n} \|N_{ji}(Z_{j+1}, \widetilde{Z}_{j+1})\zeta_i\|^2 \\
& + \sum_{j+1 \leq i \leq n+1} \|E_{ji}(z, \widetilde{z})\widetilde{z}_i\|^2 \Big]
\end{aligned} \tag{1.82}$$

for $j = 1, \ldots, n$ and

$$\begin{aligned}
& \left\| B_{j-1,1}^+ R_{j-1,1}^{-1} B_{j-1,2}^T P_{SF,j-1}^{(l)} G_{j-1}^{(l)} \widetilde{C}_{j-1,1}\Big(\widetilde{\Psi}_{j-1}(z, \widetilde{z}_{n+1}, t) - K_{j-1}\widetilde{z}_j \Big) \right\|^2 \\
& \leq \frac{1}{\gamma_{j-1}^2}\Big[\|\widetilde{N}_{jj}(z, \widetilde{z})\zeta_j\|^2 + \sum_{j+1 \leq i \leq n} \|\widetilde{N}_{ji}(Z_{j+1}, \widetilde{Z}_{j+1})\zeta_i\|^2 \\
& + \sum_{j \leq i \leq n+1} \|\widetilde{E}_{ji}(z, \widetilde{z})\widetilde{z}_i\|^2 \Big]
\end{aligned} \tag{1.83}$$

for $j = 2, \ldots, n$

(A4) *there exists $l_2^* \in (0,1]$ such that for all $l \in (0, l_2^*]$ and $j = 1, \ldots, n-1$*

$$\|R_{j1}^{-1} B_{j2}^T \sqrt{P_{SF,j}^{(l)}}\|\Big(1 + \sqrt{\frac{1}{h}}\Big)\sqrt{c_j^{(l)}} \;\leq\; \max_{V_{SF,j+1}(\zeta_{j+1}) \leq c_{j+1}^{(l)}} \{\|\zeta_{j+1}\|\} \tag{1.84}$$

$$Q_{mj}^{(l)} \geq 16\gamma_{j+1}^2 k_{j+1}^{(l)} (D_j^{(l)})^T D_j^{(l)} \tag{1.85}$$

$$Q_{SF,j}^{(s)} \geq 32\gamma_{j+1}^2 k_{j+1}^{(l)} (S_j^{(l)})^T S_j^{(l)} \tag{1.86}$$

and, in addition, for all $l \in (0, l_2^*]$ and whenever

$$V_{SF,i}(\zeta_i) \leq c_i^{(l)} \tag{1.87}$$

$$\|\widetilde{z}_{i+1}\| \leq \|R_{i1}^{-1} B_{i2}^T \sqrt{P_{SF,i}^{(l)}}\|\Big(1 + \sqrt{\frac{1}{h}}\Big)\sqrt{c_i^{(l)}} \tag{1.88}$$

for $i = 1, \ldots, n,$

$$2E_{js}^T(z, \widetilde{z}) E_{js}(z, \widetilde{z}) + \frac{4\gamma_j^2}{\gamma_{j-1}^2} \widetilde{E}_{js}^T(z, \widetilde{z}) \widetilde{E}_{js}(z, \widetilde{z})$$

$$\leq \begin{cases} \frac{k_{s-1}^{(l)} \cdots k_{j+1}^{(l)} R_{s-1,1}}{2^{n-j+1}} & \text{if } 2 \leq j \leq s-2, 4 \leq s \leq n+1 \\ \frac{R_{s-1,1}}{4} & \text{if } j = s-1, 3 \leq s \leq n+1 \end{cases} \quad (1.89)$$

$$2E_{1s}^T(z, \widetilde{z}) E_{1s}(z, \widetilde{z}) \leq \begin{cases} \frac{k_{s-1}^{(l)} \cdots k_2^{(l)} R_{s-1,1}}{2^n} & \text{if } 3 \leq s \leq n+1 \\ \frac{R_{11}}{4} & \text{if } s = 2 \end{cases}$$

$$(1.90)$$

$$\frac{16\gamma_j^2}{\gamma_{j-1}^2} \widetilde{E}_{jj}(z, \widetilde{z}) \widetilde{E}_{jj}^T(z, \widetilde{z}) \leq \frac{R_{j-1,1}}{k_j^{(l)}} \qquad j = 2, \ldots, n \quad (1.91)$$

$$\frac{16\gamma_j^2}{\gamma_{j-1}^2} \widetilde{N}_{jj}^T(z, \widetilde{z}) \widetilde{N}_{jj}(z, \widetilde{z}) \leq Q_{SF,j}^{(l)}, \qquad j = 2, \ldots, n \quad (1.92)$$

then there exists a C^0 *function* $c^{(\cdot)} : (0,1] \to \mathbb{R}^+$ *and a positive definite quadratic function* $V(z_1, \sigma_1, \ldots, z_n, \sigma_n)$ *such that for each* $\delta \in \mathbb{R}^+$ *there exist* $l^* \in (0,1]$ *such that for all* $l \in (0, l^*]$

(B1) $\Omega^{(l)} \supset \mathcal{S}_1 \times \mathcal{W}_1 \times \cdots \times \mathcal{S}_n \times \mathcal{W}_n,$ *where*

$$\Omega^{(l)} \stackrel{\text{def}}{=} \{(v_{11}, v_{12}, \ldots, v_{n1}, v_{n2}) \in \mathbb{R}^{r_1} \times \mathbb{R}^{r_1} \times \cdots \times \mathbb{R}^{r_n} \times \mathbb{R}^{r_n} : \\ V(v_{11}, v_{12}, \ldots, v_{n1}, v_{n2}) \leq c^{(l)}\} \quad (1.93)$$

(B2) $\|\widetilde{z}_{n+1}(t)\| \leq \delta$ *for all* $t \geq 0$ *and the time derivative of* $V(z_1, \sigma_1, \ldots, z_n, \sigma_n)$, *along the trajectories of (1.55)–(1.56), is negative definite whenever*

$$(z_1(t), \sigma_1(t), \ldots, z_n(t), \sigma_n(t)) \in \Omega^{(l)} \quad (1.94)$$

Remark 1.5 Theorem 1.2 gives a sufficient condition for semiglobally stabilizing (1.55) via arbitrarily bounded measurement feedback. Some key features of Theorem 1.2 consist of allowing for block–state equations and uncertainties in the outputs. As it will be shown in the proof of Theorem 1.2, our design procedures end up with *linear* controllers and *quadratic* Lyapunov functions, which can be used together with any systematic design tool such as backstepping or forwarding. It is only a matter of technicalities

([1], [2]) to see that Conditions (A1)–(A4) are satisfied for systems

$$
\begin{aligned}
\dot{z}_1 &= z_2 + p_{11}(z, u, t) \\
\dot{z}_2 &= z_3 + p_{21}(z, u, t) \\
\vdots \ &= \ \vdots \\
\dot{z}_n &= u + p_{n1}(z, u, t) \\
y_j &= z_j + p_{j2}(z, u, t), \qquad j = 1, \ldots, n
\end{aligned}
\tag{1.95}
$$

where $z_j, y, u \in \mathbb{R}$ and $z = col(z_1, \ldots, z_n)$ and $p_{j1}(z, u, t)$ and $p_{j2}(z, u, t)$, $j = 1, \ldots, n$, are *higher order* in $(z_{j+1}, \ldots, z_n, u)$, uniformly with respect to t and $z_1, \ldots, z_j$ (see [18] for state feedback), and, in addition, for systems

$$
\dot{z} = Az + Bu + \tilde{p}_1(z, u, t)u \tag{1.96}
$$
$$
y = z + \tilde{p}_2(z, u, t)u \tag{1.97}
$$

where $\tilde{p}_j(z, 0, t) = 0$, $j = 1, 2$, for all t and z, (A, B) is in Brunowski form, $z \in \mathbb{R}^n$, $y \in \mathbb{R}^n$, $u \in \mathbb{R}$ and $\tilde{p}_1(z, u, t)u$ and $\tilde{p}_2(z, u, t)u$ are *of at least order one and zero*, respectively, with respect to the dilation $\delta_l(z, u) = (l^{1-n}z_1, \ldots, l^{-1}z_{n-1}, z_n, lu)$ and uniformly w.r.t. t (see [13] for definitions and related results for state feedback).

We shortly illustrate how to satisfy Assumptions (A1)–(A4) through the following example. Let us consider

$$
\begin{aligned}
\dot{x}_1 &= x_2 \\
\dot{x}_2 &= \sin x_3 + x_1 v^2 \\
\dot{x}_3 &= v
\end{aligned}
\tag{1.98}
$$

(see the ball and beam example in [17]). Assume that only x_1 and x_3 (angular and linear positions of the ball and beam) are available for feedback. Assume also that the measure of x_1 is affected by some error, which depends on t and x_3, it is bounded with respect to t and it is zero near $x_3 = 0$, uniformly with respect to t. Our model is finally given by

$$
\begin{aligned}
\dot{x}_1 &= x_2 \\
\dot{x}_2 &= \sin x_3 + x_1 v^2 \\
\dot{x}_3 &= v \\
\tilde{y}_1 &= x_1 + \psi_1(x_3, t) \\
\tilde{y}_2 &= x_3
\end{aligned}
\tag{1.99}
$$

with $\psi_1(\cdot, \cdot)$ a C^0 (unknown) function, bounded with respect to the second argument and zero near $x_3 = 0$. Let $z_1 = (x_1, x_2)^T$, $z_2 = x_3$, $\tilde{z}_2 = x_3$ and $\tilde{z}_3 = v$. We have

$$
A_1 = \begin{pmatrix} 0 & 1 \\ 0 & 0 \end{pmatrix}, \quad B_{11} = \begin{pmatrix} 0 & 0 \\ 1 & 0 \end{pmatrix}, \quad B_{12} = \begin{pmatrix} 0 \\ 1 \end{pmatrix}
$$

$$C_{12} = \begin{pmatrix} 1 & 0 \end{pmatrix}, \; \tilde{C}_{11} = \begin{pmatrix} 0 & 1 \end{pmatrix} \tag{1.100}$$

$$A_2 = 0, \; B_{21} = \begin{pmatrix} 1 & 0 \end{pmatrix}, \; B_{22} = 1, \; C_{22} = 1, \; \tilde{C}_{21} = \begin{pmatrix} 0 & 1 \end{pmatrix} \tag{1.101}$$

$$\tilde{\Psi}_1(z, \tilde{z}, t) - K_1 \tilde{z}_2 = \begin{pmatrix} \sin x_3 - \tilde{z}_2 + x_1 \tilde{z}_3^2 \\ \psi_1(x_3, t) \end{pmatrix}$$

$$\tilde{\Psi}_2(z, \tilde{z}, t) - K_2 \tilde{z}_3 = \begin{pmatrix} 0 \\ 0 \end{pmatrix}$$

Let $\gamma_1, \gamma_2 > 1$, $0 < R_{j1} < \gamma_j^2$, $j = 1, 2$, and

$$Q_{SF,1}^{(l)} = Q_{OI,1}^{(l)} = \begin{pmatrix} l^2 & 0 \\ 0 & l \end{pmatrix}$$

Moreover, define

$$\lambda_i = \frac{1}{R_{i1}} - \frac{1}{\gamma_i^2}, \qquad i = 1, 2$$

It is to see that

$$P_{SF,1}^{(l)} = \begin{pmatrix} l\sqrt{l\left(1 + \frac{2}{\sqrt{\lambda_1}}\right)} & \frac{l}{\sqrt{\lambda_1}} \\ \frac{l}{\sqrt{\lambda_1}} & \sqrt{\frac{l\left(1 + \frac{2}{\sqrt{\lambda_1}}\right)}{\lambda_1}} \end{pmatrix}$$

$$P_{OI,1}^{(l)} = \begin{pmatrix} 2l\sqrt{3l} & -2l \\ -2l & \sqrt{3l} \end{pmatrix}, \; R_{12}^{(l)} = \frac{1}{5l^2}$$

and

$$Q_{SF,2}^{(l)} = \lambda_2 (P_{SF,2}^{(l)})^2 \tag{1.102}$$

$$Q_{OI,2}^{(l)} = (R_{22}^{(l)})^{-1} - (P_{OI,2}^{(l)})^2 \tag{1.103}$$

$$P_{OI,2}^{(l)} = \frac{2 P_{SF,2}^{(l)}}{\gamma_2^2} \tag{1.104}$$

$$R_{22}^{(l)} = \gamma_2^{-2} R_{21} (P_{OI,2}^{(l)})^{-2} \tag{1.105}$$

satisfies (A1)–(A2) with $P_{SF,2}^{(l)}$ to be specified later.

Assume that $|\psi_1(x_3, t)| \leq k|x_3|^h$, $h \geq 2$. Let $\delta_1, \delta_2 \in \mathbb{R}^+$ be such that

$$\left\| \begin{pmatrix} \sin(\zeta_2 + \tilde{z}_2) - \sin \tilde{z}_2 + \sin \tilde{z}_2 - \tilde{z}_2 + x_1 \tilde{z}_3^2 \\ \psi_1(x_3, t) \end{pmatrix} \right\|^2$$
$$\leq \frac{1}{\gamma_1^2} \left[N_{12}^2(\zeta_2, \tilde{z}_3)\zeta_2^2 + E_{12}^2 \tilde{z}_2^2 + E_{13}^2(z_1, \tilde{z}_3)\tilde{z}_3^2 \right] \tag{1.106}$$

$$\|R_{11}^{-1} P_{SF,1}^{(l)} G_1^{(l)}\|^2 \psi_1^2(x_3, t) \leq \frac{1}{\gamma_1^2} \left[\tilde{N}_{22}^2(\zeta_2, \tilde{z}_2)\zeta_2^2 + \tilde{E}_{22}^2 \tilde{z}_2^2 \right] \tag{1.107}$$

whenever $|\tilde{z}_j| \leq \delta_j$, $j = 2, 3$, for some C^0 functions $N_{12}(\cdot, \cdot)$ and $\tilde{N}_{22}(\cdot, \cdot)$, with E_{12}, independent of γ_1^2, and with $E_{13}(z_1, \tilde{z}_3) = \sqrt{2}\gamma_1|x_1 \tilde{z}_3|$. Let $c_2^{(l)} = P_{SF,2}^{(l)} h_2$, with $h_2 = 2 \sup\limits_{\zeta_2 \in \tilde{S}_2; \sigma_2 \in \tilde{W}_2} \{\zeta_2^2 + (\zeta_2 - \sigma_2)^2\}$, $\delta_2^{(l)} = \tilde{\delta}_2^{(l)} = 2R_{21}^{-1}\sqrt{h_2} P_{SF,2}^{(l)}$
and

$$k_2^{(l)} = 4 \sup\limits_{|\zeta_2| \leq \sqrt{h_2}; |\tilde{z}_2| \leq \tilde{\delta}_2^{(l)}} \left\{ \frac{\Gamma_{12}^2(\zeta_2, \tilde{z}_2)}{\lambda_2(P_{SF,2}^{(l)})^2} \right\} = \frac{4\Gamma_{12}^2}{\lambda_2(P_{SF,2}^{(l)})^2}$$

where $\Gamma_{12}(\cdot, \cdot) = N_{12}(\cdot, \cdot)$. Moreover, let $c_1^{(l)} = c_2^{(l)} k_2^{(l)}$, $\delta_1^{(l)} = 2\|R_{11}^{-1} B_{12}^T\|\sqrt{P_{SF,1}^{(l)}}\|\sqrt{c_1^{(l)}}$ and $\tilde{\delta}_1^{(l)} = \sqrt{\frac{c_2^{(l)}}{P_{SF,2}^{(l)}}}$.

(A4) can be met as long as γ_j, R_{j1}, $\delta_j^{(\cdot)}$, $R_{j2}^{(\cdot)}$, $j = 1, 2$, $\tilde{N}_{22}(\cdot, \cdot)$, $E_{13}(\cdot, \cdot)$, E_{12}, $\tilde{E}_{22}$ and $P_{SF,2}^{(l)}$ are such that

$$\gamma_j^2 > R_{j1} \tag{1.108}$$

$$0 < \gamma_1^2 P_{OI,1}^{(l)} - P_{SF,1}^{(l)} \tag{1.109}$$

$$R_{j2} > 1 \tag{1.110}$$

$$\delta_j^{(l)} \leq \min\{\delta_j, \tilde{\delta}_j^{(l)}\} \tag{1.111}$$

$$E_{13}^2(z_1, \tilde{z}_3) \leq \frac{R_{21} k_2^{(l)}}{8} \tag{1.112}$$

$$\tilde{E}_{22}^2 \leq \frac{\gamma_1^2 R_{11}}{16\gamma_2^2 k_2^{(l)}} \tag{1.113}$$

$$E_{12}^2 \leq \frac{R_{11}}{8} \tag{1.114}$$

$$\tilde{N}_{22}^2(\zeta_2, \tilde{z}_2) \leq \frac{\lambda_2 \gamma_1^2 (P_{SF,2}^{(l)})^2}{16\gamma_2^2} \tag{1.115}$$

$$Q_{m1}^{(l)} \geq 16\gamma_2^2 k_2^{(l)} (D_1^{(l)})^T D_1^{(l)} \tag{1.116}$$

$$Q_{SF,1}^{(s)} \geq 32\gamma_2^2 k_2^{(l)} (S_1^{(l)})^T S_1^{(l)} \tag{1.117}$$

for all $l \in (0, l^*]$, $l^* \in (0, 1]$, and whenever

$$|x_1| \leq \sqrt{c_1^{(l)} \frac{p_{SF,22}^{(l)}}{p_{SF,11}^{(l)} p_{SF,22}^{(l)} - (p_{SF,12}^{(l)})^2}}, \tag{1.118}$$

$|\zeta_2| \leq \sqrt{h_2}$, $|\tilde{z}_2| \leq \delta_1^{(l)}$ and $|\tilde{z}_3| \leq \delta_2^{(l)}$, with

$$p_{11}^{(l)} = l\sqrt{l\left(1 + \frac{2}{\sqrt{\lambda_1}}\right)} \tag{1.119}$$

$$p_{12}^{(l)} = \frac{l}{\sqrt{\lambda_1}} \tag{1.120}$$

$$p_{22}^{(l)} = \sqrt{\frac{l\left(1 + \frac{2}{\sqrt{\lambda_1}}\right)}{\lambda_1}} \tag{1.121}$$

Note that (1.108)–(1.117) impose that $P_{SF,2}^{(l)} = \mu\sqrt{l}$ for some $\mu \in \mathbb{R}^+$.

4   REFERENCES

[1] S. Battilotti, More results on the semiglobal stabilization of uncertain nonlinear systems via measurement feedback, *Proc. of the Conference on Decision and Control (CDC'98')*, 1998.

[2] S. Battilotti, Semiglobal stabilization of uncertain block–feedforward systems via measurement feedback, NOLCOS, 1–3 July 1998.

[3] J. C. Doyle, K. Glover, P. P. Khargonekar and B. A. Francis, State space solutions to standard $\mathcal{H}_2$ and $\mathcal{H}_\infty$ control problems, *IEEE Trans. Autom. Contr.*, **34**, 831-847, 1989.

[4] F. Esfandiari and H. K. Khalil, Output feedback stabilization of fully linearizable systems, *Internat. Journ. Contr.*, **56**, 1007-1037, 1992.

[5] H. K. Khalil and F. Esfandiari, Semiglobal stabilization of a class of nonlinear system using output feedback, *IEE Trans. Autom. Contr.*, **38**, 1412-1415, 1993.

[6] P. Khargonekar, I. R. Petersen and K. Zhou, Robust stabilization of uncertain linear systems: quadratic stabilizability and H_∞ theory *IEEE Trans. Autom. Contr.*, **35**, 356-361, 1989.

[7] P. Kokotovic and R. Marino, On vanishing stability regions in nonlinear systems with high gain feedback, *IEEE Trans. Autom. Contr.*, **31**, 967-970, 1986.

[8] Z. Lin and A. Saberi, Robust semiglobal stabilization of minimum ohase input output linearizable systems via partial state and output feedback, *IEEE Trans. Autom. Contr.*, **40**, 1029-1042, 1996.

[9] S. Morse, Structural invariants of linear multivariable systems, *SIAM Journ. Contr. Optim.*, **11**, 446–465, 1973.

[10] A. Saberi, Z. Lin and A. Teel, Control of Linear systems with saturating actuators, *IEEE Trans. Autom. Contr.*, **41**, 368-378, 1996.

[11] Semiglobal robust regulation of nonlinear systems, *Colloquium on Automatic Control, Lect. Notes in Contr. and Inform. Sciences*, **215**, 27-55, 1997.

[12] H. J. Sussmann and P. V. Kokotovic, The peaking phenomenon and the global stabilization of nonlinear systems, *IEEE Trans. Autom. Contr.*, **36**, 424-439, 1991.

[13] R. Sepulchre, Slow peaking and low–gain designs for global sstabilization of nonlinear systems, *36th IEEE Conf. Dec. Contr.*, San Diego, CA, December 1997.

[14] A. Teel and L. Praly, Tools for semiglobal stabilization by partial state and output feedback, *SIAM Journ. Contr. Optim.*, **33**, 1443-1488, 1995.

[15] A. Teel and L. Praly, Global stabilizability and observability imply semiglobal stabilizability by output feedback, *Syst. Contr. Lett.*, **22**, 313-325, 1994.

[16] A. Teel, Semiglobal stabilizability of linear null controllable systems with input nonlinearities, *IEEE Trans. Autom. Contr.*, **40**, 96-100, 1995.

[17] A. Teel, Semiglobal stabilization of the ball and beam using output feedback, *Am. Contr. Conf.*, 1993.

[18] A. Teel, Using saturation to stabilize a class of single input partially linear composite systems, *Proc. IFAC NOLCOS'92*, 232-237, 1992.

Observer-Controller Design for Global Tracking of Nonholonomic Systems

Zhong-Ping Jiang[1] and Henk Nijmeijer[2, 3]

[1]Department of Electrical Engineering, Polytechnic University, Six Metrotech Center, Brooklyn, NY 11201, U.S.A.
[2]Faculty of Mathematical Sciences, Dept. of Systems, Signals and Control, University of Twente, P.O. Box 217, 7500 AE Enschede, The Netherlands.
[3]Faculty of Mechanical Engineering, Eindhoven University of Technology, PO Box 513, 5600 MB Eindhoven, The Netherlands.

1 Introduction

Recent progress on the control of nonholonomic dynamic systems has been summarized in several excellent surveys – see, for instance, [16, 4, 2]. As it is seen from the rather complete list of numerous references in those survey papers, this nonlinear control problem has attracted attention of many researchers. Most of these papers have been devoted to the asymptotic feedback stabilization of some important classes of controllable nonholonomic systems. As it is well-known in the nonlinear control community, according to Brockett's necessary condition for feedback stabilization [1], there is no C^1 (or, even, C^0) state-dependent feedback which can asymptotically stabilize a system in these classes at the origin. As a consequence, the linearization of the system is not controllable, although the original nonlinear system itself is controllable (in the *nonlinear* sense). Several novel nonlinear control designs have been proposed ranging from time-varying smooth or nonsmooth techniques to discontinuous feedback strategies (see [2, 4, 16] for relevant references). When turning our attention to the tracking problem, the technical difficulty appears to be more tractable because the linearization around the reference trajectory often is uniformly controllable. As a matter of fact, this idea has been pursued in a number of recent work – see, for instance, [13, 22, 23, 6, 2]. Some other approaches are also proposed in [17, 5].

The purpose of this article is to continue our recent work on the state-feedback tracking control of a class of nonholonomic systems in chained form [8, 9, 10], where semiglobal and global tracking results were obtained beyond local results pertaining to the idea of linearization. Here, instead of assuming full-state information, we will study the problem of global output-feedback tracking.

The reasons why we are particularly interested in nonholonomic systems in chained form are twofold. Firstly, the chained form has proved useful in modelling the kinematics of many nonlinear mechanical systems with nonholonomic constraints. This has been shown in the pioneering paper [20], where this class of systems was originally introduced, and in papers by several other authors where the control of nonholonomic systems in chained form, or extended chained form, is the main issue (see [16] for the details). Secondly, nonholonomic systems in chained form appear to be the largest class of nonholonomic dynamic systems in the literature for which systematic controller design methods are available.

Like in our previous papers [11, 7, 8, 9, 10], the main recursive technique we will invoke in this paper is the integrator backstepping which has found applications in several control problems. We refer the reader to [18] for an expository introduction to this technique and several original references.

Our control design procedure is divided into three parts:

1. design a Luenberger-like reduced-order time-varying observer;

2. introduce an appropriate system of coordinates and apply backstepping to design an observer-based output-feedback control law u_2;

3. we invoke either Jurdjevic-Quinn method or a cascade design idea to complete the design of the other control input u_1.

It is of interest to note that we are able to allow any saturation level on u_1. In a companion paper [19] of the second author, an alternative approach was independently developed on the basis of the main results of [3, 21].

The rest of this chapter is presented as follows. Section 2 briefly formulates the problem of global output-feedback tracking. Then, we introduce a reduced-order time-varying observer in Section 3. In Section 4, two above-mentioned methods involving the application of backstepping are presented to design desired output-feedback control laws. In Section 5, an extension of our results to the extended chained-form, or the so-called dynamic model, is discussed in detail with the help of a simple benchmark nonholonomic knife-edge example. The simulation results demonstrate the effectiveness of our methodology. Finally, we close this article with some concluding remarks and follow-up problems for future work.

2   Problem Statement

The problem we deal with in this paper is the output-feedback global tracking for a class of nonholonomic mechanical systems which can be transformed into chained form

$$\dot{x}_1 \;=\; u_1$$
$$\dot{x}_2 \;=\; u_2$$
$$\dot{x}_3 \;=\; x_2 u_1$$
$$\vdots$$
$$\dot{x}_n \;=\; x_{n-1} u_1$$
$$y \;=\; (x_1, x_n)^T$$

(2.1)

where $x = (x_1, \ldots, x_n)$ is the state, y is the output, $u = (u_1, u_2)$ is the control input with u_1 satisfying the following constraint

$$|u_1(t)| \;\leq\; u_{1\,\mathrm{max}}$$

(2.2)

with $u_{1\,\mathrm{max}}$ a positive constant.

More precisely, given a desired reference path $x_d(t)$ described by

$$\dot{x}_{1d} \;=\; u_{1d}$$
$$\dot{x}_{2d} \;=\; u_{2d}$$
$$\dot{x}_{3d} \;=\; x_{2d} u_{1d}$$
$$\vdots$$
$$\dot{x}_{nd} \;=\; x_{(n-1)d} u_{1d}$$

(2.3)

with $u_d = (u_{1d}, u_{2d})$ as the bounded and continuous time-varying reference input, we wish to design a dynamic output-feedback law

$$u \;=\; \mu(t, y, \chi), \quad \dot{\chi} = \nu(t, y, \chi)$$

(2.4)

which forces $x(t)$ asymptotically track $x_d(t)$ as $t \to +\infty$.

We want to find a systematic way to construct such an output-feedback law (2.4) for any given $u_{1\,\mathrm{max}} > \sup_{t \geq 0} |u_{1d}(t)|$. It should be mentioned that the time-dependence in (2.4) comes from the reference signals $u_d(t)$ and $x_d(t)$.

In terms of differentially flat systems theory [6], the output $y = (x_1, x_n)$ is a *flat* output of system (2.1). Clearly, this is the minimal number of state components we require to know so as to solve the output-feedback tracking problem. For many physical systems which can be brought into a form of (2.1) (see, for instance, [20, 16]), u_1 takes the meaning of a linear velocity; see also the knife-edge example in Section 5. So the condition (2.2) imposes a practically meaningful constraint on the input u_1. On the other hand, the saturation condition (2.2) will also help a successful observer design, as shown in the next section.

3 Reduced-Order Observer

The purpose of this section is to introduce a Luenberger-type "observer" for the unmeasured states of the linear time-varying (LTV) $(x_2, \ldots, x_n)$-system of (2.1) if we regard u_1 as a time-varying signal and x_n as the output. Obviously, there does not exist an asymptotic observer — which asymptotically recovers the unmeasured states $(x_2, \ldots, x_{n-1})$ — irrespective of the time-varying function $u_1(t)$. Roughly speaking, this LTV system is uniformly observable only if $u_1(t)$ is persistently exciting. For instance, a null function $u_1(t)$ renders the unmeasured states unobservable. With this observation in mind, our observer yields an asymptotic estimator for the unmeasured states $(x_2, \ldots, x_{n-1})$ whenever the input u_1 is chosen appropriately. A similar idea has recently appeared in the independent work [14] on local observer designs.

In order to design a reduced-order observer, introduce the new variables

$$
\begin{aligned}
\xi_1 &= x_{n-1} - k_1(t)x_n \\
\xi_2 &= x_{n-2} - k_2(t)x_n \\
&\vdots \\
\xi_{n-2} &= x_2 - k_{n-2}(t)x_n
\end{aligned}
\tag{2.5}
$$

where every k_i is a real-valued function of time t to be chosen later.

Differentiating these variables with respect to time yields

$$
\begin{aligned}
\dot{\xi}_i &= \xi_{i+1}u_1 + k_{i+1}x_nu_1 - k_i(\xi_1 + k_1x_n)u_1 - \dot{k}_ix_n, \quad 1 \le i \le n-3 \\
\dot{\xi}_{n-2} &= u_2 - k_{n-2}(\xi_1 + k_1x_n)u_1 - \dot{k}_{n-2}x_n
\end{aligned}
\tag{2.6}
$$

This leads us to introduce the minimal-order observer of the type

$$
\begin{aligned}
\dot{\hat{\xi}}_i &= \hat{\xi}_{i+1}u_1 + k_{i+1}x_nu_1 - k_i(\hat{\xi}_1 + k_1x_n)u_1 - \dot{k}_ix_n, \quad 1 \le i \le n-3 \\
\dot{\hat{\xi}}_{n-2} &= u_2 - k_{n-2}(\hat{\xi}_1 + k_1x_n)u_1 - \dot{k}_{n-2}x_n
\end{aligned}
\tag{2.7}
$$

Denote the observation error $\tilde{\xi} = (\tilde{\xi}_1, \ldots, \tilde{\xi}_{n-2})$ with $\tilde{\xi}_i = \xi_i - \hat{\xi}_i$ for $1 \le i \le n-2$.

From (2.6) and (2.7), we have

$$
\dot{\tilde{\xi}} = u_1
\begin{pmatrix}
-k_1 & 1 & 0 & \cdots & 0 \\
-k_2 & 0 & 1 & \cdots & 0 \\
\vdots & \vdots & \vdots & \vdots & \vdots \\
-k_{n-3} & 0 & 0 & \cdots & 1 \\
-k_{n-2} & 0 & 0 & \cdots & 0
\end{pmatrix}
\tilde{\xi}
\tag{2.8}
$$

For a given C^0 bounded function $u_{1d}(t)$ (which will be understood as the reference input signal in Section 4), define $\hbar_1$ as $\hbar_1(t) = 1$ if $u_{1d}(t) \ge 0$ and

$\hbar_1(t) = -1$ otherwise. Using this notation, (2.8) is rewritten as

$$\dot{\tilde{\xi}} = (|u_{1d}| + \hbar_1(u_1 - u_{1d})) \begin{pmatrix} -k_1\hbar_1 & \hbar_1 & 0 & \cdots & 0 \\ -k_2\hbar_1 & 0 & \hbar_1 & \cdots & 0 \\ \vdots & \vdots & \vdots & \vdots & \vdots \\ -k_{n-3}\hbar_1 & 0 & 0 & \cdots & \hbar_1 \\ -k_{n-2}\hbar_1 & 0 & 0 & \cdots & 0 \end{pmatrix} \tilde{\xi}$$

$$:= (|u_{1d}| + \hbar_1(u_1 - u_{1d})) A(t)\tilde{\xi} \tag{2.9}$$

Notice that A becomes a time-invariant matrix if the sign of $u_{1d}(t)$ remains unchanged and the k_i's are constant. In this case, it is easy to select the k_i's to make A asymptotically stable with $\lambda(A)$ less than any prescribed stability margin $\lambda^* < 0$. This situation happens when the virtual reference object moves either forward or backward along the reference trajectory x_d (cf. [13, 8]).

When the sign of $u_{1d}(t)$ is not constant, we can pick the design functions k_i's appropriately so that $A(t)$ is asymptotically stable. To this purpose, notice that $\hbar_1$ takes either the value of 1 or -1. Let $K = (k_1, \ldots, k_{n-2})$ and let (K_+, K_-) be a pair of vectors which render $A(t)$ asymptotically stable corresponding to the case when $\hbar_1 = 1$ and $\hbar_1 = -1$, respectively. Given a constant matrix $Q = Q^T > 0$ and any $t \geq 0$, there is a unique matrix $P(t) = P^T(t) > 0$ such that

$$A(t)^T P(t) + P(t)A(t) = -Q \tag{2.10}$$

In particular, $P(t) = P_+$ if $u_{1d}(t) \geq 0$ and $P(t) = P_-$ if $u_{1d}(t) < 0$, with $P_+ > 0$ and $P_- > 0$ solving the Lyapunov matrix equation (2.10) for $(K, \hbar_1) = (K_+, 1)$ and $(K, \hbar_1) = (K_-, -1)$, respectively.

The following result shows that our reduced-order observer produces an asymptotic estimate for the unmeasured states $(x_2, \ldots, x_{n-1})$ of system (2.1), provided that $u_1(t)$ is persistently exciting in the sense of (2.11).

Proposition 2.1 *Let $u_{1d}(t)$ be any bounded C^0 function. If $u_1(t)$ is bounded and there exist two constants $t_0 \geq 0$ and $\gamma = \gamma(t_0) > 0$ such that*

$$\liminf_{t \to \infty} \frac{1}{t} \int_{t_0}^{t_0+t} |u_{1d}(\tau)| d\tau > \gamma \sup_{\tau \geq t_0} |u_1(\tau) - u_{1d}(\tau)| \tag{2.11}$$

and if $\gamma > \lambda_{\max}(P)\lambda_{\max}(Q)\lambda_{\min}^{-1}(P)\lambda_{\min}^{-1}(Q)$, then, for any $\tilde{\xi}(0) \in \mathbb{R}^{n-2}$, the solution $\tilde{\xi}(t)$ of (2.9) exponentially converges to 0 as $t \to \infty$.

Proof. Consider the positive-definite and radially unbounded function

$$V(t, \tilde{\xi}) = \tilde{\xi}^T P(t)\tilde{\xi} \tag{2.12}$$

Noticing that $\dot{P} = 0$ almost everywhere (a.e.), the time derivative of V along the solutions of (2.9) satisfies

$$\dot{V} = (-|u_{1d}(t)| - \hbar_1(u_1(t) - u_{1d}(t))) \, \tilde{\xi}^T Q \tilde{\xi} \qquad \text{a.e.} \qquad (2.13)$$

Then,

$$\dot{V} \le -|u_{1d}(t)| \frac{\lambda_{\min}(Q)}{\lambda_{\max}(P)} V + \frac{\lambda_{\max}(Q)}{\lambda_{\min}(P)} |u_1(t) - u_{1d}(t)| V \qquad \text{a.e.} \qquad (2.14)$$

Using the variation of constants formula and Gronwall-Bellman inequality [15], it follows from (2.14) that, for all $t \ge t_0$,

$$\begin{aligned}
V(t, \tilde{\xi}(t)) \;\le\; & V(t_0, \tilde{\xi}(t_0)) \exp \left(-\frac{\lambda_{\min}(Q)}{\lambda_{\max}(P)} \int_{t_0}^t |u_{1d}(\tau)| d\tau \right. \\
& \left. + \frac{\lambda_{\max}(Q)}{\lambda_{\min}(P)} (t - t_0) \sup_{\tau \ge t_0} |u_1(\tau) - u_{1d}(\tau)| \right) \qquad (2.15)
\end{aligned}$$

In view of (2.11), there exist two positive constants δ and T such that, for all $t \ge T > t_0$,

$$\frac{1}{t - t_0} \int_{t_0}^t |u_{1d}(\tau)| d\tau - \gamma \sup_{\tau \ge t_0} |u_1(\tau) - u_{1d}(\tau)| \ge \delta \qquad (2.16)$$

From (2.15) and (2.16), we have

$$V(t, \tilde{\xi}(t)) \le V(t_0, \tilde{\xi}(t_0)) \exp(-\delta(t - t_0)), \qquad \forall t \ge T \qquad (2.17)$$

which completes the proof. ∎

Remark 2.1 If the left-hand side of (2.11) is positive, this "persistent excitation" condition can be met via two different ways: (i) design a saturated single-input u_1 with the saturation level as small as possible; (ii) design a (not necessarily saturated) control u_1 so that $u_1(t) - u_{1d}(t)$ tends to 0 as $t \to \infty$. In both cases, $\gamma = \gamma(t_0)$ can be made large for some $t_0 > 0$. We develop this idea in the next section.

4   Output-Feedback Design

Using the observer (2.7), we will design a dynamic output-feedback law (2.4) to drive the tracking error $x(t) - x_d(t)$ to zero. To this end, we notice that

$$x_i = \hat{\xi}_{n-i} + \tilde{\xi}_{n-i} + k_{n-i}(t)x_n \qquad \forall 2 \le i \le n - 1 \qquad (2.18)$$

Introduce the new variables

$$\begin{aligned}
\zeta_1 &= x_n - x_{nd} \\
\zeta_i &= \hat{\xi}_{i-1} - (x_{(n-i+1)d} - k_{i-1}x_{nd}) \quad \forall 2 \le i \le n-1 \\
\zeta_n &= x_1 - x_{1d}
\end{aligned} \tag{2.19}$$

Let $\bar{x} = (x_2, \ldots, x_n)$, $\bar{x}_d = (x_{2d}, \ldots, x_{nd})$ and $\bar{\zeta} = (\zeta_1, \ldots, \zeta_{n-1})$. If all signals are bounded and the conditions of Proposition 2.1 hold, then (2.19) together with Proposition 2.1 implies that $\bar{x}(t) - \bar{x}_d(t)$ converges to 0 if and only if $\bar{\zeta}(t)$ converges to 0.

Notice that, when k_i is selected as in Section 3, $\dot{k}_i = 0$ a.e. for all $1 \le i \le n-2$. Differentiating the variable ζ along the solutions of (2.1)-(2.3) yields

$$\begin{aligned}
\dot{\zeta}_1 &= (\zeta_2 + k_1\zeta_1)u_{1d} + \tilde{\xi}_1 u_1 + (\zeta_2 + x_{(n-1)d} + k_1\zeta_1)(u_1 - u_{1d}) \\
&\;\;\vdots \\
\dot{\zeta}_i &= (\zeta_{i+1} + k_i\zeta_1 - k_{i-1}(\zeta_2 + k_1\zeta_1))u_{1d} + [\zeta_{i+1} + k_i\zeta_1 \\
&\quad -k_{i-1}(\zeta_2 + k_1\zeta_1) + x_{(n-i)d} - k_{i-1}x_{(n-1)d}] (u_1 - u_{1d}) \\
&\;\;\vdots \\
\dot{\zeta}_{n-1} &= u_2 - k_{n-2}(\hat{\xi}_1 + k_1 x_n)u_1 - (u_{2d} - k_{n-2}x_{(n-1)d}u_{1d}) \\
\dot{\zeta}_n &= u_1 - u_{1d}
\end{aligned} \tag{2.20}$$

In the sequel, we will apply the backstepping approach to the transformed system (2.20) in order to design two desired output-feedback control laws u subject to (2.2). The first one is based on a combined application of backstepping and Jurdjevic-Quinn methods. The second one is a mixture of backstepping and cascade designs.

4.1 Backstepping-Based Trackers

Noticing the lower-triangular structure of the $(\zeta_1, \ldots, \zeta_{n-1})$-subsystem of (2.20) with u_2 as the input, the backstepping technique will be first applied to design the control u_2. Then, the design of the single-input u_1 is carried out via the Jurdjevic-Quinn method [12].

Step 1: Begin with the ζ_1-subsystem of (2.20) with ζ_2 viewed as the virtual control.

Let $z_1 = \zeta_1$ and write the ζ_1-subsystem in more compact form

$$\dot{z}_1 = (\zeta_2 + \psi_1(\zeta_1))u_{1d} + \tilde{\xi}_1 u_1 + \phi_1(t, \zeta_1, \zeta_2)(u_1 - u_{1d}) \tag{2.21}$$

Consider the quadratic function

$$V_1 = \frac{1}{2}z_1^2 \tag{2.22}$$

We have

$$\dot{V}_1 = z_1 \left(\zeta_2 + \psi_1(\zeta_1) \right) u_{1d} + z_1 \widetilde{\xi}_1 u_1 + z_1 \phi_1(t, \zeta_1, \zeta_2)(u_1 - u_{1d}) \qquad (2.23)$$

Perform the change of variable $z_2 = \zeta_2 - \alpha_1(u_{1d}^{2l-1}, z_1)$ where

$$\alpha_1 = -c_1 u_{1d}^{2l-1} z_1 - \psi_1(\zeta_1) \qquad (2.24)$$

with $c_1 > 0$ a design parameter and $l \geq n - 2$ an integer.
Then, (2.23) implies

$$\dot{V}_1 = -c_1 u_{1d}^{2l} z_1^2 + z_1 z_2 u_{1d} + z_1 \widetilde{\xi}_1 u_1 + z_1 \phi_1(t, \zeta_1, \zeta_2)(u_1 - u_{1d}) \qquad (2.25)$$

For later reference, the z_2-dynamics satisfy

$$\begin{aligned}
\dot{z}_2 &= \left(\zeta_3 + \psi_2(u_{1d}^{2l-1}, u_{1d}^{2l-3}\dot{u}_{1d}, \zeta_1, \zeta_2) \right) u_{1d} + \varphi_2(t)\widetilde{\xi}_1 u_1 \\
&\quad + \phi_2(t, \zeta_1, \zeta_2, \zeta_3)(u_1 - u_{1d}) \qquad (2.26)
\end{aligned}$$

where $\varphi_2(t)$ is dependent on $u_{1d}(t)$ and k_1.

Step i $(2 \leq i \leq n - 2)$: Assume that, at Step $i - 1$, we have designed $i - 1$ virtual control functions α_j $(1 \leq j \leq i - 1)$ and obtained new variables $z_{j+1} = \zeta_{j+1} - \alpha_j(u_{1d}^{2l-1}, \ldots, u_{1d}^{2(l-j)+1}u_{1d}^{(j-1)}, \zeta_1, \ldots, \zeta_j)$. Furthermore, it is assumed that, for all $1 \leq j \leq i$,

$$\begin{aligned}
\dot{z}_j &= \left(\zeta_{j+1} + \psi_j(u_{1d}^{2l-1}, u_{1d}^{2l-3}\dot{u}_{1d}, \ldots, u_{1d}^{2(l-j)+1}u_{1d}^{(j-1)}, \zeta_1, \ldots, \zeta_j) \right) u_{1d} \\
&\quad + \varphi_j(t)\widetilde{\xi}_1 u_1 + \phi_j(t, \zeta_1, \ldots, \zeta_{j+1})(u_1 - u_{1d}) \qquad (2.27)
\end{aligned}$$

With respect to the solutions of (2.27), the time derivative of

$$V_{i-1} = \frac{1}{2}z_1^2 + \cdots + \frac{1}{2}z_{i-1}^2 \qquad (2.28)$$

satisfies

$$\begin{aligned}
\dot{V}_{i-1} &= -\sum_{j=1}^{i-1} c_j u_{1d}^{2l} z_j^2 + z_{i-1} z_i u_{1d} + \sum_{j=1}^{i-1} z_j \varphi_j(t)\widetilde{\xi}_1 u_1 \\
&\quad + \sum_{j=1}^{i-1} z_j \phi_j(t, \zeta_1, \ldots, \zeta_{j+1})(u_1 - u_{1d}) \qquad (2.29)
\end{aligned}$$

where $\varphi_1(t) = 1$.

We wish to prove that the above properties hold for the $(\zeta_1, \ldots, \zeta_i)$-subsystem with ζ_{i+1} considered as the virtual control.

To this end, consider the quadratic function

$$V_i = V_{i-1}(z_1, \ldots, z_{i-1}) + \frac{1}{2}z_i^2 \qquad (2.30)$$

In view of (2.27) and (2.29), differentiating V_i with respect to time yields

$$\dot{V}_i = -\sum_{j=1}^{i-1} c_j u_{1d}^{2l} z_j^2 + \sum_{j=1}^{i-1} z_j \varphi_j(t) \tilde{\xi}_1 u_1 + \sum_{j=1}^{i-1} z_j \phi_j(u_1 - u_{1d}) + z_{i-1} z_i u_{1d}$$

$$+ z_i \left(\zeta_{i+1} + \psi_i \right) u_{1d} + z_i \varphi_i \tilde{\xi}_1 u_1 + z_i \phi_i (u_1 - u_{1d}) \qquad (2.31)$$

Letting $z_{i+1} = \zeta_{i+1} - \alpha_i(u_{1d}^{2l-1}, \ldots, u_{1d}^{2(l-i)+1} u_{1d}^{(i-1)}, \zeta_1, \ldots, \zeta_i)$ where

$$\alpha_i = -c_i u_{1d}^{2l-1} z_i - z_{i-1} - \psi_i(u_{1d}^{2l-1}, \ldots, u_{1d}^{2(l-i)+1} u_{1d}^{(i-1)}, \zeta_1, \ldots, \zeta_i) \qquad (2.32)$$

with $c_i > 0$, it follows from (2.31) that

$$\dot{V}_i = -\sum_{j=1}^{i} c_j u_{1d}^{2l} z_j^2 + z_i z_{i+1} u_{1d} + \sum_{j=1}^{i} z_j \varphi_j \tilde{\xi}_1 u_1 + \sum_{j=1}^{i} z_j \phi_j (u_1 - u_{1d})$$

$$\qquad (2.33)$$

Step $n - 1$: At this step, we concentrate on the design of the true control u_2. Consider the Lyapunov function candidate

$$V_{n-1} = V_{n-2}(z_1, \ldots, z_{n-2}) + \frac{1}{2} z_{n-1}^2 \qquad (2.34)$$

From Step $n - 2$ and (2.20), it holds

$$\dot{V}_{n-1} = -\sum_{j=1}^{n-2} c_j u_{1d}^{2l} z_j^2 + z_{n-2} z_{n-1} u_{1d} + \sum_{j=1}^{n-1} z_j \varphi_j \tilde{\xi}_1 u_1$$

$$+ \sum_{j=1}^{n-1} z_j \phi_j (u_1 - u_{1d}) + z_{n-1} \left(u_2 + \psi_{n-1} \right) \qquad (2.35)$$

where ψ_{n-1} is a function dependent on $(t, u_1, \zeta_1, \ldots, \zeta_{n-1})$, ϕ_{n-1} is a function of $(t, \zeta_1, \ldots, \zeta_{n-1})$ and φ_{n-1} is a function of $u_{1d}(t)$ and its derivatives up to order $n - 2$.

By choosing the control law u_2 as

$$u_2 = -c_{n-1} u_{1d}^{2l} z_{n-1} - z_{n-2} u_{1d} - \psi_{n-1}(t, u_1, \zeta_1, \ldots, \zeta_{n-1}) \qquad (2.36)$$

we obtain

$$\dot{V}_{n-1} = -\sum_{j=1}^{n-1} c_j u_{1d}^{2l} z_j^2 + \sum_{j=1}^{n-1} z_j \varphi_j \tilde{\xi}_1 u_1 + \sum_{j=1}^{n-1} z_j \phi_j (u_1 - u_{1d}) \qquad (2.37)$$

Step n : In order to design a control law for u_1, let us consider the Lyapunov function

$$V_n = V_{n-1}(z_1, \ldots, z_{n-1}) + \frac{1}{2} \zeta_n^2 \qquad (2.38)$$

Along the solutions of (2.20), with (2.37), the time derivative of V_n satisfies

$$\dot{V}_n = -\sum_{j=1}^{n-1} c_j u_{1d}^{2l} z_j^2 + \sum_{j=1}^{n-1} z_j \varphi_j \tilde{\xi}_1 u_1 + \left(\zeta_n + \sum_{j=1}^{n-1} z_j \phi_j \right) (u_1 - u_{1d}) \quad (2.39)$$

This leads us to choose the control law u_1 as

$$u_1 = u_{1d} - \sigma \left(\zeta_n + \sum_{j=1}^{n-1} z_j \phi_j \right) \quad (2.40)$$

where σ in C^1 is a saturation function such that $\sigma(r) = r$ for small signals r, $r\sigma(r) > 0$ for all $r \in \mathbb{R} \setminus \{0\}$ and $\sup_{r \in \mathbb{R}} |\sigma(r)| = \sigma_m < \infty$. The saturation level σ_m is selected to meet the input constraint (2.2) with $u_{1\,max} > \sup_{t \geq 0} |u_{1d}(t)| := u_{1d,max}$.

Under this choice (2.40), (2.39) gives

$$\dot{V}_n = -\sum_{j=1}^{n-1} c_j u_{1d}^{2l} z_j^2 + \sum_{j=1}^{n-1} z_j \varphi_j \tilde{\xi}_1 u_1 - \left(\zeta_n + \sum_{j=1}^{n-1} z_j \phi_j \right) \sigma \left(\zeta_n + \sum_{j=1}^{n-1} z_j \phi_j \right)$$
$$(2.41)$$

Finally, we are in a position to formulate the following tracking result.

Proposition 2.2 *Assume that the reference trajectories $x_{id}(t)$ $(2 \leq i \leq n)$ and reference input $u_d(t)$ are bounded. It is further assumed that the derivatives of $u_{1d}(t)$ up to order $n-2$ are bounded on $[0, \infty)$. If there exists a constant $\ell_u > 0$ such that*

$$\liminf_{t \to \infty} |u_{1d}(t)| \geq \ell_u , \quad (2.42)$$

then, σ_m can be tuned towards any level of size $u_{1\,max} - u_{1d,max}$ so that, for any initial tracking error $x(0) - x_d(0) \in \mathbb{R}^n$ and any initial condition $\hat{\xi}(0) \in \mathbb{R}^{n-1}$, the trajectory $(x(t) - x_d(t), \hat{\xi}(t))$ of system (2.1), (2.7), (2.36) and (2.40) is bounded with the following properties

$$\lim_{t \to \infty} |x_i(t) - \hat{\xi}_{n-i}(t) - k_{n-i}(t) x_n(t)| = 0 \quad \forall 2 \leq i \leq n-1 \quad (2.43)$$

$$\lim_{t \to \infty} |x(t) - x_d(t)| = 0 \quad (2.44)$$

Furthermore, the convergence rate in (2.43) and (2.44) is exponential.

Proof. We first prove the boundedness property. By assumptions, we can choose a sufficiently small constant σ_m such that the conditions of Proposition 2.1 hold. As a consequence, the observation error $\tilde{\xi}(t)$ exponentially converges to 0 and thus the property (2.43) is satisfied. We can rewrite the ζ-system (2.20) in more compact form

$$\dot{\overline{\zeta}} = f(t,\overline{\zeta}) \tag{2.45}$$

$$\dot{\zeta}_n = -\sigma\left(\zeta_n + \sum_{j=1}^{n-1} z_j\phi_j\right) \tag{2.46}$$

It is directly checked that f is linear in $\overline{\zeta}$ for any fixed t. Therefore, the closed-loop solutions $\zeta(t) = (\overline{\zeta}(t), \zeta_n(t))$ and $x(t) - x_d(t)$ do not exhibit finite escape.

Given a positive constant ε, by means of the Schwartz inequality, (2.41) gives

$$
\begin{aligned}
\dot{V}_n \leq\ & -\sum_{j=1}^{n-1}(c_j u_{1d}^{2l} - \varepsilon)z_j^2 - \left(\zeta_n + \sum_{j=1}^{n-1} z_j\phi_j\right)\sigma\left(\zeta_n + \sum_{j=1}^{n-1} z_j\phi_j\right) \\
& + \sum_{j=1}^{n-1}\frac{1}{4\varepsilon}|\varphi_j\widetilde{\xi}_1 u_1|^2
\end{aligned} \tag{2.47}
$$

From (2.42), there exist a time instant $t_0 > 0$ and design parameters c_j's such that

$$c_j u_{1d}^{2l}(t) - \varepsilon > 0 \quad \forall t \geq t_0 \tag{2.48}$$

Then, it follows from (2.47) that

$$V_n(z(t)) \leq V_n(z(t_0)) + \sum_{j=1}^{n-1}\int_{t_0}^{t}\frac{1}{4\varepsilon}|\varphi_j\widetilde{\xi}_1 u_1|^2 d\tau \tag{2.49}$$

With the help of Proposition 2.1, (2.49) completes the proof of the first statement since the φ_j's and u_1 are bounded.

Back to the inequality (2.47), using the fact that $\alpha_i = 0$ if $\zeta_1 = \cdots = \zeta_i = 0$, a direct application of Barbălat lemma [15] and Proposition 2.1 yields the last statement (2.44). ∎

4.2   A Modification

In the first step, the Jurdjevic-Quinn method was used to design a controller u_1 to diminish the effect of the $(u_1 - u_{1d})$-related terms on the $(\zeta_1, \ldots, \zeta_{n-1})$-subsystem of (2.20). In this subsection, we pursue the line of a cascade design. That is, we design u_1 in such a way that $x_1(t) - x_{1d}(t)$ converges to zero, regardless of the $(\zeta_1, \ldots, \zeta_{n-1})$-subsystem design. For instance, looking at the ζ_n-subsystem of (2.20), we can simply choose the controller u_1 as

$$u_1 = u_{1d} - \sigma(\zeta_n) \tag{2.50}$$

where σ is a saturation function as defined above.

The global output-feedback tracking result is stated below.

Proposition 2.3 *Assume that the reference trajectories $x_{id}(t)$ ($2 \leq i \leq n$) and reference input $u_d(t)$ are bounded. It is further assumed that the derivatives of $u_{1d}(t)$ up to order $n-2$ are bounded on $[0, \infty)$. If there exist an integer $l \geq n-2$ and a constant $\beta > 0$ such that*

$$\left.\begin{array}{c} \displaystyle\liminf_{t \to \infty} \frac{1}{t} \int_{t_0}^{t_0+t} |u_{1d}(\tau)| d\tau \;>\; 0 \\[4mm] \displaystyle\liminf_{t \to \infty} \frac{1}{t} \int_{t_0}^{t_0+t} |u_{1d}(\tau)|^{2l} d\tau \;>\; 0 \end{array}\right\} \qquad \forall\, t_0 \geq \beta \tag{2.51}$$

then, for any initial tracking error $x(0) - x_d(0) \in \mathbb{R}^n$ and any initial condition $\widehat{\xi}(0) \in \mathbb{R}^{n-1}$, the trajectory $(x(t) - x_d(t), \widehat{\xi}(t))$ of system (2.1), (2.7), (2.36) and (2.50) is bounded. Furthermore, (2.43) and (2.44) hold with exponential convergence.

Proof. As in the proof of Proposition 2.2, we can prove that the closed-loop trajectories do not exhibit finite escape.

Thanks to the choice (2.50), the closed-loop solution $\zeta_n(t)$ satisfies

$$\dot{\zeta}_n = -\sigma(\zeta_n) \tag{2.52}$$

and converges to zero when $t \to \infty$. Moreover, there exist a finite time instant t^o (probably dependent on the initial condition $\zeta_n(0)$), two positive constants p_1 (dependent on the initial condition $\zeta_n(0)$) and p_2 (independent of the initial condition $\zeta_n(0)$) such that

$$|\zeta_n(t)| \leq p_1 \exp(-p_2 t) \quad \forall\, t \geq t^o \tag{2.53}$$

Notice that $t^o = 0$, $p_1 = |\zeta_n(0)|$ and $p_2 = \sigma_0$ if $\sigma(r) = \sigma_0 r$ for $\sigma_0 > 0$.

Without loss of generality, we may assume that $p_1 \exp(-p_2 t)$ is so small for $t \geq t^o$ that $\sigma(\zeta_n(t)) = \zeta_n(t)$ and

$$|u_1(t) - u_{1d}(t)| \leq p_1 \exp(-p_2 t) \quad \forall\, t \geq t^o \tag{2.54}$$

With the aid of (2.54), pick a sufficiently large $t_0 \geq t^o$ so as to check (2.11) and (2.51). As a result, by Proposition 2.1,

$$|\widetilde{\xi}(t)| \leq q_1 \exp(-q_2 t) \quad \forall\, t \geq t_0 \tag{2.55}$$

where $q_1 > 0$ is a constant which depends on $\tilde{\xi}(0)$ and $q_2 > 0$ is a constant which does not depend on $\tilde{\xi}(0)$.

Let us now look at eq. (2.37). By virtue of (2.54) and (2.55), noticing the fact that every ϕ_j is overbounded by $a_{i1} + a_{i2}|z|$ with (a_{i1}, a_{i2}) a pair of positive constants, there exist positive constants c, a_1, a_2, b_1 and b_2 such that

$$\dot{V}_{n-1} \leq -cu_{1d}^{2l}V_{n-1}(z) + a(t)V_{n-1}(z) + b(t)V_{n-1}^{\frac{1}{2}}(z) \qquad (2.56)$$

where $a(t)$ and $b(t)$ are two time-varying signals satisfying

$$\left. \begin{array}{ll} |a(t)| & \leq \ a_1 \exp(-a_2 t) \\ |b(t)| & \leq \ b_1 \exp(-b_2 t) \end{array} \right\} \quad \forall t \geq t_0 \qquad (2.57)$$

From (2.56) and (2.57), Proposition 2.3 follows readily. Indeed, (2.56) implies that $W(z) = V_{n-1}^{\frac{1}{2}}(z)$ satisfies

$$\dot{W} \leq -0.5cu_{1d}^{2l}W(z) + 0.5a(t)W + 0.5b(t) \qquad (2.58)$$

The rest of the proof goes like the proof of Proposition 2.1.   ∎

Remark 2.2 It should be mentioned that Proposition 2.3 can be extended to a simplified dynamic extension of (2.1). That is, it is composed of (2.1) and two integrators $\dot{u}_1 = v_1$, $\dot{u}_2 = v_2$, with (v_1, v_2) considered as the new control. In the next section, we illustrate this extension via a simple benchmark mechanical system with nonholonomic constraints.

5   Example: A Knife-Edge

We illustrate the presented output-feedback design methodology with the help of a simple knife-edge, which moves on the plane and has often served as an elementary illustrative example for theoretic studies on nonholonomic control systems. We refer the reader to [17] for details on this system.

The knife-edge dynamics satisfy the following differential equations [17]:

$$\begin{aligned} \ddot{x}_c &= \frac{\gamma}{m}\sin\phi + \frac{\tau_1}{m}\cos\phi \\ \ddot{y}_c &= -\frac{\gamma}{m}\cos\phi + \frac{\tau_1}{m}\sin\phi \\ \ddot{\phi} &= \frac{\tau_2}{I_c} \\ \dot{x}_c\sin\phi &= \dot{y}_c\cos\phi \end{aligned} \qquad (2.59)$$

where (x_c, y_c) denotes the coordinates for the center of mass of the knife-edge, ϕ denotes the heading angle measured from the x-axis, and τ_1 is the pushing force in the direction of the heading angle, τ_2 is the steering torque about the vertical axis through the center of mass. The constants (m, I_c) are the mass and the moment of inertia of the knife-edge respectively, and γ is the scalar constrain multiplier. Note that the fourth-equation in (2.18) represents the nonholonomic constraint on the velocity of the knife-edge system.

It was shown in [9] that, after a suitable transformation of coordinates and state feedback, the system (2.59) can be brought into

$$
\begin{aligned}
\dot{x}_1 &= x_4 \\
\dot{x}_2 &= x_5 \\
\dot{x}_3 &= x_2 x_4 \\
\dot{x}_4 &= v_1 \\
\dot{x}_5 &= v_2
\end{aligned}
\tag{2.60}
$$

As in [17], consider the reference trajectory

$$
\phi^{\mathrm{r}}(t) = t , \quad x_c^{\mathrm{r}}(t) = \sin t , \quad y_c^{\mathrm{r}}(t) = -\cos t \quad \forall t \geq 0 ,
\tag{2.61}
$$

which corresponds to the center of mass of the knife-edge moving along the circle centered at the origin of unit radius with uniform angular rate.

In the transformed x-coordinates, the desired trajectory is:

$$
\begin{aligned}
x_{1d}(t) &= t , \quad x_{2d}(t) = 0 , \quad x_{3d}(t) = 1 , \tag{2.62} \\
x_{4d}(t) &= u_{1d}(t) = 1 , \quad x_{5d}(t) = u_{2d}(t) = 0 . \tag{2.63}
\end{aligned}
$$

For this particular reference trajectory (which is a straight line in the new x-coordinates), a global *state*-feedback tracking control law has been derived in [9] via a recursive approach.

For simulation purposes, we recall that the coordinate and feedback transformation leading to (2.60) is

$$
\begin{aligned}
x_1 &= \phi , \quad x_2 = x_c \cos\phi + y_c \sin\phi \\[4pt]
x_3 &= x_c \sin\phi - y_c \cos\phi , \quad x_4 = \dot{\phi} \\[4pt]
x_5 &= \dot{x}_c \cos\phi + \dot{y}_c \sin\phi + \dot{\phi}(-x_c \sin\phi + y_c \cos\phi) \tag{2.64} \\[4pt]
v_1 &= \frac{\tau_2}{I_c} \\[4pt]
v_2 &= \frac{\tau_1}{m} + \frac{\tau_2}{I_c}(-x_c \sin\phi + y_c \cos\phi) - \dot{\phi}^2(x_c \cos\phi + y_c \sin\phi)
\end{aligned}
$$

For the new system (2.60), we consider $y = (x_1, x_3)$ as the output and assume that the other states (x_2, x_4, x_5) are unavailable to the designer.

We first introduce an observer to reconstruct the unmeasured state x_4. Introduce a new variable $\xi_3 = x_4 - k_3 x_1$ with $k_3 > 0$ a design parameter, which satisfies

$$\dot{\xi}_3 = -k_3 \xi_3 - k_3^2 x_1 + v_1 \tag{2.65}$$

Then, the reduced-order observer is introduced

$$\dot{\hat{\xi}}_3 = -k_3 \hat{\xi}_3 - k_3^2 x_1 + v_1 \tag{2.66}$$

which leads to an exponentially stable linear dynamics $\dot{\tilde{\xi}}_3 = -k_3 \tilde{\xi}_3$ where $\tilde{\xi}_3 = \xi_3 - \hat{\xi}_3$.

Consequently, the unmeasured state $x_4 = \hat{\xi}_3 + k_3 x_1 + \tilde{\xi}_3$ can be exponentially recovered via the observer (2.66).

Next, we turn to the observer design for the unmeasured states (x_2, x_5). Guided by the development in Section 3, introduce the new variables

$$\xi_1 = x_2 - k_1 x_3 , \quad \xi_2 = x_5 - k_2 x_3 \tag{2.67}$$

where $K = (k_1, k_2)$ is a vector of design parameters, which are constant here because the sign of $u_{1d} = x_{4d} = 1$ does not change.

Direct computation yields

$$\begin{aligned} \dot{\xi}_1 &= \xi_2 + k_2 x_3 - k_1(\xi_1 + k_1 x_3)x_4 \\ \dot{\xi}_2 &= v_2 - k_2(\xi_1 + k_1 x_3)x_4 \end{aligned} \tag{2.68}$$

Since x_4 is not measured in the present case of *extended* chained form (2.60), in contrast to the observer (2.7) for the chained form case, the following observer is introduced in which the estimate $\hat{x}_4 := \hat{\xi}_3 + k_3 x_1$ of x_4 is used in place of x_4:

$$\begin{aligned} \dot{\hat{\xi}}_1 &= \hat{\xi}_2 + k_2 x_3 - k_1(\hat{\xi}_1 + k_1 x_3)\hat{x}_4 \\ \dot{\hat{\xi}}_2 &= v_2 - k_2(\hat{\xi}_1 + k_1 x_3)\hat{x}_4 \end{aligned} \tag{2.69}$$

Letting $\tilde{\xi}_1 = \xi_1 - \hat{\xi}_1$ and $\tilde{\xi}_2 = \xi_2 - \hat{\xi}_2$, (2.68) and (2.69) imply

$$\begin{pmatrix} \dot{\tilde{\xi}}_1 \\ \dot{\tilde{\xi}}_2 \end{pmatrix} = \underbrace{\begin{pmatrix} -k_1 & 1 \\ -k_2 & 0 \end{pmatrix}}_{A} \begin{pmatrix} \tilde{\xi}_1 \\ \tilde{\xi}_2 \end{pmatrix} - \begin{pmatrix} k_1 \\ k_2 \end{pmatrix} \left(\hat{\xi}_1(x_4 - 1) + (\hat{\xi}_1 + k_1 x_3)\tilde{\xi}_3 \right) \tag{2.70}$$

Clearly, we can pick two parameters k_1 and k_2 such that A is an asymptotically stable matrix. For simulation use, take $k_1 = 2$ and $k_2 = 1$. The second term in the last brackets of (2.70) is new comparing with the chained form case, a special class of the so-called Chaplygin (kinematic) form. Bare in mind that the first term depends upon x_4 which is a state component and thus, unlike in the case of chained form, is not free to choose. These terms together prevent us from applying Proposition 2.1 in order to conclude the exponential convergence of the observation error $(\tilde{\xi}_1, \tilde{\xi}_2)$. Nevertheless, as we will show below, we can still design an output-feedback control law to achieve the global tracking task.

Before designing such a controller, we introduce a system of coordinates under which our synthesis is developed

$$
\begin{aligned}
\zeta_1 &= x_3 - x_{3d} , \quad \zeta_2 = \hat{\xi}_1 - (x_{2d} - k_1 x_{3d}) \\
\zeta_3 &= \hat{\xi}_2 - (x_{5d} - k_2 x_{3d}) , \quad \zeta_4 = x_1 - x_{1d} \\
\zeta_5 &= \hat{\xi}_3 - (x_{4d} - k_3 x_{1d})
\end{aligned}
\tag{2.71}
$$

Then,

$$
\begin{cases}
\dot{\zeta}_1 &= (\zeta_2 + k_1\zeta_1) + \tilde{\xi}_1 x_4 + (\zeta_2 + k_1\zeta_1)\tilde{\xi}_3 + (\zeta_2 + k_1\zeta_1)(\zeta_5 + k_3\zeta_4) \\
\dot{\zeta}_2 &= \zeta_3 + k_2\zeta_1 - k_1(\zeta_2 + k_1\zeta_1)(1 + \zeta_5 + k_3\zeta_4) \\
\dot{\zeta}_3 &= v_2 - k_2(\zeta_2 + k_1\zeta_1)(1 + \zeta_5 + k_3\zeta_4) \\
\dot{\zeta}_4 &= (\zeta_5 + k_3\zeta_4) + \tilde{\xi}_3 \\
\dot{\zeta}_5 &= v_1 - k_3\zeta_5 - k_3^2\zeta_4
\end{cases}
\tag{2.72}
$$

where

$$
x_4 = 1 + \zeta_5 + k_3\zeta_4 + \tilde{\xi}_3
\tag{2.73}
$$

Notice that the states of system (2.72) are measured and available for feedback design. If $\hat{\xi}(t) - \xi(t)$ goes to 0 as $t \to \infty$, then $\zeta(t)$ converges to 0 if and only if $x(t) - x_d(t)$ does. In other words, we have converted the global output-feedback tracking problem into a global *state-feedback regulation* issue.

In the sequel, the design of our desired dynamic output-feedback controllers v_1 and v_2 will be developed according to the second method proposed in Section 4.2.

First, we observe that the (ζ_4, ζ_5)-subsystem of (2.72) can be easily made GES (globally exponentially stable) at the origin. Indeed, a direct application of integrator backstepping generates a Lyapunov function candidate

$$
W_1 = \frac{1}{2}\zeta_4^2 + \frac{1}{2}(\zeta_5 + k_3\zeta_4 + c_4\zeta_4)^2
\tag{2.74}
$$

where $c_4 > 0$ is a design parameter.

To render $\dot{W}_1$ nonpositive when $\tilde{\xi}_3 = 0$, we are led to choose the control law

$$v_1 = -c_5(\zeta_5 + k_3\zeta_4 + c_4\zeta_4) - \zeta_4 - c_4(\zeta_5 + k_3\zeta_4), \quad c_5 > 0 \tag{2.75}$$

which gives

$$\dot{W}_1 = -c_4\zeta_4^2 - c_5(\zeta_5 + k_3\zeta_4 + c_4\zeta_4)^2 + [\zeta_4 + (k_3 + c_4)(\zeta_5 + k_3\zeta_4 + c_4\zeta_4)]\tilde{\xi}_3$$

Now, consider the quadratic Lyapunov function

$$W_2 = W_1(\zeta_4, \zeta_5) + \frac{1}{2}\tilde{\xi}_3^2 \tag{2.76}$$

Then, we have

$$\begin{aligned}\dot{W}_2 &= -c_4\zeta_4^2 - c_5(\zeta_5 + k_3\zeta_4 + c_4\zeta_4)^2 - k_3\tilde{\xi}_3^2 \\ &\quad +\zeta_4\tilde{\xi}_3 + (k_3 + c_4)(\zeta_5 + k_3\zeta_4 + c_4\zeta_4)\tilde{\xi}_3\end{aligned} \tag{2.77}$$

Hence, $\dot{W}_2$ is negative-definite if $c_4 > 0.5\varepsilon^{-1}$, $c_5 > 0.5\varepsilon^{-1}(k_3 + c_4)^2$ and $k_3 > \varepsilon$, with $\varepsilon > 0$ being arbitrary. In addition, given any $c > 0$, we can select the design parameters c_4, c_5 and k_3 appropriately such that

$$\dot{W}_2(\zeta_4, \zeta_5, \tilde{\xi}_3) \leq -cW_2(\zeta_4, \zeta_5, \tilde{\xi}_3) \tag{2.78}$$

From (2.78) and (2.76), it follows that there exist two positive constants δ_1 and δ_2 such that

$$|(\zeta_4(t), \zeta_5(t), \tilde{\xi}_3(t))| \leq \delta_1|(\zeta_4(0), \zeta_5(0), \tilde{\xi}_3(0))|\exp(-\delta_2 t) \tag{2.79}$$

Next, it remains to design a suitable control law for the input v_2. As above, we approach this goal by an application of backstepping to the $(\zeta_1, \zeta_2, \zeta_3)$-subsystem of (2.72). Without going into details, a direct application of backstepping generates a Lyapunov function

$$V_3 = \frac{1}{2}z_1^2 + \frac{1}{2}z_2^2 + \frac{1}{2}z_3^2 \tag{2.80}$$

where $z_1 = \zeta_1$, $z_2 = \zeta_2 - \alpha_1(\zeta_1)$, $z_3 = \zeta_3 - \alpha_2(\zeta_1, \zeta_2, \zeta_4, \zeta_5)$ and

$$\alpha_1 = -(c_1 + k_1)\zeta_1, \quad c_1 > 0 \tag{2.81}$$

$$\alpha_2 = -c_2 z_2 - z_1 - k_2\zeta_1 - c_1(\zeta_2 + k_1\zeta_1)(1 + \zeta_5 + k_3\zeta_4) \tag{2.82}$$

If we choose the control law

$$\begin{aligned}v_2 &= -c_3 z_3 - z_2 + (k_2 + \frac{\partial\alpha_2}{\partial\zeta_1})(\zeta_2 + k_1\zeta_1)(1 + \zeta_5 + k_3\zeta_4) \\ &\quad +\frac{\partial\alpha_2}{\partial\zeta_2}(\zeta_3 + k_2\zeta_1 - k_1(\zeta_2 + k_1\zeta_1)(1 + \zeta_5 + k_3\zeta_4)) \\ &\quad +\frac{\partial\alpha_2}{\partial\zeta_4}(\zeta_5 + k_3\zeta_4) + \frac{\partial\alpha_2}{\partial\zeta_5}(v_1 - k_3\zeta_5 - k_3^2\zeta_4)\end{aligned} \tag{2.83}$$

with $c_3 > 0$ a design parameter, it follows that

$$
\begin{aligned}
\dot{V}_3 &= -c_1 z_1^2 - c_2 z_2^2 - c_3 z_3^2 + z_1(\zeta_2 + k_1\zeta_1)(\zeta_5 + k_3\zeta_4) - z_3\frac{\partial\alpha_2}{\partial\zeta_4}\tilde{\xi}_3 \\
&\quad + (z_1 + (c_1 + k_1)z_2 - \frac{\partial\alpha_2}{\partial\zeta_1}z_3)(\tilde{\xi}_1 x_4 + (\zeta_2 + k_1\zeta_1)\tilde{\xi}_3) \quad\quad (2.84)
\end{aligned}
$$

Since the matrix A in (2.70) is asymptotically stable, there exists a unique solution $P = P^T > 0$ to the Lyapunov matrix equation

$$
PA + A^T P = -I_2 \quad\quad (2.85)
$$

where I_2 is the 2×2 unit matrix.
    Consider now the quadratic function

$$
V(z_1, z_2, z_3, \tilde{\xi}_1, \tilde{\xi}_2) = V_3(z_1, z_2, z_3) + (\tilde{\xi}_1, \tilde{\xi}_2)P(\tilde{\xi}_1, \tilde{\xi}_2)^T \quad\quad (2.86)
$$

In view of (2.71), (2.73) and (2.84), from (2.79), it follows the existence of a positive constant $\bar{c} > 0$ and two exponentially converging signals $a(t) \geq 0$, $b(t) \geq 0$ such that

$$
\dot{V} \leq -(\bar{c} - a(t))V + b(t) , \quad \forall t \geq 0 \qu\quad (2.87)
$$

From (2.87), like in the proof of Proposition 2.1, we can invoke the variation of constants formula and Gronwall-Bellman inequality to conclude the exponential convergence of $V(z_1(t), z_2(t), z_3(t), \tilde{\xi}_1(t), \tilde{\xi}_2(t))$ and, therefore of the tracking error $x(t) - x_d(t)$, to 0 as t goes to ∞.
    The simulations in Figure 1 were obtained with the following values of design parameters and initial conditions

$$
\begin{aligned}
&k_1 = 2 , \quad k_2 = k_3 = 1 , \quad c_1 = c_2 = c_3 = c_4 = 1 , \quad c_5 = 3 , \\
&\phi(0) = x_c(0) = y_c(0) = 1 , \quad \dot{\phi}(0) = \dot{x}_c(0) = \dot{y}_c(0) = 0.5 , \quad\quad (2.88) \\
&\hat{\xi}_1(0) = \hat{\xi}_2(0) = \hat{\xi}_3(0) = 1 .
\end{aligned}
$$

The responses indicate that the tracking error exponentially converges to 0 under mild control effort.

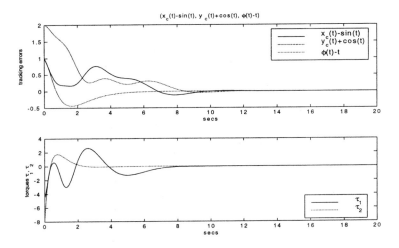

FIGURE 1. Global output-feedback tracking of the knife-edge (2.59).

6   Conclusions and Future Work

The problem of global output-feedback tracking was addressed for a class of nonholonomic systems in this paper. The presented design methodology is a natural extension of our recent state-feedback tracking algorithms proposed in [8, 9, 10]. More specifically, when considering a flat output of a chained-form system in this class as the only accessible measurements, we first design a Luenberger-like reduced-order time-varying observer in order to recover the remaining unmeasured states. Under a condition of persistant excitation on the reference input $u_{1d}(t)$, the observation error was shown to converge to zero at an exponential rate if u_1 is chosen appropriately. Then, based on this observer and using the chained form structure, two constructive methods involving the backstepping technique have been proposed to design desired output-feedback tracking controllers. An extension to the simplified dynamic model was discussed via a simple benchmark nonholonomic knife-edge system.

It is of interest to mention that an arbitrary saturation level can be imposed on the control input u_1. However, we are unable to extend our approaches to cover the case when the other control input u_2 is saturated. The chained form represents a good model for many nonholonomic mechanical systems in the ideal case, that is, when the disturbances are ignored. However, almost all physical systems are subject to some kind of disturbance. It turns out to be necessary to examine the robustness of the global tracking property which was guaranteed by our current trackers. In short, the following problems are meaningful from a practical point of view and deserve our further investigation:

1. What happens if all control inputs of a nonholonomic system in chained-form are subject to some L_∞-type constraints? In relation to the knife-edge example, further difficulties arise here in the boundedness of the controller for the dynamic extension of the considered chained models.

2. In case when uncertainties occur in nonholonomic mechanical systems, how do we give a good mathematical description of these uncertainties? If the nominal system is transformable into a chained form, how will these uncertainties affect the coordinates and feedback transformation and the stability obtained from the undisturbed chained form? We are also interested to know how to modify our proposed tracking controllers in [8, 9, 10] and in this paper in order to maintain stability properties in the presence of uncertainties.

3. Experimental work ought to be done on some laboratory-type robots so as to test the effectiveness of our proposed tracking approaches.

7   References

[1] R. W. Brockett, Asymptotic stability and feedback stabilization, in: R.W. Brockett, R.S. Millman and H.J. Sussmann, eds., *Differential Geometric Control Theory*, pp. 181-191, 1983.

[2] C. Canudas de Wit, B. Siciliano and G. Bastin (Eds), *Theory of Robot Control*. London: Springer-Verlag, 1996.

[3] M.-S. Chen, Control of linear time-varying systems by the gradient algorithm, *Proc. 36th IEEE Conf. Dec. Control*, pp. 4549-4553, San Diego, 1997.

[4] J.-M. Coron, Stabilizing time-varying feedback, *NOLCOS'95*, Tahoe City, CA, pp. 176-183, 1995.

[5] G. Escobar, R. Ortega and M. Reyhanoglu, Regulation and tracking of the nonholonomic double integrator: A field-oriented control approach, *Automatica*, **34**, pp. 125-131, 1998.

[6] M. Fliess, J. Levine, P. Martin and P. Rouchon, Flatness and defect of non-linear systems: introductory theory and examples. *Int. J. Control*, **61**, pp. 1327-1361, 1995.

[7] Z. P. Jiang, Iterative design of time-varying stabilizers for multi-input systems in chained form, *Syst. Contr. Letters*, **28**, pp. 255-262, 1996.

[8] Z. P. Jiang and H. Nijmeijer, Tracking control of mobile robots: a case study in backstepping, *Automatica*, **33**, pp. 1393-1399, 1997.

[9] Z. P. Jiang and H. Nijmeijer, A recursive technique for tracking control of nonholonomic systems in chained form, to appear in: *IEEE Trans. Automat. Control*, Feb. 1999.

[10] Z. P. Jiang and H. Nijmeijer, Backstepping-based tracking control of nonholonomic chained systems, *Proc. European Control Conference*, 1-4 July, 1997, Brussels.

[11] Z. P. Jiang and J.-B. Pomet, Combining backstepping and time-varying techniques for a new set of adaptive controllers, *Proc. 33rd IEEE Conf. Dec. Control*, pp. 2207-2212, Florida, 1994; also in: *Int. J. Adaptive Contr. Signal Processing*, vol. 10, pp. 47-59, 1996.

[12] V. Jurdjevic and J.P. Quinn, Controllability and stability, *J. Diff. Eqs.*, **28**, pp. 381-389, 1979.

[13] Y. Kanayama, Y. Kimura, F. Miyazaki and T. Noguchi, A stable tracking control scheme for an autonomous mobile robot, *Proc. IEEE 1990 Int. Conf. on Robotics and Automation*, pp. 384-389, 1990.

[14] W. Kang and A. J. Krener, Nonlinear observer design, a backstepping approach, preprint, 1998.

[15] H. K. Khalil, *Nonlinear Systems*. Prentice Hall, Upper Saddle River, NJ, 2nd edition, 1996.

[16] I. Kolmanovsky and N. H. McClamroch, Developments in nonholonomic control problems, *IEEE Control Systems Magazine*, Vol. 15, No. 6, pp. 20-36, 1995.

[17] I. Kolmanovsky and N. H. McClamroch, Hybrid feedback laws for a class of cascaded nonlinear control systems, *IEEE Trans. Automat. Control*, **41**, pp. 1271-1282, 1996.

[18] M. Krstić, I. Kanellakopoulos and P. V. Kokotović, *Nonlinear and Adaptive Control Design*. New York: John Wiley & Sons, 1995.

[19] E. Lefeber, A. Robertsson and H. Nijmeijer, Output feedback tracking of nonholonomic systems in chained form, *preprint*, October 1998.

[20] R. M. Murray and S. Sastry, Nonholonomic motion planning: steering using sinusoids, *IEEE Trans. Automat. Contr.*, **38**, pp. 700-716, 1993.

[21] E. Panteley and A. Loria, On global uniform asymptotic stability of nonlinear time-varying systems in cascade, *Systems & Control Letters*, **33**, pp. 131-138, 1998.

[22] C. Samson and K. Ait-Abderrahim, Feedback control of a nonholonomic wheeled cart in Cartesian space, *Proc. of the 1991 IEEE Int. Conf. Robotics and Automation*, Sacramento, pp. 1136-1141, 1991.

[23] G. Walsh, D. Tilbury, S. Sastry, R. Murray and J. P. Laumond, Stabilization of trajectories for systems with nonholonomic constraints, *IEEE Trans. Automat. Contr.*, **39**, pp. 216-222, 1994.

A Separation Principle for a Class of Euler-Lagrange Systems

Antonio Loría[♡] and Elena Panteley[†]

[♡]C.N.R.S., UMR 5228, Laboratoire d'Automatique de Grenoble, ENSIEG, St. Martin d'Hêres, France.
[†]I.N.R.I.A., Rhône Alpes, Projet BIP, St. Martin d'Hêres, France.

1 Introduction

The solution to the *state* feedback tracking control problem of fully damped Euler-Lagrange systems (in particular, rigid-robot manipulators) has been known from many years now – for a literature review, see e.g. [22, 27] –. Nevertheless, a drawback of many of the available results in the literature is that they require the measurement of joint *velocities* which may be contaminated by noise. An *ad hoc* solution, often taken in practice, is to numerically differentiate the joint positions. However, it has been shown experimentally [2] that this method is inefficient for high and slow velocities.

This has motivated researchers in the robotics community to solve the *global* output feedback control problem of robot manipulators. This problem has been open for many years now.

As in the regulation control problem, an approach alternative to numerical differentiation, that has been widely considered in the literature, is to design an observer that makes use of position information to reconstruct the velocity signal. Then, the controller is implemented replacing the velocity measurement by its estimate. Even though the certainty equivalence does not apply for general nonlinear systems, the rationale behind this approach is precisely that the estimate will converge to the true signal, and this should in turn entail stability of the closed loop. As far as we know, some of the earliest works on state estimation for robot manipulators are [20, 17] and some of the references therein. See also [21] for some interesting experimental results. In [17] the authors used a nonlinear observer that reproduces the robot dynamics, in a PD plus gravity compensation scheme. The authors prove the equilibrium is *locally* asymptotically stable provided the observer gain satisfies some lower bound determined by the robot parameters and the trajectories error norms. See also [7] where a sliding mode approach is taken.

The authors of [4] proposed a linear observer-computed torque scheme which exploits the feedback linearizing property of the computed torque scheme providing an efficient tuning technique. Later, using the same tuning idea [3] presented a systematic procedure that exploits the passivity properties of robot manipulators into the design of controller-observer systems to solve both the position and tracking control problems. Local asymptotic stability was proved for sufficiently high gains.

Later in [16], based on a computed torque plus PD-like controller first appeared in [27], we added an n-th order "approximative differentiation filter", to eliminate the necessity of velocity measurements. In that paper we proved semiglobal asymptotic stability of the closed loop system hence showing that the domain of attraction can be arbitrarily enlarged by increasing the filter gain. Some more recent results addressing the same problem are for instance: [11, 18], and [19]. The authors of [11] proposed the first adaptive controller for flexible joint robots by using only position measurements. Simultaneously, [18] proposed a globally asymptotically stable observer-based controller needing only link (position and *velocity*) measurements and later in [19] they extended this result to link position feedback.

The approaches mentioned above, rely on a Lyapunov design, that is, the principal aim is to design an observer and a controller such that, the total time derivative along the closed loop trajectories, be negative definite. A common drawback however, is the appearance of high order terms in the derivative of the Lyapunov function, and which can be dominated only for small states. In the best case, one can prove that the region of attraction can be enlarged for large control gains.

As an attempt to bound the cubic terms in the time derivative of the storage function we presented in [13], as far as we know, the first smooth controller which renders the *one-degree-of-freedom* (dof) EL system. Our approach relies on a computed torque plus PD structure and a nonlinear dynamic extension based on the *linear* approximate differentiation filter. The main innovation in our controller, which allows us to give explicit lower bounds for the controller gains, in order to ensure GUAS, is the use of hyperbolic trigonometric functions in a Lyapunov function with cross terms.

Global uniform asymptotic stability is ensured provided the controller and filter gains satisfy some lower bound depending on the system parameters and the reference trajectory norm. Unfortunately, the performance of our approach can be ensured only for one dof systems and nothing can be claimed for the general multivariable case.

Independently, in [6] Burkov showed by using singular perturbation techniques, that a computed torque like controller plus a linear observer is capable of making a rigid joint robot track a trajectory starting from *any* initial conditions. The main drawback of this result is that no explicit bounds for the observer and control gains can be given. Thus, the author proves in an elegant way, the *existence* of an output feedback tracking controller that ensures GUAS.

Later, A.A.J. Lefeber proposed in [10] an approach which consists on applying a global output feedback *set point* control law (for instance an EL controller) from the initial time t_o until some "switching time" t_s, at which it is supposed that the trajectories are contained in some pre-specified bounded set. At time t_s one switches to a local output feedback tracking control law (such as any among those mentioned above). The obvious drawback of this idea is that the controller is no longer smooth, furthermore, the switching time may depend on bounds on the *unmeasured* variables. The results contained in [10] concern the *existence* of the time instant t_s such that the closed loop system is GUAS.

Most recently, based on the controller of [13] the authors of [28] proposed a dynamic output feedback controller for the multivariable case. The Lyapunov stability proof for the closed loop system is carried out relying on a nonlinear change of coordinates (See Eqs. 35 and 39 of that reference). This change of coordinates is *not* invertible, and therefore the controller the authors propose in [28] is *not* implementable without velocity measurements, for *any* intial conditions of the dynamic extension.

In [5] an elegant alternative result for one-degree-of-freedom systems was reported. The controller proposed in [5] is based upon a *global* nonlinear change of coordinates which makes the system *affine* in the unmeasured velocities. This is crucial to define a very simple controller which has at most *linear* growth in the state variables, as a matter of fact the proposed controller is of a PD+ type. This must be contrasted with the *exponential* growth of the control law proposed in [13], due to the use of hyperbolic trigonometric functions. Hence, from a practical point of view, the controller of [5] supersedes by far that of [13]. The work of [5] suggests that more attention should be payed to the modelling stage of the control design.

As far as we know, the position tracking control problem stated at the beginning of this section for *any* initial conditions and for n-degrees-of-freedom EL systems still remains open. In this chapter we will present a solution to this problem, for a class of n-degrees of freedom EL systems (including robot manipulators). The systems belonging to this class, allow a factorisation which does not exhibit the Coriolis effects in the dynamic model. Inspired by [5, 9, 24], we consider a kinematics model which in other words, provides a global change of coordinates. As it will become clear later, the model considered here covers a fairly wide class of EL systems, however, in general, it is very difficult to find such factorisation.

Our main result is to prove that, for this class of EL systems, it is possible to design a state observer and a state feedback controller *independently* of each other. That is, we will establish a separation principle for a class of EL systems. Our results are an extension of [14] to the tracking problem, i.e., to the time-varying case. In a more general context, some work on separation principles for nonlinear systems has been done recently for local stabilisation of input-output linearizable systems [1] and for the case of

nonaffine systems in [12]. Our results differ from those of in the latter
references in that, neither high gains nor bounded feedbacks are required.
Moreover, we consider here *time-varying* systems.

This chapter is organised as follows. In next section we present the model
we consider here. In Section 3.1 we construct a state estimator and prove
global exponential stability in closed loop with the plant dynamics and
kinematics. In Section 3.2 we construct a state feedback controller and
prove global exponential stability. In Section 3.3 we establish our separation principle, i.e., we prove that, if the state-feedback control law is
implemented using the state estimates, the overall closed loop system is
uniformly globally asymptotically stable (UGAS). Finally, in Section 4 we
discuss our results, when applied to robot manipulators.

Notations. In this chapter, we use $\| \cdot \|$ to denote the Euclidean norm of
vectors and induced norm of matrices. The symbols k_m and k_M are used
for lower and upper bounds on $\|K\|$. The symbols := and =: mean "equal
by definition". A continuous function $\beta : \mathbb{R}_{\geq 0} \to \mathbb{R}_{\geq 0}$ is said to be of class
$\mathcal{K}$ ($\beta \in \mathcal{K}$), if $\beta(s)$ is strictly increasing and $\beta(0) = 0$

2 Model and Problem Formulation

We consider in this chapter, fully actuated Euler-Lagrange systems with
generalised coordinates $q \in \mathbb{R}^n$, and control inputs $u \in \mathbb{R}^n$, i.e.,

$$\frac{d}{dt}\left(\frac{\partial \mathcal{L}}{\partial \dot{q}}\right) - \frac{\partial \mathcal{L}}{\partial q} = u \tag{3.1}$$

where the Lagrangian $\mathcal{L}(q, \dot{q}) := T(q, \dot{q}) - V(q)$. It is assumed that the
kinetic energy function is of the quadratic form,

$$T(q, \dot{q}) = \frac{1}{2}\dot{q}^\top D(q)\dot{q}$$

where the inertia matrix $D(q) \in \mathbb{R}^{n \times n}$ is positive definite and uniformly
bounded. The potential energy function, $V(q)$, is assumed to be uniformly
bounded from below, i.e., we assume that there exists a real number c, such
that $V(q) > c$ for all $q \in \mathbb{R}^n$.

As it is well known, using the Christoffel symbols of the first kind [26, 24],
the system (3.1) can be rewritten in the form

$$D(q)\ddot{q} + C(q, \dot{q})\dot{q} + g(q) = u \tag{3.2}$$

where, in our notation, the matrix $C(q, \dot{q})$ contains the terms corresponding
to centrifugal and Coriolis effects, and the vector $g(q) := \frac{\partial V(q)}{\partial q}$.

As it is discussed in [13], a common drawback of output feedback controllers relying on Lyapunov design, is that certain 3rd order terms that

appear in the Lyapunov function derivative, cannot be dominated. These high order terms arise since the Coriolis and centrifugal forces vector in (3.2), has a quadratic growth in the generalised velocities, which are not measured.

The global change of coordinates introduced in [5] for one degree-of-freedom systems overcomes this problem by rewriting the dynamics with functions which are linear in the unmeasured velocities. A physical meaning for this "change of coordinates", which makes best sense when considering mechanical systems, is that this can be regarded as a *kinematic* model.

The result of [9] for output feedback control of boats in slow motion tasks, combined with the underlying ideas in [5] suggest that, if we could rewrite the model (3.2) in a way which exhibited these kinematic relations and a dynamic model, linear in the unmeasured states, the problem of output feedback tracking should be considerably simplified.

For setpoint control, a first step has been undertaken in [14] where a separation principle for dynamic positioning of ships was already proven. Thus, inspired by the results of [9, 5] and motivated by those in [14], in this chapter, we extend the latter to the tracking problem. For this, *we will assume* that there exists a function $J : \mathbb{R}^n \to \mathbb{R}^{n \times n}$, with the following properties

P1 $J(q)$ is invertible for all $q \in \mathbb{R}^n$ and satisfies $0 < k_{j_m} \leq \|J(q)\| \leq k_{j_M} < \infty$.

P2 $\dfrac{dJ(q)}{dt} =: \dot{J}(q, \dot{q})$ is globally Lipschitz in $\dot{q}$, uniformly in q; i.e., there exists $k'_j > 0$ such that,

$$\|\dot{J}(q, x) - \dot{J}(q, y)\| \leq k'_j \|x - y\| \qquad \forall q \in \mathbb{R}^n . \tag{3.3}$$

If a function $J(q)$ with the properties above is available, we can rewrite the model (3.2) as

$$\left[J^{-\top} M J^{-1} \right] (q) \ddot{q} + \left[-J^{-\top} M J^{-1} \dot{J} J^{-1} \right] (q, \dot{q}) \dot{q} + J^{-\top}(q) v(q) = J^{-\top}(q) \tau \tag{3.4}$$

by defining $\left[J^{-\top} M J^{-1} \right] (q) =: D(q)$, with M being positive definite and constant, $\left[-J^{-\top} M J^{-1} \dot{J} J^{-1} \right] (q, \dot{q}) =: C(q, \dot{q})$, $J^{-\top}(q) v(q) =: g(q)$ and $J^{-\top}(q) \tau := u$.

We will consider in this chapter, EL systems for which a $J(q)$, satisfying **P1** and **P2**, exists. Our motivation to consider EL systems which can be modelled by (3.4) is that these equations are equivalent to

$$M \dot{\nu} + v(q) = \tau \tag{3.5}$$

$$\dot{q} = J(q) \nu . \tag{3.6}$$

where the *dynamics* (3.5) of the system have been expressed separately from the *kinematic* model (3.6).

Notice that under the assumption that the inertia matrix $D(q)$ is positive definite, there exists a uniformly bounded function $\Delta : \mathbb{R}^n \to \mathbb{R}^{n \times n}$ such that $\Delta(q)^\top \Delta(q) > 0$ for all $q \in \mathbb{R}^n$. While this property is true in general for positive definite matrices, it is usually very hard to find such factorisation for n degrees of freedom systems.

In [8, 9, 14] the model (3.5,3.6) represents the dynamics an kinematics of a surface vessel, for small motion applications. In [5], the author proposed a "change of coordinates" similar to (3.6) for one-degree-of-freedom systems. The control problem we solve in this chapter is the following.

Definition 3.1 (Global output feedback tracking) *Let $\dot{q}_d : \mathbb{R}_{\geq 0} \to \mathbb{R}^n$ be twice continuously differentiable and assume there exists $\beta_d > 0$ such that, $\max\{\|q_d(t)\|, \|\dot{q}_d(t)\|\} \leq \beta_d$, uniformly in t. Assume that only q is available for measurement. Under these conditions, find a dynamic controller $\tau(t, q, \xi)$, $\dot{\xi} = \phi(t, \xi, q)$, such that, for any initial conditions $(t_\circ, q(t_\circ), \dot{q}(t_\circ), \xi(t_\circ)) \in \mathbb{R}_{\geq 0} \times \mathbb{R}^n \times \mathbb{R}^n \times \mathbb{R}^m$, the system (3.4) in closed loop with $\tau(t, q, \xi)$, be uniformly globally asymptotically stable (UGAS).*

3 A Cascades Approach to a Separation Principle

Our main result in this chapter is a separation principle for EL systems under the assumptions made in the previous section. Our control design relies on defining an observer and a control law, with the aim at having a cascaded closed loop system, i.e., we seek for an error dynamics of the form

$$\Sigma_1 \quad : \quad \dot{x}_1 = f_1(t, x_1) + g(t, x)x_2 \tag{3.7}$$

$$\Sigma_2 \quad : \quad \dot{x}_2 = f_2(t, x_2) \tag{3.8}$$

where $x_1 \in \mathbb{R}^{2n}$, $x_2 \in \mathbb{R}^{2n}$, the functions $f_1(t, x_1)$, $f_1(t, x_1)$, and $g(t, x)$ are continuously differentiable and, both subsystems, Σ_2 and

$$\Sigma_1^0 \quad : \quad \dot{x}_1 = f_1(t, x_1) \tag{3.9}$$

are UGAS. Our motivation for considering this class of systems, is that the sufficient conditions for UGAS of cascades, are often easier to verify than to find a Lyapunov function for the system (Σ_1, Σ_2), with a negative definite time derivative. In particular, in this chapter we will use Theorem 2 from [23].

In classic Lyapunov control design, one aims at designing a control law which yields a Lyapunov function with a negative definite derivative. In our control design for EL systems in the form (3.5, 3.6), the system Σ_2 will be the estimation error dynamics, hence, our first goal is to construct an exponentially convergent observer. The system Σ^0 will correspond to the plant in closed loop with a state feedback controller. Then, Σ_1 will correspond to the system (3.5, 3.6), in closed loop with the output feedback controller. In other words, $g(t, x)x_2$ will correspond to nonlinearities of the system that result from implementing the state feedback control law, using the state estimates, instead of the true values. Then, to analyse the stability of the overall system, we will invoke [23, Theorem 2]. Hence, our design is made with aim at verifying the conditions of that theorem.

3.1 Observer Design

The observer design is based on [9]. With respect to the result in the last reference, we relax the assumption that the dynamic model (3.5) is internally damped. Consider the observer

$$M\dot{\hat{\nu}} + v(q) = \tau + MK_{o_2}(q)\bar{q} \tag{3.10}$$

$$\dot{\hat{q}} = J(q)\hat{\nu} + K_{o_1}\bar{q}. \tag{3.11}$$

where $K_{o_1} \in \mathbb{R}^{n\times n}$ and $K_{o_2}(q) \in \mathbb{R}^{n\times n}$ are to be defined later and we denote the estimation error $\bar{q} = q - \hat{q}$, correspondingly for the other variables. The estimation error dynamics (3.5, 3.6), (3.10, 3.11) is

$$\dot{\bar{\nu}} = -K_{o_2}(q)\bar{q} \tag{3.12}$$

$$\dot{\bar{q}} = J(q)\bar{\nu} - K_{o_1}\bar{q}. \tag{3.13}$$

Proposition 3.1 (Exponentially convergent observer) *Let P_1, P_2 be positive definite, K_{o_1} be such that $P_1K_{o_1} + K_{o_1}^\top P_1$ is positive definite, and let $K_{o_2}(q) := P_2^{-1}J(q)^\top P_1$. Then, the origin $(\bar{q}, \bar{\nu}) = (0,0)$ of the system (3.13,3.12) is uniformly globally stable (UGS). Furthermore, assume that the trajectories $q(t)$ and $\nu(t)$, starting at (t_o, q_o, ν_o) are globally uniformly bounded, i.e., there exist $c > 0$ and $\beta \in \mathcal{K}$ such that $\|[q(t); \nu(t)]\| \leq \beta(\|q_o; \nu_o\|) + c$ for all $t \geq t_o \geq 0$ and all $(q_o, \nu_o) \in \mathbb{R}^{n\times n}$. Then, the origin is UGAS.*

Remark 3.1 *The assumption on the uniform boundedness of the plant trajectories is needed here to establish UGAS for (3.12,3.13), however, this condition will be relaxed later when considering the overall closed loop system. That is, when introducing the output feedback controller.*

Proof of Proposition 3.1. Consider the control Lyapunov function candidate

$$V_o(\bar{q}, \bar{\nu}) = \frac{1}{2}\left(\bar{q}^\top P_1\bar{q} + \bar{\nu}^\top P_2\bar{\nu}\right) \tag{3.14}$$

where $P_1 \in \mathbb{R}^{n \times n}$ and $P_2 \in \mathbb{R}^{n \times n}$ are positive definite matrices. The time derivative of $V_o(\bar{q}, \bar{\nu})$ along the trajectories of (3.13,3.12) yields

$$\dot{V}_o(\bar{q}, \bar{\nu}) = -\frac{1}{2}\bar{q}^\top \left(P_1 K_{o_1} + K_{o_1}^\top P_1\right) \bar{q} + \bar{q}^\top P_1 J(q)\bar{\nu} - \bar{q}^\top K_{o_2}(q)^\top P_2 \bar{\nu} \tag{3.15}$$

hence, using the definition $K_{o_2}(q) = P_2^{-1} J(q)^\top P_1$, we obtain that

$$\dot{V}_o(\bar{q}, \bar{\nu}) = -\frac{1}{2}\bar{q}^\top \left(P_1 K_{o_1} + K_{o_1}^\top P_1\right) \bar{q}. \tag{3.16}$$

Since by assumption, $P_1 K_{o_1} + K_{o_1}^\top P_1$ is positive (semi-)definite, the time derivative $\dot{V}_o(\bar{q}, \bar{\nu})$ is negative semidefinite. We conclude that the origin of the system is uniformly globally stable.

To prove global exponential stability we rely on the a theorem, from [15], which is repeated in the Appendix A for the sake of completeness. To apply Theorem 3.3, let $\xi_1 := \bar{q}$, $\xi_2 := \bar{\nu}$, $W(t, \xi_1) := \frac{1}{2}\bar{q}^\top P_1 \bar{q}$, $G(t, \xi) := J(\xi_1 + \hat{q}(t)) = J(q)$, $P := P_2$ and $h(t, \xi_1) := -K_{o_1}\bar{q}$. With these definitions, it is clear that the system (3.13,3.12) is of the form (3.51,3.52). Hence, we simply have to verify that conditions **A1** - **A2** hold. The bound (3.53) is clearly satisfied with $\rho_1(\cdot) = \max\{k_{o_{1M}}, p_{1M}\}$. The bounds (3.54,3.56) hold due to the property **P1**. Also, the inequality (3.55) is satisfied since, using $\dot{J}(q, 0) \equiv 0$ and (3.11,3.13), we can compute

$$
\begin{aligned}
\left\| \dot{J}(q, \dot{q}) \right\| &\leq \left\| \dot{J}(q, \dot{q}) - \dot{J}(q, 0) \right\| + \left\| \dot{J}(q, 0) \right\| \\
&\leq k_j' \left(\|\dot{\tilde{q}}\| + \|J(q)[\nu(t) - \bar{\nu}]\| \right) + k_{o_{1M}} \|\bar{q}\| \\
&\leq k_j' \left(\|\dot{\tilde{q}}\| + k_j \left(c + \beta(\|[q_o, \nu_o]\|)\right)\|\bar{\nu}\| \right) + k_{o_{1M}} \|\bar{q}\| =: \rho_3(\|\xi\|).
\end{aligned}
$$

Finally, it is immediate to verify that Assumption **A2** holds true with $\alpha_1(\cdot) = 0.5 p_{1m}$, $\alpha_2(\cdot) = 0.5 p_{1M}$, and $\mu = \lambda_m(P_1 K_{o_1} + K_{o_1}^\top P_1)$.

We conclude that the origin $(\bar{q}, \bar{\nu}) = (0, 0)$ of (3.13,3.12) is globally uniformly asymptotically stable and uniformly locally exponentially stable.

3.2   State Feedback Controller

Having designed an exponentially converging state observer, we proceed now to derive a state feedback controller for the system (3.5, 3.6). In the next section we will prove that this state feedback controller can be implemented using the state estimates and the observer above, leading to uniform global asymptotic stability of the closed loop.

As in [25], let us define the virtual error $s := \dot{\tilde{q}} + K_p \tilde{q}$, where $\tilde{q} := q - q_d$, and K_p is positive definite. From (3.6), we also have that

$$s = J(q)\nu - \dot{q}_d + K_p \tilde{q}. \tag{3.17}$$

The control design consists on finding $\tau(t, q, \dot{q})$ such that the closed loop system have the form

$$\dot{s} = -K_d s - \tilde{q} \tag{3.18}$$

$$\dot{\tilde{q}} = -K_p \tilde{q} + s. \tag{3.19}$$

The obvious reason we aim at such closed loop system, is that $(3.18, 3.19)$ is globally exponentially stable for any positive definite matrices K_d and K_p. This can be verified, using the Lyapunov function candidate

$$V_c(t, \tilde{q}, s) = \frac{1}{2} \left(\|\tilde{q}\|^2 + \|s\|^2 \right) \tag{3.20}$$

which is positive definite and radially unbounded in the state $(\tilde{q}, s)$.

Thus, to derive $\tau(t, q, \dot{q})$, we evaluate the time derivative on both sides of (3.17), using (3.5), to obtain

$$\dot{s} = J(q)M^{-1}[\tau - v(q)] + \dot{J}(q, \dot{q})\nu - \ddot{q}_d + K_p \dot{\tilde{q}}, \tag{3.21}$$

hence, substituting the right hand side of (3.18) in (3.21) we have that the latter is satisfied if we apply the control law $\tau = \tau^*$, with

$$\tau^* = v(q) + MJ(q)^{-1} \left[-\tilde{q} + \ddot{q}_d - K_p \dot{\tilde{q}} - \dot{J}(q, \dot{q})\nu - K_d s \right]. \tag{3.22}$$

This control law can be implemented under the assumptions that velocity and position measurements are available, $J(q)$ is invertible, and using (3.6).

We conclude that the system (3.5, 3.6) in closed loop with (3.22) is globally exponentially stable for any positive definite matrices K_p and K_d.

3.3   A Separation Principle

We show in this section that, the control law (3.22) can be implemented using the state estimates provided by the observer of Section 3.1 and UGAS can still be achieved. This is our main result.

Consider the controller

$$\hat{\tau} := v(q) + MJ(q)^{-1} \left[-\tilde{q} + \ddot{q}_d - K_p \hat{\tilde{q}} - \dot{J}(q, J(q)\hat{\nu})\hat{\nu} - K_d \hat{s} \right] \tag{3.23}$$

$$\hat{\tilde{q}} := \hat{\dot{q}} - \dot{q}_d := J(q)\hat{\nu} - \dot{q}_d \tag{3.24}$$

$$\hat{s} := (J(q)\hat{\nu} - \dot{q}_d + K_p \tilde{q}). \tag{3.25}$$

We are now ready to state our main result.

Theorem 3.1 (Separation principle) *The system (3.5,3.6) in closed loop with the observer (3.10,3.11) and the controller (3.23)–(3.25), is UGAS for any positive definite matrices K_p, K_d, P_1, P_2 and K_{o_1} such that $P_1 K_{o_1} + K_{o_1}^\top P_1$ is positive definite, and $K_{o_2}(q) := P_2^{-1} J(q) P_1$.*

Remark 3.2 *In order to prove Theorem 3.1 we will invoke [23, Theorem 2] (see Theorem 3.2) which is formulated for cascades of UGAS systems. It can be proven along the same lines as in [23], that, under the conditions of that paper, the cascade of a subsystem Σ_1 being GES, with a subsystem Σ_2, being UGAS and ULES, yields a UGAS and ULES cascade. Therefore, even though we do not state it explicitly in Theorem 3.1, the closed loop system is UGAS and ULES.*

Proof of Theorem 3.1. Comparing (3.23) to (3.22), we have that

$$\hat{\tau} \;=\; \tau^* + g_1(t, s, \tilde{q}, \bar{\nu}, \bar{q}) \tag{3.26}$$

$$g_1 \;:=\; MJ(q)^{-1}\left[(K_p + K_d)J(q)\bar{\nu} + \dot{J}(q, \dot{q})\nu - \dot{J}(q, \hat{\dot{q}})\nu + \dot{J}(q, \hat{\dot{q}})\bar{\nu}\right] \tag{3.27}$$

$$\hat{\dot{q}} \;:=\; J(q)\hat{\nu} \tag{3.28}$$

and we remark that we have abbreviated $\tilde{q} + q_d = q$ and $J(q)^{-1}[s - K_p\tilde{q} + \dot{q}_d] = \nu$ for simplicity in the notation.

Thus, the closed loop system (3.5,3.6); with $\tau = \hat{\tau}$, (3.10, 3.11) and (3.23) yields the system

$$\begin{bmatrix} \dot{\tilde{q}} \\ \dot{s} \end{bmatrix} = \begin{bmatrix} -K_p\tilde{q} + s \\ -K_d s - \tilde{q} \end{bmatrix} + g(t, s, \tilde{q}, \bar{\nu}, \bar{q}) \tag{3.29}$$

$$\begin{bmatrix} \dot{\bar{q}} \\ \dot{\bar{\nu}} \end{bmatrix} = \begin{bmatrix} -K_{o_1}\bar{q} + J(q(t))\bar{\nu} \\ -K_{o_2}(q(t))\bar{q} \end{bmatrix} \tag{3.30}$$

where the interconnection term

$$g(t, s, \tilde{q}, \bar{\nu}, \bar{q}) := \begin{bmatrix} 0 \\ g_1(t, s, \tilde{q}, \bar{\nu}, \bar{q}) \end{bmatrix}. \tag{3.31}$$

In order to study the stability of (3.29-3.31), we present below a theorem which follows from the proofs of [23, Theorems 1 and 2] for nonautonomous cascades.

Theorem 3.2 *Assume that the system (3.9) is UGAS and that the trajectories of (3.8) are globally bounded, uniformly in the initial states and the initial time (x_o, t_o). If moreover, Assumptions **A3** – **A5** below are satisfied, then the solutions $x(t, t_o, x_o)$ of the system (3.7,3.8) are globally uniformly bounded. If furthermore, the system (3.8) is UGAS, then so is the origin of the cascade (3.7,3.8).*

A3 *There exists a Lyapunov function $V(t, x_1)$ for (3.9) such that $V : \mathbb{R}_{\geq 0} \times \mathbb{R}^n \to \mathbb{R}_{\geq 0}$ is positive definite (that is $V(t, 0) = 0$ and $V(t, x_1) > 0$ for all $x_1 \neq 0$) and radially unbounded, which satisfies*

$$\left\|\frac{\partial V}{\partial x_1}\right\| \|x_1\| \leq c_1 V(t, x_1) \qquad \forall \|x_1\| \geq \eta \tag{3.32}$$

where c_1, $\eta > 0$. We also assume that $\frac{\partial V}{\partial x_1}(t, x_1)$ is bounded uniformly in t for all $\|x_1\| \leq \eta$, that is, there exists a constant $c_2 > 0$ such that for all $t \geq t_o \geq 0$

$$\left\|\frac{\partial V}{\partial x_1}\right\| \leq c_2 \qquad\qquad \forall \|x_1\| \leq \eta \qquad\qquad (3.33)$$

A4 *There exist two continuous functions θ_1, $\theta_2 : \mathbb{R}_{\geq 0} \to \mathbb{R}_{\geq 0}$, such that $g(t, x)$ satisfies*

$$\|g(t, x)\| \leq \theta_1(\|x_2\|) + \theta_2(\|x_2\|)\|x_1\| \qquad\qquad (3.34)$$

A5 *There exists a class $\mathcal{K}$ function $\phi(s)$ such that, for all $t_o \geq 0$, the trajectories of the system (3.8) satisfy*

$$\int_{t_o}^{\infty} \|x_2(t, t_o, x_2(t_o))\| dt \leq \phi(\|x_2(t_o)\|) . \qquad\qquad (3.35)$$

In order to apply Theorem 3.2, let $x_1 := \mathrm{col}(\tilde{q}, s)$ and, $x_2 := \mathrm{col}(\bar{q}, \bar{\nu})$. Notice that, $g(t, \tilde{q}, 0, 0) \equiv 0$ and, the zero input dynamics of (3.29),

$$\begin{bmatrix} \dot{\tilde{q}} \\ \dot{s} \end{bmatrix} = \begin{bmatrix} -K_p\tilde{q} + s \\ -K_d s - \tilde{q} \end{bmatrix} ,$$

corresponds to the closed loop system with the state feedback control law τ^*, which was designed to be GES for any positive definite K_p and K_d. Also, it is clear that the Lyapunov function (3.20) satisfies **A3** above.

We show now, that $g(t, \tilde{q}, s, \bar{q}, \bar{\nu})$ as defined in (3.31,3.27), satisfies Assumption **A4** of Theorem 3.2. For this, consider the following bounds:

$$\|(K_p + K_d)J(q)\bar{\nu}\| \leq (k_{pM} + k_{dM})k_{jM}\|\bar{\nu}\|$$

$$\left\|\left[\dot{J}(q, \dot{q}) - \dot{J}(q, \hat{\dot{q}})\right]\nu\right\| = \left\|\left[\dot{J}(q, J(q)\nu) - \dot{J}(q, J(q)\hat{\nu})\right]\nu\right\|$$

$$\leq k_j'\|J(q)\bar{\nu}\|\|\nu\|$$

$$\leq \frac{k_j'k_{jM}}{k_{jm}}\left(\|s\| + k_{pM}\|\tilde{q}\| + \beta_d\right)\|\bar{\nu}\|$$

$$\left\|\dot{J}(q, J(q)\hat{\nu})\bar{\nu}\right\| \leq \left(\left\|\dot{J}(q, J(q)\hat{\nu}) - \dot{J}(q, 0)\right\|\right)\|\bar{\nu}\|$$

$$\leq k_j'k_{jM}\|\nu - \bar{\nu}\|\|\bar{\nu}\|$$

$$\leq k_j'k_{jM}\left(\frac{1}{k_{jm}}(\|s\| + k_{pM}\|\tilde{q}\| + \beta_d) + \|\bar{\nu}\|\right)\|\bar{\nu}\|$$

where we have used properties **P1** and **P2** and $\dot{J}(q, 0) \equiv 0$. It follows that

$g(t, \tilde{q}, s, \bar{q}, \bar{\nu})$ satisfies (3.34) with

$$\theta_1(\|\bar{q}; \bar{\nu}\|) \quad := \quad \frac{m_M}{k_{jm}}\left[(k_{pM} + k_{dM})k_{jM} + \frac{2k'_j k_{jM}\beta_d}{k_{jm}}\right] + \frac{m_M k'_j k_{jM}}{k_{jm}}\|\bar{\nu}\|^2$$

$$\theta_2(\|\bar{q}; \bar{\nu}\|) \quad := \quad \frac{2k'_j k_{jM} m_M}{k_{jm}^2}(k_{pM} + 1)\|\bar{\nu}\|.$$

Therefore, the trajectories $(\tilde{q}(t), s(t), \bar{q}(t), \bar{\nu}(t))$ of the overall closed loop system (3.5,3.6), (3.10,3.11), $\tau = \hat{\tau}$ and (3.23), are uniformly bounded for any initial conditions. Since $q_d(t)$ and $\dot{q}_d(t)$ are uniformly bounded, so are $\dot{q}(t)$ and $q(t)$ for all initial conditions. Finally, from (3.6) and Property **P1** it follows that $\nu(t)$ is also uniformly bounded for all initial conditions.

At this point, we can invoke Proposition 3.1 to conclude that the origin $(\bar{q}, \bar{\nu}) = (0, 0)$ of the time-varying system

$$\dot{\bar{\nu}} = -K_{o_2}(q(t))\bar{q}$$
$$\dot{\bar{q}} = J(q(t))\bar{\nu} - K_{o_1}\bar{q}$$

is uniformly globally asymptotically stable. Moreover, this property is also uniform in the initial conditions of the trajectories $q(t, t_o, q_o)$ and $\nu(t, t_o, \nu_o)$ since these are globally uniformly bounded. From these properties, it is clear that there exist $\phi \in \mathcal{K}$ such that the trajectories $x_2(t) = [\bar{q}(t); \bar{\nu}(t)]$ satisfy (3.35).

We conclude that the cascaded system (3.29,3.30) is UGAS.   ∎

4 Application to Robot Manipulators

Our main result applies to EL systems (3.1) with quadratic kinetic energy function, and potential energy bounded from below. However, the control implementation involves the knowledge of a matrix $J(q)$ such that the inertia matrix can be written as $D(q) = J(q)^{-\top} M J(q)^{-1}$. While for systems, such as marine vessels, the Jacobian $J(q)$ is a simple orthogonal rotation matrix, in general, this factorisation is very hard to find (such is the case of robot manipulators). In this section we give the insight on an alternative result for robot manipulators.

We start with the expression of the kinetic energy of a robot manipulator, assuming that each articulation is actuated thru a rigid transmission. Following the notation of [24], let p_i denote the position of the centre of mass of the i-th link, expressed in the Cartesian coordinates of the base frame hence, $p_i \in \mathbb{R}^3$. Let, $\dot{p}_i$ denote the linear velocities of the centre of mass and, ω_i denote the angular velocity of the i-th link, expressed in the base frame, due to the rotations (if any) of the i-th link. Define m_i as the mass of the i-th link and I_i^i the inertia tensor of the i-th link, expressed in an inertial reference frame, with origin at the centre of mass of the i-th link. Define also, the position and orientation Jacobians for the i-th link as the maps, $J_P^i : \mathbb{R}^n \to \mathbb{R}^{3 \times n}$ and $J_O^i : \mathbb{R}^n \to \mathbb{R}^{3 \times n}$ from the space of linear and angular velocities into generalised velocities, i.e., let

$$\dot{p}_i \;=\; J_P^i(q)\dot{q} \tag{3.36}$$

$$\omega_i \;=\; J_O^i(q)\dot{q} . \tag{3.37}$$

With all these definitions, and neglecting the inertia contributions of the actuators, we have that the kinetic energy of the manipulator is given by

$$T(\dot{q}, q) = \frac{1}{2} \sum_{i=1}^{n} \left(m_i \dot{q}^\top J_P^i(q)^\top J_P^i(q)\dot{q} + \dot{q}^\top J_O^i(q)^\top R_i(q) I_i^i R_i^\top(q) J_O^i(q)\dot{q} \right) \tag{3.38}$$

where $R_i(q)$ is the rotation matrix expressing the orientation of the i-th inertial frame, in the base frame. Using, (3.36,3.37), the expression (3.38) is equivalent to

$$T(\dot{p}, \omega_i^i) = \frac{1}{2} \sum_{i=1}^{n} \left(m_i \dot{p}_i^\top \dot{p}_i + {\omega_i^i}^\top I_i^i \omega_i^i \right) \tag{3.39}$$

where $\omega_i^i = R_i(q)^\top \omega_i$ is the angular velocity of the i-th link, expressed in the coordinates of its inertial frame. At this point, let us define the vector $\nu := \mathrm{col}[\dot{p}_1, \dots, \dot{p}_n, \omega_1^1, \dots, \omega_n^n]$ and the matrix $\mathcal{J}(q) \in \mathbb{R}^{6n \times n}$ as

$$\mathcal{J}(q) \;\; := \;\; \begin{bmatrix} J_P^1(q) \\ \vdots \\ J_P^n(q) \\ R_1^\top(q) J_O^1(q) \\ \vdots \\ R_n^\top(q) J_O^n(q) \end{bmatrix}, \tag{3.40}$$

hence we have that

$$\nu := \mathcal{J}(q)\dot{q} . \tag{3.41}$$

Observe that, since there are n generalised (independent) coordinates, the column rank of $\mathcal{J}(q)$ is n. Therefore, we can also write

$$\dot{q} := [\mathcal{J}(q)^\top \mathcal{J}(q)]^{-1} \mathcal{J}(q)^\top \nu . \tag{3.42}$$

Using all these definitions, the kinetic energy function can be rewritten (with an obvious abuse of notation) as

$$T(\nu) = \nu^\top \mathcal{M} \nu \tag{3.43}$$

where $\mathcal{M} := \text{blockdiag}\{m_1 I, \ldots, m_n I, I_1^1, \ldots, I_n^n\}$ where $I \in \mathbb{R}^{3 \times 3}$ is the identity matrix and we recall that I_i^i is the *constant* inertia tensor of the i-th link, i.e., referred to the inertial frame.

Concerning the potential energy of the manipulator, this is given by

$$\mathcal{U}(p) := \sum_{i=1}^{n} m_i g_o^\top p_i \tag{3.44}$$

where $g_o \in \mathbb{R}^3$ is the vector of gravity acceleration.

Therefore, using (3.43,3.44) we can derive the dynamic model of the manipulator,

$$\mathcal{M}\nu + v = \tau \tag{3.45}$$

where the constant vector

$$v := \left[\frac{\partial \mathcal{U}(p)}{\partial p}^\top, \; 0, \; \ldots, \; 0_n \right]^\top$$

and which is subject to the holonomic constraints imposed by the kinematics equation (3.41).

Two observations are important at this point. Firstly notice that the manipulator model (3.45,3.41) is of the form (3.5,3.6) except of course, for $\mathcal{J}(q)$ which is not square. Moreover, it can be shown that $\mathcal{J}(q)$ satisfies **P1** and **P2**. Indeed, **P1** holds since $\mathcal{J}(q)$ is full column rank. This is equivalent to the assumption that the inertia matrix is positive definite and its induced norm is bounded from above.

Secondly, differentiating (3.41) with respect to time and substituting for $\dot{\nu}$, in (3.45), we obtain that

$$\left[\mathcal{J}^\top \mathcal{M} \mathcal{J}\right](q)\ddot{q} + \left[\mathcal{J}^\top \mathcal{M} \dot{\mathcal{J}}\right](q, \dot{q})\dot{q} + \mathcal{J}^\top(q)v = \mathcal{J}^\top(q)\tau. \tag{3.46}$$

The inertia matrix $D(q) := \mathcal{J}(q)^\top \mathcal{M} \mathcal{J}(q)$ is positive definite *if and only if* the "Jacobian" $\mathcal{J}(q)$ is of full column rank and $\mathcal{M}$ is positive definite. Also, it is easy to see that, defining the Coriolis and centrifugal matrix $C(q, \dot{q}) := \left[\mathcal{J}^\top \mathcal{M} \dot{\mathcal{J}}\right](q, \dot{q})$, the matrix $0.5\dot{D}(q, \dot{q}) - C(q, \dot{q})$ is skew-symmetric. Finally, the gravity forces vector

$$g(q) := \frac{\partial U(p(q))}{\partial q} = \mathcal{J}(q)^\top v.$$

We remark that this does not mean that $\frac{\partial p(q)}{\partial q} = \mathcal{J}(q)$.

Comparing the systems (3.45,3.41) to (3.5,3.6) and from Theorem 3.1, one could conjecture that the controller

$$\hat{\tau} \; := \; v + \mathcal{M}\mathcal{J}(q)\left[-\tilde{q} + \ddot{q}_d - K_p\hat{\tilde{q}} - \dot{J}(q, J(q)\hat{\nu})\hat{\nu} - K_d\hat{s}\right] \quad (3.47)$$

$$\dot{\hat{\tilde{q}}} \; := \; J(q)\hat{\nu} - \dot{q}_d \quad (3.48)$$

$$\hat{s} \; := \; (J(q)\hat{\nu} - \dot{q}_d + K_p\tilde{q}) \quad (3.49)$$

$$J(q) \; := \; \mathcal{J}(q)^{\dagger} := \left[\mathcal{J}(q)^{\top}\mathcal{J}(q)\right]^{-1}\mathcal{J}(q)^{\top} \quad (3.50)$$

together with the observer (3.10,3.11), renders the robot manipulator system (3.46), UGAS. However, it shall be remarked that the observer system was designed under the implicit assumption that the coordinates ν are linearly *independent*. In the case of the robot manipulator, $\nu \in \mathbb{R}^{6n}$ and these variables are subject to $5n$ holonomic constraints. The physical meaning of these constraints is clear if we recall that a manipulator consists of an open kinematic chain of n rigid *linked* bodies.

Therefore, strictly speaking, we cannot talk about UGAS, more specifically, of global (uniform) attractivity of the origin $(\bar{q}, \bar{\nu}, \tilde{q}, s) = (0, 0, 0, 0)$. Yet, based on the results of the previous section, we conjecture the following.

Conjecture 3.1 *Consider the closed loop system (3.46), (3.10,3.11), (3.47–3.50). Under the conditions of Theorem 3.1, the origin $(\bar{q}, \bar{\nu}, \tilde{q}, s) = (0, 0, 0, 0)$ is UGS and the signals $\bar{q}(t)$, $\dot{\bar{q}}(t) := J(q)\bar{\nu}(t)$, $\tilde{q}(t)$ and $\dot{\tilde{q}}(t)$ converge uniformly to zero, for any initial conditions.*

5 Conclusions

We have addressed in this chapter, the problem of output feedback trajectory tracking of Euler Lagrange systems. For a subclass of these systems (including manipulators), characterised by certain factorisation, we have provided a separation principle. Our main result establishes that, if a globally exponentially stabilising state feedback controller can be implemented using the state estimates provided by a globally exponentially convergent observer; the overall closed loop system, remains uniformly globally asymptotically stable. Even though our results cannot be directly applied to robot manipulators, for these systems, we have conjectured the weaker property of UGS plus global uniform convergence of part of the state variables. These include the position and velocity tracking errors. The proof of the latter is currently under investigation.

Acknowledgements

This work was realised while the authors were with the Dept. of ECE, University of California at Santa Barbara, USA, under grant NSF-9812346.

The second author is on leave from the IMPE, Academy of Sciences of Russia, St. Petersburg, Russia.

6 REFERENCES

[1] A.N. Atassi and H. Khalil. A separation principle for the stabilization of a class of nonlinear systems. In *Proc. 4th. European Contr. Conf.*, Bruxelles, BE, 1997. CD-ROM Paper no. 952.

[2] P. Bélanger. Estimation of angular velocity and acceleration from shaft encoder measurements. In *Proc. IEEE Conf. Robotics Automat.*, volume 1, pages 585–592, Nice, France, 1992.

[3] H. Berghuis and H. Nijmeijer. A passivity approach to controller-observer design for robots. *IEEE Trans. on Robotics Automat.*, 9-6:740–755, 1993.

[4] H. Berghuis, H. Nijmeijer, and P. Löhnberg. Observer design in the tracking control problem of robots. In *Proc. IFAC NOLCOS*, pages 588–593, Bordeaux, France, 1992.

[5] G. Besançon. Simple global output feedback tracking control for one-degree-of-freedom Euler-Lagrange systems. In *IFAC Conf. on Systems Structure and Control*, pp. 267–271, 1998.

[6] I. V. Burkov. Mechanical system stabilization via differential observer. In *IFAC Conference on System Structure and Control*, pages 532–535, Nantes, France, 1995.

[7] C. Canudas de Wit, K. Astrom, and N. Fixot. Computed torque control via a nonlinear observer. *Int. J. Adapt. Contr. Sign. Process.*, 4(3):443–452, 1990.

[8] T. I. Fossen. *Guidance and control of ocean vehicles*. John Wiley & Sons Ltd., 1994.

[9] T. I. Fossen and Å . Grøvlen. Nonlinear output feedback control of dynamically positioned ships using vectorial observer backstepping. *IEEE Trans. Contr. Syst. Technol.*, 6(1):121–128, 1998.

[10] A. A. J. Lefeber. (Adaptive) Control of chaotic and robot systems via bounded feedback control. Master's thesis, University of Twente, The Netherlands, 1996. Available from the author upon request to A.A.J.Lefeber@math.utwente.nl .

[11] S. Lim, D. Dawson, J. Hu, and M. S. de Queiroz. An adaptive link position tracking controller for rigid link, flexible joint robots without velocity measurements. In *Proc. 33rd. IEEE Conf. Decision Contr.*, pages 351–357, Orlando, FL., 1994.

[12] W. Lin. Bounded smooth state feedback and a global separation principle for non-affine nonlinear systems. *Syst. Contr. Letters*, 26:41–53, 1995.

[13] A. Loria. Global tracking control of one degree of freedom Euler-Lagrange systems without velocity measurements. *European J. of Contr.*, 2(2), 1996. See also: Proc. 13th IFAC World Congress 1996, vol. E, pp. 419-424.

[14] A. Loria, T. I. Fossen, and E. Panteley. Global output-feedback dynamic positioning of ships: A cascaded approach. Technical report, Norwegian University of Science and Technology, November 1997.

[15] A. Loría, T I. Fossen, and A. Teel. UGAS and ULES of nonautonomous systems: Applications to integral control of ships and manipulators. In *Proc. 5th. European Contr. Conf.*, Karlsruhe, Germany, 1999. Submitted for presentation.

[16] A. Loría and R. Ortega. On tracking control of rigid and flexible joints robots. *Appl. Math. and Comp. Sci., special issue on* Mathematical Methods in Robotics, 5(2):101–113, 1995. eds. K. Tchon and A. Gosiewsky.

[17] S. Nicosia and P. Tomei. Robot control by using only joint position measurement. *IEEE Trans. on Automat. Contr.*, 35-9:1058–1061, 1990.

[18] S. Nicosia and P. Tomei. On the control of flexible joint robots by dynamic output feedback. In *Proc. 4th. Symp. Robot Contr.*, pages 717–722, Capri, Italy, 1994.

[19] S. Nicosia and P. Tomei. A tracking controller for flexible joint robots using only link position feedback. *IEEE Trans. on Automat. Contr.*, AC-40(5):885–890, 1995.

[20] S. Nicosia and A. Tornambè. High-gain observers in the state and parameter estimation of robots having elastic joints. *Syst. Contr. Letters*, 13:331–337, 1989.

[21] S. Nicosia, A. Tornambè, and P. Valigi. Experimental results in estimation of industrial robots. In *Proc. 29th. IEEE Conf. Decision Contr.*, pages 360–365, Honolulu, HI, 1990.

[22] R. Ortega and M. Spong. Adaptive motion control of rigid robots: A tutorial. *Automatica*, 25-6:877–888, 1989.

[23] E. Panteley and A. Loria. On global uniform asymptotic stability of non linear time-varying non autonomous systems in cascade. *Syst. Contr. Letters*, 33(2):131–138, 1998.

[24] L. Sciavicco and B. Siciliano. *Modeling and control of robot manipulators*. McGraw Hill, New York, 1996.

[25] J. J. Slotine and W. Li. Adaptive manipulator control: a case study. *IEEE Trans. on Automat. Contr.*, AC-33:995–1003, 1988.

[26] M. Spong and M. Vidyasagar. *Robot Dynamics and Control*. John Wiley & Sons, New York, 1989.

[27] J. T. Wen and S. D. Bayard. New class of control laws for robot manipulators, Parts I and II. *Int. J. of Contr.*, 47-5:1288–1310, 1988.

[28] F. Zhang, D. M. Dawson, M. S. de Queiroz, and W. Dixon. Global adaptive output feedback tracking control of robot manipulators. In *Proc. 36th. IEEE Conf. Decision Contr.*, pages 3634–3639, San Diego, USA, 1997.

Appendix A: A Theorem on UGAS for Strictly Passive Systems

Theorem 3.3 *[15] Consider the system*

$$\dot{\xi}_1 = h(t,\xi_1) + G(t,x)\xi_2 \tag{3.51}$$

$$\dot{\xi}_2 = -PG(t,\xi)^\top \left(\frac{\partial W(t,\xi_1)}{\partial \xi_1} \right)^\top, \tag{3.52}$$

*where $\xi_1 \in \mathbb{R}^m$, $\xi_2 \in \mathbb{R}^m$, $P = P^\top > 0$ and $W : \mathbb{R}^n \times \mathbb{R}_{\geq 0} \to \mathbb{R}_{\geq 0}$ is a C^1 function satisfying certain properties (see **A1**). If Assumptions **A1** and **A2** below hold, then the origin of the system (3.51,3.52) is UGAS.*

A1 *There exist continuous nondecreasing functions $\rho_j : \mathbb{R}_{\geq 0} \to \mathbb{R}_{\geq 0}$, $(j = 1, 2, 3)$ such that, for all $t \geq 0$, $x \in \mathbb{R}^{n+m}$*

$$\max \left\{ \|h(t,\xi_1)\|, \left\| \frac{\partial W(t,\xi_1)}{\partial \xi_1} \right\| \right\} \leq \rho_1(\|\xi_1\|)\|\xi_1\| \tag{3.53}$$

$$\|G(t,\xi)\| \leq \rho_2(\|\xi\|) \tag{3.54}$$

$$\|\dot{G}(t,\xi)\| \leq \rho_3(\|\xi\|). \tag{3.55}$$

Furthermore, for each compact set $K \subset \mathbb{R}^{n+m}$ there exist $b_m > 0$ such that

$$G(t,\xi)^\top G(t,\xi) \geq b_m^2 I \tag{3.56}$$

for all $(t,\xi) \in K \times \mathbb{R}_{\geq 0}$.

A2 *There exist class-$\mathcal{K}_\infty$ functions α_1 and α_2 and $\mu > 0$ such that*

$$\alpha_1(\|\xi_1\|) \leq W(t, \xi_1) \leq \alpha_2(\|\xi_1\|) \tag{3.57}$$

$$\frac{\partial W(t, \xi_1)}{\partial t} + \frac{\partial W(t, \xi_1)}{\partial \xi_1} h(t, \xi_1) \leq -\mu\|\xi_1\|^2 . \tag{3.58}$$

Moreover, if $\alpha_2(s) \propto s^2$ then the origin is uniformly locally exponentially stable (ULES); i.e., there exist γ_1, γ_2 and $r > 0$ such that

$$\|\xi_0\| \leq r \implies \|\xi(t, t_0, \xi_0)\| \leq \gamma_1\|\xi_0\|e^{-\gamma_2(t-t_0)} , \tag{3.59}$$

where $\xi(t, t_0, \xi_0)$ is the solution of (3.51,3.52) starting at (t_0, ξ_0).

Remark 3.3 *It is important to mention that the inequality (3.55) used here, differs from the original condition*

$$\max\left\{\left\|\frac{\partial G(t, \xi)}{\partial t}\right\|, \left\|\frac{\partial G(t, \xi)}{\partial \xi}\right\|\right\} \leq \rho_3(\|\xi\|),$$
$$i = 1, 2 \tag{3.60}$$

with $\rho(\cdot)$ continuous nondecreasing, used in [15]. However, the condition (3.60) is used in that reference, in the proof of Theorem 3.3, to imply (3.55). We have modified this condition here for the purposes of this chapter.

High-Gain Observers in Nonlinear Feedback Control

Hassan K. Khalil

Department of Electrical and Computer Engineering
Michigan State University
East Lansing, MI 48824-1226, USA

1 Introduction

The use of high-gain observers has evolved as an important technique for the design of output feedback control of nonlinear systems. The basic ingredients of this technique are

(1) a high-gain observer that robustly estimates the derivatives of the output;

(2) a globally bounded state feedback control, usually obtained by saturating a continuous state feedback function outside a compact region of interest, that meets the design objectives. The global boundedness of the control protects the state of the plant from peaking when the high-gain observer estimates are used instead of the true states.

The technique was first introduced by Esfandiari and Khalil [14] and since then has been used in about forty papers, about two thirds of which represent the work of Khalil and coworkers while the rest represents the work of about twenty nonlinear control researchers.

This chapter is intended as a tutorial/survey paper on the use of high-gain observers in nonlinear control. In Section 2, we use a second-order example to illustrate the main ideas of the technique. In Section 3, we review some recent results by Attasi and Khalil which give a fairly general separation principle for a class of nonlinear systems. These separation results can be used to derive most of the results reported in the following sections. There are a few exceptions where the results cannot be derived using the separation approach, like the adaptive control results of Section 6 in the lack of full persistence of excitation and the sliding mode control results of Section 7. Sections 4 through 10 survey work on stabilization, nonlinear servomechanisms, adaptive control, sliding mode control, robustness to fast unmodeled dynamics, discrete-time implementation, and application to induction motors.

2   Motivating Example

Consider the second-order nonlinear system

$$
\begin{aligned}
\dot{x}_1 &= x_2 \\
\dot{x}_2 &= \phi(x, u) \\
y &= x_1
\end{aligned}
\tag{4.1}
$$

where $x = [x_1, x_2]^T$. Suppose $u = \gamma(x)$ is a state feedback control that stabilizes the origin $x = 0$ of the closed-loop system

$$
\begin{aligned}
\dot{x}_1 &= x_2 \\
\dot{x}_2 &= \phi(x, \gamma(x))
\end{aligned}
\tag{4.2}
$$

To implement this feedback control using only measurements of the output y, we use the observer

$$
\begin{aligned}
\dot{\hat{x}}_1 &= \hat{x}_2 + h_1(y - \hat{x}_1) \\
\dot{\hat{x}}_2 &= \phi_0(\hat{x}, u) + h_2(y - \hat{x}_1)
\end{aligned}
\tag{4.3}
$$

where $\phi_0(x, u)$ is a nominal model of the nonlinear function $\phi(x, u)$. The estimation error

$$
\tilde{x} = \begin{bmatrix} \tilde{x}_1 \\ \tilde{x}_2 \end{bmatrix} = \begin{bmatrix} x_1 - \hat{x}_1 \\ x_2 - \hat{x}_2 \end{bmatrix}
$$

satisfies the equation

$$
\begin{aligned}
\dot{\tilde{x}}_1 &= -h_1 \tilde{x}_1 + \tilde{x}_2 \\
\dot{\tilde{x}}_2 &= -h_2 \tilde{x}_1 + \delta(x, \tilde{x})
\end{aligned}
\tag{4.4}
$$

where $\delta(x, \tilde{x}) = \phi(x, \gamma(\hat{x})) - \phi_0(\hat{x}, \gamma(\hat{x}))$. As in any asymptotic observer, we want to design the observer gain $H = [h_1, h_2]^T$ to achieve asymptotic error convergence, that is, $\lim_{t \to \infty} \tilde{x}(t) = 0$. In the absence of the disturbance term $\delta(x, \tilde{x})$, asymptotic error convergence is achieved by designing the observer gain such that the matrix

$$
A_o = \begin{bmatrix} -h_1 & 1 \\ -h_2 & 0 \end{bmatrix}
$$

is Hurwitz; that is, its eigenvalues have negative real parts. For this second-order system, A_o is Hurwitz for any positive constants h_1 and h_2. In the presence of δ, we need to design the observer gain with the additional goal of rejecting the effect of the disturbance term δ on the estimation error $\tilde{x}$. This is ideally achieved, for any disturbance term δ, if the transfer function

$$
H_o(s) = \frac{1}{s^2 + h_1 s + h_2} \begin{bmatrix} 1 \\ s + h_1 \end{bmatrix}
$$

from δ to $\tilde{x}$ is identically zero. While this is not possible, we can design the observer gain such that the transfer function H_o is arbitrarily close to zero. By calculating the H_∞ norm of $H_o(s)$ it can be seen that the norm can be made arbitrarily small by choosing $h_2 \gg h_1 \gg 1$. In particular, taking

$$h_1 = \frac{\alpha_1}{\epsilon}, \quad h_2 = \frac{\alpha_2}{\epsilon^2} \tag{4.5}$$

for some positive constants α_1, α_2, and ϵ, with $\epsilon \ll 1$, it can be shown that

$$H_o(s) = \frac{\epsilon}{(\epsilon s)^2 + \alpha_1 \epsilon s + \alpha_2} \left[\begin{array}{c} \epsilon \\ \epsilon s + \alpha_1 \end{array} \right]$$

Hence, $\lim_{\epsilon \to 0} H_o(s) = 0$. The disturbance rejection property of the high-gain-observer design (4.5) can be also seen in the time domain by representing the error equation (4.4) in the singularly perturbed form. Towards that end, define the scaled estimation errors

$$\eta_1 = \frac{\tilde{x}_1}{\epsilon}, \quad \eta_2 = \tilde{x}_2 \tag{4.6}$$

The newly defined variables satisfy the singularly perturbed equation

$$\begin{aligned} \epsilon \dot{\eta}_1 &= -\alpha_1 \eta_1 + \eta_2 \\ \epsilon \dot{\eta}_2 &= -\alpha_2 \eta_1 + \epsilon \delta(x, \tilde{x}) \end{aligned} \tag{4.7}$$

This equation shows clearly that reducing ϵ diminishes the effect of the disturbance term δ. It shows also that, for small ϵ, the dynamics of the estimation error will be much faster than the dynamics of x. Notice, however, that the change of variables (4.6) may cause the initial condition $\eta_1(0)$ to be of order $O(1/\epsilon)$ even when $\tilde{x}_1(0)$ is of order $O(1)$. With this initial condition, the solution of (4.7) will contain a term of the form $(1/\epsilon)e^{-at/\epsilon}$ for some $a > 0$. While this exponential mode decays rapidly, it exhibits an impulsive-like behavior where the transient peaks to $O(1/\epsilon)$ values before it decays rapidly towards zero. In fact, the function $(1/\epsilon)e^{-at/\epsilon}$ approaches an impulse function as ϵ tends to zero. This behavior is known as the *peaking phenomenon*. It is important to realize that the peaking phenomenon is not a consequence of using the change of variables (4.6) to represent the error dynamics in the singularly perturbed form. It is an intrinsic feature of any high-gain-observer design that rejects the effect of the disturbance term δ in (4.4); that is, any design with $h_2 \gg h_1 \gg 1$. This point can be seen by calculating the transition matrix $e^{A_o t}$ and noting that the $(2,1)$ element is given by

$$\frac{-2h_2}{\sqrt{4h_2 - h_1^2}} e^{-h_1 t/2} \sin\left(\frac{t\sqrt{4h_2 - h_1^2}}{2} \right)$$

when $4h_2 > h_1^2$, and

$$\frac{-h_2}{\sqrt{h_1^2 - 4h_2}} \left\{ \exp\left[-\left(\frac{h_1 + \sqrt{h_1^2 - 4h_2}}{2} \right) t \right] - \exp\left[-\left(\frac{h_1 - \sqrt{h_1^2 - 4h_2}}{2} \right) t \right] \right\}$$

when $4h_2 < h_1^2$. The magnitude of the coefficient of the exponential mode is greater than $\sqrt{h_2}$ in the first case and h_2/h_1 in the second one. Thus, as we increase h_1 and h_2/h_1, we drive this coefficient toward infinity.

The peaking phenomenon was studied in the context of high-gain state feedback control of linear [39, 15, 29] and nonlinear [31, 52] systems. Its investigation for high-gain observers appeared as early as 1979 in the work of Polotski [45]. The impact of observer peaking on closed-loop stability was studied by Esfandiari and Khalil [14]. Examples 3 and 4 of [14] illustrate the peaking phenomenon and show how it could destabilize the closed-loop system, as the impulsive-like behavior is transmitted from the observer to the plant. A key contribution of [14] is the following observation. If the state feedback control $\gamma(x)$ is a globally bounded function of x, it will provide a buffer that protects the plant from peaking. In particular, during the peaking period, the control $\gamma(\hat{x})$ saturates. Since the peaking period shrinks to zero as ϵ tends to zero, for sufficiently small ϵ the peaking period becomes so small that the state of the plant x remains close to its initial value.

After the peaking period, the estimation error becomes of order $O(\epsilon)$ and the feedback control $\gamma(\hat{x})$ becomes $O(\epsilon)$ close to $\gamma(x)$. Consequently, the trajectories of the closed-loop system under output feedback asymptotically approach its trajectories under state feedback as ϵ tends to zero. This leads to recovery of the performance achieved under state feedback. The global boundedness of $\gamma(x)$ can be always achieved by saturating the state feedback control, or the state estimates, outside a compact region of interest.

The analysis of the closed-loop system under output feedback proceeds as follows. The system is represented in the singularly perturbed form

$$
\begin{aligned}
\dot{x}_1 &= x_2 \\
\dot{x}_2 &= \phi(x, \gamma(\hat{x})) \\
\epsilon\dot{\eta}_1 &= -\alpha_1\eta_1 + \eta_2 \\
\epsilon\dot{\eta}_2 &= -\alpha_2\eta_1 + \epsilon\delta(x, \tilde{x})
\end{aligned}
\tag{4.8}
$$

where $\hat{x}_1 = x_1 - \epsilon\eta_1$ and $\hat{x}_2 = x_2 - \eta_2$. The slow subsystem of (4.8), obtained by setting $\epsilon = 0$, is the closed-loop system under state feedback (4.2). The fast subsystem is

$$
\epsilon\dot{\eta} = \begin{bmatrix} -\alpha_1 & 1 \\ -\alpha_2 & 0 \end{bmatrix} \eta \overset{\text{def}}{=} A_0\eta
$$

Let $V(x)$ be a Lyapunov function for the slow subsystem (4.2), which is guaranteed to exist for any stabilizing state feedback control $\gamma(x)$, and let $W(\eta) = \eta^T P_0 \eta$ be a Lyapunov function for the fast subsystem, where P_0 is the solution of the Lyapunov equation $P_0 A_0 + A_0^T P_0^T = -I$. Define the sets Ω_c and Σ by $\Omega_c = \{V(x) \le c\}$ and $\Sigma = \{W(\eta) \le \rho\epsilon^2\}$, where $c > 0$ is chosen such that Ω_c is in the interior of the region of attraction of (4.2).

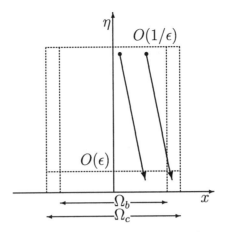

FIGURE 1. Illustration of fast convergence to the set $\Omega_c \times \Sigma$.

The analysis can be divided in two basic steps. In the first step we show that for sufficiently large ρ there is $\epsilon_1^* > 0$ such that for each $0 < \epsilon < \epsilon_1^*$ the origin of the closed-loop system is asymptotically stable and the set $\Omega_c \times \Sigma$ is a positively invariant subset of the region of attraction. The proof makes use of the fact that in $\Omega_c \times \Sigma$, η is $O(\epsilon)$. In the second step of the analysis we show that for any bounded $\hat{x}(0)$ and any $x(0) \in \Omega_b$, where $0 < b < c$, there exists $\epsilon_2^* > 0$ such that for each $0 < \epsilon < \epsilon_2^*$ the trajectory enters the set $\Omega_c \times \Sigma$ in finite time. The proof makes use of the fact that Ω_b is in the interior of Ω_c and $\gamma(\hat{x})$ is globally bounded. Hence, there exits a time $T_1 > 0$, independent of ϵ, such that any trajectory starting in Ω_b will remain in Ω_c for all $t \in [0, T_1]$. Then, using the fact that the fast variables η decay faster than an exponential mode of the form $(1/\epsilon)e^{-at/\epsilon}$, we can show that the trajectory enters the set $\Omega_c \times \Sigma$ within the time interval $[0, T(\epsilon)]$ where $\lim_{\epsilon \to 0} T(\epsilon) = 0$. Thus, by choosing ϵ small enough we can ensure that $T(\epsilon) < T_1$. Figure 1 gives a sketch that illustrates this behavior.

The full-order observer (4.3) provides estimates $(\hat{x}_1, \hat{x}_2)$ of the full state vector which are then used to replace (x_1, x_2) in the feedback control law. We can use the fact that $y = x_1$ is measured in two different ways. On one hand, we can use only $\hat{x}_2$ to replace x_2 in the control law, while using the measured x_1. This approach does not change the analysis of the closed-loop system and we obtain the same results as before. On the other hand, we can use a reduced-order observer that estimates only $\hat{x}_2$. Such an observer is given by

$$
\begin{aligned}
\dot{w} &= -h(w + hy) + \phi_o(\hat{x}, u) \\
\hat{x}_2 &= w + hy
\end{aligned}
\tag{4.9}
$$

where $h = \alpha/\epsilon$ for some positive constants α and ϵ with $\epsilon \ll 1$. It is not difficult to see that the high-gain reduced-order observer (4.9) exhibits the peaking phenomenon, and that global boundedness of the state feedback control plays the same role as in the full-order observer case.

The high-gain observer (4.3) or (4.9) is basically an approximate differentiator. This point can be easily seen in the special case when the nominal function ϕ_0 is chosen to be zero; for then the observer is linear. For the full-order observer (4.3) the transfer function from y to $\hat{x}$ is given by

$$\frac{\alpha_2}{(\epsilon s)^2 + \alpha_1 \epsilon s + \alpha_2} \left[\begin{array}{c} 1 + (\epsilon \alpha_1/\alpha_2)s \\ s \end{array} \right] \rightarrow \left[\begin{array}{c} 1 \\ s \end{array} \right] \text{ as } \epsilon \to 0$$

and for the reduced-order observer (4.9) the transfer function from y to $\hat{x}_2$ is given by

$$\frac{s}{(\epsilon/\alpha)s + 1} \rightarrow s \text{ as } \epsilon \to 0$$

Thus, on a compact frequency interval, the high-gain observer approximates $\dot{y}$ for sufficiently small ϵ.

Realizing that the high-gain observer is basically an approximate differentiator, we can see that measurement noise and unmodeled high-frequency sensor dynamics will put a practical limit on how small ϵ could be. As we will see later on, despite this limitation there are interesting applications where the range of permissible values of ϵ allowed successful application of high-gain observers in experimental testing.

It is useful to note that for low-frequency (slow) measurement noise, we can handle the effect of measurement noise as part of the state feedback control design. This idea was used by Khalil in [22, 24, 25]. To illustrate the idea, let us reconsider the system (4.1) with measurement noise $v(t)$ where $\dot{v}(t)$ and $\ddot{v}(t)$ are bounded:

$$\begin{aligned} \dot{x}_1 &= x_2 \\ \dot{x}_2 &= \phi(x, u) \\ y &= x_1 + v \end{aligned} \qquad (4.10)$$

Suppose we want y to track a reference signal r. In the error coordinates

$$e_1 = y - r, \qquad e_2 = \dot{y} - \dot{r}$$

the system is represented by

$$\begin{aligned} \dot{e}_1 &= e_2 \\ \dot{e}_2 &= \phi(x, u) + \ddot{v} - \ddot{r} \end{aligned} \qquad (4.11)$$

The effect of measurement noise now appears as a bounded disturbance in the state equation, and can usually be handled by robust control techniques. As for the observer design, with e_1 as the measured output we can use the same high-gain observer as before.

3   Separation Principle

The combination of globally bounded state feedback control with high-gain observers allows for a separation approach where the state feedback control is designed first to meet the design objectives, then the high-gain observer is designed, fast enough, to recover the performance achieved under state feedback. This separation approach is used in most of the papers that utilize high-gain observers. It is proved in a generic form in the work of Teel and Praly [53], where it is shown that global stabilizability by state feedback and uniform observability imply semiglobal stabilizability by output feedback. A more comprehensive separation principle is proved by Atassi and Khalil [10]. They consider a class of multi-input-multi-output nonlinear systems of the form

$$
\begin{aligned}
\dot{x} &= Ax + B\phi(x, z, u) \\
\dot{z} &= \psi(x, z, u) \\
y &= Cx \\
\zeta &= q(x, z)
\end{aligned}
\tag{4.12}
$$

where u is the control input, y and ζ are measured outputs, and x and z constitute the state vector. The $r \times r$ matrix A, the $r \times p$ matrix B, and the $p \times r$ matrix C, given by

$$
A = \text{block diag}[A_1, \dots, A_p], \quad A_i = \begin{bmatrix} 0 & 1 & \cdots & \cdots & 0 \\ 0 & 0 & 1 & \cdots & 0 \\ \vdots & & & & \vdots \\ 0 & \cdots & \cdots & 0 & 1 \\ 0 & \cdots & \cdots & \cdots & 0 \end{bmatrix}_{r_i \times r_i}
$$

$$
B = \text{block diag}[B_1, \dots, B_p], \quad B_i = \begin{bmatrix} 0 \\ 0 \\ \vdots \\ 0 \\ 1 \end{bmatrix}_{r_i \times 1}
$$

$$
C = \text{block diag}[C_1, \dots, C_p], \quad C_i = \begin{bmatrix} 1 & 0 & \cdots & \cdots & 0 \end{bmatrix}_{1 \times r_i}
$$

where $1 \le i \le p$ and $r = r_1 + \dots + r_p$, represent p chains of integrators. The system is assumed to satisfy appropriate regularity conditions.

The goal of [10] is to design feedback control to stabilize the origin of the closed-loop system using only the measured outputs y and ζ. A two-step approach is followed. First a partial state feedback control that uses measurements of x and ζ is designed to asymptotically stabilize the origin.

Then a high-gain observer is used to estimate x from y. The state feedback control is allowed to be a dynamic system of the form

$$\begin{aligned} \dot{\vartheta} &= \Gamma(\vartheta, x, \zeta) \\ u &= \gamma(\vartheta, x, \zeta) \end{aligned} \tag{4.13}$$

The control (4.13) is implemented using

$$\begin{aligned} \dot{\vartheta} &= \Gamma(\vartheta, \hat{x}, \zeta) \\ u &= \gamma(\vartheta, \hat{x}, \zeta) \end{aligned} \tag{4.14}$$

where the state estimate $\hat{x}$ is generated by the high-gain observer

$$\dot{\hat{x}} = A\hat{x} + B\phi_0(\hat{x}, \zeta, u) + H(y - C\hat{x}) \tag{4.15}$$

The observer gain H is chosen as

$$H = \text{block diag}[H_1, \dots, H_p], \quad H_i = \begin{bmatrix} \alpha_1^i/\epsilon \\ \alpha_2^i/\epsilon^2 \\ \vdots \\ \alpha_{r_i-1}^i/\epsilon^{r_i-1} \\ \alpha_{r_i}^i/\epsilon^{r_i} \end{bmatrix}_{r_i \times 1} \tag{4.16}$$

where ϵ is a positive constant to be specified and the positive constants α_j^i are chosen such that the roots of

$$s^{r_i} + \alpha_1^i s^{r_i-1} + \cdots + \alpha_{r_i-1}^i s^1 + \alpha_{r_i}^i = 0$$

are in the open left-half plane, for all $i = 1, \dots, p$. The function $\phi_0(x, \zeta, u)$ is a nominal model of $\phi(x, z, u)$. The function ϕ_0 is required to be locally Lipschitz in its arguments over the domain of interest and globally bounded in x.

For the purpose of analysis, the observer dynamics are replaced by the equivalent dynamics of the scaled estimation error

$$\eta_{ij} = \frac{x_{ij} - \hat{x}_{ij}}{\epsilon^{r_i-j}} \tag{4.17}$$

for $1 \le i \le p$ and $1 \le j \le r_i$. Hence, $\hat{x} = x - D(\epsilon)\eta$ where

$$\begin{aligned} \eta &= [\eta_{11}, \dots, \eta_{1r_1}, \dots, \eta_{p1}, \dots, \eta_{pr_p}]^T \\ D(\epsilon) &= \text{block diag}[D_1, \dots, D_p], \quad D_i = \text{diag}[\epsilon^{r_i-1}, \dots, 1]_{r_i \times r_i} \end{aligned}$$

The closed-loop system is represented by

$$\begin{aligned} \dot{x} &= Ax + B\phi(x, z, \gamma(\vartheta, x - D(\epsilon)\eta, \zeta)) \\ \dot{z} &= \psi(x, z, \gamma(\vartheta, x - D(\epsilon)\eta, \zeta)) \\ \dot{\vartheta} &= \Gamma(\vartheta, x - D(\epsilon)\eta, \zeta) \\ \epsilon\dot{\eta} &= A_0\eta + \epsilon B\delta(x, z, \vartheta, D(\epsilon)\eta) \end{aligned} \tag{4.18}$$

where

$$\delta(x, z, \vartheta, D(\epsilon)\eta) = \phi(x, z, \gamma(\vartheta, \hat{x}, \zeta)) - \phi_0(\hat{x}, \zeta, \gamma(\vartheta, \hat{x}, \zeta))$$

and $\frac{1}{\epsilon}A_0 = D^{-1}(\epsilon)(A - HC)D(\epsilon)$ is an $r \times r$ Hurwitz matrix.

It is shown in [10] that the output feedback controller (4.14) recovers the performance of the state feedback controller (4.13) for sufficiently small ϵ. The performance recovery manifests itself in three points. First, the origin $(x = 0, z = 0, \vartheta = 0, \hat{x} = 0)$ of the closed-loop system under output feedback is asymptotically stable. Second, the output feedback controller recovers the region of attraction of the state feedback controller in the sense that if $\mathcal{R}$ is the region of attraction under state feedback, then for any compact set $\mathcal{S}$ in the interior of $\mathcal{R}$ and any compact set $\mathcal{Q} \subseteq R^r$, the set $\mathcal{S} \times \mathcal{Q}$ is included in the region of attraction under output feedback control. Third, the trajectory of (x, z, ϑ) under output feedback approaches the trajectory under state feedback as $\epsilon \to 0$.

Performance recovery is shown in three steps. First, boundedness of trajectories starting in the specified compact set is established by regulating the parameter ϵ such that state estimation is fast enough. Of paramount importance at this stage is the global boundedness of the control function. Then, these trajectories are shown to be arbitrarily close to the origin after a finite time interval; thus a property of ultimate boundedness is established. Finally, local asymptotic stability of the origin is argued in three cases: the case where perfect knowledge of the system's nonlinearity is available ($\phi_0 = \phi$), the case where the origin under state feedback control is exponentially stable, and the case where the origin under state feedback control is asymptotically by not exponentially stable combined with an imperfect knowledge of the system's nonlinearity. In the last case certain conditions were imposed on the growth of the modeling error due to the imperfect knowledge of the system's nonlinearity.

Atassi and Khalil extended the results of [10] in two different directions. In [8] they proved the separation principle for the more general case when the state feedback control renders a certain compact set positively invariant and asymptotically attractive. This more general result allows the separation principle to be applied to a number of control tasks beyond the stabilization of an equilibrium point. Examples include finite-time convergence to a set [13], ultimate boundedness [54], servomechanisms [22, 37, 38, 18, 24], and adaptive control [23, 2, 3],

In [9], they extended the result of [10] in a different direction. They reviewed various techniques for the design of high-gain observers and classified them into three groups. First, pole-placement algorithms which lead to either a two-time scale structure as in [14] or a multiple time-scale structure as in [47]. Second, Riccati equation-based algorithms which lead to either an H_2 Riccati equation as in [12] and [48, Section 4.4.1] or to an H_∞ Riccati equation as in [44] and [48, Section 4.4.2]. Third, Lyapunov equation-based algorithm as in [16]. They showed that separation results similar to those of [10] can be obtained for any one of the other high-gain observer designs provided the state feedback control is globally bounded.

4 Stabilization and Semiglobal Stabilization

Stabilization of nonlinear systems using high-gain observers appeared before the work of Esfandiari and Khalil [14], which was the first paper to draw attention to the impact of peaking on closed-loop stability and to suggest the use of globally bounded feedback controllers. The results which were obtained without globally bounded controllers are either local results where the region of attraction shrinks with decreasing ϵ, or global results which require global Lipschitz conditions. Note that in the case of the global results peaking is present but it does not destabilize the system due to the restrictive assumptions used. It is our opinion, however, that such global results are not very useful because the presence of peaking presents a clearly unacceptable transient response. Examples of the local stabilization results can be found in the work of Tornambé [55] for a class of input-output linearizable systems and the work of Nicosia and Tornambé [41] on robots with elastic joints. Examples of the global stabilization results can be found in the work of Khalil and Saberi [27], Saberi and Sannuti [47], and Gauthier, Hammouri, and Othman [16] for different classes of input-output linearizable systems. The paper [27] does not explicitly use high-gain observers, but it uses approximate differentiators which are equivalent to linear high-gain observers, as shown in Section 2.

Esfandiari and Khalil [14] studied the stabilization of fully-linearizable uncertain systems using high-gain observers and robust state feedback control techniques. They first gave local and global stabilization results under output feedback. Then, they illustrated the peaking phenomenon and showed how it could lead to shrinking of the region of attraction and even destabilization of the system. They then suggested the use of saturation to render the state feedback control globally bounded and overcome the destabilizing effect of peaking. They used singular perturbations as the main tool of analysis, which is the same approach used later on by Khalil and coworkers in all their work. This paper has been the impetus for a number of research contributions by Khalil's group as well as by several other researchers. A key feature of the combination of globally bounded state feedback control and high-gain observers is the recovery of the region of attraction. This feature was made explicit in the follow-up work by Khalil and Esfandiari [26] where they showed that their approach can achieve semiglobal stabilization.

Teel and Praly [53, 54] picked up on the work of Esfandiari and Khalil and developed some fundamental tools for semiglobal stabilization using high-gain observers and saturation [54]. Their results covered stabilization as well as ultimate boundedness. A key technical contribution of their work is the use of Lyapunov functions to prove asymptotic stability of the closed-loop system under output feedback without resorting to singular perturbation arguments as in Khalil's work. It should be mentioned, however, that while the Lyapunov argument of [54] is more elegant than the singular perturbation argument as a way of proving asymptotic stability, the singular perturbation argument shows that trajectories under output feedback approach trajectories under state feedback as ϵ tends to zero, a property which cannot be shown using Lyapunov theory. In [53], Teel and Praly combined results from Tornambé [55] and Esfandiari and Khalil [14] to give the first non-local separation principle for nonlinear systems, which we have already discussed in the previous section.

Other stabilization results using high-gain observers include the work of Lin and Saberi [35], Praly and Jiang [46], Lin and Qian [34], and Jiang, Hill, and Guo [21]. Except for [21], the other papers combine high-gain observers with saturation to make the feedback control globally bounded, and use either the singular perturbation approach of Khalil or the tools of Teel and Praly. The paper [21] does not use saturation but avoids peaking by a special choice of the initial state of the observer. This is an option that is valid only for systems of relative degree two. The idea can be illustrated by the second-order example of Section 2. Recall from (4.6) that peaking is induced by the fact that $\eta_1(0) = [x_1(0) - \hat{x}_1(0)]/\epsilon$. Since x_1 is measured we can, in essence, choose $\hat{x}_1(0)$ equal to $x_1(0)$, or within $O(\epsilon)$ from it, thus avoiding an $O(1/\epsilon)$ initial value of $\eta_1(0)$. There may be some difficulty in implementing this idea due to measurement delays.

5 Nonlinear Servomechanisms

In a series of papers [22, 37, 38, 24, 25], Khalil and coworkers used high-gain observers in the nonlinear servomechanism problem. The paper [22] considers a class of input-output linearizable systems with no zero dynamics and with the disturbance satisfying a strict feedback structure. By using the tracking error and its derivatives as state variables, the system is represented as a chain of integrators with all the uncertainty in the state equation satisfying the matching condition. The paper then designs a robust controller which can achieve regional or semiglobal tracking. There are three basic ingredients of the approach used in [22]. First, by studying the dynamics of the system on the zero-error manifold, a linear internal model is identified. The internal model generates not only the trajectories of the exosystem but also a number of higher-order harmonics generated by the nonlinearities. A linear servo compensator is then synthesized and augmented with the plant. Second, the separation approach is used to design a robust output feedback controller, where a state feedback controller is designed first and then a high-gain observer, that estimates the derivatives of the output, is used to recover the performance achieved under state feedback. A key tool in this approach is the saturation of the state feedback control outside a compact region of interest. Third, to achieve robust regional or semiglobal stabilization of the augmented system (formed of the plant and the servo compensator), the state feedback design uses an effective strategy, whereby a robust control is designed as if the goal was to stabilize the origin. This control brings the trajectories of the system to some neighborhood of the origin in finite time. Near the origin, the robust controller acts as a high-gain feedback controller that stabilizes the disturbance-dependent zero-error manifold. It is emphasized that the output feedback controller of [22] is an error-driven controller which could not have been implemented by state feedback since the change of variables from the original states of the system to the states where the design is calculated is disturbance-dependent.

The results of [22] were extended by Mahmoud and Khalil [37, 38] to input-output linearizable systems with zero dynamics, where [37] deals with the special case of constant exogenous systems and [38] deals with the more general case of time-varying exogenous signals. In [24] and [25], Khalil extended the results of [37] and [38] by designing the controller as a universal one that uses limited information about the plant under control. In particular, in the case of constant exogenous signals [24], it is shown that the controller can be designed knowing only the relative degree of the plant and the sign of its high-frequency gain. It is shown in [24] that such universal regulator reduces to the classical PI controller followed by saturation for relative-degree-one systems, and to the classical PID controller followed by saturation for relative-degree-two systems. In the more general case of time-varying signals [25], the controller can be designed knowing only the relative degree of the plant, the sign of its high-frequency gain, and the characteristic equation of the internal model. Moreover, it is shown in [25] that a δ error in the characteristic equation of the internal model results an $O(\delta)$ error in the steady-state tracking error.

Isidori [18] unified the approach of Khalil with the general theory of nonlinear servomechanisms, as presented in [17]. He showed that for a class of systems that exhibit a triangular structure, exponential stabilizability of the plant at the origin is a sufficient condition for the existence of a globally-defined zero-error manifold. The manifold representation of [18] plays a key role in deriving the results of [25]. Scrrani and Isidori [49] extended the results of [18] removing the minimum-phase assumption.

In a different approach to the universal regulator design, Alvarez-Ramirez, Alvarez, and Suárez [7] used high-gain observers in the design of output feedback control for a class of nonlinear systems of relative degree one. They showed that their controller is equivalent to a standard PI controller with antirest windup structure. The result has some similarities with the universal regulator of [24].

6 Adaptive Control

The use of high-gain observers in adaptive control of nonlinear systems simplifies the analysis considerably because it reduces the Lyapunov-based adaptive design to a design under state feedback. In effect, it reduces a high-relative-degree problem to a relative-degree-one problem. This line of research was pursued independently by Khalil [23] and Jankovic [19, 20]. Khalil [23] considers a single-input–single-output minimum phase nonlinear system which can be represented globally by an input-output model. The model depends linearly on unknown parameters which belong to a known compact convex set. The paper designs a semiglobal adaptive output feedback controller which ensures that the output of the system tracks any given reference signal which is bounded and has bounded derivatives up to the order of the system. The reference signal and its derivatives are assumed to belong to a known compact set. They are also assumed to be sufficiently rich to satisfy a persistence of excitation condition. The design process is simple. First it is assumed that the output and its derivatives are available for feedback and the adaptive controller is designed as a state feedback controller in appropriate coordinates. Then, the controller is saturated outside a compact region of interest and a high-gain observer is used to estimate the derivatives of the output. It is shown via asymptotic analysis that, for sufficiently small ϵ, the adaptive output feedback controller recovers the performance achieved under state feedback. One drawback of [23] is that persistence of excitation is required not only for parameter convergence, but even for tracking error convergence. Without persistence of excitation, it is only shown that the mean square tracking error is of order $O(\epsilon)$. This drawback was removed in the work of Aloliwi and Khalil [2, 3] where tracking error convergence is shown without persistence of excitation. To arrive at this result, [2] and [3] analyze the closed-loop system under output feedback directly, rather than the separation approach used in [23]. This is one of the few results reported in this paper which cannot be proved using the separation principle of Section 3 because convergence under state feedback cannot be represented as asymptotic stability of a positively invariant set.

Robustness of the adaptive controller of [23] is studied in [2, 3], where two robustness results are shown. First, it is shown that the adaptive controller is robust to sufficiently small bounded disturbance. Second, for a wide class of not-necessarily-small bounded disturbance, a robustifying control component is added to achieve a small ultimate bound on the tracking error, provided an upper bound on the disturbance is known. Similar results are obtained for unmodeled dynamics in [4].

Jankovic [20] derives an output feedback adaptive controller for a class of nonlinear systems using a high-gain observer combined with saturation. The class of systems is similar to the one used in [23], but the approach differs from [23] in two aspects. First, he uses a modified version of the observer-based identification scheme of [32]. Second, the design is not based on a separation approach as in [23]. The result of [20] is semiglobal and requires a persistence of excitation condition for tracking error and parameter error convergence. The persistence of excitation condition is relaxed in [19] by choosing the adaptation gain sufficiently large.

The early work on the use of high-gain observers in adaptive control, especially [23], has inspired a number of interesting extensions. Lee and Khalil [33] apply the adaptive controller of [23] to the control of an n-link robot manipulator with unknown load, using only joint position measurements. High-gain observers are used to estimate joint velocities and the control inputs are saturated outside a compact region of interest. Seshagiri and Khalil [50] design an adaptive output feedback controller for a nonlinear system where Radial Basis Function (RBF) neural networks are used to model the system's nonlinearities. This is essentially an application of the results of [3] since RBF networks are linearly parameterized in the weights and the approximation error can be treated as bounded disturbance. Zhang, Ge, and Hang [56] carry the neural network application one step further by working with multilayer neural networks which depend nonlinearly on the weights. Miyasato [40] shows how the use of high-gain observers can simplify the traditional model reference adaptive control of linear systems.

7   Sliding Mode Control

Oh and Khalil [42, 43] used high-gain observers in sliding mode control. The first paper [42] deals with stabilization of an input-output linearizable system with no zero dynamics, while the second paper [43] deals with tracking in the presence of disturbances and allows zero dynamics. Due to discontinuity of the sliding mode control, the separation approach of Section 3 cannot be used. The approach used in [42, 43] is to design the high-gain observer first; then design the sliding mode control as a globally bounded function of the state estimates to ensure attractivity of the sliding manifold. The results are semiglobal and show ultimate boundedness with an ultimate bound of order $O(\epsilon)$.

8   Unmodeled Fast Dynamics

Since high-gain observers extend the bandwidth of the controller, it is important to study the robustness of the control design to unmodeled fast (high-frequency) dynamics. Such a study was initiated by Aldhaheri and Khalil [1] who studied the robustness of the stabilizing controller of [14] to unmodeled actuator dynamics. Their analysis confirms the intuition gained from singular perturbation theory [30] that the design will be robust provided the actuator dynamics are sufficiently fast relative to the dynamics of the nominal closed-loop system. An interesting finding of [1] is the fact that the actuator dynamics need not be faster than the observer dynamics. Mahmoud and Khalil [36] extend the results of [1] in two directions. First, they consider sensor dynamics in addition to actuator dynamics. Second, they work with a general stabilizing state feedback controller in the spirit of the separation principle of Atassi and Khalil [10]. They show that, given any globally-bounded stabilizing state-feedback control, the closed-loop system performance can be recovered by a sufficiently fast high-gain observer in the presence of sufficiently fast actuator and sensor dynamics. The actuator dynamics needs not be faster than the observer dynamics, but the sensor dynamics should be sufficiently faster than the observer dynamics.

9   Discrete-Time Implementation

Dabroom and Khalil [11] studied discrete-time implementation of linear high-gain observers and their use as numerical differentiators. It is shown in [11] that discretization using the bilinear transformation method gives better results than other discretization methods. Moreover, many of the available numerical differentiators are special cases of the bilinear discrete-time equivalents of full-order or reduced-order high-gain observers.

10   Application to Induction Motors

One application of high-gain observers which has been carried successfully to experimental testing is its use in the control of induction motors. There are two distinct applications of high-gain observers in induction motors: first, their use to estimate the rotor speed from the rotor position measurement; second, their use to estimate the derivative of the stator current in senesorless control, i.e., control without measurement of the rotor position. The first application appears in the work of Aloliwi, Strangas and Khalil [28, 5, 6] and the second one in the work of Strangas et al [51].

Acknowledgement

This work was supported by the National Science Foundation under grant number ECS-9703742.

11 REFERENCES

[1] R. W. Aldhaheri and H. K. Khalil. Effect of unmodeled actuator dynamics on output feedback stabilization of nonlinear systems. *Automatica*, 32(9):1323–1327, 1996.

[2] B. Aloliwi and H.K. Khalil. Adaptive output feedback regulation of a class of nonlinear systems: convergence and robustness. *IEEE Trans. Automat. Contr.*, 42:1714–1716, 1997.

[3] B. Aloliwi and H.K. Khalil. Robust adaptive output feedback control of nonlinear systems without persistence of excitation. *Automatica*, 33:2025–2032, 1997.

[4] B. Aloliwi and H.K. Khalil. Robust adaptive control of nonlinear systems with unmodeled dynamics. In *Proc. IEEE Conf. on Decision and Control*, pages 2872–2873, Tampa, FL, December 1998.

[5] B. Aloliwi, H.K. Khalil, and E.G. Strangas. Robust speed control of induction motors. In *Proc. American Control Conf.*, Albuquerque, NM, June 1997. WP16:4.

[6] B. Aloliwi, H.K. Khalil, and E.G. Strangas. Robust speed control of induction motors: application to a benchmark example. 1998. Submitted for publication.

[7] J. Alvarez-Ramirez, J. Alvarez, and R. Suárez. Robust PI control of a class of nonlinear systems. 1998. Submitted for publication.

[8] A. N. Atassi and H. K. Khalil. A separation principle for the control of a class of nonlinear systems. In *Proc. IEEE Conf. on Decision and Control*, pages 855–860, Tampa, FL, December 1998.

[9] A. N. Atassi and H. K. Khalil. Separation results for the stabilization of nonlinear systems using different high-gain observer designs. 1998. Submitted for publication.

[10] A.N. Atassi and H.K. Khalil. A separation principle for the stabilization of a class of nonlinear systems. *IEEE Trans. Automat. Contr.*, 44, 1999. To appear. See also *Proc. European Control Conf.*, Brussels, July 1997. WE-A-A-4.

[11] A. Dabroom and H.K. Khalil. Discrete-time implementation of high-gain observers for numerical differentiation. *Int. J. Contr.*, 1999. To appear.

[12] J.C. Doyle and G. Stein. Robustness with observers. *IEEE Trans. Automat. Contr.*, AC-24(4):607–611, 1979.

[13] F. Esfandiari and H.K. Khalil. Observer-based design of uncertain systems: recovering state feedback robustness under matching conditions. In *Proc. Allerton Conf.*, pages 97–106, Monticello, IL, September 1987.

[14] F. Esfandiari and H.K. Khalil. Output feedback stabilization of fully linearizable systems. *Int. J. Contr.*, 56:1007–1037, 1992.

[15] B.A. Francis and K. Glover. Bounded peaking in the optimal linear regulator with cheap control. *IEEE Trans. Automat. Contr.*, AC-23(4):608–617, 1978.

[16] J.P. Gauthier, H. Hammouri, and S. Othman. A simple observer for nonlinear systems application to bioreactors. *IEEE Trans. Automat. Contr.*, 37(6):875–880, 1992.

[17] A. Isidori. *Nonlinear Control Systems*. Springer-Verlag, New York, 3rd edition, 1995.

[18] A. Isidori. A remark on the problem of semiglobal nonlinear output regulation. *IEEE Trans. Automat. Contr.*, 42(12):1734–1738, 1997.

[19] M. Jankovic. Adaptive output feedback control of nonlinear feedback linearizable systems. *Int. J. Adaptive Control and Signal Processing*, 10:1–18, 1996.

[20] M. Jankovic. Adaptive nonlinear output feedback tracking with a partial high-gain observer and backstepping. *IEEE Trans. Automat. Contr.*, 42(1):106–113, 1997.

[21] Z.P. Jiang, D.J. Hill, and Y. Guo. Semi-global output feedback stabilization for the nonlinear benchmark example. In *Proc. European Control Conf.*, Brussels, July 1997. FR-A-K-8.

[22] H.K. Khalil. Robust servomechanism output feedback controllers for a class of feedback linearizable systems. *Automatica*, 30(10):1587–1599, 1994.

[23] H.K. Khalil. Adaptive output feedback control of nonlinear systems represented by input-output models. *IEEE Trans. Automat. Contr.*, 41(2):177–188, 1996.

[24] H.K. Khalil. Universal regulators for minimum phase nonlinear systems. In *Proc. American Control Conf.*, Philadelphia, PA, June 1998.

[25] H.K. Khalil. On the design of robust servomechanisms for minimum phase nonlinear systems. In *Proc. IEEE Conf. on Decision and Control*, pages 3075–3080, Tampa, FL, December 1998.

[26] H.K. Khalil and F. Esfandiari. Semiglobal stabilization of a class of nonlinear systems using output feedback. *IEEE Trans. Automat. Contr.*, 38(9):1412–1415, 1993.

[27] H.K. Khalil and A. Saberi. Adaptive stabilization of a class of nonlinear systems using high-gain feedback. *IEEE Trans. Automat. Contr.*, AC-32(11):1031–1035, 1987.

[28] H.K. Khalil and E.G. Strangas. Robust speed control of induction motors using position and current measurement. *IEEE Trans. Automat. Contr.*, 41:1216–1220, 1996.

[29] H. Kimura. A new approach to the perfect regulation and the bounded peaking in linear multivariable control systems. *IEEE Trans. Automat. Contr.*, AC-26(1):253–270, 1981.

[30] P.V. Kokotovic, H.K. Khalil, and J. O'Reilly. *Singular Perturbations Methods in Control: Analysis and Design*. Academic Press, New York, 1986.

[31] P.V. Kokotovic and R. Marino. On vanishing stability regions in nonlinear systems with high-gain feedback. *IEEE Trans. Automat. Contr.*, AC-31(10):967–970, 1986.

[32] M. Krstic, P.V. Kokotovic, and I. Kanellakopoulos. Adaptive nonlinear output-feedback control with an observer-based identifier. In *Proc. American Control Conf.*, pages 2821–2825, San Francisco, CA, June 1993.

[33] K.W. Lee and H.K. Khalil. Adaptive output feedback control of robot manipulators using high-gain observers. *Int. J. Contr.*, 67(6):869–886, 1997.

[34] W. Lin and C. Qian. Semiglobal robust stabilization of nonlinear systems by partial state and output feedback. In *Proc. IEEE Conf. on Decision and Control*, pages 3105–3110, Tampa, FL, December 1998.

[35] Z. Lin and A. Saberi. Robust semi-global stabilization of minimum-phase input-output linearizable systems via partial state and output feedback. *IEEE Trans. Automat. Contr.*, 40(6):1029–1041, 1995.

[36] M.S. Mahmoud and H.K. Khalil. Robustness of high-gain observer-based nonlinear controllers to unmodeled actuators and sensors. 1998. Submitted for publication.

[37] N.A. Mahmoud and H.K. Khalil. Asymptotic regulation of minimum phase nonlinear systems using output feedback. *IEEE Trans. Automat. Contr.*, 41(10):1402–1412, 1996.

[38] N.A. Mahmoud and H.K. Khalil. Robust control for a nonlinear servomechanism problem. *Int. J. Contr.*, 66(6):779–802, 1997.

[39] T. Mita. On zeros and responses of linear regulators and linear observers. *IEEE Trans. Automat. Contr.*, AC-22(3):423–428, 1977.

[40] Y. Miyasato. A simple redesign of model reference adaptive control system and its robustness. In *Proc. IEEE Conf. on Decision and Control*, pages 2880–2885, Tampa, FL, December 1998.

[41] S. Nicosia and A. Tornambé. High-gain observers in the state and parameter estimation of robots having elastic joints. *Systems Contr. Lett.*, 19:331–337, 1993.

[42] S. Oh and H.K. Khalil. Output feedback stabilization using variable structure control. *Int. J. Contr.*, 62:831–848, 1995.

[43] S. Oh and H.K. Khalil. Nonlinear output feedback tracking using high-gain observer and variable structure control. *Automatica*, 33:1845–1856, 1997.

[44] I.R. Petersen and C.V. Holot. High-gain observers applied to problems in disturbance attenuation, H-infinity optimization and the stabilization of uncertain linear systems. In *Proc. American Control Conf.*, pages 2490–2496, Atlanta, GA, June 1988.

[45] V.N. Polotskii. On the maximal errors of an asymptotic state identifier. *Automation and Remote Control*, 11:1116–1121, 1979.

[46] L. Praly and Z.P. Jiang. Further results on robust semiglobal stabilization with dynamic input uncertainties. In *Proc. IEEE Conf. on Decision and Control*, pages 891–896, Tampa, FL, December 1998.

[47] A. Saberi and P. Sannuti. Observer design for loop transfer recovery and for uncertain dynamical systems. *IEEE Trans. Automat. Contr.*, 35(8):878–897, 1990.

[48] A. Saberi B.M. Chen and P. Sannuti. *Loop Transfer Recovery: Analysis and Design*. Springer-Verlag, New York, 1993.

[49] A. Scrrani and A. Isidori. Robust output regulation for a class of non-minimum phase systems. In *Proc. IEEE Conf. on Decision and Control*, pages 867–872, Tampa, FL, December 1998.

[50] S. Seshagiri and H.K. Khalil. Output feedback control of nonlinear systems using RBF neural networks. 1998. Submitted for publication.

[51] E.G. Strangas, H.K. Khalil, B. Aloliwi, L. Laubinger, and J. Miller. Robust tracking controllers for induction motors without rotor position sensor: analysis and experimental results. *IEEE Trans. Energy Conversion* 1999. To appear.

[52] H.J. Sussmann and P.V. Kokotovic. The peaking phenomenon and the global stabilization of nonlinear systems. *IEEE Trans. Automat. Contr.*, 36(4):424–440, 1991.

[53] A. Teel and L. Praly. Global stabilizability and observability imply semi-global stabilizability by output feedback. *Systems Contr. Lett.*, 22:313–325, 1994.

[54] A. Teel and L. Praly. Tools for semiglobal stabilization by partial state and output feedback. *SIAM J. Control & Optimization*, 33:1443–1488, 1995.

[55] A. Tornambé. Output feedback stabilization of a class of non-minimum phase nonlinear systems. *Systems Contr. Lett.*, 19:193–204, 1992.

[56] T. Zhang, S.S. Ge, and C.C. Hang. Adaptive output feedback control for general nonlinear systems using multilayer neural networks. In *Proc. American Control Conf.*, pages 520–524, Philadelphia, June 1998.

Output-Feedback Control of Stochastic Nonlinear Systems
Stabilization, Disturbance Attenuation, and Adaptation

Miroslav Krstić and Hua Deng

Department of Applied Mechanics and Engineering Sciences
University of California at San Diego
La Jolla, CA 92093-0411, USA

1 Introduction

Despite huge popularity of the LQG control problem, the stabilization problem for *nonlinear stochastic* systems has been receiving relatively little attention until recently. Efforts towards (global) *stabilization* of stochastic nonlinear systems have been initiated in the work of Florchinger [6, 7, 8] who, among other things, extended the concept of control Lyapunov functions and Sontag's stabilization formula [24] to the stochastic setting. A breakthrough towards arriving at *constructive* methods for stabilization of broader classes of stochastic nonlinear systems came with the result of Pan and Başar [21] who derived a backstepping design for strict-feedback systems motivated by a risk-sensitive cost criterion [1, 11, 19, 23]. In [2, 3], for the same class of systems as in [21], we designed inverse optimal control laws, which, unlike those in [21], can be designed in an automated manner (via symbolic software). Important extensions were reported by Tsinias [27, 28].

In this chapter, we address the *output-feedback* problem for stochastic nonlinear systems. The output-feedback problem has received considerable attention in the recent robust and adaptive nonlinear control literature [18, 17, 22, 26, 14, 12]. The present chapter is the first to address the output-feedback problem in the stochastic setting.

We present three results. First, in Section 3, we design an output-feedback (observer based) backstepping control law which guarantees global asymptotic stability in probability. The result in Section 3 solves only the equilibrium stabilization problem under the assumption that the noise vector field is vanishing (which preserves the equilibrium) and the assumption that a bound on the noise covariance is known. These assumptions allow some interesting nonlinear systems but exclude linear systems with additive noise! For this reason, in [4] we addressed systems with nonvanishing noise vector field and unknown bound on covariance, and derived both "robust" and "adaptive" controllers for the class of strict-feedback systems. In Sections 4 and 5, we extend these results to the class of output feedback systems. The class of systems that we consider is the stochastic version of the *output-feedback form*, which is the broadest class for which *global* output-feedback controllers currently exist in the deterministic setting.

2   Preliminaries on Stochastic Stability

In this section, we briefly review some stochastic stability concepts and theorems from [16], which will be used in the following sections. Consider the nonlinear stochastic system

$$dx = f(x)dt + g(x)dw, \tag{5.1}$$

where $x \in \mathbb{R}^n$ is the state, w is an r-dimensional Wiener process with incremental covariance $\Sigma(t)\Sigma(t)^T dt$, i.e., $E\left\{dwdw^T\right\} = \Sigma(t)\Sigma(t)^T dt$, where $\Sigma(t)$ is a bounded function taking values in the set of nonnegative definite matrices, $f : \mathbb{R}^n \to \mathbb{R}^n$ and $g : \mathbb{R}^n \to \mathbb{R}^{n \times r}$ are locally Lipschitz, and $f(0) = 0$.

We first state notation which will be used in the sequel. For a matrix $X = [x_1, x_2, \cdots, x_n]$,

$$|X|_{\mathcal{F}} \triangleq \left(\text{Tr}\left\{X^T X\right\}\right)^{1/2} = \left(\text{Tr}\left\{XX^T\right\}\right)^{1/2} \tag{5.2}$$

denotes the Frobenius norm, and obviously,

$$|X|_{\mathcal{F}} = |\text{col}(X)| \tag{5.3}$$

where $\text{col}(X) = [x_1^T, x_2^T, \cdots, x_n^T]^T$.

Definition 5.1 *The system (5.1) is* noise-to-state stable (NSS) *if* $\forall \epsilon > 0$, *there exists a class* $\mathcal{KL}$ *function* $\beta(\cdot, \cdot)$ *and a class* $\mathcal{K}$ *function* $\gamma(\cdot)$, *such that*

$$P\left\{|(x(t)| < \beta(|x_0|, t) + \gamma\left(\sup_{t \geq s \geq t_0} |\Sigma(s)\Sigma(s)^T|_{\mathcal{F}}\right)\right\} \geq 1 - \epsilon, \ \forall t \geq 0, \forall x_0 \in \mathbb{R}^n \backslash \{0\}. \tag{5.4}$$

NSS is a stochastic analog of input-to-state stability [25].

Theorem 5.1 ([16]) *Consider system (5.1) and suppose there exists a* $\mathcal{C}^2$ *function* $V(x)$, *class* $\mathcal{K}_\infty$ *functions* α_1, α_2 *and* ρ, *and a class* $\mathcal{K}$ *function* α_3, *such that*

$$\alpha_1(|x|) \leq V(x) \leq \alpha_2(|x|), \tag{5.5}$$

$$|x| \geq \rho\left(|\Sigma\Sigma^T|_{\mathcal{F}}\right) \Rightarrow \mathcal{L}V(x, \Sigma) = \frac{\partial V}{\partial x}f(x) + \frac{1}{2}\text{Tr}\left\{\Sigma^T g^T \frac{\partial^2 V}{\partial x^2} g\Sigma\right\} \leq -\alpha_3(|x|). \tag{5.6}$$

Then the system (5.1) is NSS.

Consider the nonlinear stochastic system (5.1) with additional assumptions that $g(0) = 0$ and $\Sigma(t) \equiv I$.

Definition 5.2 *The equilibrium $x = 0$ of the system (5.1) is*

- globally stable in probability *if $\forall \epsilon > 0$ there exists a class $\mathcal{K}$ function $\gamma(\cdot)$ such that*

$$P\{|x(t)| < \gamma(|x_0|)\} \geq 1 - \epsilon, \qquad \forall t \geq 0, \forall x_0 \in \mathbb{R}^n \backslash \{0\}, \qquad (5.7)$$

- globally asymptotically stable in probability *if $\forall \epsilon > 0$ there exists a class $\mathcal{KL}$ function $\beta(\cdot, \cdot)$ such that*

$$P\{|x(t)| < \beta(|x_0|, t)\} \geq 1 - \epsilon, \qquad \forall t \geq 0, \forall x_0 \in \mathbb{R}^n \backslash \{0\}. \qquad (5.8)$$

Theorem 5.2 ([16]) *Consider system (5.1) and suppose there exists a C^2 function $V(x)$ and class $\mathcal{K}_\infty$ functions α_1 and α_2, such that*

$$\alpha_1(|x|) \leq V(x) \leq \alpha_2(|x|) \qquad (5.9)$$

$$\mathcal{L}V(x) = \frac{\partial V}{\partial x} f(x) + \frac{1}{2} \text{Tr}\left\{ g^T \frac{\partial^2 V}{\partial x^2} g \right\} \leq -W(x), \qquad (5.10)$$

where $W(x)$ is continuous and nonnegative. Then the equilibrium $x = 0$ is globally stable in probability and

$$P\left\{ \lim_{t \to \infty} W(x) = 0 \right\} = 1. \qquad (5.11)$$

This theorem is a stochastic analog of LaSalle's theorem, and the main Lyapunov theorem comes as a corollary.

Theorem 5.3 *Consider system (5.1) and suppose there exists a C^2 function $V(x)$, class $\mathcal{K}_\infty$ functions α_1 and α_2, and a class $\mathcal{K}$ function α_3, such that*

$$\alpha_1(|x|) \leq V(x) \leq \alpha_2(|x|), \qquad (5.12)$$

$$\mathcal{L}V(x) = \frac{\partial V}{\partial x} f(x) + \frac{1}{2} \text{Tr}\left\{ g^T \frac{\partial^2 V}{\partial x^2} g \right\} \leq -\alpha_3(|x|). \qquad (5.13)$$

Then the equilibrium $x = 0$ is globally asymptotically stable in probability.

3   Output-Feedback Stabilization in Probability

In this section we deal with nonlinear *output-feedback* systems driven by white noise. This class of systems is given by the following nonlinear stochastic differential equations:

$$
\begin{aligned}
dx_i &= x_{i+1}dt + \varphi_i(y)^\mathrm{T} dw, \qquad i = 1, \cdots, n-1 \\
dx_n &= u\,dt + \varphi_n(y)^\mathrm{T} dw \\
y &= x_1,
\end{aligned}
\tag{5.14}
$$

where $\varphi_i(y)$ are r-vector-valued smooth functions with $\varphi_i(0) = 0$, and w is an independent r-dimensional standard Wiener process.

Since the states $x_2, \cdots, x_n$ are not measured, we first design an observer which would provide exponentially convergent estimates of the unmeasured states in the absence of noise. The observer is designed as

$$
\dot{\hat{x}}_i = \hat{x}_{i+1} + k_i\,(y - \hat{x}_1), \qquad i = 1, \cdots, n
\tag{5.15}
$$

where $\hat{x}_{n+1} = u$. The observation error $\tilde{x} = x - \hat{x}$ satisfies

$$
\begin{aligned}
d\tilde{x} &=
\left[
\begin{array}{cccc}
-k_1 & & & \\
\vdots & & I & \\
-k_n & 0 & \cdots & 0
\end{array}
\right]
\tilde{x}\,dt + \varphi(y)^\mathrm{T} dw \\
&= A_0 \tilde{x}\,dt + \varphi(y)^\mathrm{T} dw,
\end{aligned}
\tag{5.16}
$$

where A_0 is designed to be asymptotically stable. Now, the entire system can be expressed as

$$
\begin{aligned}
d\tilde{x} &= A_0 \tilde{x}\,dt + \varphi(y)^\mathrm{T} dw \\
dy &= (\hat{x}_2 + \tilde{x}_2)\,dt + \varphi_1(y)^\mathrm{T} dw \\
d\hat{x}_2 &= [\hat{x}_3 + k_2\,(y - \hat{x}_1)]\,dt \\
&\;\;\vdots \\
d\hat{x}_n &= [u + k_n\,(y - \hat{x}_1)]\,dt.
\end{aligned}
\tag{5.17}
$$

Our output-feedback design will consist in applying a backstepping procedure to the system $(y, \hat{x}_2, \cdots, \hat{x}_n)$, while also taking care of the feedback connection through the $\tilde{x}$ system.

In the standard backstepping method for deterministic systems [9] (where dw/dt would be a bounded deterministic disturbance), a sequence of stabilizing functions $\alpha_i(\bar{\hat{x}}_i, y)$, where $\bar{\hat{x}}_i = [\hat{x}_2, \cdots, \hat{x}_i]^\mathrm{T}$, is constructed recursively to build a Lyapunov function of the form

$$
V = \sum_{i=1}^{n} \frac{1}{2} z_i^2 + \tilde{x}^\mathrm{T} P \tilde{x},
\tag{5.18}
$$

where P is a positive definite matrix which satisfies $A_0^\mathrm{T} P + P A_0 = -I$, and the error variables z_i are given by

$$
\begin{aligned}
z_1 &= y \\
z_i &= \hat{x}_i - \alpha_{i-1}\left(\bar{\hat{x}}_{i-1}, y\right), \qquad i = 2, \cdots, n.
\end{aligned}
\tag{5.19}
\tag{5.20}
$$

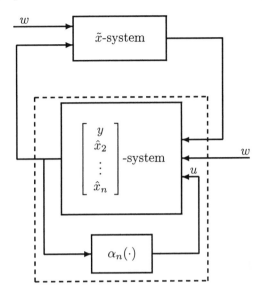

FIGURE 1. Feedback structure of the system (5.17)

The Lyapunov design for stochastic systems cannot be performed using the quadratic Lyapunov function (5.18) because of the term

$$\frac{1}{2}\mathrm{Tr}\left\{g^{\mathrm{T}}\frac{\partial^2 V}{\partial x^2}g\right\}$$

in (5.10). We instead employ a *quartic* (fourth order) Lyapunov function

$$V = \sum_{i=1}^{n}\frac{1}{4}z_i^4 + \left(\tilde{x}^{\mathrm{T}}P\tilde{x}\right)^2. \tag{5.21}$$

Our presentation of the backstepping procedure here is very concise: instead of introducing the stabilizing functions α_i in a step-by-step fashion, we derive them simultaneously. A reader who is a novice to the technique of backstepping is referred to [17].

We start by an important preparatory comment. Since $\varphi_i(0) = 0$, the α_i's will vanish at $\bar{\tilde{x}}_i = 0$, $y = 0$, as well as at $\bar{z}_i = 0$, where $\bar{z}_i = [z_1, \cdots, z_i]^{\mathrm{T}}$. Thus, by the mean value theorem, $\alpha_i(\bar{\tilde{x}}_i, y)$ and $\varphi(y)$ can be expressed respectively as

$$\alpha_i(\bar{\tilde{x}}_i, y) = \sum_{l=1}^{i} z_l \alpha_{il}(\bar{\tilde{x}}_i, y), \tag{5.22}$$

$$\varphi(y) = y\psi(y) \tag{5.23}$$

where $\alpha_{il}(\bar{\tilde{x}}_i, y)$ and $\psi(y)$ are smooth functions.

Now, we are ready to start the backstepping design procedure. According to Itô's differentiation rule [20], we have

$$dz_1 = (\hat{x}_2 + \tilde{x}_2)\, dt + \varphi_1(y)^{\mathrm{T}} dw \tag{5.24}$$

$$
\begin{aligned}
dz_i = {} & \left[\hat{x}_{i+1} + k_i \tilde{x}_1 - \sum_{l=2}^{i-1} \frac{\partial \alpha_{i-1}}{\partial \hat{x}_l} (\hat{x}_{l+1} + k_l \tilde{x}_1) - \frac{\partial \alpha_{i-1}}{\partial y} (\hat{x}_2 + \tilde{x}_2) \right. \\
& \left. - \frac{1}{2} \left(\frac{\partial^2 \alpha_{i-1}}{\partial y^2} \right) \varphi_1(y)^{\mathrm{T}} \varphi_1(y) \right] dt - \frac{\partial \alpha_{i-1}}{\partial y} \varphi_1(y)^{\mathrm{T}} dw \tag{5.25}
\end{aligned}
$$

$$i = 2, \cdots, n.$$

As we announced previously, we employ a Lyapunov function of a quartic form

$$V(z, \tilde{x}) = \frac{1}{4} y^4 + \frac{1}{4} \sum_{i=2}^{n} z_i^4 + \frac{b}{2} \left(\tilde{x}^{\mathrm{T}} P \tilde{x} \right)^2, \tag{5.26}$$

where b is a positive constant.

This form of the Lyapunov function clearly indicates that we view the system as a feedback connection in Figure 1. The first two terms in (5.26) constitute a Lyapunov function for the $(y, \hat{x}_2, \cdots, \hat{x}_n)$-system, while the third term in (5.26) is a Lyapunov function for the $\tilde{x}$-system. Even though not obvious from the calculations that follow, we achieve a nonlinear small-gain global stabilization (in probability) in the style of [13].

Now we start the process of selecting the functions $\alpha_i(\hat{\bar{x}}_i, y)$ to make $\mathcal{L}V$ negative definite. Along the solutions of (5.16), (5.24) and (5.25), we have

$$
\begin{aligned}
\mathcal{L}V = {} & y^3 (\hat{x}_2 + \tilde{x}_2) + \frac{3}{2} y^2 \varphi_1(y)^{\mathrm{T}} \varphi_1(y) + \sum_{i=2}^{n} z_i^3 \left[\hat{x}_{i+1} + k_i \tilde{x}_1 \right. \\
& - \sum_{l=2}^{i-1} \frac{\partial \alpha_{i-1}}{\partial \hat{x}_l} (\hat{x}_{l+1} + k_l \tilde{x}_1) - \frac{\partial \alpha_{i-1}}{\partial y} (\hat{x}_2 + \tilde{x}_2) \\
& - \frac{1}{2} \left(\frac{\partial^2 \alpha_{i-1}}{\partial y^2} \right) \varphi_1(y)^{\mathrm{T}} \varphi_1(y) \right] + \frac{3}{2} \sum_{i=2}^{n} z_i^2 \left(\frac{\partial \alpha_{i-1}}{\partial y} \right)^2 \varphi_1(y)^{\mathrm{T}} \varphi_1(y) \\
& - b\tilde{x}^{\mathrm{T}} P \tilde{x} |\tilde{x}|^2 + 2b \mathrm{Tr} \left\{ \varphi(y) \left(2P \tilde{x} \tilde{x}^{\mathrm{T}} P + \tilde{x}^{\mathrm{T}} P \tilde{x} P \right) \varphi(y)^{\mathrm{T}} \right\} \\
= {} & - b\tilde{x}^{\mathrm{T}} P \tilde{x} |\tilde{x}|^2 + 2b \mathrm{Tr} \left\{ \varphi(y) \left(2P \tilde{x} \tilde{x}^{\mathrm{T}} P + \tilde{x}^{\mathrm{T}} P \tilde{x} P \right) \varphi(y)^{\mathrm{T}} \right\} \\
& + y^3 (\alpha_1 + z_2 + \tilde{x}_2) + \frac{3}{2} y^2 \varphi_1(y)^{\mathrm{T}} \varphi_1(y) \\
& + \sum_{i=2}^{n} z_i^3 \left[\alpha_i + z_{i+1} + k_i \tilde{x}_1 - \sum_{l=2}^{i-1} \frac{\partial \alpha_{i-1}}{\partial \hat{x}_l} (\hat{x}_{l+1} + k_l \tilde{x}_1) - \frac{\partial \alpha_{i-1}}{\partial y} (\hat{x}_2 + \tilde{x}_2) \right.
\end{aligned}
$$

$$-\frac{1}{2}\left(\frac{\partial^2 \alpha_{i-1}}{\partial y^2}\right)\varphi_1(y)^{\mathrm{T}}\varphi_1(y)\right] + \frac{3}{2}\sum_{i=2}^{n} z_i^2 \left(\frac{\partial \alpha_{i-1}}{\partial y}\right)^2 \varphi_1(y)^{\mathrm{T}}\varphi_1(y)$$

$$\leq -\left[b\lambda - 3bn\sqrt{n}\epsilon_2^2|P|^4 - \frac{1}{4}\sum_{i=2}^{n}\frac{1}{\eta_i^4} - \frac{1}{4\epsilon_1^4}\right]|\tilde{x}|^4 + y^3\left[\alpha_1 + \frac{3}{2}\psi_1(y)^{\mathrm{T}}\psi_1(y)y\right.$$

$$\left. +\frac{3}{4}\delta_1^{\frac{4}{3}}y + \frac{3}{4}\epsilon_1^{\frac{4}{3}}y + \frac{3}{4}\sum_{i=2}^{n}\xi_i^2\left(\psi_1(y)^{\mathrm{T}}\psi_1(y)\right)^2 y + \frac{3bn\sqrt{n}}{\epsilon_2^2}|\psi(y)|^4 y\right]$$

$$+\sum_{i=2}^{n-1} z_i^3\left[\alpha_i + k_i\tilde{x}_1 - \sum_{l=2}^{i-1}\frac{\partial \alpha_{i-1}}{\partial \hat{x}_l}(\hat{x}_{l+1} + k_l\tilde{x}_1) - \frac{\partial \alpha_{i-1}}{\partial y}\hat{x}_2\right.$$

$$-\frac{1}{2}\frac{\partial^2 \alpha_{i-1}}{\partial y^2}\varphi_1(y)^{\mathrm{T}}\varphi_1(y) + \frac{3}{4}\delta_i^{\frac{4}{3}}z_i + \frac{1}{4\delta_{i-1}^4}z_i + \frac{3}{4}\eta_i^{\frac{4}{3}}\left(\frac{\partial \alpha_{i-1}}{\partial y}\right)^{\frac{4}{3}}z_i$$

$$\left. +\frac{3}{4\xi_i^2}\left(\frac{\partial \alpha_{i-1}}{\partial y}\right)^4 z_i\right] + z_n^3\left[u + k_n\tilde{x}_1 - \sum_{l=2}^{n-1}\frac{\partial \alpha_{n-1}}{\partial \hat{x}_l}(\hat{x}_{l+1} + k_l\tilde{x}_1) - \frac{\partial \alpha_{n-1}}{\partial y}\hat{x}_2\right.$$

$$-\frac{1}{2}\frac{\partial^2 \alpha_{n-1}}{\partial y^2}\varphi_1(y)^{\mathrm{T}}\varphi_1(y) + \frac{1}{4\delta_{n-1}^4}z_n + \frac{3}{4}\eta_n^{\frac{4}{3}}\left(\frac{\partial \alpha_{n-1}}{\partial y}\right)^{\frac{4}{3}}z_n$$

$$\left. +\frac{3}{4\xi_n^2}\left(\frac{\partial \alpha_{n-1}}{\partial y}\right)^4 z_n\right]$$
(5.27)

where $\lambda > 0$ is the smallest eigenvalue of P. The second equality comes from substituting $\hat{x}_i = z_i + \alpha_{i-1}$, and the inequality comes from Young's inequalities in Appendix A. At this point, we can see that all the terms can be cancelled by u and α_i. If we choose ϵ_1, ϵ_2 and η_i to satisfy

$$b\lambda - 3bn\sqrt{n}\epsilon_2^2|P|^4 - \frac{1}{4}\sum_{i=2}^{n}\frac{1}{\eta_i^4} - \frac{1}{4\epsilon_1^4} = p > 0,$$
(5.28)

and α_i and u as

$$\alpha_1 = -c_1 y - \frac{3}{2}\psi_1(y)^{\mathrm{T}}\psi_1(y)y - \frac{3}{4}\delta_1^{\frac{4}{3}}y - \frac{3}{4}\epsilon_1^{\frac{4}{3}}y - \frac{3}{4}\sum_{i=2}^{n}\xi_i^2\left(\psi_1(y)^{\mathrm{T}}\psi_1(y)\right)^2 y$$

$$- \frac{3bn\sqrt{n}}{\epsilon_2^2}|\psi(y)|^4 y \tag{5.29}$$

$$\alpha_i = -c_i z_i - k_i \tilde{x}_1 + \sum_{l=2}^{i-1}\frac{\partial \alpha_{i-1}}{\partial \hat{x}_l}(\hat{x}_{l+1}+k_l\tilde{x}_1) + \frac{\partial \alpha_{i-1}}{\partial y}\hat{x}_2 + \frac{1}{2}\frac{\partial^2 \alpha_{i-1}}{\partial y^2}\varphi_1(y)^{\mathrm{T}}\varphi_1(y)$$

$$- \frac{3}{4}\delta_i^{\frac{4}{3}}z_i - \frac{1}{4\delta_{i-1}^4}z_i - \frac{3}{4}\eta_i^{\frac{4}{3}}\left(\frac{\partial \alpha_{i-1}}{\partial y}\right)^{\frac{4}{3}}z_i - \frac{3}{4\xi_i^2}\left(\frac{\partial \alpha_{i-1}}{\partial y}\right)^4 z_i \tag{5.30}$$

$$u = -c_n z_n - k_n \tilde{x}_1 + \sum_{l=2}^{n-1}\frac{\partial \alpha_{n-1}}{\partial \hat{x}_l}(\hat{x}_{l+1}+k_l\tilde{x}_1)$$

$$+ \frac{\partial \alpha_{n-1}}{\partial y}\hat{x}_2 + \frac{1}{2}\frac{\partial^2 \alpha_{n-1}}{\partial y^2}\varphi_1(y)^{\mathrm{T}}\varphi_1(y)$$

$$- \frac{1}{4\delta_{n-1}^4}z_n - \frac{3}{4}\eta_n^{\frac{4}{3}}\left(\frac{\partial \alpha_{n-1}}{\partial y}\right)^{\frac{4}{3}}z_n - \frac{3}{4\xi_n^2}\left(\frac{\partial \alpha_{n-1}}{\partial y}\right)^4 z_n, \tag{5.31}$$

where $c_i > 0$, then the infinitesimal generator of the closed-loop system (5.16), (5.24), (5.25) and (5.31) is negative definite:

$$\mathcal{L}V \leq -\sum_{i=1}^{n}c_i z_i^4 - p|\tilde{x}|^4. \tag{5.32}$$

With (5.32), we have the following stability result.

Theorem 5.4 *The equilibrium at the origin of the closed-loop stochastic system (5.17), (5.31) is globally asymptotically stable in probability.*

4 Output-Feedback Noise-to-State Stabilization

In this section we deal with nonlinear *output-feedback* systems driven by white noise with bounded but unknown covariance. This class of systems is given by the following nonlinear stochastic differential equations:

$$\begin{aligned}
dx_i &= x_{i+1}dt + \varphi_i(y)^{\mathrm{T}}dw, & i = 1,\ldots,n-1 \\
dx_n &= u\,dt + \varphi_n(y)^{\mathrm{T}}dw \\
y &= x_1,
\end{aligned} \tag{5.33}$$

where $\varphi_i(y)$ are r-vector-valued smooth functions, and w is an r-dimensional Wiener process with incremental covariance $E\left\{dwdw^{\mathrm{T}}\right\} = \Sigma(t)\Sigma(t)^{\mathrm{T}}dt$.

The observer is designed as in (5.15), and the entire system can be expressed as (5.17). The error variables z_i are defined as in (5.19), (5.20). With Itô's differentiation rule, we have

$$dz_1 = (\hat{x}_2 + \tilde{x}_2)\, dt + \varphi_1(y)^{\mathrm{T}} dw \tag{5.34}$$

$$
\begin{aligned}
dz_i = {} & \left[\hat{x}_{i+1} + k_i\tilde{x}_1 - \sum_{l=2}^{i-1} \frac{\partial \alpha_{i-1}}{\partial \hat{x}_l} (\hat{x}_{l+1} + k_l\tilde{x}_1) - \frac{\partial \alpha_{i-1}}{\partial y}(\hat{x}_2 + \tilde{x}_2) \right. \\
& \left. - \frac{1}{2}\left(\frac{\partial^2 \alpha_{i-1}}{\partial y^2}\right)\varphi_1(y)^{\mathrm{T}}\Sigma\Sigma^{\mathrm{T}}\varphi_1(y) \right] dt - \frac{\partial \alpha_{i-1}}{\partial y}\varphi_1(y)^{\mathrm{T}} dw
\end{aligned} \tag{5.35}
$$

$$i = 2, \ldots, n.$$

As in Section 3, we employ a quartic Lyapunov function

$$V(z, \tilde{x}) = \frac{1}{4}y^4 + \frac{1}{4}\sum_{i=2}^{n} z_i^4 + \frac{b}{2}\left(\tilde{x}^{\mathrm{T}} P\tilde{x}\right)^2. \tag{5.36}$$

Now we start the process of selecting the functions $\alpha_i(\bar{\hat{x}}_i, y)$ to make $\mathcal{L}V$ in the form

$$\mathcal{L}V \le -\rho(\tilde{x}, y, \hat{x}) + \gamma(|\Sigma|) \tag{5.37}$$

where ρ is positive definite, radially unbounded, and γ is a class $\mathcal{K}$ function. Since $\varphi(y)$ is a smooth function, according to mean value theorem, we can write it as

$$\varphi(y) = \varphi(0) + y\psi(y) \tag{5.38}$$

where $\psi(y)$ is a smooth function. Along the solutions of (5.16), (5.34) and (5.35), we have

$$
\begin{aligned}
\mathcal{L}V = {} & y^3(\hat{x}_2 + \tilde{x}_2) + \frac{3}{2}y^2\varphi_1(y)^{\mathrm{T}}\Sigma\Sigma^{\mathrm{T}}\varphi_1(y) + \sum_{i=2}^{n} z_i^3\left[\hat{x}_{i+1} + k_i\tilde{x}_1 - \sum_{l=2}^{i-1}\frac{\partial \alpha_{i-1}}{\partial \hat{x}_l} \right. \\
& \left. (\hat{x}_{l+1} + k_l\tilde{x}_1) - \frac{\partial \alpha_{i-1}}{\partial y}(\hat{x}_2 + \tilde{x}_2) - \frac{1}{2}\left(\frac{\partial^2 \alpha_{i-1}}{\partial y^2}\right)\varphi_1(y)^{\mathrm{T}}\Sigma\Sigma^{\mathrm{T}}\varphi_1(y) \right] \\
& + \frac{3}{2}\sum_{i=2}^{n} z_i^2\left(\frac{\partial \alpha_{i-1}}{\partial y}\right)^2\varphi_1(y)^{\mathrm{T}}\Sigma\Sigma^{\mathrm{T}}\varphi_1(y) \\
& - b\tilde{x}^{\mathrm{T}} P\tilde{x}|\tilde{x}|^2 + b\mathrm{Tr}\left\{\varphi(y)\Sigma^{\mathrm{T}}\left(2P\tilde{x}\tilde{x}^{\mathrm{T}}P + \tilde{x}^{\mathrm{T}} P\tilde{x}P\right)\Sigma\varphi(y)^{\mathrm{T}}\right\}
\end{aligned} \tag{5.39}
$$

Therefore

$$
\begin{aligned}
\mathcal{L}V \le\ & -b\tilde{x}^{\mathrm{T}} P\tilde{x}|\tilde{x}|^2 + 3bn\sqrt{n}|P|^2\left(\frac{1}{2\epsilon_2^2}|\varphi(0)^{\mathrm{T}}\varphi(0)| + \frac{1}{\epsilon_3^2} + \frac{1}{2\epsilon_4^2}\right)|\tilde{x}|^4 \\
& + \frac{3bn\sqrt{n}|P|^2\epsilon_2^2}{2}|\varphi(0)^{\mathrm{T}}\varphi(0)||\Sigma|^4 + 3bn\sqrt{n}|P|^2\left(\frac{\epsilon_3^2}{2} + \frac{\epsilon_4^2}{4}\right)|\Sigma|^8 \\
& + 3bn\sqrt{n}|P|^2\left(\frac{\epsilon_2^2}{2}|\varphi(0)^{\mathrm{T}}\psi(y)|^4 y + \frac{\epsilon_4^2}{4}|\psi(y)^{\mathrm{T}}\psi(y)|^4 y^5\right)y^3 \\
& + y^3\left(\alpha_1 + z_2 + \tilde{x}_2\right) + \frac{3}{2}y^2\varphi_1(y)^{\mathrm{T}}\Sigma\Sigma^{\mathrm{T}}\varphi_1(y) \\
& + \sum_{i=2}^{n} z_i^3\left[\alpha_i + z_{i+1} + k_i\tilde{x}_1 - \sum_{l=2}^{i-1}\frac{\partial\alpha_{i-1}}{\partial\hat{x}_l}(\hat{x}_{l+1} + k_l\tilde{x}_1) - \frac{\partial\alpha_{i-1}}{\partial y}(\hat{x}_2 + \tilde{x}_2)\right. \\
& \left. - \frac{1}{2}\left(\frac{\partial^2\alpha_{i-1}}{\partial y^2}\right)\varphi_1(y)^{\mathrm{T}}\Sigma\Sigma^{\mathrm{T}}\varphi_1(y)\right] + \frac{3}{2}\sum_{i=2}^{n} z_i^2\left(\frac{\partial\alpha_{i-1}}{\partial y}\right)^2\varphi_1(y)^{\mathrm{T}}\Sigma\Sigma^{\mathrm{T}}\varphi_1(y) \\[4pt]
\le\ & -\left[b\lambda - 3bn\sqrt{n}|P|^2\left(\frac{1}{2\epsilon_2^2}|\varphi(0)^{\mathrm{T}}\varphi(0)| + \frac{1}{\epsilon_3^2} + \frac{1}{2\epsilon_4^2}\right) - \frac{1}{4}\sum_{i=2}^{n}\frac{1}{\eta_i^4} - \frac{1}{4\epsilon_1^4}\right]|\tilde{x}|^4 \\
& + y^3\left[\alpha_1 + \frac{3}{4}\left(\varphi_1(y)^{\mathrm{T}}\varphi_1(y)\right)^2 y + \frac{3}{4}\delta_1^{\frac{4}{3}}y + \frac{3}{4}\epsilon_1^{\frac{4}{3}}y\right. \\
& \left. + 3bn\sqrt{n}|P|^2\left(\frac{\epsilon_2^2}{2}|\varphi(0)^{\mathrm{T}}\psi(y)|^4 y + \frac{\epsilon_4^2}{4}|\psi(y)^{\mathrm{T}}\psi(y)|^4 y^5\right)\right] \\
& + \sum_{i=2}^{n} z_i^3\left[\alpha_i + k_i\tilde{x}_1 - \sum_{l=2}^{i-1}\frac{\partial\alpha_{i-1}}{\partial\hat{x}_l}(\hat{x}_{l+1} + k_l\tilde{x}_1) - \frac{\partial\alpha_{i-1}}{\partial y}\hat{x}_2\right. \\
& + \frac{1}{4}\left(\frac{\partial^2\alpha_{i-1}}{\partial y^2}\varphi_1(y)^{\mathrm{T}}\varphi_1(y)\right)^2 z_i^3 + \frac{3}{4}\delta_i^{\frac{4}{3}}z_i + \frac{1}{4\delta_{i-1}^4}z_i \\
& \left. + \frac{3}{4}\eta_i^{\frac{4}{3}}\left(\frac{\partial\alpha_{i-1}}{\partial y}\right)^{\frac{4}{3}}z_i + \frac{3}{4\varsigma_i^2}\left(\frac{\partial\alpha_{i-1}}{\partial y}\right)^4 z_i\left(\varphi_1(y)^{\mathrm{T}}\varphi_1(y)\right)^2\right] \\
& + \left(\frac{3bn\sqrt{n}|P|^2\epsilon_2^2}{2}|\varphi(0)^{\mathrm{T}}\varphi(0)| + \frac{3}{4}\sum_{i=2}^{n}\varsigma_i^2 + 1\right)|\Sigma|^4 \\
& + 3bn\sqrt{n}|P|^2\left(\frac{\epsilon_3^2}{2} + \frac{\epsilon_4^2}{4}\right)|\Sigma|^8 \tag{5.40}
\end{aligned}
$$

where $\lambda > 0$ is the smallest eigenvalue of P. The inequalities come from substituting $\hat{x}_i = z_i + \alpha_{i-1}$, and Young's inequalities in Appendix B and (5.68), (5.69), (5.70), (5.71) in Appendix A. At this point, we can see that all the terms can be cancelled by u and α_i. If we choose ϵ_1, ϵ_2, ϵ_3, ϵ_4 and η_i to satisfy

$$b\lambda - 3bn\sqrt{n}|P|^2\left(\frac{1}{2\epsilon_2^2}|\varphi(0)^{\mathrm{T}}\varphi(0)| + \frac{1}{\epsilon_3^2} + \frac{1}{2\epsilon_4^2}\right) - \frac{1}{4}\sum_{i=2}^{n}\frac{1}{\eta_i^4} - \frac{1}{4\epsilon_1^4} = p > 0,$$

$$(5.41)$$

and α_i and u as

$$\alpha_1 = -c_1 y - \frac{3}{4}\left(\varphi_1(y)^{\mathrm{T}}\varphi_1(y)\right)^2 y - \frac{3}{4}\delta_1^{\frac{4}{3}}y - \frac{3}{4}\epsilon_1^{\frac{4}{3}}y$$

$$-3bn\sqrt{n}|P|^2\left(\frac{\epsilon_2^2}{2}|\varphi(0)^{\mathrm{T}}\psi(y)|^4 y + \frac{\epsilon_4^2}{4}|\psi(y)^{\mathrm{T}}\psi(y)|^4 y^5\right) \quad (5.42)$$

$$\vdots$$

$$\alpha_i = -c_i z_i - k_i \tilde{x}_1 + \sum_{l=2}^{i-1}\frac{\partial\alpha_{i-1}}{\partial\hat{x}_l}(\hat{x}_{l+1} + k_l\tilde{x}_1) + \frac{\partial\alpha_{i-1}}{\partial y}\hat{x}_2$$

$$-\frac{1}{4}\left(\frac{\partial^2\alpha_{i-1}}{\partial y^2}\varphi_1(y)^{\mathrm{T}}\varphi_1(y)\right)^2 z_i^3 - \frac{3}{4}\delta_i^{\frac{4}{3}}z_i - \frac{1}{4\delta_{i-1}^4}z_i$$

$$-\frac{3}{4}\eta_i^{\frac{4}{3}}\left(\frac{\partial\alpha_{i-1}}{\partial y}\right)^{\frac{4}{3}}z_i - \frac{3}{4\xi_i^2}\left(\frac{\partial\alpha_{i-1}}{\partial y}\right)^4\left(\varphi_1(y)^{\mathrm{T}}\varphi_1(y)\right)^2 z_i(5.43)$$

$$u = \alpha_n \quad (5.44)$$

where $c_i > 0$ and $\delta_n = 0$, then the infinitesimal generator of the closed-loop system (5.16), (5.34), (5.35) and (5.44) satisfies:

$$\mathcal{L}V \leq -\sum_{i=1}^{n}c_i z_i^4 - p|\tilde{x}|^4 + \left(\frac{3bn\sqrt{n}|P|^2\epsilon_2^2}{2}|\varphi(0)^{\mathrm{T}}\varphi(0)| + \frac{3}{4}\sum_{i=2}^{n}\xi_i^2 + 1\right)|\Sigma|^4$$

$$+3bn\sqrt{n}|P|^2\left(\frac{\epsilon_3^2}{2} + \frac{\epsilon_4^2}{4}\right)|\Sigma|^8 \quad (5.45)$$

With (5.45), according to Theorem 5.1, we have the following stability result.

Theorem 5.5 *The closed-loop stochastic system (5.33), (5.15), (5.44) is NSS.*

5   Output-Feedback Adaptive Stabilization

In this section, we deal with output-feedback systems (5.33) with an additional assumption that $\varphi_i(0) = 0$. Since $\varphi_i(0) = 0$, by the mean value theorem, $\varphi(y)$ can be expressed as

$$\varphi(y) \;=\; y\psi(y) \tag{5.46}$$

where $\psi(y)$ is a smooth function. As we will see in the sequel, to achieve adaptive stabilization in the presence of unknown Σ, it is not necessary to estimate the entire matrix Σ. Instead, we will estimate only one unknown parameter $\theta = \| \Sigma\Sigma^{\mathrm{T}} \|_\infty^2$ using an estimate $\hat\theta$. Employing the same observer (5.15), the entire system is:

$$
\begin{aligned}
d\tilde{x} &= A_0\tilde{x}dt + \varphi(y)^{\mathrm{T}}dw \\
dy &= (\hat{x}_2 + \tilde{x}_2)\,dt + \varphi_1(y)^{\mathrm{T}}dw \\
d\hat{x}_2 &= [\hat{x}_3 + k_2\,(y - \hat{x}_1)]\,dt \\
&\;\;\vdots \\
d\hat{x}_n &= \Big[\alpha_n(\hat{x}, y, \hat\theta) + k_n\,(y - \hat{x}_1)\Big]\,dt \\
\dot{\hat\theta} &= \gamma\tau_n(\hat{x}, y, \hat\theta),
\end{aligned}
\tag{5.47}
$$

where α_n and τ_n are functions to be designed.

In the *adaptive* backstepping method, the error variables z_i are given by

$$z_1 \;=\; y \tag{5.48}$$

$$z_i \;=\; \hat{x}_i - \alpha_{i-1}\left(\bar{\hat{x}}_{i-1}, y, \hat\theta\right), \qquad i = 2, \ldots, n. \tag{5.49}$$

According to Itô's differentiation rule, we have

$$dz_1 \;=\; (\hat{x}_2 + \tilde{x}_2)\,dt + \varphi_1(y)^{\mathrm{T}}dw \tag{5.50}$$

$$
\begin{aligned}
dz_i \;=\; & \Bigg[\hat{x}_{i+1} + k_i\tilde{x}_1 - \sum_{l=2}^{i-1}\frac{\partial\alpha_{i-1}}{\partial\hat{x}_l}(\hat{x}_{l+1} + k_l\tilde{x}_1) - \frac{\partial\alpha_{i-1}}{\partial y}(\hat{x}_2 + \tilde{x}_2) \\
& -\frac{1}{2}\!\left(\frac{\partial^2\alpha_{i-1}}{\partial y^2}\right)\!\varphi_1(y)^{\mathrm{T}}\Sigma\Sigma^{\mathrm{T}}\varphi_1(y) - \frac{\partial\alpha_{i-1}}{\partial\hat\theta}\dot{\hat\theta}\Bigg]dt - \frac{\partial\alpha_{i-1}}{\partial y}\varphi_1(y)^{\mathrm{T}}dw
\end{aligned}
$$

$$i \;=\; 2, \ldots, n. \tag{5.51}$$

As in the previous sections, we employ a Lyapunov function of a quartic form. In this case it also includes the parameter estimation error $\tilde\theta = \| \Sigma\Sigma^{\mathrm{T}} \|_\infty^2 - \hat\theta$,

$$V(z, \tilde{x}, \hat\theta) = \frac{1}{4}y^4 + \frac{1}{4}\sum_{i=2}^{n} z_i^4 + \frac{b}{2}\left(\tilde{x}^{\mathrm{T}}P\tilde{x}\right)^2 + \frac{1}{2\gamma}\tilde\theta^2, \tag{5.52}$$

where b is a positive constant and P satisfies

$$A_0^T P + P A_0 = -I.$$

Now we start the process of selecting the functions $\alpha_i(\bar{\hat{x}}_i, y, \hat{\theta})$ to make $\mathcal{L}V$ in the form

$$\mathcal{L}V \leq -\rho(\tilde{x}, y, \hat{x}, \hat{\theta}) \tag{5.53}$$

where ρ is a positive definite function in $\tilde{x}$, y, $\hat{x}$ for each value of $\hat{\theta}$. Along the solutions of (5.16), (5.50) and (5.51), we have

$$
\begin{aligned}
\mathcal{L}V &= y^3 (\alpha_1 + z_2 + \tilde{x}_2) + \frac{3}{2} y^2 \varphi_1(y)^T \Sigma \Sigma^T \varphi_1(y) \\
&\quad + \sum_{i=2}^{n} z_i^3 \left[\alpha_i + z_{i+1} + k_i \tilde{x}_1 - \sum_{l=2}^{i-1} \frac{\partial \alpha_{i-1}}{\partial \hat{x}_l} (\hat{x}_{l+1} + k_l \tilde{x}_1) - \frac{\partial \alpha_{i-1}}{\partial y} (\hat{x}_2 + \tilde{x}_2) \right. \\
&\quad \left. - \frac{1}{2} \left(\frac{\partial^2 \alpha_{i-1}}{\partial y^2} \right) \varphi_1(y)^T \Sigma \Sigma^T \varphi_1(y) - \frac{\partial \alpha_{i-1}}{\partial \hat{\theta}} \dot{\hat{\theta}} \right] \\
&\quad + \frac{3}{2} \sum_{i=2}^{n} z_i^2 \left(\frac{\partial \alpha_{i-1}}{\partial y} \right)^2 \varphi_1(y)^T \Sigma \Sigma^T \varphi_1(y) \\
&\quad - b \tilde{x}^T P \tilde{x} |\tilde{x}|^2 + b \text{Tr} \left\{ \varphi(y) \Sigma^T \left(2 P \tilde{x} \tilde{x}^T P + \tilde{x}^T P \tilde{x} P \right) \Sigma \varphi(y)^T \right\} - \frac{\tilde{\theta} \dot{\hat{\theta}}}{\gamma} \\
&\leq - \left[b\lambda - \frac{3bn\sqrt{n}}{2\epsilon_2^2} |P|^2 - \frac{1}{4} \sum_{i=2}^{n} \frac{1}{\eta_i^4} - \frac{1}{4\epsilon_1^4} \right] |\tilde{x}|^4 + y^3 \left[\alpha_1 + \frac{3}{4} \left(\psi_1(y)^T \psi_1(y) \right)^2 y \right. \\
&\quad + \frac{3}{4} \delta_1^{\frac{4}{3}} y + \frac{3}{4} \epsilon_1^{\frac{4}{3}} y + \frac{3bn\sqrt{n}|P|^2 \epsilon_2^2}{2} |\psi(y)|^4 y \| \Sigma \Sigma^T \|_\infty^2 + \frac{3}{4} y \| \Sigma \Sigma^T \|_\infty^2 \\
&\quad \left. + \frac{1}{4} \left(\psi_1(y)^T \psi_1(y) \right)^2 y + \frac{3}{4} (n-1) \left(\psi_1(y)^T \psi_1(y) \right)^2 y \right] \\
&\quad + \sum_{i=2}^{n} z_i^3 \left[\alpha_i + k_i \tilde{x}_1 - \sum_{l=2}^{i-1} \frac{\partial \alpha_{i-1}}{\partial \hat{x}_l} (\hat{x}_{l+1} + k_l \tilde{x}_1) - \frac{\partial \alpha_{i-1}}{\partial y} \hat{x}_2 \right. \\
&\quad + \frac{1}{4} z_i^3 \left(\frac{\partial^2 \alpha_{i-1}}{\partial y^2} \right)^2 \| \Sigma \Sigma^T \|_\infty^2 + \frac{3}{4} \delta_i^{\frac{4}{3}} z_i + \frac{1}{4\delta_{i-1}^4} z_i + \frac{3}{4} \eta_i^{\frac{4}{3}} \left(\frac{\partial \alpha_{i-1}}{\partial y} \right)^{\frac{4}{3}} z_i \\
&\quad \left. + \frac{3}{4} \left(\frac{\partial \alpha_{i-1}}{\partial y} \right)^4 z_i \| \Sigma \Sigma^T \|_\infty^2 - \frac{\partial \alpha_{i-1}}{\partial \hat{\theta}} \dot{\hat{\theta}} \right] - \frac{\tilde{\theta} \dot{\hat{\theta}}}{\gamma}
\end{aligned}
$$

$$
= -\left[b\lambda - \frac{3bn\sqrt{n}}{2\epsilon_2^2}|P|^2 - \frac{1}{4}\sum_{i=2}^{n}\frac{1}{\eta_i^4} - \frac{1}{4\epsilon_1^4}\right]|\tilde{x}|^4 + y^3\left[\alpha_1 + \frac{3n+1}{4}\left(\psi_1(y)^\mathrm{T}\psi_1(y)\right)^2 y\right.
$$

$$
\left. + \frac{3}{4}\delta_1^{\frac{4}{3}}y + \frac{3}{4}\epsilon_1^{\frac{4}{3}}y + \frac{3bn\sqrt{n}|P|^2\epsilon_2^2}{2}|\psi(y)|^4 y\hat{\theta} + \frac{3}{4}y\hat{\theta}\right]
$$

$$
+ \sum_{i=2}^{n}z_i^3\left[\alpha_i + k_i\tilde{x}_1 - \sum_{l=2}^{i-1}\frac{\partial\alpha_{i-1}}{\partial\hat{x}_l}(\hat{x}_{l+1} + k_l\tilde{x}_1) - \frac{\partial\alpha_{i-1}}{\partial y}\hat{x}_2 + \frac{1}{4}z_i^3\left(\frac{\partial^2\alpha_{i-1}}{\partial y^2}\right)^2\hat{\theta}\right.
$$

$$
\left. + \frac{3}{4}\delta_i^{\frac{4}{3}}z_i + \frac{1}{4\delta_{i-1}^4}z_i + \frac{3}{4}\eta_i^{\frac{4}{3}}\left(\frac{\partial\alpha_{i-1}}{\partial y}\right)^{\frac{4}{3}}z_i + \frac{3}{4}\left(\frac{\partial\alpha_{i-1}}{\partial y}\right)^4 z_i\hat{\theta} - \frac{\partial\alpha_{i-1}}{\partial\hat{\theta}}\dot{\hat{\theta}}\right]
$$

$$
- \tilde{\theta}\left[\frac{\dot{\hat{\theta}}}{\gamma} - \frac{3bn\sqrt{n}|P|^2\epsilon_2^2}{2}|\psi(y)|^4 y^4 - \frac{3}{4}y^4 - \frac{1}{4}\sum_{i=2}^{n}z_i^6\left(\frac{\partial^2\alpha_{i-1}}{\partial y^2}\right)^2\right.
$$

$$
\left. - \frac{3}{4}\sum_{i=2}^{n}z_i^4\left(\frac{\partial\alpha_{i-1}}{\partial y}\right)^4\right] \tag{5.54}
$$

where $\lambda > 0$ is the smallest eigenvalue of P, $x_{n+1} = u$, $z_{n+1} = 0$, $\alpha_n = u$. The inequalities come from substituting $\hat{x}_i = z_i + \alpha_{i-1}$, Young's inequalities in Appendix C and (5.68), (5.69), (5.70), (5.71) in Appendix A. Let

$$
\tau_1 = \frac{3bn\sqrt{n}|P|^2\epsilon_2^2}{2}|\psi(y)|^4 y^4 + \frac{3}{4}y^4 \tag{5.55}
$$

$$
\tau_i = \tau_{i-1} + z_i^3\omega_i, \qquad i = 2,\cdots,n \tag{5.56}
$$

$$
\dot{\hat{\theta}} = \gamma\tau_n \tag{5.57}
$$

where

$$
\omega_i = \frac{1}{4}z_i^3\left(\frac{\partial^2\alpha_{i-1}}{\partial y^2}\right)^2 + \frac{3}{4}z_i\left(\frac{\partial\alpha_{i-1}}{\partial y}\right)^4. \tag{5.58}
$$

Then

$$\mathcal{L}V \;\leq\; -\left[b\lambda - \frac{3bn\sqrt{n}}{2\epsilon_2^2}|P|^2 - \frac14\sum_{i=2}^{n}\frac{1}{\eta_i^4} - \frac{1}{4\epsilon_1^4}\right]|\tilde{x}|^4$$

$$+y^3\left[\alpha_1 + \frac{3n+1}{4}\big(\psi_1(y)^{\mathrm T}\psi_1(y)\big)^2 y\right.$$

$$\left. + \frac34\delta_1^{\frac43}y + \frac34\epsilon_1^{\frac43}y + \frac{3bn\sqrt{n}|P|^2\epsilon_2^2}{2}|\psi(y)|^4 y\hat\theta + \frac34 y\hat\theta\right]$$

$$+\sum_{i=2}^{n}z_i^3\left[\alpha_i + k_i\tilde{x}_1 - \sum_{l=2}^{i-1}\frac{\partial\alpha_{i-1}}{\partial\hat{x}_l}(\hat{x}_{l+1}+k_l\tilde{x}_1) - \frac{\partial\alpha_{i-1}}{\partial y}\hat{x}_2\right.$$

$$+\frac34\delta_i^{\frac43}z_i + \frac{1}{4\delta_{i-1}^4}z_i + \frac34\eta_i^{\frac43}\left(\frac{\partial\alpha_{i-1}}{\partial y}\right)^{\frac43}z_i - \omega_i\hat\theta - \frac{\partial\alpha_{i-1}}{\partial\hat\theta}\sum_{j=2}^{i}\gamma z_j^3\omega_j$$

$$\left.-\sum_{j=2}^{i-1}\gamma z_j^3\frac{\partial\alpha_{j-1}}{\partial\hat\theta}\omega_i\right] \tag{5.59}$$

If we choose ϵ_1, ϵ_2 and η_i to satisfy

$$b\lambda - \frac{3bn\sqrt{n}}{2\epsilon_2^2}|P|^2 - \frac14\sum_{i=2}^{n}\frac{1}{\eta_i^4} - \frac{1}{4\epsilon_1^4} = p > 0, \tag{5.60}$$

and α_i and u as

$$\alpha_1 \;=\; -c_1 y - \frac{3n+1}{4}\big(\psi_1(y)^{\mathrm T}\psi_1(y)\big)^2 y - \frac34\delta_1^{\frac43}y - \frac34\epsilon_1^{\frac43}y$$

$$-\frac{3bn\sqrt{n}|P|^2\epsilon_2^2}{2}|\psi(y)|^4 y\hat\theta - \frac34 y\hat\theta \tag{5.61}$$

$$\alpha_i \;=\; -c_i z_i - k_i\tilde{x}_1 + \sum_{l=2}^{i-1}\frac{\partial\alpha_{i-1}}{\partial\hat{x}_l}(\hat{x}_{l+1}+k_l\tilde{x}_1) + \frac{\partial\alpha_{i-1}}{\partial y}\hat{x}_2$$

$$-\frac34\delta_i^{\frac43}z_i - \frac{1}{4\delta_{i-1}^4}z_i - \frac34\eta_i^{\frac43}\left(\frac{\partial\alpha_{i-1}}{\partial y}\right)^{\frac43}z_i$$

$$+\omega_i\hat\theta + \frac{\partial\alpha_{i-1}}{\partial\hat\theta}\sum_{j=2}^{i}\gamma z_j^3\omega_j + \sum_{j=2}^{i-1}\gamma z_j^3\frac{\partial\alpha_{j-1}}{\partial\hat\theta}\omega_i \tag{5.62}$$

$$u \;=\; \alpha_n \tag{5.63}$$

where $c_i > 0$ and $\delta_n = 0$, then the infinitesimal generator of the closed-loop system (5.16), (5.50), (5.51) and (5.63) satisfies:

$$\mathcal{L}V \leq -\sum_{i=1}^{n} c_i z_i^4 - p|\tilde{x}|^4. \tag{5.64}$$

Since $z = 0$ and $\tilde{x} = 0$ implies $x = 0$, by Theorem 5.2, we have the following result.

Theorem 5.6 *The equilibrium $x = 0$, $\hat{\theta} = \parallel \Sigma\Sigma^T \parallel_\infty^2$ of the closed-loop system (5.33), (5.15), (5.57) and (5.63) is globally stable in probability and*

$$P\left\{ \lim_{t \to \infty} x(t) = 0 \text{ and } \lim_{t \to \infty} \hat{x}(t) = 0 \right\} = 1. \tag{5.65}$$

Remark 5.1 Since $\mathcal{L}V$ is nonpositive, $EV(t)$ is nonincreasing. Since V is also bounded from below by zero, $EV(t)$ has a limit. Since $z(t)$ and $\tilde{x}(t)$ converge to zero with probability one, thus $E\left\{ \tilde{\theta}(t)^2 \right\}$ has a limit. Denote this limit by $\Delta\theta^2$. Thus

$$\underset{t \to \infty}{\text{l.i.m.}}\ \hat{\theta}(t) = \parallel \Sigma\Sigma^T \parallel_\infty^2 \pm \Delta\theta. \tag{5.66}$$

It is well known [5] that *convergence in the mean* implies *convergence in probability*, i.e.

$$\lim_{t \to \infty} P\left\{ \left| \hat{\theta}(t) - \left(\parallel\Sigma\Sigma^T\parallel_\infty^2 + \Delta\theta \right) \right| > \epsilon \right\} P\left\{ \left| \hat{\theta}(t) - \left(\parallel\Sigma\Sigma^T\parallel_\infty^2 - \Delta\theta \right) \right| > \epsilon \right\} = 0, \ \forall \epsilon > 0.$$

6   References

[1] T. Başar and P. Bernhard, H^∞-*Optimal Control and Related Minimax Design Problems: A Dynamic Game Approach*, Boston, MA: Birkhäuser, 2nd ed., 1995.

[2] H. Deng and M. Krstić, "Stochastic nonlinear stabilization—Part I: A Backstepping Design," *Systems & Control Letters*, vol. 32, pp. 143–150, 1997.

[3] H. Deng and M. Krstić, "Stochastic nonlinear stabilization—Part II: Inverse optimality," *Systems & Control Letters*, vol. 32, pp. 151–159, 1997.

[4] H. Deng and M. Krstić, "Stabilization of stochastic nonlinear systems driven by noise of unknown covariance," *Proceedings of ACC*, Philadelphia, 1998.

[5] J. L. Doob, *Stochastic Processes*, New York: John Wiley & Sons, 1953.

[6] P. Florchinger, "Lyapunov-like techniques for stochastic stability," *SIAM Journal of Control and Optimization,* vol. 33, pp. 1151–1169, 1995.

[7] P. Florchinger, "Global stabilization of cascade stochastic systems," *Proceedings of the 34th Conference on Decision & Control,* New Orleans, LA, pp. 2185–2186, 1995.

[8] P. Florchinger, "A universal formula for the stabilization of control stochastic differential equations," *Stochastic Analysis and Applications,* vol. 11, pp. 155–162, 1993.

[9] R. A. Freeman and P. V. Kokotović, *Robust Nonlinear Control Design: State-Space and Lyapunov Techniques,* Boston, MA: Birkhäuser, 1996.

[10] G. Hardy, J. E. Littlewood, and G. Polya, *Inequalities,* 2nd Edition, Cambridge University Press, 1989.

[11] M. R. James, J. Baras and R. J. Elliott, "Risk-sensitive control and dynamic games for partially observed discrete-time nonlinear systems," *IEEE Transactions on Automatic Control,* vol. 39, pp. 780–792, 1994.

[12] M. Jankovic, "Adaptive nonlinear output feedback tracking with a partial high-gain observer and backstepping", *IEEE Transactions on Automatic Control,* vol.42, pp. 106–113, Jan. 1997.

[13] Z. P. Jiang, A. R. Teel, and L. Praly, "Small-gain theorem for ISS systems and applications," *Mathematics of Control, Signals, and Systems,* vol. 7, pp. 95–120, 1995.

[14] H. K. Khalil, "Adaptive output feedback control of nonlinear systems represented by input-output models", *IEEE Transactions on Automatic Control,* vol.41, pp. 177–188, Feb. 1996.

[15] R. Z. Khas'minskii, *Stochastic Stability of Differential Equations,* Rockville, Maryland: S & N International publisher, 1980.

[16] M. Krstić and H. Deng, *Stabilization of Nonlinear Uncertain Systems,* Springer, 1998.

[17] M. Krstić, I. Kanellakopoulos, and P. V. Kokotović, *Nonlinear and Adaptive Control Design,* New York: Wiley, 1995.

[18] R. Marino and P. Tomei, *Nonlinear Control Design : Geometric, Adaptive, and Robust,* New York : Prentice Hall, c1995.

[19] H. Nagai, "Bellman equations of risk-sensitive control," *SIAM Journal of Control and Optimization,* vol. 34, pp. 74–101, 1996.

[20] B. Øksendal, *Stochastic Differential Equations–An Introduction with Applications,* New York: Springer-Verlag, 1995.

[21] Z. Pan and T. Başar, "Backstepping controller design for nonlinear stochastic systems under a risk-sensitive cost criterion," submitted to *SIAM Journal of Control and Optimization,* 1996.

[22] L. Praly and Z. P. Jiang, "Stabilization by output feedback for systems with ISS inverse dynamics", *Systems & Control Letters,* vol. 21, pp. 19–33, July 1993.

[23] T. Runolfsson, "The equivalence between infinite horizon control of stochastic systems with exponential-of-integral performance index and stochastic differential games," *IEEE Transactions on Automatic Control,* vol. 39, pp. 1551–1563, 1994.

[24] E. D. Sontag, "A 'universal' construction of Artstein's theorem on nonlinear stabilization," *Systems & Control Letters,* vol. 13, pp. 117–123, 1989.

[25] E. D. Sontag, "Smooth stabilization implies coprime factorization," *IEEE Transactions on Automatic Control,* vol. 34, pp. 435–443, 1989.

[26] A. Teel and L. Praly, "Tools for semiglobal stabilization by partial state and output feedback", *SIAM Journal on Control and Optimization,* vol. 33, pp. 1443–1488, Sept. 1995.

[27] J. Tsinias, "The concept of 'exponential ISS' for stochastic systems and applications to feedback stabilization", preprint, 1997.

[28] J. Tsinias, "Stochastic input-to-state stability and applications to global feedback stabilization", submitted to *International Journal on Control* for special issue on *Breakthrough in the control of nonlinear systems,* 1998.

Appendix A

In this and the following appendix, we use Young's inequality [10, Theorem 156]:

$$xy \le \frac{\epsilon^p}{p}|x|^p + \frac{1}{q\epsilon^q}|y|^q, \tag{5.67}$$

where $\epsilon > 0$, the constants $p > 1$ and $q > 1$ satisfy $(p-1)(q-1) = 1$, and $(x, y) \in \mathbb{R}^2$. Applying these inequalities leads to

$$y^3 z_2 \le \frac{3}{4}\delta_1^{\frac{4}{3}} y^4 + \frac{1}{4\delta_1^4} z_2^4 \tag{5.68}$$

$$y^3 \tilde{x}_2 \le \frac{3}{4}\epsilon_1^{\frac{4}{3}} y^4 + \frac{1}{4\epsilon_1^4} \tilde{x}_2^4 \le \frac{3}{4}\epsilon_1^{\frac{4}{3}} y^4 + \frac{1}{4\epsilon_1^4}|\tilde{x}|^4 \tag{5.69}$$

$$\sum_{i=2}^{n} z_i^3 z_{i+1} \le \frac{3}{4}\sum_{i=2}^{n-1}\delta_i^{\frac{4}{3}} z_i^4 + \frac{1}{4}\sum_{i=3}^{n}\frac{1}{\delta_{i-1}^4} z_i^4 \tag{5.70}$$

$$-\sum_{i=2}^{n} z_i^3 \frac{\partial \alpha_{i-1}}{\partial y}\tilde{x}_2 \le \frac{3}{4}\sum_{i=2}^{n}\eta_i^{\frac{4}{3}}\left(\frac{\partial \alpha_{i-1}}{\partial y}\right)^{\frac{4}{3}} z_i^4 + \frac{1}{4}\sum_{i=2}^{n}\frac{1}{\eta_i^4}\tilde{x}_2^4$$

$$\le \frac{3}{4}\sum_{i=2}^{n}\eta_i^{\frac{4}{3}}\left(\frac{\partial \alpha_{i-1}}{\partial y}\right)^{\frac{4}{3}} z_i^4 + \frac{1}{4}\sum_{i=2}^{n}\frac{1}{\eta_i^4}|\tilde{x}|^4 \tag{5.71}$$

$$\frac{3}{2}\sum_{i=2}^{n} z_i^2 \left(\frac{\partial \alpha_{i-1}}{\partial y}\right)^2 \varphi_1(y)^{\mathrm{T}}\varphi_1(y)$$

$$\le \frac{3}{4}\sum_{i=2}^{n}\frac{1}{\xi_i^2}\left(\frac{\partial \alpha_{i-1}}{\partial y}\right)^4 z_i^4 + \frac{3}{4}\sum_{i=2}^{n}\xi_i^2\left(\varphi_1(y)^{\mathrm{T}}\varphi_1(y)\right)^2 \tag{5.72}$$

$$2b\mathrm{Tr}\left\{\varphi(y)\left(2P\tilde{x}\tilde{x}^{\mathrm{T}}P + \tilde{x}^{\mathrm{T}}P\tilde{x}P\right)\varphi(y)^{\mathrm{T}}\right\}$$

$$\le 2bn\left|\varphi(y)\left(2P\tilde{x}\tilde{x}^{\mathrm{T}}P + \tilde{x}^{\mathrm{T}}P\tilde{x}P\right)\varphi(y)^{\mathrm{T}}\right|_{\infty}$$

$$\le 2bn\sqrt{n}\left|\varphi(y)\left(2P\tilde{x}\tilde{x}^{\mathrm{T}}P + \tilde{x}^{\mathrm{T}}P\tilde{x}P\right)\varphi(y)^{\mathrm{T}}\right|$$

$$\le 6bn\sqrt{n}y^2|\psi(y)|^2|P|^2|\tilde{x}|^2 \quad \text{(cf. (2.10))}$$

$$\le \frac{3bn\sqrt{n}}{\epsilon_2^2}y^4|\psi(y)|^4 + 3bn\sqrt{n}\epsilon_2^2|P|^4|\tilde{x}|^4, \tag{5.73}$$

where the ϵ's, δ's, η's and ξ's are positive constants to be chosen.

Appendix B

Similar to Appendix A, in the following inequalities, ϵ's and ξ's are constants to be chosen.

$$\frac{3}{2}\sum_{i=2}^{n} z_i^2 \left(\frac{\partial \alpha_{i-1}}{\partial y}\right)^2 \varphi_1(y)^{\mathrm{T}}\Sigma\Sigma^{\mathrm{T}}\varphi_1(y)$$

$$\leq \frac{3}{4}\sum_{i=2}^{n}\frac{1}{\xi_i^2}\left(\frac{\partial \alpha_{i-1}}{\partial y}\right)^4 z_i^4 \left(\varphi_1(y)^{\mathrm{T}}\varphi_1(y)\right)^2 + \frac{3}{4}\sum_{i=2}^{n}\xi_i^2|\Sigma\Sigma^{\mathrm{T}}|^2 \qquad (5.74)$$

$$b\mathrm{Tr}\left\{\varphi(y)\left(2P\tilde{x}\tilde{x}^{\mathrm{T}}P + \tilde{x}^{\mathrm{T}}P\tilde{x}P\right)\varphi(y)^{\mathrm{T}}\right\}$$

$$\leq 3bn\sqrt{n}|\varphi(y)^{\mathrm{T}}\varphi(y)||P|^2|\tilde{x}|^2|\Sigma|^2$$

$$= 3bn\sqrt{n}|P|^2|\varphi(0)^{\mathrm{T}}\varphi(0)||\tilde{x}|^2|\Sigma|^2 + 6bn\sqrt{n}|P|^2|y\varphi(0)^{\mathrm{T}}\psi(y)||\tilde{x}|^2|\Sigma|^2$$

$$+ 3bn\sqrt{n}|P|^2 y^2|\psi(y)^{\mathrm{T}}\psi(y)||\tilde{x}|^2|\Sigma|^2$$

$$\leq \frac{3bn\sqrt{n}|P|^2}{2\epsilon_2^2}|\varphi(0)^{\mathrm{T}}\varphi(0)||\tilde{x}|^4 + \frac{3bn\sqrt{n}|P|^2\epsilon_2^2}{2}|\varphi(0)^{\mathrm{T}}\varphi(0)||\Sigma|^4$$

$$+ 3bn\sqrt{n}|P|^2\epsilon_3^2|\varphi(0)^{\mathrm{T}}\psi(y)|^2 y^2|\Sigma|^4 + \frac{3bn\sqrt{n}|P|^2}{\epsilon_3^2}|\tilde{x}|^4 + \frac{3bn\sqrt{n}|P|^2}{2\epsilon_4^2}|\tilde{x}|^4$$

$$+ \frac{3bn\sqrt{n}|P|^2\epsilon_4^2}{2}y^4|\psi(y)^{\mathrm{T}}\psi(y)|^2|\Sigma|^4$$

$$\leq 3bn\sqrt{n}|P|^2\left(\frac{1}{2\epsilon_2^2}|\varphi(0)^{\mathrm{T}}\varphi(0)| + \frac{1}{\epsilon_3^2} + \frac{1}{2\epsilon_4^2}\right)|\tilde{x}|^4$$

$$+ \frac{3bn\sqrt{n}|P|^2\epsilon_2^2}{2}|\varphi(0)^{\mathrm{T}}\varphi(0)||\Sigma|^4 + 3bn\sqrt{n}|P|^2\left(\frac{\epsilon_3^2}{2} + \frac{\epsilon_4^2}{4}\right)|\Sigma|^8$$

$$+ 3bn\sqrt{n}|P|^2\left(\frac{\epsilon_2^2}{2}|\varphi(0)^{\mathrm{T}}\psi(y)|^4 y + \frac{\epsilon_4^2}{4}|\psi(y)^{\mathrm{T}}\psi(y)|^4 y^5\right)y^3 \qquad (5.75)$$

$$\frac{3}{2}y^2\varphi_1(y)^{\mathrm{T}}\Sigma\Sigma^{\mathrm{T}}\varphi_1(y) \leq \frac{3}{4}y^4\left(\varphi_1(y)^{\mathrm{T}} varphi_1(y)\right)^2 + \frac{3}{4}|\Sigma|^4 \quad (5.76)$$

$$-\frac{1}{2}\frac{\partial^2 \partial_{i-1}}{\partial y^2}z_i^3\varphi_1(y)^{\mathrm{T}}\Sigma\Sigma^{\mathrm{T}}\varphi_1(y) \leq \frac{1}{4}z_i^6\left(\frac{\partial^2 \partial_{i-1}}{\partial y^2}\varphi_1(y)^{\mathrm{T}}\varphi_1(y)\right)^2 + \frac{1}{4}|\Sigma|^4 \quad (5.77)$$

Appendix C

Similar to Appendix B, in the following inequalities, ϵ_2 is a constant to be chosen.

$$\frac{3}{2} \sum_{i=2}^{n} z_i^2 \left(\frac{\partial \alpha_{i-1}}{\partial y}\right)^2 \varphi_1(y)^{\mathrm{T}} \Sigma \Sigma^{\mathrm{T}} \varphi_1(y)$$

$$\leq \frac{3}{4} \sum_{i=2}^{n} \left(\frac{\partial \alpha_{i-1}}{\partial y}\right)^4 z_i^4 \| \Sigma \Sigma^{\mathrm{T}} \|_\infty^2 + \frac{3}{4} \sum_{i=2}^{n} y^4 \left(\psi_1(y)^{\mathrm{T}} \psi_1(y)\right)^2 (5.78)$$

$$b \mathrm{Tr} \left\{ \varphi(y) \left(2 P \tilde{x} \tilde{x}^{\mathrm{T}} P + \tilde{x}^{\mathrm{T}} P \tilde{x} P\right) \varphi(y)^{\mathrm{T}} \right\}$$

$$\leq 3bn\sqrt{n} y^2 |\psi(y)|^2 |P|^2 |\tilde{x}|^2 \| \Sigma \Sigma^{\mathrm{T}} \|_\infty$$

$$\leq \frac{3bn\sqrt{n}\epsilon_2^2 |P|^2}{2} y^4 |\psi(y)|^4 \| \Sigma \Sigma^{\mathrm{T}} \|_\infty^2 + \frac{3bn\sqrt{n}|P|^2}{2\epsilon_2^2} |\tilde{x}|^4 \qquad (5.79)$$

$$\frac{3}{2} y^2 \varphi_1(y)^{\mathrm{T}} \Sigma \Sigma^{\mathrm{T}} \varphi_1(y)$$

$$\leq \frac{3}{4} \left(\varphi_1(y)^{\mathrm{T}} \varphi_1(y)\right)^2 + \frac{3}{4} y^4 \| \Sigma \Sigma^{\mathrm{T}} \|_\infty^2$$

$$= \frac{3}{4} y^4 \left(\psi_1(y)^{\mathrm{T}} \psi_1(y)\right)^2 + \frac{3}{4} y^4 \| \Sigma \Sigma^{\mathrm{T}} \|_\infty^2 \qquad (5.80)$$

$$-\frac{1}{2} \frac{\partial^2 \alpha_{i-1}}{\partial y^2} z_i^3 \varphi_1(y)^{\mathrm{T}} \Sigma \Sigma^{\mathrm{T}} \varphi_1(y)$$

$$\leq \frac{1}{4} z_i^6 \left(\frac{\partial^2 \alpha_{i-1}}{\partial y^2}\right)^2 \| \Sigma \Sigma^{\mathrm{T}} \|_\infty^2 + \frac{1}{4} y^4 \left(\psi_1(y)^{\mathrm{T}} \psi_1(y)\right)^2 \qquad (5.81)$$

Output Feedback Control of Food–Chain Systems

Romeo Ortega[1], Alessandro Astolfi[2], Georges Bastin[3] and Hugo Rodrigues–Cortes[1]

[1]Laboratoire des Signaux et Systèmes, Ecole Supérieure d'Electricité, Paris, France.
[2]Centre for Process Systems Engineering, Imperial College of Science, London, UK.
[3]Centre for Systems Engineering and Applied Mechanics, Universite Catholique de Louvain, Louvain–La–Neuve, Belgium.

1   Introduction

In this chapter, we consider the problem of output feedback control of a class of non–linear mass balance models that describe the behavior of certain food–chain systems. These models are of interest, amoung other fields, in environmental engineering.

The control approach we use to solve the stabilization problem builds upon some recent developments on passivity–based stabilization of port–controlled Hamiltonian systems reported in [7], [5]. Since the design procedure is applicable to a broad class of mass–balance systems of similar structure (such as comportamental systems and stirred tank reactors, see [1] and the references therein), we present it in a rather general form. In this technique the original Hamiltonian structure of the system is preserved in closed–loop, and only the energy function and the dissipation are modified via the control. Preservation of the Hamiltonian structure allows stabilization to be understood in terms of energy. These feature makes the method very appealing in applications, since the action of the control has a clear physical interpretation that simplifies its comissioning. This task is particularly difficult in mass–balance systems where the control (and the system state) should be positive.

One further advantage of the method, central for the developments in this paper, is that the restriction of disposing only of output–feedback (as opposed to full–state feedback) can be naturally incorporated into the controller design. In particular, we show here that to obtain an output–feedback control strategy, some of the natural damping of the mass–balance equations should be removed, leaving only the damping of the measurable coordinate, which is necessary to ensure asymptotic stability. To better explain this modification we present first a state–feedback solution for the simplest second order model. In this case we leave untouched the natural damping of the system and apply *verbatim* the method proposed in [7]. A careful observation of the energy–shaping plus damping injection conditions of [7] reveals that with a, rather unusual, injection of positive damping we can easily obtain an output–feedback solution. Furthermore, the new solution is a simple linear controller, while the state–feedback controller is nonlinear, and rather involved. It is interesting to note that, the injection of positive damping allows us to obtain a stabilizing controller for the n–th order model, while the solution without removal of damping cannot be extended beyond the second order case.

Some simulation results are presented to illustrate the properties of the controller, and we conclude the chapter with some open questions and final remarks.

2 Controller Design Procedure

In this section we review the basic material of [7] presented in a form suitable for the problem considered in this chapter. Even though we deal with mass–balances instead of energy–balances, to keep up with the standard notation we will use throughout the word "energy".

We consider, so–called port–controlled Hamiltonian models of the form [6], [11], [10]

$$\Sigma \ : \ \dot{x} = [J(x) - R(x)]\frac{\partial H}{\partial x}(x) + g(x)u, \tag{6.1}$$

where $x \in \Re_+^n \subset \Re^n$, $u \in \Re_+^m \subset \Re^m$, are the mass variables, and the control, respectively. The set $\Re_+^n$ is the n–dimensional positive orthant. The smooth function $H(x) : \Re^n \to \Re$, which typically represents the total stored energy, will denote for our mass–balance systems the *total mass*, and it will be non–negative. The matrices

$$J(x) = -J^\top(x), \ \ R(x) = R^\top(x) \geq 0, \ \ \forall\, x \in \Re_+^n,$$

capture the internal interconnections and the natural damping of the system, respectively, while $g(x)$ defines the interconnection of the system with its environment. We assume measurable the q–dimensional output vector function $y = h(x)$. This output should not be confused with the natural outputs associated to the port–controlled Hamiltonian system Σ defined as $g^\top(x)\frac{\partial H}{\partial x}(x)$ [7].

The control objective is to stabilize, via output–feedback, an equilibrium $\bar{x} \in \Re_+^n$ preserving in closed–loop the Hamiltonian structure. The latter property allows us to provide an *energy* interpretation of the control action.

We will consider only static controllers, but as shown in [7] the procedure can be easily modified to incorporate controller dynamics.

Following the principles of passivity–based control [8], [10], we will achieve the stabilization objective by the standard energy–shaping plus damping injection stages. That is:

1. Assigning to the closed–loop an energy function $H_d(x)$, which should have a strict local minimum at $\bar{x}$. (That is, there exists an open neighbourhood $\mathcal{B}$ of $\bar{x}$ such that $H_d(x) > H_d(\bar{x})$ for all $x \in \mathcal{B}$.) We will define

$$H_d(x) \triangleq H(x) + H_a(x) \tag{6.2}$$

where $H_a(x)$ is a function to be defined.

2. Injecting some additional damping $R_a(x)$ to get

$$R_d(x) \triangleq R(x) + R_a(x) \geq 0, \quad \forall\, x \in \Re_+^n \tag{6.3}$$

That is, we look for an output–feedback control $u(h(x))$ such that

$$[J(x) - R(x)]\frac{\partial H}{\partial x}(x) + g(x)u(h(x)) = [J(x) - R_d(x)]\frac{\partial H_d}{\partial x}(x)$$

holds $\forall\, x \in \Re_+^n$, with $H_d(x)$, $R_d(x)$ defined by (6.2) and (6.3), respectively. In this way, the closed-loop dynamics will be defined as

$$\dot{x} = [J(x) - R_d(x)]\frac{\partial H_d}{\partial x}(x), \tag{6.4}$$

and along the trajectories of (6.4) we will have

$$\frac{d}{dt}H_d = -\left[\frac{\partial H_d}{\partial x}(x)\right]^{\mathsf{T}} R_d(x)\frac{\partial H_d}{\partial x}(x) \leq 0, \; \forall x \in \Re_+^n \tag{6.5}$$

Thus, $\bar{x}$ will be a *stable* equilibrium.

For ease of presentation we will assume throughout the following:

Assumption A $[J(x) - R_d(x)]$ is invertible for every $x \in \Re_+^n$.

□□□

It is important to remark that this does not imply that the closed–loop system is fully damped. That is, we do not require $R_d(x) > 0$, $\forall x \in \Re_+^n$. Actually, it is shown in [7] that Assumption A is not needed for the proof of the proposition below.

We have the following basic result.

Proposition 6.1 *[7]   Given $J(x), R(x), H(x), g(x)$. Assume we can find and output–feedback control $u(h(x))$ and a matrix $R_a(x)$ such that $R(x) + R_a(x) \geq 0$, Assumption A hold, and the vector function $K(x)$, defined as,*

$$K(x) \triangleq [J(x) - (R(x) + R_a(x))]^{-1} [R_a(x) \frac{\partial H}{\partial x}(x) + g(x)u(h(x))] \quad (6.6)$$

satisfies

- *(Integrability) $K(x)$ is the gradient of a scalar function. That is,*

$$\frac{\partial K}{\partial x}(x) = \left[\frac{\partial K}{\partial x}(x) \right]^{\mathsf{T}} \quad (6.7)$$

- *(Equilibrium assignment) $K(x)$, at $\bar{x}$, verifies*

$$K(\bar{x}) = -\frac{\partial H}{\partial x}(\bar{x}) \quad (6.8)$$

- *(Lyapunov stability) The Jacobian of $K(x)$, at $\bar{x}$, satisfies the bound*

$$\frac{\partial K}{\partial x}(\bar{x}) > -\frac{\partial^2 H}{\partial x^2}(\bar{x}) \quad (6.9)$$

Then, $\bar{x}$ will be a locally stable equilibrium of the closed–loop. It will be asymptotically stable if, furthermore, the largest invariant set under the closed–loop dynamics contained in

$$\left\{ x \in \Re_+^n \cap \mathcal{B} \mid \left[\frac{\partial H_d}{\partial x}(x) \right]^{\mathsf{T}} R_d(x) \frac{\partial H_d}{\partial x}(x) = 0 \right\} \quad (6.10)$$

equals $\{\bar{x}\}$, where $H_d(x)$ is given by (6.2). The latter condition will be automatically satisfied if we can achieve full damping, that is, if $R_d(x) > 0$ for every $x \in \Re_+^n$.

Proof
First, notice that, using (6.2), (6.3) and Assumption A, the identity (6.4) may be equivalently written as

$$\frac{\partial H_a}{\partial x}(x) = [J(x) - R_d(x)]^{-1} [R_a(x) \frac{\partial H}{\partial x}(x) + g(x)u(h(x))] \quad (6.11)$$

For every given $u(h(x)), R_a(x)$, this is a linear PDE. A necessary and sufficient condition for the solvability of this PDE (on every contractible neighbourhood of $\Re_+^n$) is that the gradient of the right hand side of (6.11) is a symmetric matrix. From (6.3), (6.6) and (6.11) we see that

$$K(x) = \frac{\partial H_a}{\partial x}(x) \quad (6.12)$$

Henceforth, the matrix mentioned above will be symmetric iff the integrability condition (6.7) of the proposition is satisfied.

The stability proof is concluded invoking standard Lyapunov stability arguments [4]. Namely, from (6.5), we conclude that, under the standing assumptions, $H_d(x)$ qualifies as a Lyapunov function. Asymptotic stability follows from a direct application of La Salle's invariance principle and (6.10).

□□□

Remark 6.1 *Notice that the construction above* does not require *the explicit derivation of the Lyapunov function $H_d(x)$. This can be obtained, though, as a by–product integrating $K(x) = \frac{\partial H_a}{\partial x}(x)$.*

Remark 6.2 *Port-controlled Hamiltonian models (6.1) encompass a very large class of physical nonlinear systems, strictly containing the class of Euler–Lagrange models considered, for instance, in [8]. They result from the network modeling of energy–conserving lumped–parameter physical systems with independent storage elements, and have been advocated in a series of recent papers [6], [11] as an alternative to more classical Euler–Lagrange (or standard Hamiltonian) models.*

3   State–Feedback Control of a Simple Prey–Predator System

As pointed out in the introduction, to motivate our output–fedback control (which is given in the next section) we present first a state–feedback stabilizer for a simple second order food–chain system. The controller is obtained from a *verbatim* application of the method described above. This is a systematic technique that can be efficiently combined with symbolic computation. See, for instance, the simple Maple code given in Appendix A.

System Model

We consider the normalized second order prey–predator system (see e.g. [3])

$$\begin{aligned}
\dot{x}_1 &= f(x) - x_1 \\
\dot{x}_2 &= -f(x) - x_2 + u
\end{aligned} \tag{6.13}$$

The state variables x_1, x_2 represent the amount of mass of the two species (preys and predators) involved in the system. The function $f(x)$ describes the predation mechanism, we consider here the classical Lotka–Volterra mechanism $f(x) = x_1 x_2$. The terms $-x_1, -x_2$ in (6.13) represent the natural mortality of the species, while the control action u is a feeding inflow rate of preys. For the output feedback case, we will consider that the variable available for measurement is the last one in the chain, in this case, x_2.

The evolution of the system is clearly restricted to the positive orthant with $u \geq 0$. That is,

$$x_i(0) \geq 0, \text{ and } u(t) \geq 0, \ \forall t \geq 0 \ \Rightarrow \ x_i(t) \geq 0, \ \forall t \geq 0$$

It is possible to show that any equilibrium of the open–loop system with a lit *constant* input $\bar{u} \geq 0$ is globally asymptotically stable. The *control objective* is, then, to asymptotically stabilize a *given* non–zero equilibrium $\bar{x} \in \mathbb{R}^2_+$ with a positive control. The *achievable* equilibria are $\bar{x} = [\bar{x}_1, \bar{x}_2]^\top = [x_1^*, 1]^\top$, with $x_1^* > 0$ the reference for x_1.

If we define the total mass

$$H(x) = x_1 + x_2$$

the system (6.13) may be written in the form (6.1) with

$$J(x) = \begin{bmatrix} 0 & x_1 x_2 \\ -x_1 x_2 & 0 \end{bmatrix}, \quad R(x) = \begin{bmatrix} x_1 & 0 \\ 0 & x_2 \end{bmatrix} \quad g(x) = g = \begin{bmatrix} 0 \\ 1 \end{bmatrix}$$

The skew–symmetry of $J(x)$ captures the mass–conservative feature of the system without inflows and outflows.

State–Feedback Stabilization

Since the system is already fully damped, i.e., $R(x) > 0$, $\forall x \neq 0, x \in \mathbb{R}^n_+$, it seems reasonable (as our first try) to set $R_a(x) = 0$. That is, we will not inject additional damping, but rely instead on the natural damping of the system to ensure the attractivity. In this case, the vector function (6.6) reduces to

$$K(x) = \begin{bmatrix} k_1(x) \\ k_2(x) \end{bmatrix} = \frac{-u(x)}{1 + x_1 x_2} \begin{bmatrix} 1 \\ \frac{1}{x_2} \end{bmatrix}$$

From which we immediately conclude that

$$x_2 k_2(x) = k_1(x) \tag{6.14}$$

The integrability condition (6.7) in this two–dimensional case reduces to

$$\frac{\partial k_1}{\partial x_2}(x) = \frac{\partial k_2}{\partial x_1}(x),$$

which, combined with (6.14), yields the linear PDE

$$\frac{\partial k_1}{\partial x_1}(x) - x_2 \frac{\partial k_1}{\partial x_2}(x) = 0 \tag{6.15}$$

A family of solutions of this PDE is easily obtained as

$$
\begin{aligned}
k_1(x) &= \Phi(\zeta(x)) \\
\zeta(x) &= x_1 + \log x_2,
\end{aligned}
$$

for all differentiable functions $\Phi(\cdot)$. From (6.14) we also obtain

$$
k_2(x) = \frac{1}{x_2}\Phi(\zeta(x))
$$

The equilibrium condition (6.8) imposes

$$
\begin{bmatrix} k_1(\bar{x}) \\ k_2(\bar{x}) \end{bmatrix} = - \begin{bmatrix} 1 \\ 1 \end{bmatrix} \tag{6.16}
$$

Hence, $\Phi(\cdot)$ must be such that $\Phi(\zeta(\bar{x})) = -1$, where $\zeta(\bar{x}) = \bar{x}_1 + \log \bar{x}_2 = x_1^*$. It is clear then that we cannot take $\Phi(\zeta) = \zeta$. We propose the function

$$
\Phi(\zeta) = c_1 \exp^{c_2 \zeta},
$$

with c_1, c_2 constants to be defined. (Although this choice of function might seem a bit contrived, we should note that this is the function that results if we directly apply the method of undetermined coefficients to the PDE (6.15). See Appendix A). The equilibrium condition $\Phi(\zeta(\bar{x})) = -1$ fixes the first constant as

$$
c_1 = -\exp^{-c_2 x_1^*}
$$

We will now verify the Hessian condition (6.9). Some simple calculations yield

$$
\frac{\partial K}{\partial x}(x) = c_1 c_2 \exp^{c_2 \zeta} \begin{bmatrix} 1 & \frac{1}{x_2} \\ \frac{1}{x_2} & \frac{1}{x_2^2}(1 - \frac{1}{c_2}) \end{bmatrix} \left(= \frac{\partial^2 H_d}{\partial x^2}(x) \right),
$$

which evaluated in the equilibrium point gives

$$
\frac{\partial K}{\partial x}(\bar{x}) = -c_2 \begin{bmatrix} 1 & 1 \\ 1 & \frac{c_2-1}{c_2} \end{bmatrix}
$$

The determinant of this matrix is 1, hence it is positive definite iff $c_2 < 0$.

We will investigate now the asymptotic stability properties. To this end, we see that the ω–limit set (6.10) is defined as

$$
\left\{ x \in \Re_+^2 \cap \mathcal{B} \mid -x_1(1 + k_1(x))^2 - x_2(1 + k_2(x))^2 = 0 \right\},
$$

which consists only of the points $x = 0$ and $x = \bar{x}$. But, it can be easily shown, that $x = 0$ is an unstable equilibrium of the closed loop dynamics.

We have established the following result.

Proposition 6.2 *Consider the system (6.13), with $f(x) = x_1 x_2$, in closed–loop with the positive control*

$$u(x) = (1 + x_1 x_2) x_2^c \exp^{c(x_1 - x_1^*)} \tag{6.17}$$

with $x_1^* > 0$ the reference for x_1, and $c < 0$. Then, all trajectories starting in $x(0) \in \Re_+^2$, will converge asymptotically to the desired equilibrium point $(x_1^*, 1)$.

□□□

Let us summarize the calculations carried out above:

1. Fix the added damping $R_a(x)$ – to 0 in this case, since the open–loop system is fully damped –;

2. Define the vector $K(x)$, (6.6), as a function of $u(x)$;

3. Use the integrability conditions (6.7) to eliminate the control and obtain a linear PDE (6.15) to be solved for $K(x)$;

4. Find a solution of this PDE that satisfies the equilibrium (6.8) and Lyapunov stability conditions (6.9);

5. Derive the control law (6.17) from the definition of $K(x)$.

Remark 6.3 *As pointed out in Remark 2 as a by–product of our analysis we can get a Lyapunov function, which in our case is*

$$H_d(x) = \underbrace{x_1 + x_2}_{H(x)} - \underbrace{\frac{1}{k} x_2^k \exp^{k(x_1 - x_1^*)} + \frac{1}{k}}_{H_a(x)} - (1 + x_1^*)$$

where the third and fourth right hand constant terms are added to enforce $H_d(\bar{x}) = 0$. It is worth noting that $H_d(x)$ above is the classical Lyapunov function for the stability analysis of Lotka–Volterra ecologies (see e.g. [3] and [9] among many other references). The design procedure of this paper allows to rediscover this Lyapunov function in a very natural way.

Remark 6.4 *There is an easier way to derive the structural constraint (6.14) that does not require the inversion of the matrix $J(x) - R_a(x)$. To this end, rewrite (6.6) as*

$$[J(x) - (R(x) + R_a(x))]K(x) = [R_a(x)\frac{\partial H}{\partial x}(x) + g(x)u(x)] \tag{6.18}$$

The first equation of (6.18) for this example yields

$$-x_1 k_1(x) + x_1 x_2 k_2(x) = 0$$

which, upon division by x_1, is precisely (6.14). The second equation simply defines the control law, in terms of $K(x)$, as

$$u(x) = -x_2 k_2(x) - x_1 x_2 k_1(x) \tag{6.19}$$

It is precisely this observation that will motivate the modification, introduced in the next section, that yields an output–feedback stabilizer.

4   Output–Feedback Stabilization

There are two important drawbacks of the solution proposed in the previous section. First, it requires measurement of all the state. Second, it can not be extended to treat the general food–chain system model, which is of the form

$$
\begin{aligned}
\dot{x}_1 &= x_1 x_2 - x_1 \\
\dot{x}_2 &= x_2 x_3 - x_1 x_2 - x_2 \\
\dot{x}_3 &= x_3 x_4 - x_2 x_3 - x_3 \\
&\vdots \qquad \vdots \\
\dot{x}_n &= -x_{(n-1)} x_n - x_n + u \\
y &= x_n
\end{aligned}
\tag{6.20}
$$

To prove the second statement, let us write the model in the form (6.1) with $H(x) = \Sigma_{i=1}^{n} x_i$ and

$$
J(x) = \begin{bmatrix}
0 & x_1 x_2 & 0 & \cdots & 0 \\
-x_1 x_2 & 0 & x_2 x_3 & \cdots & 0 \\
\vdots & \vdots & \vdots & \cdots & \vdots \\
0 & 0 & 0 & \cdots & 0
\end{bmatrix} = -J^{\mathsf{T}}(x)
$$

$$
R(x) = \begin{bmatrix}
x_1 & 0 & \cdots & 0 \\
0 & x_2 & \cdots & 0 \\
\vdots & \vdots & \cdots & \vdots \\
0 & 0 & \cdots & x_n
\end{bmatrix} = R^{\mathsf{T}}(x) \geq 0,\; g(x) = g = \begin{bmatrix} 0 \\ 0 \\ \vdots \\ 1 \end{bmatrix}
$$

Then, notice that the distribution spanned by the vector fields defined by the column vectors obtained from the first $n-1$ rows of $J(x) - R(x)$ is not involutive. Consequently, the key PDE

$$[J(x) - R(x)] \frac{\partial H_a}{\partial x}(x) = gu(x)$$

can not be solved.

In this section we show how, for our second order example (6.13), these limitations can be overcome modifying the damping of the closed–loop. In the next section we extend this result to the general n–th order model (6.20).

Towards this end, let us remove the damping from the first coordinate. That is, define $R_a(x)$ like

$$R_a(x) = \begin{bmatrix} -x_1 & 0 \\ 0 & 0 \end{bmatrix}$$

Notice the negative sign. With this choice, the vector function (6.6) becomes now

$$K(x) = \begin{bmatrix} k_1(x) \\ k_2(x) \end{bmatrix} = \begin{bmatrix} -\frac{1}{x_1 x_2}(u-1) \\ -\frac{1}{x_2} \end{bmatrix}$$

Choosing the control law as the simple output–feedback

$$u(x_2) = cx_2 + 1,$$

with c some constant to be defined, yields

$$K(x) = \begin{bmatrix} -\frac{c}{x_1} \\ -\frac{1}{x_2} \end{bmatrix} \tag{6.21}$$

which is clearly the gradient of a scalar function. Hence, the integrability condition (6.7) is satisfied. We will now verify if we can find a constant c such that the remaining stability conditions of Proposition 6.1 are also satisfied. The equilibrium condition (6.16) imposes $c = x_1^*$. For the Hessian condition (6.9) we first observe from (6.21) and $\frac{\partial^2 H}{\partial x^2}(x) = 0$, that

$$\frac{\partial^2 H_d}{\partial x^2}(x) = \frac{\partial K}{\partial x}(x) = \begin{bmatrix} \frac{x_1^*}{x_1^2} & 0 \\ 0 & \frac{1}{x_2^2} \end{bmatrix}$$

Evaluated in the equilibrium point gives

$$\frac{\partial^2 H_d}{\partial x^2}(\bar{x}) = \frac{\partial K}{\partial x}(\bar{x}) = \begin{bmatrix} \frac{1}{x_1^*} & 0 \\ 0 & 1 \end{bmatrix},$$

which will be positive definite for any $x_1^* > 0$.

Finally, asymptotic stability is ensured because the ω–limit set (6.10) is now defined as

$$\left\{ x \in \Re_+^2 \cap B \mid \frac{x_2 - 1}{x_2} = 0 \right\},$$

which consists only of the point $x = \bar{x}$.

The new Lyapunov function is

$$H_d(x) = \underbrace{x_1 + x_2}_{H(x)} - \underbrace{x_1^* \ln(x_1) - \ln(x_2)}_{H_a(x)} - (x_1^* + 1 - x_1^* \ln(x_1^*)),$$

where the third right hand constant term is, again, added to enforce $H_d(\bar{x}) = 0$.

We have established the following result.

Proposition 6.3 *Consider the system (6.13), with $f(x) = x_1 x_2$, in closed–loop with the positive output–feedback control*

$$u(x_2) = 1 + x_1^* x_2 \tag{6.22}$$

with $x_1^* > 0$ the reference for x_1. Then, all trajectories starting in $x(0) \in \mathcal{R}_+^2$, will converge asymptotically to the desired equilibrium point $(x_1^*, 1)$.

□□□

Remark 6.5 *To increase the speed of convergence it is possible to inject some additional damping on the actuated coordinate x_2. To this end, we choose*

$$R_a(x) = \begin{bmatrix} -x_1 & 0 \\ 0 & (r-1)x_2 \end{bmatrix},$$

with the desired damping a constant $1 < r < 1 + x_1^*$. Going through the calculations we get the control law

$$u(x_2) = r + (x_1^* - r + 1)x_2 \tag{6.23}$$

It can be shown that this control law is also globally asymptotically stabilizing. Notice that with $r = 1$ we recover the controller (6.22).

5 Main Result

In this section we present the generalization of the previous result to the n–th order case.

Theorem 6.1 *Consider the general food chain system (6.20) in closed-loop with the output–feedback positive control*

$$u(x_n) = mx_n + m + x_1^*, \quad m = \frac{n-1}{2}$$

for n odd, and

$$u(x_n) = (m + x_1^*)x_n + \frac{n}{2}, \quad m = \frac{n}{2} - 1$$

for n even, with $x_1^* > 0$ the reference for x_1. Then, all trajectories starting in $x(0) \in \mathfrak{R}_+^n$ will converge asymptotically to the desired equilibrium point $\bar{x} = [x_1^*, \bar{x}_2, ..., \bar{x}_n]$.

□□□

Proof

Motivated by the developments of the second order case above we propose to remove the damping from all non–actuated coordinates. That is, we choose

$$R_a(x) = \begin{bmatrix} -x_1 & 0 & \cdots & 0 \\ 0 & -x_2 & \cdots & 0 \\ . & . & \cdots & . \\ 0 & 0 & \cdots & 0 \end{bmatrix}$$

We will now verify the three conditions of Proposition 6.1.

- *Integrability*

The key equation (6.11) becomes then

$$\begin{bmatrix} 0 & x_1 x_2 & 0 & \cdots & 0 & 0 \\ -x_1 x_2 & 0 & x_2 x_3 & \cdots & 0 & 0 \\ . & . & . & \cdots & . & . \\ 0 & 0 & 0 & \cdots & 0 & x_{n-1} x_n \\ 0 & 0 & 0 & \cdots & -x_{n-1} x_n & x_n \end{bmatrix} \begin{bmatrix} k_1(x) \\ k_2(x) \\ . \\ . \\ k_{n-1}(x) \\ k_n(x) \end{bmatrix}$$

$$= \begin{bmatrix} -x_1 \\ -x_2 \\ . \\ . \\ -x_{n-1} \\ u(x) \end{bmatrix},$$

which can be compactly written as $\bar{J}(x) K(x) = \bar{g}(x)$. Now, $\bar{J}(x)$ admits a factorization of the form

$$\bar{J}(x) = \text{diag}\{x_i\} \begin{bmatrix} 0 & 1 & 0 & \cdots & 0 & 0 \\ -1 & 0 & 1 & \cdots & 0 & 0 \\ . & . & . & \cdots & . & . \\ 0 & 0 & 0 & \cdots & 0 & 1 \\ 0 & 0 & 0 & \cdots & -1 & -\frac{1}{x_n} \end{bmatrix} \text{diag}\{x_i\}$$

This leads to

$$
\begin{bmatrix}
0 & 1 & 0 & \cdots & 0 & 0 \\
-1 & 0 & 1 & \cdots & 0 & 0 \\
\vdots & \vdots & \vdots & \cdots & \vdots & \vdots \\
0 & 0 & 0 & \cdots & 0 & 1 \\
0 & 0 & 0 & \cdots & -1 & -\frac{1}{x_n}
\end{bmatrix}
\begin{bmatrix}
x_1 k_1(x) \\
x_2 k_2(x) \\
\cdot \\
\cdot \\
\cdot \\
x_{n-1} k_{n-1}(x) \\
x_n k_n(x)
\end{bmatrix}
=
\begin{bmatrix}
-1 \\
-1 \\
\cdot \\
\cdot \\
\cdot \\
-1 \\
\frac{u(x)}{x_n}
\end{bmatrix}
$$

From which we obtain a system of equations of the form

$$
\begin{aligned}
x_2 k_2(x) &= -1 \\
-x_1 k_1(x) + x_3 k_3(x) &= -1 \\
-x_2 k_2(x) + x_4 k_4(x) &= -1 \\
&\;\;\vdots \qquad\qquad \vdots \\
-x_{n-2} k_{n-2}(x) + x_n k_n(x) &= -1 \\
-x_{n-1} k_{n-1}(x) - k_n(x) &= \frac{u(x)}{x_n}
\end{aligned}
\qquad (6.24)
$$

Notice that from the first equation of (6.24) we have

$$
k_2(x) = -\frac{1}{x_2}
$$

Subsequently, the functions $k_i(x)$, for i even, have a unique solution, which is furthermore of the form $k_i(x) = k_i(x_i)$. Now, choosing

$$
k_1(x) = -\frac{c}{x_1}
$$

we can also obtain a unique solution $k_i(x_i)$, for i odd. The vector function $K(x)$ is finally given by

$$
K(x) = \begin{bmatrix} -\frac{1}{x_1} & -\frac{1}{x_2} & \cdots & -\frac{m}{x_{n-1}} & -\frac{m+c}{x_n} \end{bmatrix}^T, \quad m = \frac{n-1}{2},
$$

for n odd, and

$$
K(x) = \begin{bmatrix} -\frac{c}{x_1} & -\frac{1}{x_2} & \cdots & -\frac{m+c}{x_{n-1}} & -\frac{\frac{n}{2}}{x_n} \end{bmatrix}^T, \quad m = \frac{n}{2} - 1,
$$

for n even. It is clear that, in both cases, the integrability conditions are satisfied.

Also, from the last equation of (6.24) we compute the control law

$$
u(x) = -x_n[x_{n-1} k_{n-1}(x) + k_n(x)]
$$

- *Equilibrium Assignment*

The equilibrium condition is

$$K(\bar{x}) = -\frac{\partial H}{\partial x}(\bar{x}) = -\begin{bmatrix} 1 & \cdots & 1 & 1 \end{bmatrix}^T = \begin{bmatrix} -\frac{c}{x_1^*} & \cdots & -\frac{m+c}{x_{n-1}^*} & -\frac{\frac{n}{2}}{x_n^*} \end{bmatrix}^T,$$

which is satisfied with $c = x_1^*$.

- *Lyapunov Stability*

We will now verify the Hessian condition. Some simple calculations yield

$$\frac{\partial K}{\partial x}(x) = \begin{bmatrix} \frac{x_1^*}{x_1^2} & 0 & 0 & 0 & \cdots & 0 & 0 \\ 0 & \frac{1}{x_2^2} & 0 & 0 & \cdots & 0 & 0 \\ 0 & 0 & \frac{1+x_1^*}{x_3^2} & 0 & \cdots & 0 & 0 \\ 0 & 0 & 0 & & \cdots & & \cdot \\ \cdot & \cdot & \cdot & \cdot & \cdots & \cdot & \cdot \\ \cdot & \cdot & \cdot & \cdot & \cdots & 0 & 0 \\ 0 & 0 & 0 & 0 & \cdots & 0 & \frac{m+x_1^*}{x_n^2} \end{bmatrix} \qquad (6.25)$$

This matrix will be positive definite for any $x \in \Re_+^n$ and any $x_1^* > 0$. Finally, the ω limit set for n odd is defined as $\{x \in \Re_+^n \cap B \mid \frac{x_n - (m + x_1^*)}{x_n} = 0\}$ and $\{x \in \Re_+^n \cap B \mid \frac{x_n - (\frac{n}{2})}{x_n} = 0\}$ for n even. In both cases the ω limit set consists only of the point $x_n = \bar{x}_n$. This, together with uniqueness of the equilibrium, completes the proof of asymptotic stability.

Remark 6.6 *The proposed control design can be easily applied to the more general class of Lotka–Volterra ecologies defined as follows:*

$$\dot{x}_i = x_i\left(-k_i + \sum_{j \neq i} a_{ij} x_j\right) \quad i = 1, ..., n-1$$

$$\dot{x}_n = x_n\left(-k_n + \sum_{j \neq n} a_{nj} x_j\right) + u$$

with $k_i > 0$ the natural mortality rates, $a_{ij} = -a_{ij}, \forall i \neq j$, the predation coefficients and u the feeding rate of species x_n, with $u(t) \geq 0$ $\forall t$.

The procedure yields the classical Lyapunov function for Lotka–Volterra ecologies

$$\sum_{i=1}^n x_i - \bar{x}_i \ln(x_i),$$

and we obtain the following output feedback control law

$$u(x_n) = \bar{u} + \lambda(x_n - \bar{x}_n)$$

with $\bar{u}$ the constant control that assigns the desired equilibrium, and $0 < \lambda < \frac{\bar{u}}{\bar{x}_n}$ an arbitrary design parameter.

6 Simulations

Numerical simulations of the second order model (6.13) were carried out in order to show the performance of the proposed controllers. The parameters used in the simulations were, $c = -0.2$ for the state feedback controller (6.17), and $r = 1, 2.1$, for the output feedback controller (6.23). The desired equilibrium of the system is $\bar{x} = [1.2, 1]^T$. The initial conditions in all the simulations are $x_1(0) = 2$ and $x_2(0) = 2$.

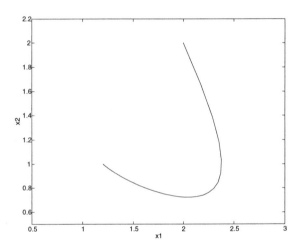

FIGURE 1. Open–loop trajectory

For the sake of comparison, in Fig. 1 we present the behaviour of the open loop trajectory in the state space with a constant input $\bar{u} = 2.2$, while Fig. 2 depicts the behaviour of the state and ouput feedback controllers. Finally, the control signals are shown in Fig. 3. As seem from the Figs. 2, 3 the addition of damping effectively increases the convergence rate with the additional advantage of reducing the control effort.

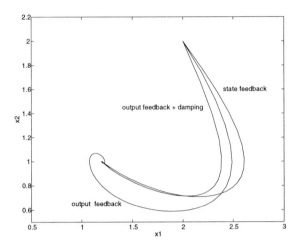

FIGURE 2. State space of the closed–loop trajectory

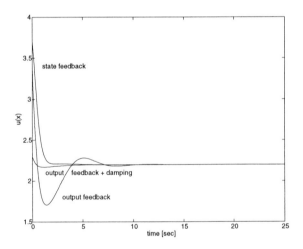

FIGURE 3. Control signals

7   Concluding Remarks

We have illustrated in this chapter how the application of the passivity–based controller design technique of [7] allows us to solve output–feedback stabilization problems for a class of mass–balance systems. The procedure is illustrated in detail with an n–th order food–chain model. It can, *mutatae–mutandi*, be applied also to other mass–balance models studied in [1], [3],

[9]. For instance, it can be shown that for the compartmental model of Section 4 in [1] the technique yields also asymptotically stabilizing controllers. However, we require in this case the knowledge of the full state.

We have not stressed here the advantages of taking a physically–based approach for controller design, see e.g. [8], [7], [10] for a detailed discussion. We should underscore, however, that the preservation of a physical interpretation to the control action (in terms of damping injection) was instrumental for our result. Finally, we bring to the readers attention the simplicity of the resulting control law. This important feature is a characteristic of passivity–based controllers.

As shown in this chapter the approach of [7] provides a flexible methodology to design controllers for physical systems. As discussed in that paper, we can also aim at modifying the internal interconnection structure $\mathcal{J}(x)$. In this way, we recover some of the results obtained with the technique of controlled Lagrangians, reported in [2]. Current research is under way to explore this interesting possibility for mass–balance systems.

Acknowledgements

The first author would like to express his deep gratitude to Bernhard Maschke and Arjan van der Schaft, with whom the basic principles underlying the developments reported here were obtained.

8 References

[1] G. Bastin and L. Praly. Feedback stabilization with positive control of a class of dissipative mass balance systems. accepted *IFAC World Congress*, Beijing, 1999.

[2] M. Bloch, E. Leonhard and J. Marsden. Controlled Lagrangians and the Stabilization of Mechanical Systems I: The First Matching Theorem. *IEEE Conf. Decision and Control*, Tampa, FL, 1998.

[3] J. Hofbaner and K. Sigmund. *Evolutionary Games and Population Dynamics*. Cambridge University Press, 1998.

[4] H. Khalil. *Nonlinear systems*. Prentice–Hall, 2nd edition, 1996. ISBN 0-13-22824-8.

[5] B. Maschke, R. Ortega and A. van der Schaft. Energy–based Lyapunov functions for forced Hamiltonian systems with dissipation. *IEEE Conf. Dec. and Control*, Tampa, FL, 1998.

[6] B.M. Maschke and A.J. van der Schaft. Port controlled Hamiltonian systems: modeling origins and system theoretic properties. *Proc. 2nd IFAC Symp. on Nonlinear Control Systems design*, NOLCOS'92, pp.282-288, Bordeaux, 1992.

[7] R. Ortega, B. Maschke, A. van der Schaft and G. Escobar. Passivity–based control of port–controlled Hamiltonian systems, *LSS–SUPELEC*, France, Int. Rep., 1998.

[8] R. Ortega and A. Loria, P. J. Nicklasson and H. Sira–Ramirez. *Passivity–Based Control of Euler–Lagrange Systems*. Springer-Verlag, Berlin, Communications and Control Engineering, 1998.

[9] F. Sendo and J. Ziegler. *The Golden Age of theoretical Ecology.* Lecture notes in Biomathematics , Springer Verlag, 1978.

[10] A. van der Schaft. L_2–*Gain and Passivity Techniques in Nonlinear Control.* Lect. Notes in Contr. and Inf. Sc., Vol. 218, Springer–Verlag, Berlin, 1996.

[11] A. van der Schaft and B. Maschke. The Hamiltonian formulation of energy–conserving physical systems with external ports. *Archiv für Elektronik und Übertragungstechnik*, 49, pp. 362–371, 1995.

Appendix A: Maple Code

In this appendix we present a Maple code that guides us in the solution of the example of Section 2. The calculations proceed as follows

1. Definition of the system (with $JmR \overset{\triangle}{=} J(x) - R(x)$ and $gu \overset{\triangle}{=} gu(x)$) :

```
> with(linalg):
> JmR := matrix(2,2,[-x1,x1*x2,-x1*x2,-x2]);
                        [ -x1        x1 x2]
              JmR := [              ]
                        [-x1 x2      -x2 ]

> gu := vector([0,u(x1,x2)]);
                    gu := [0, u(x1, x2)]
```

2. Computation of $K(x)$ and its Jacobian

```
> K := multiply(inverse(JmR),gu);
                    [  u(x1, x2)         u(x1, x2)    ]
            K := [-  ----------,  -  --------------]
                    [  1 + x1 x2     x2 (1 + x1 x2)]
```

```
> Jac := jacobian(K,[x1,x2]);

  Jac :=
```

```
[                  d                           d              ]
[              --- u(x1, x2)               --- u(x1, x2)]
[[u(x1, x2) x2     dx1          u(x1, x2) x1    dx2          ]
[------------ - -------------  , ------------ - -------------]
[          2     1 + x1 x2              2       1 + x1 x2   ]
[(1 + x1 x2)                   (1 + x1 x2)                   ]
```

```
[                  d
[              --- u(x1, x2)
[ u(x1, x2)        dx1
[----------- - --------------   ,
[          2    x2 (1 + x1 x2)
[(1 + x1 x2)
```

```
                                   d              ]
                               --- u(x1, x2) ]
     u(x1, x2)         u(x1, x2) x1    dx2          ]
  --------------- + --------------- - --------------]
       2                  2       x2 (1 + x1 x2)]
  x2  (1 + x1 x2)    x2 (1 + x1 x2)              ]
```

3. Definition of the term

$$eq12 \triangleq \frac{\partial k_2}{\partial x_1}(x) - \frac{\partial k_1}{\partial x_2}(x)$$

```
> eq12 := Jac[2,1]-Jac[1,2];
```

```
                       d
                   --- u(x1, x2)
         u(x1, x2)     dx1                  u(x1, x2) x1
eq12 := ------------ - --------------- - ------------
               2        x2 (1 + x1 x2)             2
         (1 + x1 x2)                        (1 + x1 x2)
```

```
           d
       --- u(x1, x2)
          dx2
     + -------------
         1 + x1 x2
```

4. Determination of the control $u(x)$ which solves $eq12 = 0$, i.e., which ensures the integrability condition.

```
> u_star:=rhs(pdesolve(eq12=0,u(x1,x2)));
```

$$u\_star := \_F1\left(\frac{x2}{\exp(-x1)}\right)(1 + x1\ x2)$$

Notice that in the line above $\_F1(\cdot)$ is any differentiable function.

5. Evaluation of the Hessian for the given control expression.

```
> subs(u(x1,x2)=u_star,evalm(Jac));
```

$$
\left[
\begin{array}{l}
\dfrac{\_F1\left(\dfrac{x2}{\exp(-x1)}\right)x2 \quad \dfrac{d}{dx1}\%1}{1 + x1\ x2} - \dfrac{\dfrac{d}{dx1}\%1}{1 + x1\ x2} \quad , \quad \dfrac{\_F1\left(\dfrac{x2}{\exp(-x1)}\right)x1}{1 + x1\ x2} - \dfrac{\dfrac{d}{dx2}\%1}{1 + x1\ x2}
\end{array}
\right]
$$

$$
\left[
\dfrac{\_F1\left(\dfrac{x2}{\exp(-x1)}\right)}{1 + x1\ x2} - \dfrac{\dfrac{d}{dx1}\%1}{x2\ (1 + x1\ x2)} \quad , \right.
$$

$$
\left.
\dfrac{\_F1\left(\dfrac{x2}{\exp(-x1)}\right)}{x2^2} + \dfrac{\_F1\left(\dfrac{x2}{\exp(-x1)}\right)x1}{x2\ (1 + x1\ x2)} - \dfrac{\dfrac{d}{dx2}\%1}{x2\ (1 + x1\ x2)}
\right]
$$

$$
\%1 := \_F1\left(\frac{x2}{\exp(-x1)}\right)(1 + x1\ x2)
$$

6. The design can be concluded selecting a function $\_F1(\cdot)$ that satisfies the equilibrium assignment and Lyapunov stability conditions of Proposition 6.1. In Section 2 we have chosen $\_F1(\zeta) = \zeta^k$.

Output Feedback Tracking Control for Ships

K. Y. Pettersen[1] and H. Nijmeijer[2]

[1]Department of Engineering Cybernetics, Norwegian University of Science and Technology,Trondheim, Norway
[2] Faculty of Mathematical Sciences, University of Twente, Enschede, The Netherlands
[2] Faculty of Mechanical Engineering, Eindhoven University of Technology, Eindhoven, The Netherlands

1 Introduction

For most ships, measurements of the ship velocities are not available. For feedback control of the ship, estimates of the velocities must therefore be computed from the position and heading measurements. The ship position is typically measured using the Navstar differential global positioning system (DGPS), while the heading is usually measured by a gyro compass. As the position measurements are quite corrupted by noise, numerical position differentiation is not desirable. Instead, an observer should be used to obtain velocity estimates from the position measurements. In conventional ship control systems, the estimation problem is solved using a linear Kalman filter. A linearized ship model is then used. The kinematic equations of motion are typically linearized about a set of 36 constant yaw angles (separated by 10 deg in order to cover the whole operating area of 360 deg). For each of these linearized models Kalman filters and feedback control gains have to be computed. The control and filter gains are then modified on-line using gain-scheduling techniques. The drawback of this approach is the considerable amount of tuning work, and the *ad hoc* nature of the approach which does not guarantee the desired stability and convergence properties.

In [8] a nonlinear observer is developed and is proven to be globally exponentially stable. Hence, only one set of observer gains is needed to cover the whole state space. The observer is developed independently of the ship control scheme. In [1] a feedback control law for dynamic positioning of ships is developed based on the estimates from the observer in [8], giving a globally exponentially stable closed-loop system. In both these works, the dynamic positioning (DP) problem for ships is considered, and the observer and the control law are thus developed based on a ship model not including Coriolis and centripetal forces and moments. For the DP-problem it is a valid assumption to disregard the Coriolis and centripetal forces and moments acting on the ship, while for tracking control where the velocities of the ship cannot be assumed to be close to zero, these forces and moments must be considered in both the observer and the controller design.

In this work we consider the output feedback tracking control problem for ships. The Coriolis and centripetal forces and moments must thus be included in the ship model, leading to quadratic velocity terms in the dynamics. Moreover, instead of designing an open-loop observer, we seek to combine the observer and controller design such that the controller exploits the underlying observer structure and vice versa, in order to find a computationally simple control law and observer. The observer-controller scheme is designed using a passivity-based approach.

The idea of passivity-based control methods is to reshape the system's natural energy via state feedback, in order to achieve the control objective. In this way the passivity property of the system is preserved in the closed loop, and therefore the approach has been named the passivity-based approach. This approach has gained much attention, and based on this approach [17] proposed a solution to the robot position control problem, and [13] solved the problem of robot motion control. Also, for adaptive robot control the passivity-based approach has been studied extensively, see for instance [15],[9] and [12].

The output feedback control problem for robots has been considered by several authors, for instance [4] and [16] where observers based on the sliding mode concept were proposed, and in [10] where a linear high-gain strategy was proposed. The observers proposed were developed independently of the robot control scheme. In [5] a modified version of the computed-torque controller was proposed, and local exponential stability of the closed-loop system was proven. In [11] some known state-feedback controllers were considered, using velocity feedback from a nonlinear observer, and local asymptotic stability of the closed-loop systems were proven. In [2],[3] the output feedback control problem for robot manipulators was solved using the passivity-based approach. A key point in [3] was the fine tuning of the controller and observer structure to each other, providing solutions of the output feedback control problem that were conceptually simple and easily implementable in industrial applications.

The output feedback tracking control problem for ships has been addressed in [14] and [18]. Both these works consider a 1 degree of freedom (DOF) nonlinear model and address the yaw angle tracking control problem (autopilot design). In [14] a passive control law without velocity feedback is proposed and proved to asymptotically stabilize the desired yaw angle. In [18], based on the ideas presented in [3] an observer-controller structure is proposed and proved to semi-globally exponentially stabilize the desired yaw angle. In this chapter we use the same ideas as presented in [3] to design a 3 DOF output feedback tracking controller for ships. However, in this chapter we use the same ideas as presented in [3] for the controller-observer design for ships. However, the control law proposed in [3] uses feedback from the position measurements together with the estimated velocities. For the ship, the gyro compass measurement noise will typically be less than 0.1 deg. However, the position measurements are quite corrupted by noise, as the DGPS measurement noise will be in the range of 1-3 m. Therefore, filtering of the position measurements is necessary, and we seek to find a tracking control law that uses feedback from the filtered position variables.

In Section 2 the ship model is presented. In Section 3 we develop a controller-observer combination for output feedback tracking control of ships, and prove that the closed-loop system is semi-globally exponentially stable. If the Coriolis and centripetal forces and moments are negligible, as for the special case where the desired trajectory is a constant position and orientation, the system is globally exponentially stable. In Section 4 simulation results for this output feedback tracking control scheme are presented. Then the problem of bias estimation is addressed in Section 5, and simulations for the output feedback tracking control scheme including bias estimation are presented in Section 6. Finally, conclusions are given in Section 7.

2 The Ship Model

The ship model is based on [7, 6]. We use the earth-fixed vector representation

$$M(\psi)\ddot{\eta} + C(\psi, \dot{\eta})\dot{\eta} + D(\psi)\dot{\eta} = \tau_e \qquad (7.1)$$

where $\eta = [x, y, \psi]^T$. The variables x and y are the position variables, while ψ is the yaw angle. The vector $\tau_e \in \Re^3$ is the control vector. The model has the following properties: The matrix of inertia including hydrodynamic added inertia effects, $M(\psi)$, is symmetric and positive definite. The symmetry property is based on the ship having starboard and port symmetries together with the assumption of low speed, as opposed to high-speed applications, as we assume that the ships considered are conventional ships, not high-speed crafts. The matrix is bounded with respect to ψ

$$0 < M_m \le ||M(\psi)|| \le M_M \qquad \forall \psi \in S^1 \qquad (7.2)$$

The Coriolis and centripetal matrix, also including added inertia effects, $C(\psi, \dot{\eta})$, satisfies the properties

$$C(\psi, v)w = C(\psi, w)v \qquad \forall \psi \in S^1, \forall v, w \in \Re^3 \quad (7.3)$$

$$C(\psi, \alpha v + \beta w) = \alpha C(\psi, v) + \beta C(\psi, w) \quad \forall \psi \in S^1, \forall v, w \in \Re^3 \quad (7.4)$$

$$\exists C_M > 0 \qquad ||C(\psi, v)|| \leq C_M ||v|| \qquad \forall \psi \in S^1, \forall v \in \Re^3 \qquad (7.5)$$

Furthermore, the matrix $\dot{M} - 2C$, where $\{\dot{M}\}_{ij} = \frac{\partial \{M\}_{ij}}{\partial \psi} \dot{\psi}$, is skew symmetric

$$s^T (\dot{M}(\psi) - 2C(\psi, \dot{\eta}))s = 0 \qquad \forall \psi \in S^1, \forall \dot{\eta}, s \in \Re^3 \qquad (7.6)$$

The damping is assumed to be linear in $\dot{\eta}$, which is a good assumption for low-speed applications and for cruising at a constant speed. The hydrodynamic damping matrix, $D(\psi)$, is in general non-symmetric. The hydrodynamic damping is due to wave drift damping and laminar skin friction, and it is dissipative

$$s^T D(\psi)s > 0 \qquad \forall \psi \in S^1, \forall s \in \Re^3 \backslash \{0\} \qquad (7.7)$$

Moreover, the damping matrix is bounded with respect to ψ

$$0 < D_m \leq ||D(\psi)|| \leq D_M \qquad \forall \psi \in S^1 \qquad (7.8)$$

3   Design of an Output Feedback Tracking Control Law

In this section we develop a combined observer-controller structure for tracking control of a ship when only the position and heading measurements are available. The observer-controller design is based on passivity in the sense that both the controller and observer are designed such that the closed-loop system matches a predefined desired energy function. This approach was proposed for the control of robot manipulators in [3]. The controller proposed by [3] uses the position measurements and the estimated velocities for feedback. For the ship, using a gyro compass for the heading measurement, the measurement of the yaw angle ψ is quite accurate. The measurement noise will typically be less than 0.1 deg. However, the measurements of the position variables x and y will be corrupted by measurement noise, typically in the range of $1 - 3$ m. Therefore, filtering of these measurements is necessary and we want to develop a controller that uses filtered position variables. To this end, we consider the following Lyapunov function candidate

$$V(s_1, \hat{e}, s_2, \tilde{\eta}) = \frac{1}{2} s_1^T M(\psi) s_1 + \frac{1}{2} \hat{e}^T K_{p1} \hat{e} + \frac{1}{2} s_2^T M(\psi) s_2 + \frac{1}{2} \tilde{\eta}^T K_{p2} \tilde{\eta} \qquad (7.9)$$

where K_{p1} and K_{p2} are symmetric, positive definite matrices. The variable $\tilde{\eta} = \eta - \hat{\eta}$, where $\hat{\eta}$ is the estimate of the variable η. Define the error variable $\hat{e} = \hat{\eta} - \eta_d$, where $\eta_d(t)$ is the reference. We further define

$$\dot{\eta}^1 \quad = \quad \dot{\eta}_d - \Lambda_1 \hat{e} \tag{7.10}$$

$$\dot{\eta}^2 \quad = \quad \frac{d}{dt}\hat{\eta} - \Lambda_2 \tilde{\eta} \tag{7.11}$$

$$s_1 \quad = \quad \frac{d}{dt}\hat{\eta} - \dot{\eta}^1 = \frac{d}{dt}\hat{e} + \Lambda_1 \hat{e} \tag{7.12}$$

$$s_2 \quad = \quad \dot{\eta} - \dot{\eta}^2 = \frac{d}{dt}\tilde{\eta} + \Lambda_2 \tilde{\eta} \tag{7.13}$$

We propose the following observer-controller scheme

$$\frac{d}{dt}\hat{\eta} \quad = \quad z + K_d \tilde{\eta} \tag{7.14}$$

$$M(\psi)\dot{z} \quad = \quad -C(\psi, \dot{\eta}^2)s_1 - K_1 s_1 - K_{p1}\hat{e} + M(\psi)\ddot{\eta}^1$$
$$+ M(\psi)K_d\Lambda_2\tilde{\eta} \tag{7.15}$$

$$\tau_e \quad = \quad C(\psi, \dot{\eta}^2)(\dot{\eta}^2 - s_1) + D(\psi)\dot{\eta}^2 - K_1 s_1 - K_{p1}\hat{e} + M(\psi)\ddot{\eta}^1$$
$$- (K_{p2} - M(\psi)\Lambda_2^2)\tilde{\eta} \tag{7.16}$$

We then find that the time-derivative of V along the system trajectories is

$$\dot{V} \quad = \quad -s_1^T(K_1 - C(\psi, s_2))s_1 + s_1^T M(\psi)K_d s_2 - s_2^T(M(\psi)(K_d - \Lambda_2)$$
$$+ C(\psi, \dot{\eta} - s_2))s_2 - s_2^T D(\psi)s_2 - \hat{e}^T K_p \Lambda_1 \hat{e} - \tilde{\eta}^T K_{p2}\Lambda_2\tilde{\eta} \tag{7.17}$$

The observer and controller matrices $K_{p1}, K_{p2}, \Lambda_1, \Lambda_2, K_d$ and K_1 are constant, symmetric and positive definite. In the following we choose

$$\Lambda_1 \quad = \quad \mathrm{diag}\{\lambda_{11}, \lambda_{12}, \lambda_{13}\} \tag{7.18}$$

$$\Lambda_2 \quad = \quad \mathrm{diag}\{\lambda_{21}, \lambda_{22}, \lambda_{23}\} \tag{7.19}$$

$$K_{p1} \quad = \quad \Lambda_1 \tag{7.20}$$

$$K_{p2} \quad = \quad \Lambda_2 \tag{7.21}$$

$$K_d \quad = \quad \Lambda_2 + l_d I_{3\times 3} \tag{7.22}$$

$$K_1 \quad = \quad l_1 I_{3\times 3} \tag{7.23}$$

where

$$l_d \quad > \quad \Lambda_{2,M} + \frac{2C_M V_M}{M_m} \tag{7.24}$$

$$l_1 \quad > \quad \frac{1}{2}\frac{M_M^2}{M_m}(\Lambda_{2,M} + l_d) \tag{7.25}$$

where $\Lambda_{2,M}$ is the maximum eigenvalue of the matrix Λ_2, and V_M is the maximum of the reference velocity $\dot{\eta}_d$. The time-derivative of the Lyapunov function candidate is then

$$
\begin{aligned}
\dot{V} &= -s_1^T(l_1 I_{3\times3} - C(\psi, s_2))s_1 + s_1^T(M(\psi)\Lambda_2 + l_d M(\psi))s_2 \\
&\quad -s_2^T(l_d M(\psi) + C(\psi, \dot{\eta} - s_2))s_2 - s_2^T D(\psi)s_2 \\
&\quad -\hat{e}^T\Lambda_1^T\Lambda_1\hat{e} - \tilde{\eta}^T\Lambda_2^T\Lambda_2\tilde{\eta}
\end{aligned}
\tag{7.26}
$$

Using the matrix properties given in Section 2 we find that this is upper bounded by

$$
\begin{aligned}
\dot{V} &\leq -(l_1 - C_M||s_2||)||s_1||^2 + M_M(l_d + \Lambda_{2,M})||s_1||\,||s_2|| \\
&\quad -(l_d M_m - C_M||\dot{\eta} - s_2||)||s_2||^2 - D_m||s_2||^2 \\
&\quad -||\Lambda_1\hat{e}||^2 - ||\Lambda_2\tilde{\eta}||^2
\end{aligned}
\tag{7.27}
$$

By completing the squares, we find that

$$
\begin{aligned}
\dot{V} &\leq -(l_1 - C_M||s_2|| - \frac{1}{2}\frac{M_M^2}{M_m}(l_d + \Lambda_{2,M}))||s_1||^2 \\
&\quad -(\frac{1}{2}l_d M_m - C_M||\dot{\eta} - s_2|| - \frac{1}{2}M_m\Lambda_{2,M})||s_2||^2 \\
&\quad -D_m||s_2||^2 - ||\Lambda_1\hat{e}||^2 - ||\Lambda_2\tilde{\eta}||^2
\end{aligned}
\tag{7.28}
$$

Noting that

$$
\dot{\eta} - s_2 = s_1 + \dot{\eta}_d - \Lambda_1\hat{e} - \Lambda_2\tilde{\eta}
\tag{7.29}
$$

we see that if (7.24–7.25) are satisfied, then there exists a region Ω in which for some $\alpha > 0$

$$
\dot{V} \leq -\alpha||\mathbf{y}||^2 \qquad \forall \mathbf{y} \in \Omega
\tag{7.30}
$$

where $\mathbf{y} = [s_1, \Lambda_1\hat{e}, s_2, \Lambda_2\tilde{\eta}]^T$. By Lyapunov theory we thus have that $\mathbf{y} = \mathbf{0}$ is an exponentially stable equilibrium of the system (7.1)–(7.14–7.16).

We find an estimate of the region of attraction as follows. If the inequality

$$
(1 + \sqrt{2})||\mathbf{y}|| < \frac{M_m}{2C_M}(l_d - \Lambda_{2,M}) - V_M
\tag{7.31}
$$

is satisfied, then by Cauchy-Schwartz

$$
C_M(||s_1|| + ||\Lambda_1\hat{e}|| + ||\Lambda_2\tilde{\eta}||) < \frac{M_m}{2}(l_d - \Lambda_{2,M}) - C_M V_M
\tag{7.32}
$$

If the inequality

$$
||\mathbf{y}|| < \frac{l_1}{C_M} - \frac{1}{2}\frac{M_M^2}{M_m C_M}(l_d + \Lambda_{2,M})
\tag{7.33}
$$

is satisfied, then

$$C_M||s_2|| < l_1 - \frac{1}{2}\frac{M_M^2}{M_m}(l_d + \Lambda_{2,M}) \tag{7.34}$$

Define

$$\delta = \min\{\frac{1}{(1+\sqrt{2})}(\frac{M_m}{2C_M}(l_d - \Lambda_{2,M}) - V_M), \frac{l_1}{C_M} - \frac{1}{2}\frac{M_M^2}{M_m C_M}(l_d + \Lambda_{2,M})\} \tag{7.35}$$

Then (7.30) holds for Ω defined by

$$\{\mathbf{y} \in \Re^{12} | \ ||\mathbf{y}|| < \delta\} \tag{7.36}$$

Furthermore, note that

$$\frac{1}{2}P_m||\mathbf{y}||^2 \le V(\mathbf{y}) \le \frac{1}{2}P_M||\mathbf{y}||^2 \tag{7.37}$$

where

$$P_m = \min\{M_m, \Lambda_{1,M}^{-1}, \Lambda_{2,M}^{-1}\} \tag{7.38}$$
$$P_M = \max\{M_M, \Lambda_{1,m}^{-1}, \Lambda_{2,m}^{-1}\} \tag{7.39}$$

Thus if

$$||\mathbf{y}(0)|| < \sqrt{\frac{P_m}{P_M}}\delta \tag{7.40}$$

then

$$\dot{V} \le -\alpha||\mathbf{y}(t)||^2 \qquad \forall t > 0 \tag{7.41}$$

We have thus proved the following proposition

Proposition 7.1 *Consider the ship (7.1) and the observer-controller scheme*

$$\frac{d}{dt}\hat{\eta} = z + (\Lambda_2 + l_d I_{3\times3})\tilde{\eta} \tag{7.42}$$
$$M(\psi)\dot{z} = -C(\psi, \dot{\eta}^2)s_1 - l_1 s_1 - \Lambda_1 \hat{e} + M(\psi)\ddot{\eta}^1 \\ + M(\psi)(\Lambda_2 + l_d I_{3\times3})\Lambda_2 \tilde{\eta} \tag{7.43}$$
$$\tau_e = C(\psi, \dot{\eta}^2)(\dot{\eta}^2 - s_1) + D(\psi)\dot{\eta}^2 - l_1 s_1 - \Lambda_1 \hat{e} + M(\psi)\ddot{\eta}^1 \\ - (\Lambda_2 - M(\psi)\Lambda_2^2)\tilde{\eta} \tag{7.44}$$

where $\Lambda_1 = diag\{\lambda_{11}, \lambda_{12}, \lambda_{13}\} > 0$ and $\Lambda_2 = diag\{\lambda_{21}, \lambda_{22}, \lambda_{23}\} > 0$. Under the conditions

$$l_d > \Lambda_{2,M} + \frac{2C_M V_M}{M_m} \tag{7.45}$$

$$l_1 > \frac{1}{2}\frac{M_M^2}{M_m}(\Lambda_{2,M} + l_d) \tag{7.46}$$

where V_M is the maximum value of the reference velocity $\dot{\eta}_d$, $\Lambda_{2,M} = \max\{\lambda_{21}, \lambda_{22}, \lambda_{23}\}$ and C_M, M_M, M_m are defined in (7.2–7.5), the closed-loop system is locally exponentially stable.

A region of attraction is given by

$$A = \{\mathbf{y} \in \Re^{12}|\ ||\mathbf{y}|| < \sqrt{\frac{P_m}{P_M}}\delta\} \tag{7.47}$$

where $\mathbf{y} = [s_1, \Lambda_1 \hat{e}, s_2, \Lambda_2 \tilde{\eta}]^T$

$$\delta = \min\{\frac{1}{(1+\sqrt{2})}(\frac{M_m}{2C_M}(l_d - \Lambda_{2,M}) - V_M), \frac{l_1}{C_M} - \frac{1}{2}\frac{M_M^2}{M_m C_M}(l_d + \Lambda_{2,M})\}$$

$$P_m = \min\{M_m, \Lambda_{1,M}^{-1}, \Lambda_{2,M}^{-1}\} \text{ and } P_M = \max\{M_M, \Lambda_{1,m}^{-1}, \Lambda_{2,m}^{-1}\}.$$

Remark 7.1 *The exponential convergence of $\mathbf{y}$ implies the exponential convergence of $\mathbf{x} = [s_1, \hat{e}, s_2, \tilde{\eta}]^T$ as $\mathbf{y}$ and $\mathbf{x}$ are linearly related.*

Remark 7.2 *Note that the region of attraction can be made arbitrarily large by choosing the control parameters l_d and l_1 large enough. This means that the closed-loop system is semi-globally exponentially stable.*

Remark 7.3 *Note that for the special case where the desired trajectory is a constant position $\eta_d = $ constant (dynamic positioning), then the closed-loop system is semi-globally exponentially stable if*

$$l_d > \Lambda_{2,M} \tag{7.48}$$

$$l_1 > \frac{1}{2}\frac{M_M^2}{M_m}(\Lambda_{2,M} + l_d) \tag{7.49}$$

Remark 7.4 *>From (7.28) we see that it is due to the Coriolis and centripetal term that the result is semi-global as opposed to global. For the dynamic positioning problem it can be assumed that the Coriolis and centripetal forces and moments are zero, and then $C_M = 0$. Then, under the conditions (7.48–7.49) the closed-loop system is globally exponentially stable.*

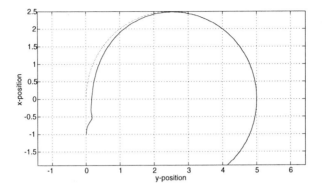

FIGURE 1. The ship trajectory (–) and the desired trajectory (- -) in the xy-plane.

Remark 7.5 *Due to the combined observer-controller design, in which the observer and controller structure are chosen in accordance with each other, we have obtained a computationally simple observer and controller. The tuning of the system will therefore be quite simple. We may choose Λ_1 and Λ_2 to give the desired rate of convergence of $\hat{e}$ and $\tilde{\eta}$ ($\Lambda_1 < \Lambda_2$ as we want the estimation error to converge to zero faster than $\hat{e}$). Then, we choose the control parameters l_d and l_1 satisfying the inequalities (7.45–7.46). The choice of l_d and l_1 will be a trade-off between the size of the region of attraction on the one side and the thruster limits and amplification of measurement noise (since these parameters can be viewed as the derivative gains) on the other side. The observer-controller design does however not take thruster limitations into consideration, and the closed-loop system does not respect such bounds if the initial conditions are poorly chosen.*

4   Simulations

The simulations were performed using the mathematical model of Cybership I, a model ship of scale 1:70 of an offshore supply vessel. Cybership I has a mass of 17.6 kg and a length of 1.2 m. The centre of gravity is located at $x_G = -0.04$ m aft of midships, and this is the origin of the body-fixed coordinate system. Assuming that the Froude number is constant, we have

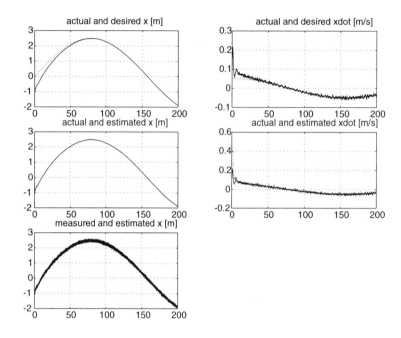

FIGURE 2. The position variable x (–) and the desired position x_d (- -) [m], the velocity variable $\dot{x}$ (–) and the desired velocity $\dot{x}_d$(- -)[m/s], the position variable x (–) and its estimate $\hat{x}$ (- -), the velocity $\dot{x}$ (–) and its estimate $\frac{d}{dt}\hat{x}$ (- -), the measurement of x (–) and the estimate $\hat{x}$ (- -).

the following relationship between the speed of the ship and the model ship

$$U_S \approx 8.37 U_M \,(\text{m/s}) \qquad (7.50)$$

where the subscripts S and M denote the ship and the model respectively. In the simulations, we introduced input magnitude saturation $\tau_{1,2}^{\max} = 10$ N and $\tau_3^{\max} = 10$ Nm. The measurement frequency was 50 Hz.

In the simulations, the measurement noise in the position variables x and y were in the magnitude of 10 cm, corresponding to measurement noise of magnitude 7 m for the original supply vessel. The heading measurement noise was of magnitude 0.1 deg. No apriori information of the state variables was assumed.

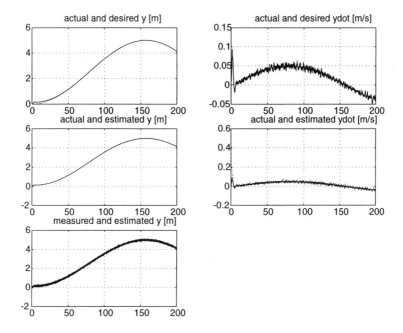

FIGURE 3. The position variable y (–) and the desired position y_d (- -) [m], the velocity variable $\dot{y}$ (–) and the desired velocity $\dot{y}_d$(- -)[m/s], the position variable y (–) and its estimate $\hat{y}$ (- -), the velocity $\dot{y}$ (–) and its estimate $\frac{d}{dt}\hat{y}$ (- -),the measurement of y (–) and the estimate $\hat{y}$ (- -).

The desired trajectory was

$$x_d = 2.5\sin(0.02t) \tag{7.51}$$
$$y_d = 2.5(1 - \cos(0.02t)) \tag{7.52}$$
$$r_d = 0.02t \tag{7.53}$$

corresponding to the model ship moving at 5 cm/s along a circle of radius 2.5 m. The initial value of the desired trajectory was at the origin of the earth-fixed coordinate system.

The initial values of Cybership I were

$$[x, y, \psi, \dot{x}, \dot{y}, \dot{\psi}]^T = [-1, 0, \frac{\pi}{4}, 0, 0, 0]^T \tag{7.54}$$

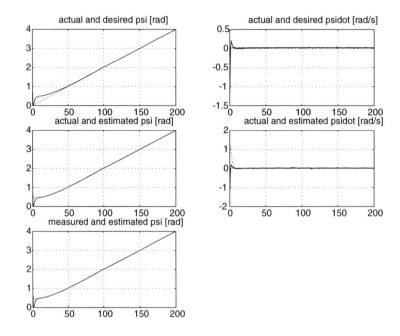

FIGURE 4. The yaw angle ψ (–) and the desired yaw angle ψ_d (- -) [rad], the yaw rate $\dot{\psi}$ (–) and the desired yaw rate $\dot{\psi}_d$(- -)[rad/s], the yaw angle ψ (–) and its estimate $\hat{\psi}$ (- -), the yaw rate $\dot{\psi}$ (–) and its estimate $\frac{d}{dt}\hat{\psi}$ (- -), the measurement of ψ (–) and the estimate $\hat{\psi}$ (- -).

The observer and controller parameters were chosen as

$$\Lambda_1 = \begin{bmatrix} 0.05 & 0 & 0 \\ 0 & 0.05 & 0 \\ 0 & 0 & 0.05 \end{bmatrix} \tag{7.55}$$

$$\Lambda_2 = \begin{bmatrix} 0.5 & 0 & 0 \\ 0 & 0.5 & 0 \\ 0 & 0 & 0.5 \end{bmatrix} \tag{7.56}$$

$$l_d = 2.5 \tag{7.57}$$

$$l_1 = 10 \tag{7.58}$$

Note that the conditions (7.45–7.46) and the estimate of the region of attraction in (7.47) are conservative. In order to reduce the thruster forces commanded by the controller, we chose l_d and l_1 below the bounds given in (7.45–7.46) in the simulations. We see in Figure 5 that the natural logarithm of the norm of $[\tilde{\eta}^T, \hat{e}^T]^T$ was upper bounded by a decreasing straight line, showing that the convergence was still exponential. The natural logarithm

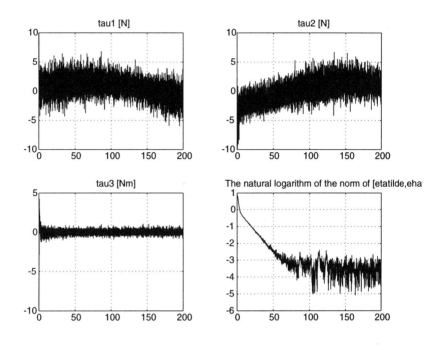

FIGURE 5. The surge control force τ_1, the sway control force τ_2 [N], the yaw control torque τ_3 [Nm] and the natural logarithm of the norm of $[\tilde{\eta}^T, \hat{e}^T]^T$.

of the norm became quite influenced by noise as it took values lower than approximately -2, corresponding to the norm value being lower than 0.15. This corresponds to the fact that the measurement noise was of magnitude 0.1, and thus the influence of the noise became significant as the norm reached such low values. The simulation results are shown in Figures 1–5. In Figure 1 we see how the ship moves in the xy−plane. We see in Figures 2c–4c that the position estimates $\hat{x}, \hat{y}$ and the yaw angle estimate $\hat{\psi}$ converge to the position variables x, y and the yaw angle ψ respectively, despite the measurement noise. The estimated position and yaw angle together with the noise-corrupted measurements which are the inputs to the observer, are shown in Figures 2e–4e. In Figures 2d–4d we see that the velocity estimates are quite noise-corrupted. The impact of the noise on the velocity estimates will depend on the "derivative gains" l_d and l_1. This can be seen from (7.42), where we see that the measurement noise is fed directly into the velocity estimate, and the choice of the control parameters Λ_2 and l_d, and by (7.46) also l_1, will therefore be a trade-off between the size of the region of attraction (7.47) and the influence of the measurement noise on the velocity estimate $\frac{d}{dt}\hat{\eta}$. The influence of the measurement noise on the velocity estimate is reflected in the control force and moment, Figure 5.

(In practice the actuator dynamics will give low-pass filtering, removing the high-frequent signals from the actual control input.) In Figures 2a–4a and 2b–4b we see that despite the influence of the measurement noise, the controller performance is good. This may be due to the second-order filtering of the ship from force and moment to position and yaw angle. In particular the position and yaw angle tracking control is good, and for ship control this is more important than the velocity tracking.

5   Bias Estimation

The ship will be subject to environmental disturbances due to wind, waves and currents. We do not want the control law to react to the high-frequent components of the disturbances, because this would give increased wear and tear on the actuators. The controller should instead compensate for the slowly-varying bias of the disturbance. We therefore want to include bias estimation in the control scheme. A frequently used bias model for maritime control applications is the 1st order Markov process [8]

$$\dot{b} = -T^{-1}b + Bn \tag{7.59}$$

where $b \in \Re^3$ is the vector of bias forces and moments and T is a diagonal matrix of positive bias time-constants. The bias model is driven by zero-mean Gaussian white noise $n \in \Re^3$, and B is a diagonal matrix scaling the amplitude of the white noise. This model can be used to describe slowly varying environmental forces and moments due to second-order wave drift, ocean currents, wind and also to describe unmodeled dynamics. For the analysis we assume that $n = 0$, since the bias estimator model will be driven by estimation errors. The ship model is then

$$M(\psi)\ddot{\eta} + C(\psi, \dot{\eta})\dot{\eta} + D(\psi)\dot{\eta} = \tau_e + b \tag{7.60}$$

We use the observer developed in Section 3 together with a bias estimator proposed by [8], and we use the bias estimate in the controller

$$\frac{d}{dt}\hat{\eta} = z + (\Lambda_2 + l_d I_{3\times 3})\tilde{\eta} \tag{7.61}$$

$$M(\psi)\dot{z} = -C(\psi, \dot{\eta}^2)s_1 - l_1 s_1 - \Lambda_1 \hat{e} + M(\psi)\ddot{\eta}^1$$
$$+ M(\psi)(\Lambda_2 + l_d I_{3\times 3})\Lambda_2 \tilde{\eta} \tag{7.62}$$

$$\frac{d}{dt}\hat{b} = -T^{-1}\hat{b} + K_b \tilde{\eta} \tag{7.63}$$

$$\tau_e = -\hat{b} + C(\psi, \dot{\eta}^2)(\dot{\eta}^2 - s_1) + D(\psi)\dot{\eta}^2 - l_1 s_1 - \Lambda_1 \hat{e} + M(\psi)\ddot{\eta}^1$$
$$-(\Lambda_2 - M(\psi)\Lambda_2^2)\tilde{\eta} \tag{7.64}$$

where $K_b = K_b^T > 0$. We cannot use the analysis of [8] to prove stability of the system, as the analysis is based on an open-loop model of the ship, and also the ship model does not include Coriolis and centripetal forces and moments as it is modeled for dynamic positioning purposes, not for tracking control. To prove that our closed-loop system is semi-globally exponentially stable, we use the Lyapunov function candidate

$$V(s_1, \hat{e}, s_2, \tilde{\eta}, \tilde{b}) = \frac{1}{2}s_1^T M(\psi)s_1 + \frac{1}{2}\hat{e}^T \Lambda_1 \hat{e} + \frac{1}{2}s_2^T M(\psi)s_2$$
$$+ \frac{1}{2}\tilde{\eta}^T \Lambda_2 \tilde{\eta} + \frac{1}{2}\gamma \tilde{b}^T \tilde{b} \tag{7.65}$$

where $\tilde{b} = b - \hat{b}$, and where γ is a positive constant. The time-derivative of V is then

$$\dot{V} = -s_1^T(l_1 I_{3\times3} - C(\psi, s_2))s_1 + s_1^T(M(\psi)\Lambda_2 + l_d M(\psi))s_2$$
$$-s_2^T(l_d M(\psi) + C(\psi, \dot{\eta} - s_2))s_2 - s_2^T D(\psi)s_2 - s_2^T \tilde{b}$$
$$-\hat{e}^T \Lambda_1^T \Lambda_1 \hat{e} - \tilde{\eta}^T \Lambda_2^T \Lambda_2 \tilde{\eta} - \gamma \tilde{b}^T K_b \tilde{b} - \gamma \tilde{b}^T T^{-1} \tilde{b} \tag{7.66}$$

Using the matrix properties we find that this is upper bounded by

$$\dot{V} \leq -(l_1 - C_M||s_2||)||s_1||^2 + M_M(l_d + \Lambda_{2,M})||s_1|| \, ||s_2|| \tag{7.67}$$
$$-(l_d M_m - C_M||\dot{\eta} - s_2||)||s_2||^2 - D_m||s_2||^2 + \frac{1}{\gamma}||s_2|| \, ||\gamma\tilde{b}||$$
$$-||\Lambda_1\hat{e}||^2 - ||\Lambda_2\tilde{\eta}||^2 + K_{bM}\Lambda_{2m}^{-1}||\gamma\tilde{b}|| \, ||\Lambda_2\tilde{\eta}|| - \frac{1}{\gamma T_M}||\gamma\tilde{b}||^2$$

where K_{bM} and T_M are the maximal eigenvalues of the matrices K_b and T respectively. By completing the squares, we find that

$$\dot{V} \leq -(l_1 - C_M||s_2|| - \frac{1}{2}\frac{M_M^2}{M_m}(l_d + \Lambda_{2,M}))||s_1||^2 \tag{7.68}$$
$$-(\frac{1}{2}l_d M_m - C_M||\dot{\eta} - s_2|| - \frac{1}{2}M_m\Lambda_{2,M} - 2\frac{T_M}{\gamma})||s_2||^2$$
$$-D_m||s_2||^2 - ||\Lambda_1\hat{e}||^2 - \frac{1}{2}||\Lambda_2\tilde{\eta}||^2 - (\frac{1}{2T_M\gamma} - 2K_{bM}^2\Lambda_{2m}^{-2})||\gamma\tilde{b}||^2$$

We see then that if the following conditions are satisfied

$$K_{bM} < \frac{\Lambda_{2m}}{2\sqrt{\gamma T_M}} \tag{7.69}$$

$$l_d > \Lambda_{2,M} + \frac{2C_M V_M}{M_m} + \frac{4T_M}{M_m\gamma} \tag{7.70}$$

$$l_1 > \frac{1}{2}\frac{M_M^2}{M_m}(\Lambda_{2,M} + l_d) \tag{7.71}$$

for some $\gamma > 0$, then there exists a region Ω in which for some $\alpha > 0$

$$\dot{V} \leq -\alpha ||\mathbf{y}||^2 \qquad\qquad \forall \mathbf{y} \in \Omega \qquad\qquad (7.72)$$

where $\mathbf{y} = [s_1, \Lambda_1 \hat{e}, s_2, \Lambda_2 \tilde{\eta}, \gamma \tilde{b}]^T$. By Lyapunov theory we thus have that $\mathbf{y} = \mathbf{0}$ is an exponentially stable equilibrium of the system (7.60–7.64). For tuning purposes, we can interpret the condition in (7.69) as an upper bound on the inverse of the integral time constant of the estimation error $\tilde{\eta}$, cf. (7.63–7.64). Furthermore, the condition (7.70) can be interpreted as a lower bound on the derivative gain of the estimation error $\tilde{\eta}$, and (7.71) as a lower bound on the derivative gain of the error $\hat{e}$. We can find an estimate of the region of attraction along the same lines as in Section 3, and we then have the following proposition

Proposition 7.2 *Consider the ship (7.60) and the observer-controller with bias estimator*

$$\frac{d}{dt}\hat{\eta} = z + (\Lambda_2 + l_d I_{3\times 3})\tilde{\eta} \qquad\qquad (7.73)$$

$$M(\psi)\dot{z} = -C(\psi, \dot{\eta}^2)s_1 - l_1 s_1 - \Lambda_1 \hat{e} + M(\psi)\ddot{\eta}^1$$
$$+ M(\psi)(\Lambda_2 + l_d I_{3\times 3})\Lambda_2 \tilde{\eta} \qquad\qquad (7.74)$$

$$\frac{d}{dt}\hat{b} = -T^{-1}\hat{b} + K_b \tilde{\eta} \qquad\qquad (7.75)$$

$$\tau_e = -\hat{b} + C(\psi, \dot{\eta}^2)(\dot{\eta}^2 - s_1) + D(\psi)\dot{\eta}^2 - l_1 s_1 - \Lambda_1 \hat{e} + M(\psi)\ddot{\eta}^1$$
$$-(\Lambda_2 - M(\psi)\Lambda_2^2)\tilde{\eta} \qquad\qquad (7.76)$$

where $\Lambda_1 = diag\{\lambda_{11}, \lambda_{12}, \lambda_{13}\} > 0$ and $\Lambda_2 = diag\{\lambda_{21}, \lambda_{22}, \lambda_{23}\} > 0$. Under the conditions

$$K_{bM} < \frac{\Lambda_{2m}}{2\sqrt{\gamma T_M}} \qquad\qquad (7.77)$$

$$l_d > \Lambda_{2,M} + \frac{2C_M V_M}{M_m} + \frac{4T_M}{M_m \gamma} \qquad\qquad (7.78)$$

$$l_1 > \frac{1}{2}\frac{M_M^2}{M_m}(\Lambda_{2,M} + l_d) \qquad\qquad (7.79)$$

where K_{bM} is the maximum eigenvalue of the symmetric positive matrix K_b, V_M is the maximum value of the reference velocity $\dot{\eta}_d$, T_M is the maximum of the bias time constants, $\Lambda_{2,M} = \max\{\lambda_{21}, \lambda_{22}, \lambda_{23}\}$, $\Lambda_{2,m} = \min\{\lambda_{21}, \lambda_{22}, \lambda_{23}\}$, where C_M, M_M, M_m are defined in (7.2–7.5) and γ is some positive constant, the closed-loop system is locally exponentially stable.

A region of attraction is given by

$$A = \{\mathbf{y} \in \Re^{12} | \ ||\mathbf{y}|| < \sqrt{\frac{P_m}{P_M}}\delta\} \tag{7.80}$$

where $\mathbf{y} = = [s_1, \Lambda_1 \hat{e}, s_2, \Lambda_2 \tilde{\eta}, \gamma \tilde{b}]^T$

$$\delta = \min\{\frac{1}{(1 + \sqrt{2})}(\frac{M_m}{2C_M}(l_d - \Lambda_{2,M}) - \frac{2T_M}{\gamma C_M} - V_M),$$

$$\frac{l_1}{C_M} - \frac{1}{2}\frac{M_M^2}{M_m C_M}(l_d + \Lambda_{2,M})\} \tag{7.81}$$

$P_m = \min\{M_m, \Lambda_{1,M}^{-1}, \Lambda_{2,M}^{-1}, \gamma^{-1}\}$ *and* $P_M = \max\{M_M, \Lambda_{1,m}^{-1}, \Lambda_{2,m}^{-1}, \gamma^{-1}\}.$

Remark 7.6 *For dynamic positioning the desired trajectory is a constant position,* $\eta_d = $ *constant, and then the closed-loop system is semi-globally exponentially stable if*

$$K_{bM} < \frac{\Lambda_{2m}}{2\sqrt{\gamma T_M}} \tag{7.82}$$

$$l_d > \Lambda_{2,M} + \frac{4T_M}{M_m \gamma} \tag{7.83}$$

$$l_1 > \frac{1}{2}\frac{M_M^2}{M_m}(\Lambda_{2,M} + l_d) \tag{7.84}$$

Moreover, for dynamic positioning purposes the Coriolis and centripetal forces and moments can be assumed to be zero. Then, from (7.68) it is seen that under the conditions (7.82–7.84) the closed-loop system is globally exponentially stable.

6   Simulations with an Environmental Disturbance

The simulations were performed with a disturbance bias that initially was

$$b(0) = \begin{bmatrix} 0.1 \ \text{N} \\ 0 \ \text{N} \\ 0 \ \text{Nm} \end{bmatrix} \tag{7.85}$$

which was of size $\frac{1}{100}$ of the control force magnitude saturation. The matrix of bias time constants was

$$T = \begin{bmatrix} 1000 & 0 & 0 \\ 0 & 1000 & 0 \\ 0 & 0 & 1000 \end{bmatrix} \tag{7.86}$$

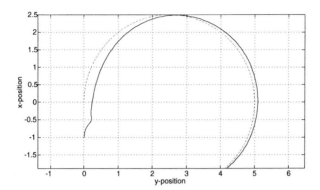

FIGURE 6. The ship trajectory (–) and the desired trajectory (- -) in the xy-plane.

The bias estimator gain matrix was chosen as

$$K_b = \begin{bmatrix} 0.2 & 0 & 0 \\ 0 & 0.2 & 0 \\ 0 & 0 & 0.2 \end{bmatrix} \qquad (7.87)$$

Furthermore the parameter $\gamma = T_M = 1000$. The other observer-controller parameters, and also the noise conditions and the reference trajectory, were chosen equal to those given in Section 4. Note that the conditions given in (7.69–7.71) are conservative. The parameter K_b chosen for the simulations is chosen above the bound of (7.69) in order to obtain faster convergence of the bias estimate. Furthermore, as in Section 4 the parameters l_d and l_1 are chosen below the bounds of (7.70–7.71) in order to reduce the thruster forces commanded by the controller. The simulation results are shown in Figures 6–11. We see in Figure 11 that the natural logarithm of the norm of $[\tilde{\eta}^T, \hat{e}^T, \tilde{b}]^T$ was upper bounded by a decreasing straight line, showing that the convergence was exponential. From Figures 7–9 we see how the position estimates follow the position variables x, y and ψ, despite the measurement noise. The velocity estimates are quite noise-corrupted, giving control inputs that are quite influenced by the noise as seen in Figure 10, but despite this the controller performance is good. In particular, the position and yaw angle tracking control is good, despite both measurement noise and the environmental disturbance. In Figure 11 we see that the bias estimates converge quite slowly, but the convergence of the vector $[\tilde{\eta}^T, \hat{e}^T, \tilde{b}]^T$ is still exponential. Simulations with a longer time-scale show that the bias error

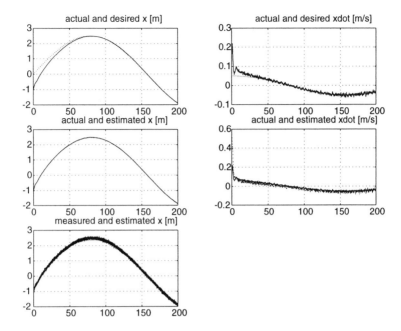

FIGURE 7. The position variable x (–) and the desired position x_d (- -) [m], the velocity variable $\dot{x}$ (–) and the desired velocity $\dot{x}_d$(- -)[m/s], the position variable x (–) and its estimate $\hat{x}$ (- -), the velocity $\dot{x}$ (–) and its estimate $\frac{d}{dt}\hat{x}$ (- -), the measurement of x (–) and the estimate $\hat{x}$ (- -).

$\tilde{b}$ converges to zero.

7 Conclusions and Future Work

In this work a nonlinear observer and feedback control law was proposed for output feedback tracking control of ships. As the measurements were quite corrupted with noise, the control law used the filtered measurements together with the estimated velocities for feedback. The observer and controller design were combined in order to utilize the observer structure in the controller design and vice versa, in order to develope a computationally simple observer and control law. The resulting system was proved to be semi-globally exponentially stable. If the Coriolis and centripetal forces and moments were negligible, as for the special case where the desired trajectory was a constant position and orientation, the system was globally exponentially stable. Furthermore, bias estimation was introduced in order to compensate for the bias of environmental forces, and the output feedback

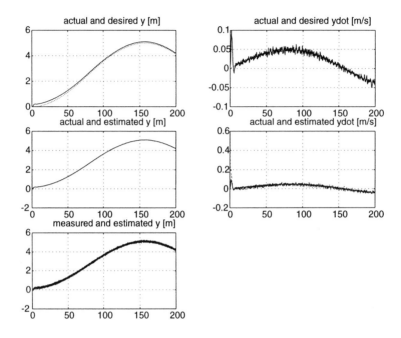

FIGURE 8. The position variable y (–) and the desired position y_d (- -) [m], the velocity variable $\dot{y}$ (–) and the desired velocity $\dot{y}_d$(- -)[m/s], the position variable y (–) and its estimate $\hat{y}$ (- -), the velocity $\dot{y}$ (–) and its estimate $\frac{d}{dt}\hat{y}$ (- -), the measurement of y (–) and the estimate $\hat{y}$ (- -).

tracking control scheme including the bias estimator was proved to give a semi-globally exponentially stable system. The results were illustrated by simulations.

The position and heading measurements of the ship will include the oscillatory wave motion. It is not desirable that the controller reacts to this wave motion, because this gives increased wear and tear on the actuators and increased fuel consumption. Therefore wave filtering should be included in future work.

Acknowledgments

The authors would like to thank T. I. Fossen and A. A. J. Lefeber for the interesting discussions regarding the topic.

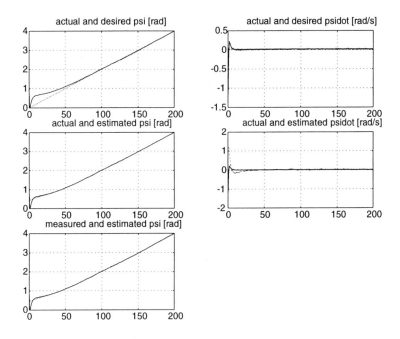

FIGURE 9. The yaw angle ψ (–) and the desired yaw angle ψ_d (- -) [rad], the yaw rate $\dot{\psi}$ (–) and the desired yaw rate $\dot{\psi}_d$(- -)[rad/s], the yaw angle ψ (–) and its estimate $\hat{\psi}$ (- -), the yaw rate $\dot{\psi}$ (–) and its estimate $\frac{d}{dt}\hat{\psi}$ (- -), the measurement of ψ (–) and the estimate $\hat{\psi}$ (- -).

8 REFERENCES

[1] M. F. Aarset, J. P. Strand and T. I. Fossen, Nonlinear Vectorial Observer Backstepping with Integral Action and Wave Filtering for Ships, *Proceedings of the IFAC Conference on Control Applications in Marine Systems (CAMS)*, Fukuoka, Japan, October, 1998.

[2] H. Berghuis, Model-based Robot Control: from Theory to Practice, Ph.D. dissertation, Univ. Twente, Enschede, The Netherlands, 1993.

[3] H. Berghuis and H. Nijmeijer, A Passivity Approach to Controller-Observer Design for Robots, *IEEE Transactions on Robotics and Automation*, Vol. 9, No. 6, pp. 740-754, 1993.

[4] C. Canudas de Wit and J.J.-E. Slotine, Sliding Observers for Robot Manipulators, *Automatica*, Vol. 27, pp. 859-864, 1991.

[5] C. Canudas de Wit, N. Fixot and K. J. Åström, Trajectory Tracking in Robot Manipulators via Nonlinear Estimated State Feedback, *IEEE*

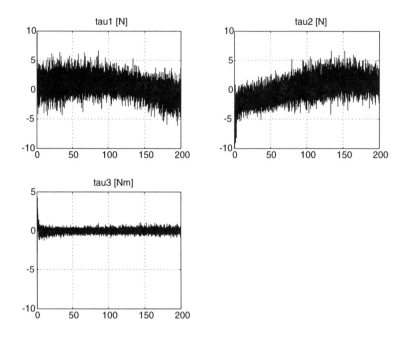

FIGURE 10. The surge control force τ_1, the sway control force τ_2 [N] and the yaw control torque τ_3 [Nm].

Trans. Robotics Automat., Vol. 8, pp. 138-144, 1992.

[6] T. I. Fossen, *Guidance and Control of Ocean Vehicles,* John Wiley & Sons Ltd., Chichester, 1994.

[7] T. I. Fossen and O.-E. Fjellstad, Nonlinear Modelling of Marine Vehicles in 6 Degrees of Freedom, *International Journal of Mathematical Modelling of Systems,* Vol. 1, No. 1, pp. 17-27, 1995.

[8] T. I. Fossen and J. P. Strand, Passive Nonlinear Observer Design for Ships Using Lyapunov Methods: Full-Scale Experiments with a Supply Vessel, *Automatica,* Vol. 35 No. 1, 1999.

[9] I. D. Landau and R. Horowitz, Applications of the Passive Systems Approach to the Stability Analysis of Adaptive Controllers for Robot Manipulators, *Int. J. Adaptive Control and Signal Processing,* Vol. 3, pp. 23-38, 1989.

[10] S. Nicosia, A. Tornambé and P. Valigi, Experimental Results in State Estimation of Industrial Robots, in *Proc. Conf. Decision and Control,* Honolulu, HI, Dec. 1990, pp. 360-365.

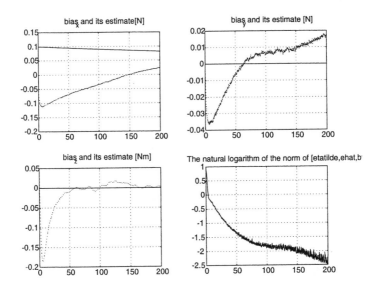

FIGURE 11. The disturbance bias (–) and its estimate (- -) in the $x-$ direction and the $y-$ direction [N], the disturbance bias about the $z-$ axis (–) and its estimate (- -) [Nm] and the natural logarithm of the norm of $[\tilde{\eta}^T, \hat{e}^T, \tilde{b}]^T$.

[11] S. Nicosia and P. Tomei, Robot Control by Using only Joint Position Measurements, *IEEE Trans. Automat. Contr.*, Vol. 35, pp. 1058-1061, 1990.

[12] R. Ortega and M. W. Spong, Adaptive Motion Control of Rigid Robots: A Tutorial, *Automatica*, Vol. 25, pp. 877-888, 1989.

[13] B. Paden and R. Panja, Globally asymptotically stable 'PD+' controller for robot manipulators, *Int. J. Control*, Vol. 47, pp. 1697-1712, 1988.

[14] M. Paulsen, O. Egeland and T. I. Fossen, A Passive Feedback Controller With Wave Filter for Marine Vehicles, International Journal of Robust and Nonlinear Control, vol.8,no.15, pp.1239-1253,1998.

[15] J.J.-E. Slotine and W. Li, On the Adaptive Control of Robot Manipulators, *Int. J. Robotics Res.*, Vol. 6, pp. 49-59, 1987.

[16] J.J.-E. Slotine, J. K. Hedrick and E. A. Misawa, Sliding Observers for Nonlinear Systems, *ASME J. Dynam. Syst., Measurement, Control,*

Vol. 109, pp. 245-252, 1987.

[17] M. Takegaki and S. Arimoto, A New Feedback Method for Dynamic Control of Manipulators, *ASME J. Dynam. Syst., Measurement, Control,* Vol. 102, pp. 119-125, 1981.

[18] B. Vik and T. I. Fossen, Semi-Global Exponential Output Feedback Control of Ships, IEEE Transactions on Control Systems Technology, TCST-5(3):360-370, 1997.

Dynamic UCO Controllers and Semiglobal Stabilization of Uncertain Nonminimum Phase Systems by Output Feedback

A. Isidori[1], A. R. Teel[2] and L. Praly[3]

[1]Department of Systems Science and Mathematics, Washington University, St. Louis, MO 63130 and Dipartimento di Informatica e Sistemistica, Università di Roma "La Sapienza", 00184 Rome, ITALY.
[2]Department of Electrical and Computer Engineering, University of California, Santa Barbara, CA 93106, USA.
[3]Centre Automatique et Systèmes, École des Mines de Paris, 35 rue St. Honoré, 77305 Fontainebleau cédex, FRANCE.

1 Introduction

One of the most active research issues in nonlinear feedback theory is the synthesis of feedback laws which robustly stabilize an uncertain system with limited measurement information. In the case of output feedback without uncertainty, one of the major achievements in this area of research has been the "nonlinear separation principle" proved in [6], where it is shown that (semi)global stabilizability via state feedback and a property of uniform observability imply the possibility of semiglobal stabilization via output feedback. To cope with the restricted information structure, the stabilization of [6] includes an approximate state observer (whose role is actually that of producing approximate estimates of a number of "higher order" derivatives of the output) earlier developed in [3] to cope with a similar (though more restricted) stabilization problem. A "robust" version of this stabilization result was given in [5], where it was shown that, in the presence of parameter uncertainties, semiglobal stabilization via output feedback is still possible if a state feedback law is known which robustly globally stabilizes the system and its value, at any time, can be expressed as a (fixed) function of the values, at this time, of a fixed number of derivatives of input and output (a *uniformly completely observable* (UCO) state feedback, in the terminology of [5]).

The design tools introduced in [3] and [5] have been recently used in [2], where a new (iterative) procedure has been proposed for the robust stabilization of certain classes of nonlinear systems. This procedure is not based

on the idea of solving separately a problem of state feedback stabilization
and a problem of asymptotic state reconstruction. Rather, it is based on the
recursive update of a sequence of "dynamic" output feedback stabilizers:
specifically, the basic result of [2] is that if a suitable subsystem of lower
dimension is robustly stabilizable by dynamic output feedback, so is the
entire system. From the point of view of the approach of [5], the condition
on which the result of [2] relies (that happens to be necessary in the case of
linear systems) can be viewed as a condition for the existence of a *dynamic*
feedback driven by functions that are expressible in terms of the output
and its derivatives, i.e., driven by UCO functions.

In this chapter we review and extend the result of [2] and we show how
this result can also be obtained as a special case of a general stabilization result based on the existence of a *dynamic* feedback driven by UCO
functions. More specifically, after some preliminary definitions in Section
2 including our definition of uniform semiglobal practical asymptotic stability, we discuss stabilization of nonminimum phase nonlinear systems by
output feedback in Section 3. This discussion is split into two parts: the
relative degree one case in Section 3.1, and the higher relative degree case
in Section 3.2. The main results of these sections are that if a reduced order,
auxiliary system can be stabilized by dynamic output feedback then the
original nonminimum phase system can be stabilized by dynamic output
feedback. In Section 4 we show how the results of Section 3 can be viewed
as special cases of a general result on semiglobal practical asymptotic stabilization by output feedback. In Section 4.1 we present some additional
definitions, including the notions of *uniformly completely observable* (UCO)
functions and uniform semiglobal practical asymptotic *stabilizability* by dynamic UCO feedback, and a general output feedback stabilization result
which expands on the ideas in [5]. This result is specialized to the case of
nonminimum phase nonlinear systems in Section 4.2. In this section, we
compare and contrast the controllers developed in Section 3 explicitly for
the nonminimum phase nonlinear system case to the controllers that result
from following the synthesis steps presented in [5].

2   Preliminaries

- For simplicity all nonlinear functions in this chapter will be assumed
  to be sufficiently smooth so that all needed derivatives exist and are
  continuous, all differential equations have solutions, etc.

- We will use $\overline{\mathcal{B}}_n(r)$, with $r > 0$, to denote a closed ball of radius r in
  $\mathbb{R}^n$.

- Unless otherwise noted, $\mu(t)$ is a measurable function taking values
  in a compact set $\mathcal{P} \subset \mathbb{R}^p$. The set of such functions is denoted $\mathcal{M}_{\mathcal{P}}$.

- The origin of a nonlinear dynamical system

$$\dot{x} = f(x, \mu(t), k) \ , \tag{8.1}$$

with $x \in \mathbb{R}^n$ and $k \in \mathbb{R}^c$, is said to be *uniformly semiglobally practically asymptotically stable in the parameter k* if

for each pair of strictly positive real numbers $0 < r < R < \infty$ there exist $\bar{k} \in \mathbb{R}^c$, an open set $\mathcal{O} \supset \overline{\mathcal{B}}_n(R)$, a function $V : \mathcal{O} \to \mathbb{R}_{\geq 0}$ that is proper on $\mathcal{O}$ and strictly positive real numbers $0 < q < Q < \infty$ such that

i.) $\overline{\mathcal{B}}_n(R) \subset \{\xi \in \mathcal{O} : V(\xi) \leq Q\}$,

ii.) $\overline{\mathcal{B}}_n(r) \supset \{\xi \in \mathcal{O} : V(\xi) \leq q\}$,

iii.) and

$$\frac{\partial V}{\partial x} f(x, \mu, \bar{k}) < 0 \quad \forall \mu \in \mathcal{P} \, , \forall x \in \{\xi \in \mathcal{O} : q \leq V(\xi) \leq Q\} \, .$$

Uniform semiglobal practical asymptotic stability implies:

for each pair of strictly positive real numbers $0 < r < R < \infty$, there exist $\bar{k} \in \mathbb{R}^c$ and $T > 0$ such that, for all initial conditions in $\overline{\mathcal{B}}_n(R)$, all resulting trajectories $x(t)$ of (8.1) with $k = \bar{k}$ are such that $x(t) \in \overline{\mathcal{B}}_n(r)$ for all $t \geq T$.

It also can be shown to imply:

for each pair of strictly positive real numbers $0 < r < R < \infty$, there exist $\bar{k} \in \mathbb{R}^c$, a compact set $\mathcal{A} \subseteq \overline{\mathcal{B}}_n(r)$ and an open set $\mathcal{G} \supset \overline{\mathcal{B}}_n(R)$ such that, for the system (8.1) with $k = \bar{k}$, the set $\mathcal{A}$ is uniformly asymptotically stable with basin of attraction $\mathcal{G}$.

By this we mean:

- for each $\epsilon > 0$ there exists $\delta > 0$ such that all trajectories starting in a δ-neighborhood of $\mathcal{A}$ remain in an ϵ-neighborhood of $\mathcal{A}$ for all time, and
- for each $\epsilon > 0$ and each compact subset of $\mathcal{G}$ there exists $T > 0$ such that all trajectories starting in the compact subset enter within T seconds and remain thereafter in an ϵ-neighborhood of $\mathcal{A}$.

In fact, due to recent converse Lyapunov function results (see [4], [1], [7]), these latter properties are equivalent characterizations of uniform semiglobal practical asymptotic stability. However, we are using the Lyapunov formulation here so that we can more directly appeal to the results on semiglobal practical asymptotic stabilization like [5, Proposition 3.1] where a Lyapunov formulation was used.

3   Stabilization of Nonminimum Phase Systems by Output Feedback

3.1   The Relative Degree One Case

Most methods for robust stabilization of a nonlinear system by relative degree one output feedback rely on the hypothesis that the system has an asymptotically stable zero dynamics. The main reason why this hypothesis is assumed is that most of the methods in question use "high-gain" feedback in order to keep the output small, thereby enforcing a behavior whose asymptotic properties are essentially determined by the asymptotic properties of the zero dynamics. In particular, asymptotic stabilization occurs only if the latter is asymptotically stable, i.e., if the system is minimum phase. Consider robust (with respect to disturbances $\mu(t)$) stabilization of the origin for the system

$$
\begin{aligned}
\dot{z} &= f_0(z, y, \mu(t)) \\
\dot{y} &= q(z, y, \mu(t)) + b(y)u
\end{aligned}
\tag{8.2}
$$

where $z \in I\!\!R^{n-1}$, $y \in I\!\!R$, $u \in I\!\!R$, $\mu(\cdot) \in \mathcal{M}_{\mathcal{P}}$ and $b(y) \neq 0$ for all y. In the case of uniformly globally asymptotically stable zero dynamics, i.e. (see [4]) when there exists a smooth, positive definite and proper function $V(z)$ such that

$$
\frac{\partial V}{\partial z} f_0(z, 0, \mu) < 0 \qquad \forall z \neq 0, \quad \forall \mu \in \mathcal{P},
$$

the control law

$$
u = -\frac{1}{b(y)} \, k \, y,
$$

where k is a sufficiently large number, solves the problem of semiglobal practical asymptotic stabilization of the origin. This follows from the fact that, given a compact set in (z, y) not containing the origin, for large enough k the negative definite term $\frac{\partial V}{\partial z} f_0(z, 0, \mu) - ky^2$ in the derivative of the composite Lyapunov function

$$
U(z, y) = V(z) + y^2,
$$

i.e., in

$$
\frac{\partial V}{\partial z} f_0(z, y, \mu) + 2y[q(z, y, \mu) - ky],
$$

is able to dominate all nonnegative terms on the given compact set.

In the case where the original output does not yield an asymptotically stable zero dynamics, one approach is to look for a new output function, of the form $y - y^*(z)$, for which the resulting system is uniformly minimum phase. Then, by following the reasoning above, the control

$$u = -\frac{1}{b(y)}\, k\,(y - y^*(z))$$

may be used to achieve robust semiglobal practical stabilization of the origin. The potential drawback to this approach is that it requires the measurement, or at least the robust observability via the actual measured output y and the input u, of the term $y^*(z)$.

Looking at the structure of the system (8.2), we see that the main information about the z subsystem that is robustly observable through the measurement y and the input u is the term $q(z, y, \mu(t))$ and perhaps its derivatives. The discussion that follows, in this and the next subsection, describes one very efficient way, suggested in [2], to use the information contained in $q(z, y, \mu(t))$ to design a stabilizing feedback law without actually requiring a measurement of $q(z, y, \mu(t))$. We will suppose

Assumption 8.1 *For the* auxiliary system

$$\begin{aligned}
\dot{z} &= f_0(z, \bar{u}, \mu(t)) \\
\bar{y} &= q(z, \bar{u}, \mu(t)) \,,
\end{aligned} \tag{8.3}$$

the controller

$$\begin{aligned}
\dot{\varphi} &= L(\varphi) + M\bar{y} \\
\bar{u} &= N(\varphi) \,,
\end{aligned} \tag{8.4}$$

with $N(0) = 0$, is such that the origin of the system (8.3),(8.4) is uniformly globally asymptotically stable.

Under this assumption, we can state the following result for the system (8.2) under the action of the controller

$$\begin{aligned}
\dot{\varphi} &= L(\varphi) + Mk[y - N(\varphi)] \\
u &= \frac{1}{b(y)}\left[\frac{\partial N}{\partial \varphi}[L(\varphi) + Mk[y - N(\varphi)]] - k[y - N(\varphi)]\right] .
\end{aligned} \tag{8.5}$$

Note that this is simply a dynamic feedback of the original (nonminimum phase) output y.

Theorem 8.2 *Under Assumption 8.1, the origin of the system (8.2), (8.5) is uniformly semiglobally practically asymptotically stable in the control parameter k.*

Proof. The result is established by noting, with the help of the input transformation

$$u = \frac{1}{b(y)}\left[\frac{\partial N}{\partial \varphi}L(\varphi) + (1 - \frac{\partial N}{\partial \varphi}M)v\right] , \tag{8.6}$$

that the system

$$
\begin{aligned}
\dot{z} &= f_0(z, y, \mu(t)) \\
\dot{\varphi} &= L(\varphi) - Mv \\
\dot{y} &= q(z, y, \mu(t)) + \frac{\partial N}{\partial \varphi} L(\varphi) + \left(1 - \frac{\partial N}{\partial \varphi} M\right) v
\end{aligned}
\tag{8.7}
$$

with output $\theta = y - N(\varphi)$ has relative degree one with high-frequency gain identically equal to one, is minimum phase and can be written, globally, in a form that matches (8.2). Specifically, in the coordinates (z, ξ, θ) where $\xi := \varphi + M\theta$, we have:

$$
\begin{aligned}
\dot{z} &= f_0(z, N(\xi - M\theta) + \theta, \mu(t)) \\
\dot{\xi} &= L(\xi - M\theta) + Mq(z, N(\xi - M\theta) + \theta, \mu(t)) \\
\dot{\theta} &= q(z, N(\xi - M\theta) + \theta, \mu(t)) + v \ .
\end{aligned}
\tag{8.8}
$$

By Assumption 8.1, when θ is set to zero, the origin of the (z, ξ) dynamics is uniformly globally asymptotically stable. It follows from the discussion above that the choice $v = -k\theta$ is semiglobally practically stabilizing for the origin of (8.8). And, since $N(0) = 0$, the origin of (8.8) corresponds to the origin of (8.2),(8.5). Moreover, with this choice for v we see from (8.6) and the $\dot{\varphi}$ equation in (8.7) that we recover the control law (8.5). $\triangle$

Remark 8.1 If a controller of a form more general than (8.4) like

$$
\begin{aligned}
\dot{\tilde{\varphi}} &= \tilde{L}(\tilde{\varphi}, \bar{y}) \\
\bar{u} &= \tilde{N}(\tilde{\varphi}, \bar{y})
\end{aligned}
$$

exists (in the case where $\bar{y}$ depends on $\bar{u}$ we would need an assumption that guarantees a solution $\bar{u}$ to the second equation), a controller of the form (8.4) can be obtained by dynamic extension as

$$
\begin{aligned}
\dot{\tilde{\varphi}} &= \tilde{L}(\tilde{\varphi}, \tilde{\varphi}_{\nu+1}) \\
\dot{\tilde{\varphi}}_{\nu+1} &= -m(\tilde{\varphi}_{\nu+1} - \bar{y}) \\
\bar{u} &= \tilde{N}(\tilde{\varphi}, \tilde{\varphi}_{\nu+1})
\end{aligned}
$$

with m a positive number. Instead of achieving uniform *global* asymptotic stability for the auxiliary system, this controller would, in general, achieve uniform semiglobal practical asymptotic stability in the parameter m, at least in the case where the functions $\mu(t)$ are restricted to have uniformly bounded derivatives. While this would complicate the above discussion, the conclusion of the theorem would still be the same. $\triangle$

Remark 8.2 As discussed in [5], various local conditions can be imposed on the system (8.8) to guarantee uniform semiglobal asymptotic stability, as opposed to only uniform semiglobal *practical* asymptotic stability. $\triangle$

3.2   The Relative Degree Greater than One Case

The result of the previous section, on stabilization by dynamic output feedback, can be extended to the case of outputs with relative degree greater than one.

Consider a nonlinear system modeled by equations of the form

$$
\begin{aligned}
\dot{z} &= f(z, \zeta_1, \dots, \zeta_r, \mu(t)) \\
\dot{\zeta}_1 &= \zeta_2 \\
\dot{\zeta}_2 &= \zeta_3 \\
&\;\;\vdots \\
\dot{\zeta}_r &= q(z, \zeta_1, \dots, \zeta_r, \mu(t)) + b(\zeta)u \\
y &= \zeta_1
\end{aligned}
\tag{8.9}
$$

in which $z \in \mathbb{R}^{n-r}$, $\mu(\cdot) \in \mathcal{M}_P$ and $b(\zeta) \neq 0$ for all ζ. This normal form may result from applying a globally defined, perhaps μ dependent, coordinate transformation to a nonlinear system given in some other form.

The only measurement that we will assume is available is the output y. What we will show is that if a particular reduced system can be stabilized with measurements of ζ and $q(z, \zeta_1, \dots, \zeta_r, \mu(t))$ then the system (8.9) can be stabilized with measurement of y only.

With the system (8.9), we associate an *auxiliary system*

$$
\begin{aligned}
\dot{x}_\mathrm{a} &= f_\mathrm{a}(x_\mathrm{a}, u_\mathrm{a}, \mu(t)) \\
y_\mathrm{a} &= h_\mathrm{a}(x_\mathrm{a}, u_\mathrm{a}, \mu(t))
\end{aligned}
\tag{8.10}
$$

in which

$$
x_\mathrm{a} = \begin{pmatrix} x_{\mathrm{a},1} \\ \hline x_{\mathrm{a},2} \end{pmatrix} := \begin{pmatrix} z \\ \hline \zeta_1 \\ \vdots \\ \zeta_{r-2} \\ \zeta_{r-1} \end{pmatrix},
$$

and

$$
f_\mathrm{a}(x_\mathrm{a}, u_\mathrm{a}, \mu(t)) = \begin{pmatrix} f_{\mathrm{a},1}(x_\mathrm{a}, u_\mathrm{a}) \\ \hline f_{\mathrm{a},2}(x_{\mathrm{a},2}, u_\mathrm{a}) \end{pmatrix} := \begin{pmatrix} f(z, \zeta_1, \dots, \zeta_{r-1}, u_\mathrm{a}, \mu(t)) \\ \hline \zeta_2 \\ \vdots \\ \zeta_{r-1} \\ u_\mathrm{a} \end{pmatrix},
$$

and

$$h_a(x_a, u_a, \mu) := q(z, \zeta_1, \ldots, \zeta_{r-1}, u_a, \mu(t)) \,.$$

About this system, we assume the following:

Assumption 8.3 *The controller*

$$\begin{aligned} \dot{\varphi} &= L(\varphi, x_{a,2}) + M y_a \\ u_a &= N(\varphi, x_{a,2}) \,, \end{aligned} \tag{8.11}$$

with $N(0,0) = 0$, *is such that the origin of the system* (8.10),(8.11) *is uniformly globally asymptotically stable.*

Under this assumption, we can state the following result for the system (8.9) under the action of the controller

$$\begin{aligned} \dot{\varphi} &= L(\varphi, x_{a,2}) + M k[\zeta_r - N(\varphi, x_{a,2})] \\ u &= \frac{1}{b(\zeta)} \left[\frac{\partial N}{\partial \varphi} [L(\varphi, x_{a,2}) + M k[\zeta_r - N(\varphi, x_{a,2})]] + \right. \\ &\qquad \left. \frac{\partial N}{\partial x_{a,2}} f_{a,2}(x_{a,2}, \zeta_r) - k[\zeta_r - N(\varphi, x_{a,2})] \right] \,. \end{aligned} \tag{8.12}$$

Note that this is a dynamic feedback of the output y and its first $r - 1$ derivatives.

Lemma 8.1 *Under Assumption 8.3, the origin of the system* (8.9), (8.12) *is uniformly semiglobally practically asymptotically stable in the control parameter* k.

Proof. The proof is the same as the proof of Theorem 8.2. With the input transformation

$$u = \frac{1}{b(\zeta)} \left[\frac{\partial N}{\partial \varphi} L(\varphi, x_{a,2}) + \frac{\partial N}{\partial x_{a,2}} f_{a,2}(x_{a,2}, \zeta_r) + \left(1 - \frac{\partial N}{\partial \varphi} M \right) v \right] \tag{8.13}$$

we get the system

$$\begin{aligned} \dot{x}_a &= f_a(x_a, \zeta_r, \mu(t)) \\ \dot{\varphi} &= L(\varphi, x_{a,2}) - M v \\ \dot{\zeta}_r &= h_a(x_a, \zeta_r, \mu(t)) + \\ &\qquad \frac{\partial N}{\partial \varphi} L(\varphi, x_{a,2}) + \frac{\partial N}{\partial x_{a,2}} f_{a,2}(x_{a,2}, \zeta_r) + \left(1 - \frac{\partial N}{\partial \varphi} M \right) v \end{aligned} \tag{8.14}$$

that, with output $\theta = \zeta_r - N(\varphi, x_{a,2})$, has relative degree one with high-frequency gain identically equal to one, is minimum phase and can be written, globally, in a form that matches (8.2). Specifically, in the coordinates

(x_a, ξ, θ) where $\xi := \varphi + M\theta$, we have:

$$\dot{x}_a = f_a\left(x_a, N(\xi - M\theta, x_{a,2}) + \theta, \mu(t)\right)$$
$$\dot{\xi} = L(\xi - M\theta, x_{a,2}) + Mh_a\left(x_a, N(\xi - M\theta, x_{a,2}) + \theta, \mu(t)\right) \quad (8.15)$$
$$\dot{\theta} = h_a\left(x_a, N(\xi - M\theta, x_{a,2}) + \theta, \mu(t)\right) + v \; .$$

By Assumption 8.3, when θ is set to zero, the origin of the (x_a, ξ) dynamics is uniformly globally asymptotically stable. It follows, as before, that the choice $v = -k\theta$ is semiglobally practically stabilizing for the origin of (8.15). And since $N(0,0) = 0$, the origin of (8.15) corresponds to the origin of (8.9),(8.12). Moreover, with this choice for v we see from (8.13) and the $\dot{\varphi}$ equation in (8.14) that we recover the control law (8.12). △

The dynamic controller (8.12) uses the state variables $\zeta_1, \ldots, \zeta_r$, i.e., the derivatives up to order $r - 1$ of the output y of system (8.9), as input. Thus, in order to find an output feedback controller, these variables must be replaced by appropriate estimates, which can be provided by a dynamical system of the form

$$\dot{\eta} = P\eta + Qy \qquad (8.16)$$

in which the matrices Q and P have the form

$$P = \begin{pmatrix} -gc_{r-1} & 1 & 0 & \cdots & 0 \\ -g^2 c_{r-2} & 0 & 1 & \cdots & 0 \\ . & . & . & \cdots & . \\ -g^{r-1}c_1 & 0 & 0 & \cdots & 1 \\ -g^r c_0 & 0 & 0 & \cdots & 0 \end{pmatrix}, \qquad Q = \begin{pmatrix} -gc_{r-1} \\ -g^2 c_{r-2} \\ . \\ -g^{r-1}c_1 \\ -g^r c_0 \end{pmatrix} . \qquad (8.17)$$

As shown in [3], it is convenient to saturate the resulting control law, at least where the estimates of ζ appear, so as to avoid the occurrence of finite escape times for large values of g. For example, we can replace the controller (8.12), which for ease of notation we now write as

$$\dot{\varphi} = C(\varphi, \zeta)$$
$$u = K(\varphi, \zeta) , \qquad (8.18)$$

with the controller

$$\dot{\varphi} = \sigma_\ell\left(C(\varphi, \eta)\right)$$
$$u = \sigma_\ell\left(K(\varphi, \eta)\right) \qquad (8.19)$$

where $\sigma_\ell(\cdot)$ is a (by abuse of notation both a scalar and vector) saturation function

$$\sigma_\ell(v) = v \cdot \min\left\{1, \frac{\ell}{|v|}\right\} .$$

A controller of this type is able to robustly semiglobally practically asymptotically stabilize the plant (8.9). In fact, using the methods of [5] for example, it is possible prove the following result.

Theorem 8.4 *(See also [2]) Under Assumption 8.3, the origin of the system (8.9), (8.16), (8.19) [with $C(\cdot,\cdot)$ and $K(\cdot,\cdot)$ defined by the identification between (8.12) and (8.18)] is uniformly semiglobally practically stable in the control parameters (k, g, ℓ).*

4   On Dynamic UCO Feedback

The basic observation of [2], summarized in Section 3.2 and on which the result of Lemma 8.1 rests, is that the term $q(z, \zeta_1, \ldots, \zeta_{r-1}, \zeta_r, \mu(t))$ in the system (8.9) can be (and, in a nonminimum phase system, has to be) "isolated" from the rest of the system, using measurements only of the output and its first $r - 1$ derivatives, and treated as a separate source of information for feedback. Then, having a dynamic controller driven by the output its first $r-1$ derivatives, as in Lemma 8.1, it is straightforward using ideas initially developed in [3] to find a dynamic output feedback controller that induces the desired properties, as in Theorem 8.4.

From this point of view, the contribution in [2] is the identification of a natural (in fact, for linear systems it can be shown to be necessary) condition (Assumption 8.3) that guarantees the existence of a dynamic feedback that is expressible in terms of the output and its derivatives. Then Theorem 8.4 can be viewed as a special case of a more general result that is essentially contained in [5] (see [5, Proposition 3.1 and footnote 5]), namely that semiglobal practical stabilization by dynamic *uniformly completely observable* (UCO) feedback implies semiglobal practical stabilization by dynamic output feedback. We make this result explicit below.

4.1   General Results

Consider multi-input, multi-output nonlinear control systems

$$\begin{aligned}
\dot{x} &= f(x, u, \mu(t)) \\
y &= h(x, u, \mu(t))
\end{aligned} \tag{8.20}$$

with $\mu(\cdot) \in \mathcal{M}_\mathcal{P}$. The definition of uniformly completely observable (UCO) dynamic feedback, given next, at times implicitly constrains $\mu(t)$ to be sufficiently smooth, where sufficiently smooth has to do with the number of times the output needs to be differentiated to reconstruct the UCO function.

Definition 8.1 *A function $\varphi(x, u, \mu)$ is said to be* uniformly completely observable *(UCO) with respect to the system (8.20) if it can be expressed as a function of a finite number of derivatives of the output y and the input u, i.e., if there exist two integers n_y and n_u and a function Ψ such that, for each solution of*

$$
\begin{aligned}
\dot{x} &= f(x, u, \mu(t)) \\
u^{(n_u+1)} &= v \\
y &= h(x, u, \mu(t))
\end{aligned} \tag{8.21}
$$

we have, for all t where the solution makes sense,

$$
\varphi(x(t), u(t), \mu(t)) = \Psi\left(y(t), \ldots, y^{(n_y)}(t), u(t), \ldots, u^{(n_u)}(t)\right) \tag{8.22}
$$

where $y^{(i)}$ denotes the ith time derivative of y at time t (and similarly for $u^{(i)}$).

Remark 8.3 As in [5, Footnote 6], note the strong requirement that Ψ is independent of $\mu(t)$. On the other hand, note that the functions

$$
\zeta_i \ , \qquad q(\zeta_1, \ldots, \zeta_r, \mu)
$$

for the system (8.9) are UCO since we can write

$$
\zeta_i = y^{(i-1)} \ , \qquad q(\zeta_1, \ldots, \zeta_r, \mu(t)) = y^{(r)} - b(y)u \ .
$$

$\triangle$

Our next definitions, on uniform semiglobal practical asymptotic *stabilizability* by dynamic UCO or output feedback, are closely related to our definition of uniform semiglobal practical asymptotic stability. However, as was the case in [5], we don't insist that the states of the dynamic compensator eventually become small in the closed-loop. We formulate the definition in Lyapunov function terms but, again, the definition could be formulated in terms of trajectories.

Definition 8.2 *The origin of (8.20) is said to be* uniformly semiglobally practically asymptotically stabilizable *by dynamic UCO feedback if for each pair of strictly positive real numbers $0 < r < R < \infty$ there exist:*

- *a UCO function $\alpha(x, u, \mu)$*

- *functions θ and κ,*

- *compact sets $C_{\eta s}$ and $C_{\eta l}$, with $C_{\eta s}$ a subset of the interior of $C_{\eta l}$,*

- *an open set $\mathcal{O} \supset \bar{B}_n(R) \times C_{\eta l}$,*

- *a function $V : \mathcal{O} \to \mathbb{R}_{\geq 0}$ that is proper on $\mathcal{O}$, and*

- *strictly positive real numbers $0 < q < Q < \infty$*

such that

i.) $(\overline{\mathcal{B}}_n(R) \times \mathcal{C}_{\eta l}) \subset \{\xi \in \mathcal{O} : V(\xi) \leq Q\},$

ii.) $(\overline{\mathcal{B}}_n(r) \times \mathcal{C}_{\eta s}) \supset \{\xi \in \mathcal{O} : V(\xi) \leq q\},$

iii.) *and*

$$\frac{\partial V}{\partial X} F(X, \mu) < 0 \qquad \forall \mu \in \mathcal{P} \ , \ \forall X \in \{\xi \in \mathcal{O} : q \leq V(\xi) \leq Q\} \ (8.23)$$

where X and $F(X, \mu)$ are defined by

$$\dot{X} = \frac{d}{dt} \begin{pmatrix} x \\ \eta \end{pmatrix} = \begin{pmatrix} f(x, u, \mu(t)) \\ \theta\big(\eta, \alpha(x, u, \mu(t))\big) \end{pmatrix} =: F(X, \mu(t)) \quad (8.24)$$

with

$$u = \kappa\Big(\eta, \alpha(x, u, \mu(t))\Big) \tag{8.25}$$

(and where, for simplicity, we assume the right-hand side of (8.25) is independent of u).

Definition 8.3 *The origin of (8.20) is said to be* uniformly semiglobally practically asymptotically stabilizable by dynamic output feedback *if, in the previous definition, we can always take $\alpha(x, u, \mu) = h(x, u, \mu)$.*

Remark 8.4 In these definitions, we could allow the right-hand side of (8.25) to depend on u if we impose an extra condition that guarantees a solution to (8.25). △

It will follow from the proof of [5, Proposition 3.1] (much like what is suggested by [5, Footnote 5]) that we have:

Theorem 8.5 *Let $\mu(\cdot) \in \mathcal{M}_{\mathcal{P}}$ be sufficiently smooth with a uniform bound on an appropriate number of derivatives. If the origin of the system (8.20) is uniformly semiglobally practically asymptotically stabilizable by dynamic UCO feedback then it is uniformly semiglobally practically asymptotically stabilizable by dynamic output feedback.*

Sketch of Proof. Fix $0 < r < R < \infty$. From the assumption of uniform semiglobal practical asymptotic stabilizability by dynamic UCO feedback, this fixes a UCO function $\alpha(x, u, \mu)$, a corresponding function Ψ that is used to reconstruct α from derivatives of y and u, functions θ and κ, compact sets $\mathcal{C}_{ns}$ and $\mathcal{C}_{nl}$, an open set $\mathcal{O}$, a function V and strictly positive real numbers $0 < q < Q < \infty$. Now we apply the proof of [5, Proposition 3.1] to the control system

$$\begin{pmatrix} \dot{x} \\ \dot{\eta} \end{pmatrix} = \begin{pmatrix} f(x, u_1, \mu(t)) \\ \theta(\eta, u_2) \end{pmatrix}$$

$$\begin{pmatrix} y_1 \\ y_2 \end{pmatrix} = \begin{pmatrix} h(x, u_1, \mu(t)) \\ \eta \end{pmatrix}$$

where the UCO feedback

$$\begin{pmatrix} u_1 \\ u_2 \end{pmatrix} = \begin{pmatrix} \kappa(\eta, \alpha(x, u_1, \mu(t))) \\ \alpha(x, u_1, \mu(t)) \end{pmatrix}$$

induces the properties for the function V that are assumed in the proof of [5, Proposition 3.1] if we define the objects $\mathcal{K}_{zs}$, $\mathcal{K}_{zl}$, ν_l, c_s and c_l used in the proof of [5, Proposition 3.1] as

$$\mathcal{K}_{zs} := \overline{\mathcal{B}}_n(r) \times \mathcal{C}_{ns}, \quad \mathcal{K}_{zl} := \overline{\mathcal{B}}_n(R) \times \mathcal{C}_{nl}$$

and

$$\nu_l := q, \quad c_s := Q, \quad c_l := Q + 1.$$

From here we follow the proof of [5, Proposition 3.1], but noting that dynamic extension is only needed on the input u_1 and no estimates of the derivatives of $y_2 = \eta$ are needed. $\triangle$

4.2 Application to Nonminimum Phase Systems

We now apply this general result to the problem considered in Section 3.2. We start with an assumption that is a combination of Assumption 8.3 and Remark 8.1.

Assumption 8.6 *The controller*

$$\begin{aligned} \dot{\varphi} &= L(\varphi, x_{a,2}, y_a) \\ u_a &= N(\varphi, x_{a,2}, y_a), \end{aligned} \tag{8.26}$$

is such that

1. *$N(0, 0, 0) = 0$ and, for simplicity, $N(\varphi, x_{a,2}, h_a(x_a, u_a, \mu))$ independent of u_a,*

2. *the origin of the system (8.10),(8.26) is uniformly globally asymptotically stable;*

3. *the functions $\mu(\cdot)$ are restricted so that*

$$\left| \frac{\partial N}{\partial h_a} \frac{\partial h_a}{\partial \mu} \dot{\mu}(t) \right|$$

is bounded in $t \geq 0$, uniformly in $\mu(\cdot)$, on each compact subset of the state-space.

Under this assumption, we can state the following result for the system (8.9) under the action of the controller

$$
\begin{aligned}
\dot{\varphi} &= L\left(\varphi, x_{a,2}, h_a(x_a, \zeta_r, \mu(t))\right) \\
u &= -k\mathrm{sgn}(b(\zeta))\left(\zeta_r - N(\varphi, x_{a,2}, h_a(x_a, \zeta_r, \mu(t)))\right) .
\end{aligned}
\tag{8.27}
$$

Note that this is a dynamic UCO feedback for the system (8.9) since, as noted in Remark 8.3, $x_{a,2}$, ζ_r and $h_a(x_a, \zeta_r, \mu(t))$ are UCO with respect to the system (8.9).

Lemma 8.2 *Under Assumption 8.6, the origin of the system (8.9), (8.27) is semiglobally practically asymptotically stable in the parameter k.*

Proof. Follows from the discussion in Section 3.1. (See also [5, Lemma 2.2 (Semiglobal backstepping I)].)   △

The final result then follows from Theorem 8.5 and Lemma 8.2.

Corollary 8.1 *Under Assumption 8.6, the origin of the system (8.9) is semiglobally practically stabilizable by dynamic output feedback.*

The controller given by Corollary 8.1, which is constructed following the proof of Theorem 8.5, is different from the one given by Theorem 8.4 together with Remark 8.1. In particular, the controller of Corollary 8.1 has the form of an observer

$$
\dot{\eta} = P\eta + Qy ,
\tag{8.28}
$$

like in (8.16) but with $\eta \in \mathbb{R}^{r+1}$, where η_{r+1} is an estimate of ζ_r, plus an estimated and saturated dynamic UCO feedback

$$
\begin{aligned}
\dot{\varphi} &= \sigma_\ell\left(L\left(\varphi, \hat{x}_{a,2}, \hat{\zeta}_r - b(\hat{\zeta})u\right)\right) \\
\dot{u} &= \sigma_\ell(k_2(v - u)) \\
v &= -k_1\mathrm{sgn}(b(\hat{\zeta}))\left(\hat{\zeta}_r - N(\varphi, \hat{x}_{a,2}, \hat{\zeta}_r - b(\hat{\zeta})u)\right) ,
\end{aligned}
\tag{8.29}
$$

like in (8.19). Compared to the controller (8.28),(8.29), the controller (8.16), (8.19) together with remark 8.1 has one less state and can be interpreted as using a reduced-order observer structure to accomplish the goal of robust semiglobal practical asymptotic stabilization.

In [5, Section 6.2], a particular nonminimum phase nonlinear system, whose auxiliary system (using the terminology of the present chapter) is semiglobally asymptotically stabilizable by (static) UCO feedback, was considered as an illustration of the result that semiglobal practical asymptotic stabilization by (static) UCO feedback implies semiglobal practical asymptotic stabilization by dynamic output feedback. The controller used in that section is the type of controller suggested by Corollary 8.1.

5 Conclusions

This chapter presented a simple design method by which it is possible to robustly stabilize, using output feedback, a significant class of uncertain nonlinear systems whose zero dynamics are unstable. The assumption made for such systems was shown to imply the existence of a stabilizing dynamic feedback that is driven by functions that are *uniformly completely observable* (UCO). In this light, the result for nonminimum phase nonlinear systems was shown to be a special case of the more general result that semiglobal practical asymptotic stabilization by dynamic UCO feedback implies semiglobal practical asymptotic stabilization by dynamic output feedback. The controllers developed in this chapter specifically for nonminimum phase nonlinear systems were compared and contrasted to the controllers that prove the general stabilization result.

Acknowledgements

The first author was supported in part by NSF under grant ECS-9707891, by AFOSR under grant F49620-95-1-0232, and by MURST.

The second author was supported by in part by the NSF under grant ECS-9896140 and by the AFOSR undergrant F49620-98-1-0087.

6 REFERENCES

[1] A. N. Atassi and H. K. Khalil. A separation principle for the control of a class of nonlinear systems. In *Proc. of the 37th IEEE Conf. on Dec. and Contr.*, pp. 855-860, Dec. 16-18, 1998, Tampa, FL.

[2] A. Isidori. A tool for semiglobal stabilization of uncertain nonminimum phase nonlinear systems via output feedback. preprint, November 1997.

[3] H. K. Khalil and F. Esfandiari. Semiglobal stabilization of a class of nonlinear systems using output feedback. *IEEE Trans. on Automatic Control*, 38 (1993), pp. 1412–1415.

[4] Y. Lin, E. D. Sontag and Y. Wang. A smooth converse Lyapunov theorem for robust stability. In *SIAM J. Control and Optimization*, 34(1996), pp. 124–160.

[5] A. R. Teel and L. Praly. Tools for semiglobal stabilization by partial state and output feedback. *SIAM J. Control Optim.*, vol. 33, no. 5, pp. 1443–1488, September, 1995.

[6] A. R. Teel and L. Praly. Global stabilizability and observability imply semi-global stabilizability by output feedback. *Systems & Control Letters*, vol. 22 (1994) 313–325.

[7] A. R. Teel and L. Praly. A smooth Lyapunov function from a class-$\mathcal{KL}$ estimate involving two positive semidefinite functions. In preparation.

Part III
Fault Detection and Isolation

In this part a slightly different notion of observer is exploited in the study of fault detection and isolation. Basically, in fault detection and isolation one faces the questions of how to detect a faulty operation of a system and, in case of a fault, what is the cause of the bad system's operation. There are several ways to approach these problems but in the frame of this book an observer perspective is taken. Although there are some clear distinctions with the standard observer problem, the usual formalism can serve as a promising starting point. A fundamental difference with the normal observer problem, however, is that certain inputs -"fault indicators"- are unknown and as long as they are zero, no fault occurs. As soon as some of these inputs become nonzero, a fault occurs and the observer should provide a rapid indication for the specific faults. At the same token, it may be less important to estimate the full state of the system. The chapters in this part of the book all present facets of observer based fault detection and isolation. Together they form an indication of today's achievements with respect to fault detection and isolation using an observer approach.

Fault Detection Observer for a Class of Nonlinear Systems

S. A. Ashton and D. N. Shields

School of MIS, Coventry University, Coventry, United Kingdom.

1 Introduction

Different approaches for fault diagnosis, based on both hardware and software, have been given, as partially described in [7] [2]. Related work on various classes of observers has been proposed, including bilinear observers [10] [3], quasi-linear observers [2] [4], nonlinear high-gain observers [6] [9], nonlinear canonical-form observers [6] [11] and nonlinear observers, based on the existence of linearizing transformations [9] [5]. Many design methods for the robust residual generation, involving observers, rely on the solution of complex partial differential equations [2] [8] or the solution of a set of algebraic-differential (or algebraic-difference) equations [7] [4]. Observer designs obtained in [10] [3] [4], based on bilinear system models, involve only the solution of nonlinear algebraic equations or Ricatti equations, and on-line residual generation is possible for low-dimensional systems. Similar propositions are reported in [1] for systems modelled by quadratic polynomial models. However, a complete theoretical analysis is lacking of the problems associated with observer stability, fault-sensitivity, disturbance decoupling and numerical tractability of the combined observer and residual generator.

This chapter extends the work in [10] and [1] and provides a design for a nonlinear fault detection observer for a class of polynomial systems such that the corresponding observer error dynamical system is linear for the fault-free case, but otherwise, nonlinear with respect to inputs, outputs and faults. A set of theoretical propositions are given for establishing the robustness, stability, existence and detectability properties of the observer, called a robust fault detection observer (RFDO). Many practical systems fall into the class of systems considered here [10] [3] [4] [1].

This chapter is organized as follows. A system description is given followed by the main section, Section 3, on observer design. Section 4 gives general sufficient conditions for detectability and Section 5 gives testable conditions for detectability including a special class of faults. Concluding remarks are given in Section 6.

2 System Description

A continuous-time system is considered with polynomial nonlinearities up to degree three. The system model considered is that of the form

$$\dot{x}(t) = Ax(t) + E_a d(t) + K_a f(t) + Bu(t) + \sum_{i=1}^{m} u^i(t) A_{ux}^i x(t)$$

$$+ \sum_{i=1}^{n} x^i(t) \left[A_{xx}^i x(t) + E^i d(t) + K^i f(t) \right]$$

$$+ \sum_{i=1}^{n} \sum_{j=1}^{n} x^i(t) x^j(t) \left[A_{xxx}^{ij} x(t) + E^{ij} d(t) + K^{ij} f(t) \right], \qquad (2.1a)$$

$$y(t) = Cx(t) + K_s f(t) + E_s d(t)$$
$$= x_1(t) + K_s f(t) + E_s d(t), \qquad (2.1b)$$

where $x(t) \in \mathbb{R}^n$ is the state vector, partitioned as,

$$x(t) = [x_1(t)', x_2(t)']',$$

where $x_1(t) \in \mathbb{R}^p$, $x_2(t) \in \mathbb{R}^{n-p}$, $u(t) \in \mathbb{R}^m$ is the input vector, $y(t) \in \mathbb{R}^p$ is the output vector and $d(t) \in \mathbb{R}^q$ is a disturbance. Here, the fault vector is partitioned as

$$f(t) = [f_a(t)', f_s(t)']' \in \mathbb{R}^v,$$

where $v = v_1 + v_2$, $f_a(t) \in \mathbb{R}^{v_1}$ represents component or actuator faults and $f_s(t) \in \mathbb{R}^{v_2}$ represents sensor faults. For convenience, without loss of generality, the constant matrices in (2.1a) and (2.1b) are partitioned as

$$C = [I_p, 0_{p \times (n-p)}], \quad A = [A_1, A_2], \quad A_{ux}^i = \left[A_{1ux}^i, A_{2ux}^i \right],$$

$$A_{xx}^i = \left[A_{1xx}^i, A_{2xx}^i \right], \quad A_{xxx}^{ij} = \left[A_{1xxx}^{ij}, A_{2xxx}^{ij} \right] \qquad (2.2)$$

where $C \in \mathbb{R}^{p \times n}$, A_1, $A_{1xx}^i (i = 1, \cdots, n)$, $A_{1xxx}^{ij} (i, j = 1, \cdots, n)$, $A_{1ux}^i (i = 1, \cdots, m) \in \mathbb{R}^{n \times p}$ and A_2, $A_{2xx}^i (i = 1, \cdots, n)$, $A_{2xxx}^{ij} (i, j = 1, \cdots, n)$, $A_{2ux}^i (i = 1, \cdots, m) \in \mathbb{R}^{n \times (n-p)}$. Also, $B \in \mathbb{R}^{n \times m}$, $E_a \in \mathbb{R}^{n \times q}$, $E_s \in \mathbb{R}^{p \times q}$ and K_a, $K_s \in \mathbb{R}^{p \times (v)}$ are all constant matrices.

It is assumed (2.1a) is controlled by the input $u(t)$ so that:

Assumption 1.1 $u(t)$ *is bounded and such that for bounded* $d(t)$ *and* $f(t) = 0$ *(all t)* $x(t)$ *exists and is bounded.*

3 Observer Design

For the system described by (2.1a)-(2.1b) an observer is proposed. The observer is linear in $z(t)$ and involves bilinear, quadratic and cubic terms in $y(t)$ and $u(t)$. The candidate observer is given by

$$\dot{z}(t) = Fz(t) + Ju(t) + \sum_{i=1}^{m} u^i(t) H_{ux}^i y(t) + Hy(t)$$

$$+ \sum_{i=1}^{p} y^i(t) H_{xx}^i y(t) + \sum_{i=1}^{p}\sum_{j=1}^{p} y^i(t) y^j(t) H_{xxx}^{ij} y(t), \qquad (3.1)$$

where $z(t) \in \mathbb{R}^d$, is a linear estimate of $Tx(t)$. A fault detection signal is defined as

$$\epsilon(t) = L_1 z(t) + L_2 y(t), \qquad (3.2)$$

where $\epsilon(t) \in \mathbb{R}^{d_o} (d_o, d \geq 1)$.

Definition 1.1 *The system (3.1)-(3.2) is a robust fault detection observer (RFDO) with respect to the system (2.1a)-(2.1b) for a class of faults, C_f, if*

1. *for all bounded $u(t)$, $d(t)$, and $z(0)$ ($t \geq 0$), and with $f(t) = 0$, the error dynamic for $e(t)$ is assymptotically stable, so that*

$$\lim_{t \to \infty} e(t) = 0,$$

$$\lim_{t \to \infty} \epsilon(t) = 0,$$

where $e(t) = z(t) - Tx(t)$ and $\epsilon(t)$ is given by (3.2).

2. *for all bounded $u(t)$, $d(t)$ and $y(t)$ ($t \geq 0$) there exists at least one fault vector $f(t) \neq 0$, $f(t) \in C_f$, such that*

$$\epsilon(t) \neq 0, \quad (t \geq t_0)$$

with $e(t_0) = 0$.

Remark 1.1 *Assumption 1.1 and condition Definition 1.1(1) ensure that $z(t)$ exists and is bounded for zero faults and bounded $z(0)$. (Note C_f is not restricted here.) Condition Definition 1.1(1) also ensures that both the observer error and fault signal converge to zero for any disturbance when no faults are present. Condition Definition 1.1(2) ensures that at least one fault exists, which can be detected for all bounded $u(t)$, $d(t)$ and $y(t)$, given that at a specific time the error signal was zero.*

Definition 1.2 *The RFDO (3.1)-(3.2) is called a strict RFDO (SRFDO) if for a class $f(t) \in C_f$, with $f(t) \neq 0$, and any bounded set $\{u(t), d(t), y(t)\}$, there exists some $t_0 \geq 0$ such that, given $e(t_0) = 0$,*

$$\epsilon(t) \neq 0, \quad (t \geq t_0).$$

Remark 1.2 *An SRFDO ensures that all $f(t) \in C_f$ can be detected, although not necessarily distinguishable.*

Design Problem:

Find matrices F, J, H, T, $H_{ux}^i(i, \cdots, m)$, $H_{xx}^i(i, \cdots, p)$, $H_{xxx}^i(i, \cdots, p)$, L_1 and L_2 and parameters d and d_0 so that Definition 1.1 is satisfied.

First condition Definition 1.1(1) is considered. Condition Definition 1.1(2) is addressed in Section 4. Consider the observer error

$$e(t) = z(t) - Tx(t). \tag{3.3}$$

Using (2.1a)-(2.1b), (3.1) and (3.3) there obtains

$$\dot{e}(t) = LI(t) + BI(t) + QU(t) + CU(t), \tag{3.4}$$

where

$$
\begin{aligned}
LI(t) = {} & Fz(t) + Ju(t) + Hy(t) \\
& - T\left[Ax(t) + Bu(t) + E_a d(t) + K_a f(t)\right],
\end{aligned} \tag{3.5a}
$$

$$BI(t) = \sum_{i=1}^{m} u^i(t) H_{ux}^i y(t) - T \sum_{i=1}^{m} u^i(t) A_{ux}^i x(t), \tag{3.5b}$$

$$
\begin{aligned}
QU(t) = {} & \sum_{i=1}^{p} y^i(t) H_{xx}^i y(t) \\
& - T \sum_{i=1}^{n} x^i(t) \left[A_{xx}^i x(t) + E^i d(t) + K^i f(t)\right],
\end{aligned} \tag{3.5c}
$$

$$
\begin{aligned}
CU(t) = {} & \sum_{i=1}^{p}\sum_{j=1}^{p} y^i(t) y^j(t) H_{xxx}^{ij} y(t) \\
& - T \sum_{i=1}^{n}\sum_{j=1}^{n} x^i(t) x^j(t) \left[A_{xxx}^{ij} x(t) + E^{ij} d(t) + K^{ij} f(t)\right], \tag{3.5d}
\end{aligned}
$$

where (3.5a)-(3.5d) correspond to linear, bilinear, quadratic and cubic terms, respectively, in $x(t)$. Using (3.4) and (3.2) the following proposition holds true.

Proposition 1.1 *If T ($T \neq 0$), J, H, F, L_1 ($L_1 \neq 0$), L_2, $H_{1ux}^i(i = 1, \cdots, m)$, $H_{xx}^i(i = 1, \cdots, p)$ and $H_{xxx}^i(i = 1, \cdots, p)$ can be found such* *that the following conditions are satisfied (for some $d_0, d \geq 1$)*

$$0 > \mathbb{R}e\left(\lambda_i(F)\right); i = 1, \cdots, d \tag{3.6a}$$

$$H = [TA - FT]\Phi \tag{3.6b}$$

$$J = TB \tag{3.6c}$$

$$L_2 = -L_1 T\Phi \tag{3.6d}$$

$$0_{d,(n-p+q)} = [TA - FT]\Omega - TE_a\Gamma \tag{3.6e}$$

$$0_{d,(n-p+q)} = L_1 T\Omega \tag{3.6f}$$

$$H_{ux}^i = TA_{ux}^i\Phi; i = 1, \cdots, m \tag{3.6g}$$

$$0_{d,(n-p+q)} = TA_{ux}^i\Omega; i = 1, \cdots, m \tag{3.6h}$$

$$2H_{xx}^i = TB_{xx}^i\Phi; i = 1, \cdots, p \tag{3.6i}$$

$$0_{d,(n-p+q)} = T\left(B_{xx}^i\Omega - E^i\Gamma\right); i = 1, \cdots, p \tag{3.6j}$$

$$0_{d,(n-p)} = TB_{xx}^{i+p}\Psi; i = 1, \cdots, n - p \tag{3.6k}$$

$$0_{d,v} = TK^{i+p}; i = 1, \cdots, n - p \tag{3.6l}$$

$$0_{d,q} = TE^{i+p}; i = 1, \cdots, n - p \tag{3.6m}$$

$$0_{d,q} = T\left[K^1, \cdots, K^p\right][I_p \otimes l_v(i)] E_s; i = 1, \cdots, v \tag{3.6n}$$

$$0_{d,q} = T\left[E^1, \cdots, E^p\right] D_{E_s}^i; i = 1, \cdots, q \tag{3.6o}$$

$$6H_{xxx}^{ij} = TB_{xxx}^{ij}\Phi; i, j = 1, \cdots, p \tag{3.6p}$$

$$0_{d,(n-p+q)} = T\left(B_{xxx}^{ij}\Omega - [E^{ij} + E^{ji}]\Gamma\right); i, j = 1, \cdots, p \tag{3.6q}$$

$$0_{d,n} = TB_{xxx}^{i+pj+p}; i, j = 1, \cdots, n - p \tag{3.6r}$$

$$0_{d,q} = T[E^{ij+p} + E^{j+pi}]; i = 1, \cdots, n; j = 1, \cdots, n - p \tag{3.6s}$$

$$0_{d,q} = T[K^{i1}, \cdots, K^{ip}][I_p \otimes l_v(j)] E_s; i, j = 1, \cdots, v \tag{3.6t}$$

$$0_{d,v} = T[K^{ij+p} + K^{j+pi}]; i = 1, \cdots, n; j = 1, \cdots, n - p \tag{3.6u}$$

$$0_{d,q} = T[E^{i1} + E^{1i}, \cdots, E^{ip} + E^{1p}] D_{E_s}^j; i, j = 1, \cdots, q, \tag{3.6v}$$

where $l_n(i) = [0, \cdots, 0, 1, 0, \cdots, 0]' \in \mathbb{R}^n$ is the i-th unit vector, $\otimes$ is the Kronecker product and

$$B_{xx}^i = A_{xx}^i + \left[A_{xx}^1, \cdots, A_{xx}^n\right][I_n \otimes l_n(i)],$$

$$B_{xxx}^{ij} = \left(A_{xxx}^{ij} + A_{xxx}^{ji} + \left[\left[A_{xxx}^{1j} + A_{xxx}^{j1}\right], \cdots, \left[A_{xxx}^{nj} + A_{xxx}^{jn}\right]\right][I_n \otimes l_n(i)]\right.$$

$$\left. + \left[\left[A_{xxx}^{1i} + A_{xxx}^{i1}\right], \cdots, \left[A_{xxx}^{ni} + A_{xxx}^{in}\right]\right][I_n \otimes l_n(j)]\right),$$

$$D_{E_s}^i = [I_p \otimes l_q(i)] E_s + [[I_p \otimes l_q(1)], \cdots, [I_p \otimes l_q(q)]][I_q \otimes E_s l_q(i)] \tag{3.7}$$

and where

$$\Phi = \begin{bmatrix} I_p \\ 0_{(n-p) \times p} \end{bmatrix}, \quad \Psi = \begin{bmatrix} 0_{p \times (n-p)} \\ I_{n-p} \end{bmatrix}, \quad \Omega = \begin{bmatrix} E_s & 0_{p \times (n-p)} \\ 0_{(n-p) \times q} & I_{n-p} \end{bmatrix},$$

$$\Gamma = \begin{bmatrix} I_q & 0_{q,(n-p)} \end{bmatrix}, \tag{3.8}$$

then $e(t)$ and $\epsilon(t)$ are implicitly decoupled from $d(t)$ and satisfy, respectively,

$$\dot{e}(t) = Fe(t) + W(t)f(t), \tag{3.9}$$

where $W(t)$ is of the form

$$W(t) = LI^* + BI^*(t) + QU^*(t) + CU^*(t),$$

where

$$LI^* = [HK_s - TK_a], \tag{3.10a}$$

$$BI^*(t) = \sum_{i=1}^{m} u^i(t) H_{ux}^i K_s, \tag{3.10b}$$

$$QU^*(t) = \sum_{i=1}^{p} \left(y^i(t) - l_p(i)' K_s f(t) \right) \left[2H_{xx}^i K_s - TK^i \right]$$

$$+ \sum_{i=1}^{p} l_p(i)' K_s f(t) H_{xx}^i K_s, \tag{3.10c}$$

$$CU^*(t) = \sum_{i=1}^{p} \sum_{j=1}^{p} l(i)' K_s f(t) l(j)' K_s f(t) \left[H_{xxx}^{ij} K_s - TK^{ij} \right]$$

$$+ \sum_{i=1}^{p} \sum_{j=1}^{p} y^i(t) l(j)' K_s f(t) \left[T[K^{ij} + K^{ji}] - 3H_{xxx}^{ij} K_s \right]$$

$$+ \sum_{i=1}^{p} \sum_{j=1}^{p} y^i(t) y^j(t) \left[3H_{xxx}^{ij} K_s - TK^{ij} \right] \tag{3.10d}$$

and

$$\epsilon(t) = L_1 \left[e(t) - T\Phi K_s f(t) \right]. \tag{3.11}$$

Proof. Firstly, consider the fault detection signal, $\epsilon(t)$, given by (3.2). Using the partition $T = [T_1, T_2]$, where $T_1 \in \mathbb{R}^{d \times p}$ and $T_2 \in \mathbb{R}^{d \times (n-p)}$, (2.1b) and (3.3) the signal $\epsilon(t)$ can be expressed as

$$\epsilon(t) = L_1 e(t) + L_2 K_s f(t) + L_2 E_s d(t)$$
$$+ [L_1 T_1 + L_2] x_1(t) + L_1 T_2 x_2(t).$$

Now, consider $LI(t)$ given in (3.5a). Using the partitions given in (2.1b) and (2.2), (3.5a) expands to

$$LI(t) = Fe(t) + [FT_1 - TA_1 + H] x_1(t) + [FT_2 - TA_2] x_2(t)$$
$$+ [J - TB] u(t) + [HK_s - TK] f(t) + [HE_s - TE_a] d(t).$$

The sufficient conditions for $LI(t)$ and $\epsilon(t)$ to be independent of $x(t)$, $d(t)$ and $u(t)$ are then given in (3.6b)-(3.6f). When these conditions hold true $LI(t)$ and $\epsilon(t)$ become respectively

$$LI(t) = Fe(t) + LI^* f(t),$$

where LI^* is given in (3.10a) and

$$\epsilon(t) = L_1 e(t) + L_2 K_s f(t).$$

Now consider the bilinear terms in (3.5b). Using (2.1b), (3.3) and the partition for A^i_{ux}, in (2.2), there obtains

$$BI(t) = \sum_{i=1}^{m} u^i(t) \left[H^i_{ux} - TA^i_{1ux} \right] x_1(t) + \sum_{i=1}^{m} u^i(t) H^i_{ux} E_s d(t)$$
$$- T \sum_{i=1}^{m} u^i(t) A^i_{2ux} x_2(t) + BI^*(t) f(t),$$

where $BI^*(t)$ is given in (3.10b). The sufficient conditions for $BI^*(t)$ to be independent of $x(t)$ and $d(t)$ are given in (3.6g)-(3.6h). When these conditions hold true $BI(t)$ reduces to $BI^*(t)f(t)$.

To obtain the most general conditions for the observer to exist, given in Proposition 1.1, the non-unique structure of the polynomial forms used in (2.1a) and (3.1) must be considered. It is assumed, without loss, that H^i_{xx} and H^{ij}_{xxx} have unique forms which satisfy the conditions

$$H^i_{xx} = [H^1_{xx}, \cdots, H^p_{xx}][I \otimes l_p(i)]; i = 1, \cdots, p,$$
$$H^{ij}_{xxx} = H^{ji}_{xxx} = [H^{1i}_{xxx}, \cdots, H^{pi}_{xxx}][I \otimes l_p(j)]$$
$$= [H^{1j}_{xxx}, \cdots, H^{pj}_{xxx}][I \otimes l_p(i)] = [H^{i1}_{xxx}, \cdots, H^{ip}_{xxx}][I \otimes l_p(j)]$$
$$= [H^{j1}_{xxx}, \cdots, H^{jp}_{xxx}][I \otimes l_p(i)]; i, j = 1, \cdots, p.$$

Next, consider (3.5c) which can be expanded partially in terms of $x_1(t)$, $x_2(t)$, $d(t)$ and $f(t)$ using (2.1b), (2.2) and

$$\sum_{i=1}^{n} x^i(t) A^i_{xx} \equiv \sum_{i=1}^{p} x_1^i(t) A^i_{xx} + \sum_{i=1}^{n-p} x_2^i(t) A^{i+p}_{xx},$$

$$\sum_{i=1}^{n-p} x_2^i(t) A^{i+p}_{1xx} x_1(t) \equiv \sum_{i=1}^{p} x_1^i(t) [A^{1+p}_{1xx}, \cdots, A^n_{1xx}][I \otimes l_p(i)] x_2(t),$$

where $x_1(t)$ and $x_2(t)$ are independent. Hence, (3.5c) can be written as

$$QU(t) = QU_x(t) + QU_{xd}(t) + QU_{xdf}(t) + QU^*(t)f(t),$$

where $QU^*(t)$ is given in (3.10c) and where

$$QU_x(t) = \sum_{i=1}^{p} x_1^i(t)[H_{xx}^i - TA_{1xx}^i]x_1(t) - T\sum_{i=1}^{n-p} x_2^i(t)A_{2xx}^{i+p}x_2(t)$$

$$- T\sum_{i=1}^{p} x_1^i(t)[A_{2xx}^i + [A_{1xx}^{1+p}, \cdots, A_{1xx}^n][I_{n-p} \otimes l_p(i)]]x_2(t),$$

$$QU_{xd}(t) = \sum_{i=1}^{p} x_1^i(t)\left[2H_{xx}^i E_s - TE^i\right]d(t) - T\sum_{i=1}^{n-p} x_2^i(t)E^{i+p}d(t)$$

$$+ \sum_{i=1}^{p} l_p(i)'E_sd(t)H_{xx}^i E_sd(t),$$

$$QU_{xdf}(t) = T\sum_{i=1}^{p} l_p(i)'E_sd(t)K^i f(t) - T\sum_{i=1}^{n-p} x_2^i(t)K^{i+p}f(t).$$

Sufficient conditions for $QU(t)$ to be independent of $x(t)$ and $d(t)$ are thus

$$2H_{xx}^i = TB_{xx}^i\Phi; i = 1, \cdots, p \tag{3.12a}$$

$$0_{d,(n-p)} = TB_{xx}^{i+p}\Psi; i = 1, \cdots, n-p \tag{3.12b}$$

$$0_{d,(n-p+q)} = TB_{xx}^i\Psi; i = 1, \cdots, p \tag{3.12c}$$

$$0_{d,v} = TK^{i+p}; i = 1, \cdots, n-p \tag{3.12d}$$

$$2H_{xx}^i E_s = TE^i; i = 1, \cdots, p \tag{3.12e}$$

$$0_{d,q} = TE^{i+p}; i = 1, \cdots, n-p \tag{3.12f}$$

$$0_{d,q} = T\left[K^1, \cdots, K^p\right][I_p \otimes l_v(i)] E_s; i = 1, \cdots, v \tag{3.12g}$$

$$0_{d,q} = \left[H_{xx}^1, \cdots, H_{xx}^p\right][I_p \otimes E_s] D_{E_s}^i; i = 1, \cdots, q, \tag{3.12h}$$

where Φ and Ψ are given in (3.8) and $D_{E_s}^i$ is given in (3.7). In obtaining the conditions (3.12a), (3.12b) and (3.12h) the following result is used:

$$0 \equiv \sum_{i=1}^{n} x^i(t)A^i x(t), \quad \text{iff} \quad 0 = A^i + [A^1, \cdots, A^n][I_n \otimes l_n(i)]. \tag{3.13}$$

Using (3.12a), (3.12e) can be combined with (3.12c) to give (3.6j). Substituting (3.12e) into (3.12h) gives (3.6o). Hence, sufficient conditions (3.12a)-(3.12h) reduce to (3.6i)-(3.6o) which, if true, imply that $QU(t)$ reduces to $QU^*(t)f(t)$.

Next, consider the expansion of (3.5d) in terms of $x_1(t)$, $d(t)$ and $f(t)$ by using (2.1b), (2.2) and the equivalences

$$\sum_{i=1}^{p}\sum_{j=1}^{n-p} x_1^i(t)x_2^j(t)A_{1xxx}^{ij+p}x_1(t)$$

$$\equiv \sum_{i=1}^{p}\sum_{j=1}^{p} x_1^i(t)x_1^j(t)[A_{1xxx}^{il+p}, \cdots, A_{1xxx}^{in}][I \otimes l_p(j)]x_2(t),$$

$$\sum_{i=1}^{p}\sum_{j=1}^{p} x_1^i(t)l_p(j)'E_sd(t)H_{xxx}^{ij}E_sd(t)$$

$$\equiv \sum_{i=1}^{p}\sum_{j=1}^{q} x_1^i(t)d^j(t)\left[H_{xxx}^{i1}, \cdots, H_{xxx}^{ip}\right][I \otimes E_s l_q(j)]d(t),$$

$$\sum_{i=1}^{n}\sum_{j=1}^{n} x^i(t)x^j(t)A_{xxx}^{ij}$$

$$\equiv \sum_{i=1}^{p}\sum_{j=1}^{p} x_1^i(t)x_1^j(t)A_{xxx}^{ij} + \sum_{i=1}^{p}\sum_{j=1}^{n-p} x_1^i(t)x_2^j(t)A_{xxx}^{ij+p}$$

$$+ \sum_{i=1}^{n-p}\sum_{j=1}^{p} x_2^i(t)x_1^j(t)A_{xxx}^{i+pj} + \sum_{i=1}^{n-p}\sum_{j=1}^{n-p} x_2^i(t)x_2^j(t)A_{xxx}^{i+pj+p},$$

where $x_1(t)$ and $x_2(t)$ are independent. There obtains

$$CU(t) = CU_x(t) + CU_{xd}(t) + CU_{xdf}(t) + CU^*(t)f(t),$$

where $CU^*(t)$ is given in (3.10d) and where

$$CU_x(t) = \sum_{i=1}^{p}\sum_{j=1}^{p} x_1^i(t)x_1^j(t)[H_{xxx}^{ij} - TA_{1xxx}^{ij}]x_1(t)$$

$$- T\sum_{i=1}^{n-p}\sum_{j=1}^{n-p} x_2^i(t)x_2^j(t)A_{2xxx}^{i+pj+P}x_2(t)$$

$$- T\sum_{i=1}^{n-p}\sum_{j=1}^{n-p} x_2^i(t)x_2^j(t)F_1^{ij}x_1(t) - T\sum_{i=1}^{p}\sum_{j=1}^{p} x_1^i(t)x_1^j(t)F_2^{ij}x_2(t),$$

where

$$F_1^{ij} = \left[A_{1xxx}^{i+pj+p} + [[A_{2xxx}^{1j+p} + A_{2xxx}^{j+p1}], \cdots, [A_{2xxx}^{nj+p} + A_{2xxx}^{j+pn}]][I_n \otimes l_{n-p}(i)]\right],$$

$$F_2^{ij} = \left[A_{2xxx}^{ij} + [[A_{1xxx}^{il+p} + A_{1xxx}^{1+pi}], \cdots, [A_{1xxx}^{in} + A_{1xxx}^{ni}]][I_n \otimes l_p(j)]\right]$$

and, also, where

$$CU_{xd}(t) = \sum_{i=1}^{p}\sum_{j=1}^{p} x_1^i(t)x_1^j(t)\left[3H_{xxx}^{ij}E_s - TE^{ij}\right]d(t)$$

$$- T\sum_{i=1}^{p}\sum_{j=1}^{n-p} x_1^i(t)x_2^j(t)\left[E^{ij+p} + E^{j+pi}\right]d(t)$$

$$- T\sum_{i=1}^{n-p}\sum_{j=1}^{n-p} x_2^i(t)x_2^j(t)E^{i+pj+p}d(t)$$

$$+ \sum_{i=1}^{p}\sum_{j=1}^{p}\left[3x_1^i(t) + l(i)'E_sd(t)\right]l(j)'E_sd(t)H_{xxx}^{ij}E_sd(t),$$

$$CU_{xdf}(t) = T\sum_{i=1}^{p}\sum_{j=1}^{p}\left[2x_1^i(t) + l(i)'E_sd(t)\right]l(j)'E_sd(t)K^{ij}f(t)$$

$$- T\sum_{i=1}^{p}\sum_{j=1}^{n-p} x_1^i(t)x_2^j(t)\left[K^{ij+p} + K^{j+pi}\right]f(t)$$

$$- T\sum_{i=1}^{n-p}\sum_{j=1}^{n-p} x_2^i(t)x_2^j(t)K^{i+pj+p}f(t).$$

A set of sufficient conditions for $CU(t)$ to be independent of $x(t)$ and $d(t)$ is then

$$6H_{xxx}^{ij} = TB_{xxx}^{ij}\Phi; i,j = 1,\cdots,p \tag{3.14a}$$

$$0_{d,n} = TB_{xxx}^{i+pj+p}\Psi; i,j = 1,\cdots,n-p \tag{3.14b}$$

$$0_{d,(n-p+q)} = TB_{xxx}^{ij}\Psi; i,j = 1,\cdots,p \tag{3.14c}$$

$$0_{d,n} = TB_{xxx}^{i+pj+p}\Phi; i,j = 1,\cdots,n-p \tag{3.14d}$$

$$6H^{ij}E_s = T[E^{ij} + E^{ji}]; i,j = 1,\cdots,p \tag{3.14e}$$

$$0_{d,q} = T[E^{ij+p} + E^{j+pi}]; i = 1,\cdots,p; j = 1,\cdots,n-p \tag{3.14f}$$

$$0_{d,q} = T[E^{i+pj+p} + E^{j+pi+p}]; i,j = 1,\cdots,n-p \tag{3.14g}$$

$$0_{d,q} = T[K^{i1},\cdots,K^{ip}][I_p \otimes l_v(j)]E_s; i,j = 1,\cdots,v \tag{3.14h}$$

$$0_{d,(v)} = T[K^{ij+p} + K^{j+pi}]; i = 1,\cdots,p; j = 1,\cdots,n-p \tag{3.14i}$$

$$0_{d,(v)} = T[K^{i+pj+p} + K^{j+pi+p}]; i,j = 1,\cdots,n-p \tag{3.14j}$$

$$0_{d,q} = [H^{i1},\cdots,H^{ip}][I_p \otimes E_s]D_{E_s}^j; i,j = 1,\cdots,q, \tag{3.14k}$$

where Φ and Ψ are given in (3.8) and $D_{E_s}^i$ is given in (3.7). In obtaining (3.14c)-(3.14e), (3.14g) and (3.14j) the following result is used:

$$0 = \sum_{i=1}^{n} \sum_{j=1}^{n} x_1^i(t) x_1^j(t) A^{ij} x_2(t), \quad \text{iff} \quad 0 = A^{ij} + A^{ji},$$

where $x_1(t)$ and $x_2(t)$ are independent. Also, for deriving (3.14a) and (3.14b) the following is used

$$0 = \sum_{i=1}^{n} \sum_{j=1}^{n} x^i(t) x^j(t) A^{ij} x(t), \quad \text{iff}$$

$$0 = A^{ij} + A^{ji} + [[A^{i1} + A^{1i}], \cdots, [A^{in} + A^{ni}]][I_n \otimes l_n(j)]$$
$$+ [[A^{j1} + A^{1j}], \cdots, [A^{jn} + A^{nj}]][I_n \otimes l_n(i)] \tag{3.15}$$

and (3.14k) is obtained by using (3.13). Using (3.14a), (3.14e) can be combined with (3.14c) to give (3.6q). The conditions (3.14b) and (3.14d), (3.14f) and (3.14g), and (3.14i) and (3.14j) can be combined to give the conditions (3.6r), (3.6s) and (3.6u), respectively. Finally, using (3.14e), (3.14k) can be written as (3.6v). Thus, sufficient conditions (3.14a)-(3.14k) reduce to (3.6p)-(3.6v), which, if true, imply that $CU(t)$ reduces to $CU^*(t)f(t)$.$\square$

Remark 1.3 *If (3.6a)-(3.6v) hold, then $\dot{e}(t)$ by (3.4), is independent of $d(t)$, and $x(t)$. A subset of these conditions have been used in [10] [3] [4] for bilinear systems and in [1] for quadratic systems. The full set is compact, much more general and the non-uniqueness of polynomial forms has been addressed.*

4   General Detectability Conditions

A general set of sufficient conditions will be given for a RFDO and a SRFDO to exist for $f(t) \in C_f$, where C_f is defined as the restricted class

$$C_f = \{f(t); f(t) = \underline{\alpha} g(t)\}, \quad (t_0 \le t \le t_0 + h), \tag{4.1}$$

where $g(t) \neq 0$ is a scalar function and $\underline{\alpha} \in \mathbb{R}^v$, $\underline{\alpha} \neq 0$.

Proposition 1.2 *Assume (3.6a)-(3.6v) are satisfied. Then for class C_f $(h = \infty)$,*

1. *(3.1)-(3.2) is a RFDO if for some $f(t) \in C_f$*

   (a) *$H_1(s, \underline{\alpha}) X(s) \neq 0$ for all s*
       and

   (b) *$rank(H_1(s, \underline{\alpha}) X(s), H_2(s, \underline{\alpha})) \neq rank(H_2(s, \underline{\alpha}))$ for all s,* *where $\mathcal{L}$ is the Laplace operator with respect to time τ, $\tau = t - t_0$, and*

$$H_1(s, \underline{\alpha}) = L_1 \left[(sI - F)^{-1}[J_g, J_{gg}, J_{ggg}] - T\Phi K_s \underline{\alpha}[I, 0, 0] \right], \quad (4.2a)$$

$$X(s) = \mathcal{L}[g(t), g^2(t), g^3(t)]', \quad (t = \tau + t_0) \quad (4.2b)$$

$$H_2(s, \underline{\alpha}) = L_1(sI - F)^{-1}[J_u, J_y, J_{yy}, \bar{J}_y] \quad (4.2c)$$

and where

$$J_g = [HK_s - TK_a]\underline{\alpha}, \quad (4.3a)$$

$$J_{gg} = \sum_{i=1}^{p} l_p(i)' K_s \underline{\alpha} \left[TK^i - H^i K_s \right] \underline{\alpha}, \quad (4.3b)$$

$$J_{ggg} = \sum_{i=1}^{p} \sum_{j=1}^{p} l_p(i)' K_s \underline{\alpha} l_p(j)' K_s \underline{\alpha} \left[H^{ij} K_s - TK^{ij} \right] \underline{\alpha}, \quad (4.3c)$$

$$J_u = [H_{ux}^1, \cdots, H_{ux}^m][I_m \otimes K_s \underline{\alpha}], \quad (4.3d)$$

$$J_y = \left(2[H_{xx}^1, \cdots, H_{xx}^p][I_p \otimes K_s] \right.$$
$$\left. - T[K^1, \cdots, K^p] \right) [I_p \otimes \underline{\alpha}], \quad (4.3e)$$

$$J_{yy} = \left[3[H_{xxx}^{11}, \cdots, H_{xxx}^{1p} + H_{xxx}^{p1}, \cdots, H_{xxx}^{pp}][I_{\frac{p}{2}(p+2)} \otimes K_s] \right.$$
$$\left. - T[K^{11}, \cdots, K^{1p} + K^{p1}, \cdots, K^{pp}] \right] [I_{\frac{p}{2}(p+2)} \otimes \underline{\alpha}], \quad (4.3f)$$

$$\bar{J}_y = \sum_{j=1}^{p} l_p(j)' K_s \underline{\alpha} \left(T \left[[K^{1j} + K^{j1}], \cdots, [K^{pj} + K^{jp}] \right] \right.$$
$$\left. - 3[H_{xxx}^{1j}, \cdots, H_{xxx}^{pj}][I_p \otimes K_s] \right) [I_p \otimes \underline{\alpha}]. \quad (4.3g)$$

2. A SRFDO exists if (1a) and (1b) hold for any $f(t) \in C_f$.

Proof. Considering the class C_f and taking the Laplace transform of the residual in (3.11) there obtains $(t = \tau + t_0)$

$$\bar{\epsilon}(s) = L_1 \left[\bar{e}(s) - T\Phi K_s \underline{\alpha} g(s) \right], \quad (4.4)$$

where $\bar{\epsilon}(s) = \mathcal{L}\epsilon(\tau)$ and $\bar{e}(s) = \mathcal{L}e(\tau)$. From (3.9),

$$\bar{e}(s) = (sI - F)^{-1}\mathcal{L}W(t)\underline{\alpha} g(t).$$

Expanding the summations in $W(t)$, (4.4) can be written in the matix form

$$\bar{\epsilon}(s) = H_1(s, \underline{\alpha})X(s) + H_2(s, \underline{\alpha})Y(s),$$

where $H_1(s, \underline{\alpha})$, $X(s)$ and $H_2(s, \underline{\alpha})$ are given in (4.2a)-(4.2c) and where

$$Y(s) = \mathcal{L}[u'(t)g(t), y'(t)g(t), yy(t)g(t), y'(t)g^2(t)]'$$

and where

$$yy(t) = \left[(y^1(t))^2, \cdots, y^1(t)y^p(t), (y^2(t))^2, \cdots, y^2(t)y^p(t), \cdots, (y^p(t))^2 \right].$$

□

Remark 1.4 *Condition Proposition 1.2(1a) is only testable if $g(t)$ is known which is usually not the case (only $\underline{\alpha}$ is known).*

5   Testable Detectability Conditions

A set of testable (numerically tractable) sufficient conditions will be given in this section for C_f defined in (4.1) for fixed $h \geq 0$.

Proposition 1.3 *Assume (3.6a)-(3.6v) are satisfied. Then*

1. *system (3.1)-(3.2) is a RFDO if there exists at least one $f(t) \in C_f$ such that*

   (a) $L_1 T\Phi K_s\underline{\alpha} \neq 0$

   or

   (b)   i. $J_1^* G^* \neq 0$

      and

      ii. $Rank(J_1^* G^*, J_2^*) \neq Rank(J_2^*)$,

      where $J_1^* \in \mathbb{R}^{d_o \times (3d+1)}$ and $J_2^* \in \mathbb{R}^{d_o \times N}$ are constant and $G^* \in \mathbb{R}^{3d+1}$ depends only on $g(t)$, where

$$J_1^* = L_1 \left[-T\Phi K_s\underline{\alpha}, F^* \left([I_d \otimes J_g], [I_d \otimes J_{gg}], [I_d \otimes J_{ggg}] \right) \right], \quad (5.1a)$$

$$G^* = [g(t_0 + h), G_g, G_{gg}, G_{ggg}]', \quad (5.1b)$$

$$J_2^* = L_1 F^* \left[[I_d \otimes J_u], [I_d \otimes J_y], [I_d \otimes \bar{J}_y], [I_d \otimes J_{yy}] \right], \quad (5.1c)$$

$$N = d \left(m + \frac{p}{2}(p+5) \right), \quad (5.1d)$$

and where J_g, J_{gg}, J_{ggg}, J_u, J_y, $\bar{J}_y$ and J_{yy} are given in (4.3a)-(4.3g) and

$$F^* = e^{Fh} \left[F^0, \cdots, F^{d-1} \right], \quad (5.2a)$$

$$G_g = \left[G_g^0, \cdots, G_g^{d-1} \right]', \quad (5.2b)$$

$$G_{gg} = \left[G_{gg}^0, \cdots, G_{gg}^{d-1} \right]', \quad (5.2c)$$

$$G_{ggg} = \left[G_{ggg}^0, \cdots, G_{ggg}^{d-1} \right]', \quad (5.2d)$$

and where

$$G_g^k = \int_0^h a_k(\tau)g(\tau + t_0)d\tau, \qquad (5.3a)$$

$$G_{gg}^k = \int_0^h a_k(\tau)g^2(\tau + t_0)d\tau, \qquad (5.3b)$$

$$G_{ggg}^k = \int_0^h a_k(\tau)g^3(\tau + t_0)d\tau. \qquad (5.3c)$$

2. *system (3.1)-(3.2) is a SRFDO for the class C_f if (1a) or (1b) holds true for any $f(t) \in C_f$.*

Proof. By assumption, for $f(t) \in C_f$, the residual from (3.11), at $t = t_0 + h$, is

$$\epsilon(t_0 + h) = L_1 \left[e(t_0 + h) - T\Phi K_s \underline{\alpha} g(t_0 + h) \right]. \qquad (5.4)$$

Solving (3.9) with $e(t_0) = 0$ and using the Cayley-Hamilton theorem [6], $a_k(\tau)$ $(k = 1, \cdots, d-1)$ exist such that

$$e^{-F\tau} = \sum_{k=0}^{d-1} a_k(\tau)F^k,$$

and then $e(t_0 + h)$ can be written as

$$e(t_0 + h) = e^{Fh} \int_0^h \sum_{k=0}^{d-1} a_k(\tau)F^k W(\tau + t_0)\underline{\alpha}g(\tau + t_0)d\tau. \qquad (5.5)$$

The error, (5.5), can be written as

$$\begin{aligned}
e(t_0 + h) = \sum_{k=0}^{d-1} e^{Fh}F^k \Bigg(& J_g G_g^k + J_{gg}G_{gg}^k + J_{ggg}G_{ggg}^k \\
& + \sum_{i=1}^{m} H_{ux}^i K_s \underline{\alpha} U_g^{ik} + \sum_{i=1}^{p} \left[2H_{xx}^i K_s - TK^i \right] \underline{\alpha} Y_g^{ik} \\
& + \sum_{i=1}^{p}\sum_{j=1}^{p} \left[3H_{xxx}^{ij}K_s - TK^{ij} \right] \underline{\alpha} Y_g^{ijk} \\
& + \sum_{i=1}^{p}\sum_{j=1}^{p} l(j)' K_s \underline{\alpha} \left(T[K^{ij} + K^{ji}] - 3H_{xxx}^{ij}K_s \right) \underline{\alpha} Y_{gg}^{ik} \Bigg), (5.6)
\end{aligned}$$

where G_g^k, G_{gg}^k and G_{ggg}^k are given in (5.3a)-(5.3c), J_g, J_{gg} and J_{ggg} are given in (4.3a)-(4.3c), and where

$$U_g^{ik} = \int_0^h a_k(\tau)u^i(\tau + t_0)g(\tau + t_0)d\tau, \tag{5.7a}$$

$$Y_g^{ik} = \int_0^h a_k(\tau)y^i(\tau + t_0)g(\tau + t_0)d\tau, \tag{5.7b}$$

$$Y_g^{ijk} = \int_0^h a_k(\tau)y^i(\tau + t_0)y^j(\tau + t_0)g(\tau + t_0)d\tau, \tag{5.7c}$$

$$Y_{gg}^{ik} = \int_0^h a_k(\tau)y^i(\tau + t_0)g^2(\tau + t_0)d\tau. \tag{5.7d}$$

The summation signs can be eliminated in (5.6) giving

$$\epsilon(t_0 + h) = J_1^* G^* + J_2^* Y^*, \tag{5.8}$$

where J_1^*, G^* and J_2^* are given in (5.1a)-(5.1c) and $Y^* \in \mathbb{R}^N$ depends upon $\{u(t), y(t), g(t)\}$, where N is given in (5.1d) and where

$$Y^* = \left[U_g, Y_g, Y_{gg}, \bar{Y}_g\right]' \tag{5.9}$$

and where

$$U_g = \left[U_g^{10}, \cdots, U_g^{m0}, \cdots, U_g^{1(d-1)}, \cdots, U_g^{m(d-1)}\right]', \tag{5.10a}$$

$$Y_g = \left[Y_g^{10}, \cdots, Y_g^{p0}, \cdots, Y_g^{1(d-1)}, \cdots, Y_g^{p(d-1)}\right]', \tag{5.10b}$$

$$Y_{gg} = \left[Y_{gg}^{10}, \cdots, Y_{gg}^{p0}, \cdots, Y_{gg}^{1(d-1)}, \cdots, Y_{gg}^{p(d-1)}\right]', \tag{5.10c}$$

$$\bar{Y}_g = \left[Y_g^{110}, \cdots, Y_g^{1p0} + Y_g^{p10}, \cdots, Y_g^{pp0}, \cdots, Y_g^{11(d-1)}, \cdots, \right.$$
$$\left. Y_g^{1p(d-1)} + Y_g^{p1(d-1)}, \cdots, Y_g^{pp(d-1)}\right]', \tag{5.10d}$$

where U_g^{ik}, Y_g^{ik}, Y_g^{ijk} and Y_{gg}^{ik} are given in (5.7a)-(5.7d) above. □

Corollary 1.1 *Let the assumptions of Proposition 1.3 hold so that (5.8) holds true.*

1. *If $J_1^* \equiv 0$, for all $\underline{\alpha}$, then system (3.1)-(3.2) is not a RFDO for the class C_f.*

2. $J_1^* \equiv 0$ for any $\underline{\alpha} \neq 0$ iff all the following hold true:

$$0 = L_1 T \Phi K_s, \tag{5.11a}$$

$$0 = L_1^* [H K_s - T K_a], \tag{5.11b}$$

$$0 = L_1^* \left(T[K^1, \cdots, K^p] - [H_{xx}^1, \cdots, H_{xx}^p][I_p \otimes K_s] \right) D_{K_s}^i, \tag{5.11c}$$

$$0 = L_1^* \hat{K} \left(\left[I_{p^2} \otimes l_v(i) \right] \left[I_p \otimes K_s \right] D_{K_s}^j \right.$$
$$+ \left[I_{p^2} \otimes l_v(j) \right] \left[I_p \otimes K_s \right] D_{K_s}^i$$
$$+ \left. \left[\left[I_{p^2} \otimes l_v(1) \right], \cdots, \left[I_{p^2} \otimes l_v(v) \right] \right] \left[G_{K_s}^{ij} + G_{K_s}^{ji} \right] \right), \tag{5.11d}$$

where (5.11a)-(5.11d) hold for $k = 1, \cdots, d - 1$, where

$$L_1^* = L_1 e^{Fh} F^k, \tag{5.12}$$

$$\hat{K} = \left[H_{xxx}^{11}, \cdots, H_{xxx}^{1p}, \cdots, H_{xxx}^{p1}, \cdots, H_{xxx}^{pp} \right] \left[I_{p^2} \otimes K_s \right]$$
$$- T \left[K^{11}, \cdots, K^{1p}, \cdots, K^{p1}, \cdots, K^{pp} \right]$$

and where

$$D_{K_s}^i = [I_p \otimes l_v(i)] K_s,$$
$$+ [[I_p \otimes l_v(1)], \cdots, [I_p \otimes l_v(v)]] [I \otimes K_s l_v(i)], \tag{5.13a}$$

$$G_{K_s}^{ij} = [I_{vp} \otimes K_s l_v(i)] [I_v \otimes K_s l_v(j)], \tag{5.13b}$$

where $i, j = 1, \cdots, v$.

Proof. From (5.8) if $Y^* = 0$ then $\epsilon(t_0 + h) = J_1^* G^*$ and hence, Corollary 1.1(1) holds true. Now, also, $J_1^* \equiv 0$ for any $\underline{\alpha} \neq 0$ iff

$$0 \equiv L_1 T \Phi K_s \underline{\alpha},$$

$$0 \equiv L_1^* [J_g, J_{gg}, J_{ggg}],$$

for $k = 1, \cdots, d - 1$, where L_1^*, J_g, J_{gg} and J_{ggg} are given in (5.12), (4.3a)-(4.3c). Using the equivalences given in (3.13) and (3.15) these equations can be written as (5.11a)-(5.11d) and hence, Corollary 1.1(2) holds true. □

5.1   A Special Class (Step-Faults)

Consider the class of faults C_f described in (4.1) where $g(t) = 1$ ($t_0 \leq t \leq t_0 + h$). Thus, step-type faults are considered along direction $\underline{\alpha}$.

Proposition 1.4 *Assume (3.6a)-(3.6v) are satisfied. When $g(t) = 1$ Proposition 1.3 can be simplified to*

1. *system (3.1)-(3.2) is a RFDO if there exists at least one $f(t) \in C_f$ such that*

(a) $L_1 T \Phi K_s \underline{\alpha} \neq 0$
   or

(b)   i. $\hat{J}_1^* \neq 0$
      and

   ii. $Rank(\hat{J}_1^*, \hat{J}_2^*) \neq Rank(\hat{J}_2^*),$
   where $\hat{J}_1^* \in \mathbb{R}^{d_o}$ and $\hat{J}_2^* \in \mathbb{R}^{d_o \times M}$, where

$$\hat{J}_1^* = \hat{L}_1^* [J_g + J_{gg} + J_{ggg}] - L_1 T \Phi K_s \underline{\alpha}, \tag{5.14a}$$

$$\hat{J}_2^* = L_1 F^* \left[[I_d \otimes J_u], [I_d \otimes [J_y + \bar{J}_y]], [I_d \otimes J_{yy}] \right], \tag{5.14b}$$

$$M = d \left(m + \frac{p}{2}(p+3) \right), \tag{5.14c}$$

and where

$$\hat{L}_1^* = L_1 F^{-1} \left[e^{Fh} - I_d \right] \tag{5.15}$$

and J_g, J_{gg}, J_{ggg}, J_u, J_y, $\bar{J}_y$ and J_{yy} are given in (4.3a)-(4.3g).

2. system (3.1)-(3.2) is a SRFDO for the class C_f, if for any $f(t) \in C_f$, (1a) or (1b) holds true.

Proof. From proof of Proposition 1.3, when $g(t) = 1$, the residual in (5.4) can be reduced to

$$\epsilon(t_0 + h) = L_1 \left[e(t_0 + h) - T \Phi K_s \underline{\alpha} \right], \tag{5.16}$$

where

$$e(t_0 + h) = e^{Fh} \int_0^h \sum_{k=0}^{d-1} a_k(\tau) F^k W (\tau + t_0) \underline{\alpha} d\tau. \tag{5.17}$$

From Proposition 1.3, when $g(t) = 1$, there follows

$$G_g^k = G_{gg}^k = G_{ggg}^k = \int_0^h a_k(\tau) d\tau$$

and this leads to

$$\sum_{k=0}^{d-1} \int_0^h a_k(\tau) F^k d\tau = \int_0^h e^{-F\tau} d\tau$$

$$= F^{-1} \left[I - e^{-Fh} \right]. \tag{5.18}$$

Also, note that when $g(t) = 1$

$$Y^{ik} = Y_g^{ik} = Y_{gg}^{ik}. \tag{5.19}$$

Using the equivalences in (5.18) and (5.19), (5.17) becomes

$$e(t_0 + h) = F^{-1} \left[e^{Fh} - I_d \right] \left[J_g + J_{gg} + J_{ggg} \right]$$

$$+ \sum_{k=0}^{d-1} e^{Fh} F^k \left(\sum_{i=1}^{m} H_{ux}^i K_s \underline{\alpha} U^{ik} + \sum_{i=1}^{p} \left[2H_{xx}^i K_s - TK^i \right. \right.$$

$$+ \sum_{j=1}^{p} l(j)' K_s \underline{\alpha} \left(T[K^{ij} + K^{ji}] - 3H_{xxx}^{ij} K_s \right) \Big] \underline{\alpha} Y^{ik}$$

$$+ \sum_{i=1}^{p} \sum_{j=1}^{p} \left[3H_{xxx}^{ij} K_s - TK^{ij} \right] \underline{\alpha} Y^{ijk} \Bigg) ,$$

where

$$U^{ik} = \int_0^h a_k(\tau) u^i(\tau + t_0) d\tau, \qquad (5.20a)$$

$$Y^{ik} = \int_0^h a_k(\tau) y^i(\tau + t_0) d\tau, \qquad (5.20b)$$

$$Y^{ijk} = \int_0^h a_k(\tau) y^i(\tau + t_0) y^j(\tau + t_0) d\tau \qquad (5.20c)$$

and J_g, J_{gg} and J_{ggg} are given in (4.3a)-(4.3c). These results can then be used to write (5.16) in the form

$$\epsilon(t_0 + h) = \hat{J}_1^* + \hat{J}_2^* \hat{Y}^*, \qquad (5.21)$$

where $\hat{J}_1^*$ and $\hat{J}_2^*$ are given in (5.1a) and (5.1c) and $\hat{Y}^* \in \mathbb{R}^M$ is defined as

$$\hat{Y}^* = [U, Y, \bar{Y}]' ,$$

where

$$U = \left[U^{10}, \cdots, U^{m0}, \cdots, U^{1(d-1)}, \cdots, U^{m(d-1)} \right]' , \qquad (5.22a)$$

$$Y = \left[Y^{10}, \cdots, Y^{p0}, \cdots, Y^{1(d-1)}, \cdots, Y^{p(d-1)} \right]' , \qquad (5.22b)$$

$$\bar{Y} = \left[Y^{110}, \cdots, Y^{1p0} + Y^{p10}, \cdots, Y^{pp0}, \cdots, \right.$$

$$\left. Y^{11(d-1)}, \cdots, Y^{1p(d-1)} + Y^{p1(d-1)}, \cdots, Y^{pp(d-1)} \right]' , \qquad (5.22c)$$

where U^{ik}, Y^{ik} and Y^{ijk} are given in (5.20a)-(5.20c) above. $\square$

Corollary 1.2 *Let the assumptions of Proposition 1.4 hold so that (5.21) holds true.*

1. If $\hat{J}_1^* \equiv 0$, all $\underline{\alpha}$, then system (3.1)-(3.2) is not a RFDO for class, C_f of faults.

2. $\hat{J}_1^* \equiv 0$ for all $\underline{\alpha} \neq 0$ iff (5.11c), (5.11d) hold true, with L_1^* replaced by $\hat{L}_1^*$ in both, and

$$0 = \hat{L}_1^* \left[HK_s - TK_a \right] - L_1 T\Phi K_s \tag{5.23}$$

holds true, where $\hat{L}_1^*$ is given in (5.15).

Proof. Corollary 1.2(1) holds by letting $Y^* = 0$ in (5.21), for which $\epsilon(t_0 + h) = \hat{J}_1^*$. If $\hat{J}_1^* \equiv 0$ for all $\underline{\alpha} \neq 0$ iff

$$0 \equiv \hat{L}_1^* J_g - L_1 T\Phi K_s \underline{\alpha},$$

$$0 \equiv \hat{L}_1^* \left[J_{gg}, J_{ggg} \right],$$

where J_g, J_{gg} and J_{ggg} are given in (4.3a)-(4.3c). >From the proof of Corollary 1.1 these equations can be written as (5.11c) and (5.11d) (replacing L_1^* by $\hat{L}_1^*$ in both) and (5.23), thus proving Corollary 1.2(2). $\square$

5.2   Numerical Calculation Procedure

The gain matrices in the design of (3.6a)-(3.6v) can be calculated efficiently. The equations given in (3.6h), (3.6j)-(3.6o) and (3.6q)-(3.6v) can be arranged to give the form

$$0_{d,\mathcal{N}} = TZ, \tag{5.24}$$

where the order of the contribution to the equation in (5.24) is not important and where

$$\mathcal{N} = \frac{n-p}{2}(2m + 3v + 3q + (p+n)(q+v+1) + p(p-n) + n(2+n))$$

$$+ \frac{q}{2}(2m + p(3+p) + 2v(v+1) + 2q(q+1)).$$

Equation (3.6e) and (5.24) are combined to give the form

$$0_{d,(5q+7n-6p)} = \left[[TA - FT]\Omega - TE_a \left[I_q, 0_{q,(n-p)} \right], TZ \right]. \tag{5.25}$$

By splitting the two terms on the right hand side of (5.25) there follows two equations in F and T which are equivalent to (5.25)

$$0_{d,(5q+7n-6p)} = FTX_1 + TX_2 \text{ or}$$

$$0_{d,(5q+7n-6p)} = [FT, T]X,$$

where

$$X_1 = \left[\Omega, 0_{n,(4q+6n-5p)}\right],$$
$$X_2 = \left[E_a\left[I_q, 0_{q,(n-p)}\right] - A\Omega, Z\right],$$
$$X = [X_1, X_2]'.$$

Using the algorithms developed in [10] [1], F, T, L_1 and L_2 (and the other gains) can now be calculated using SVD decompositions. Then conditions (1a) and (1b) of Proposition 1.4 can be tested (or similar conditions in Proposition 1.3 if $g(t)$ is known). A recursive algorithm for the complete design of a RFDO (or SRFDO) can be given along the lines developed in [10] [1].

6 Concluding Remarks

A nonlinear fault detection observer has been proposed in this chapter for a nonlinear system involving polynomial nonlinearities of bilinear, quadratic and cubic forms. Proposition 1.1 gives sufficient conditions for the error dynamics and fault detection signal to be robust with respect to a disturbance and Propositions 1.2-1.4 give conditions for a fault to be detectable (RFDO and SRFDO). Fault isolation can be performed by using a bank of RFDO's [7]. The design procedure here involves only efficient linear matrix calculations, thus ensuring easy assessment of fault detectability.

7 REFERENCES

[1] S. A. Ashton, D. N. Shields and S. Daley. Application of a Fault Detection Method for Pipelines, *System Science*, Vol. 23, No. 2, pp. 97-109, 1997.

[2] P. M. Frank. On-line Fault Detection in Uncertain Nonlinear Systems Using Diagnostic Observers : A Survey, *Int.J.Systems Sci*, Vol. 25, No 12, pp. 2129-2154, 1994.

[3] A. Hac, Design of Disturbance Decoupled Observer for Bilinear Systems, *ASME, J. Dynamic Syst. Measure. Control*, Vol. 114, N0. 12, pp. 556-562, 1992.

[4] M. Kinnaert, Y. Peng and H. Hammouri. The Fundamental Problem of Residual Generation for Bilinear Systems up to Input Injection, *Proc. IFAC conf. ECC'95*, Rome, Italy, pp. 3777-3782, 1995.

[5] A. J. Krener and A. Isidori. Linearization by Output Injection and Nonlinear Observers, *Systems and Control Letters*, Vol. 3, pp. 47-52, 1983.

[6] H. Nijmeijer and A. Van der Schaft. Nonlinear Dynamical Control Systems, Springer Verlag, 1990.

[7] R. Patton, P. Frank and R. Clark., Fault Diagnosis in Dynamic Systems, Theory and Applications, *Prentice Hall*, 1989.

[8] R. Seliger and P. M. Frank. Robust Component Fault Detection and Isolation in Nonlinear Dynamic Systems using Nonlinear Unknown Input Observers, *Preprints of SAFEPROCESS '91*, Sept. 10-13, Baden-Baden, FRG. Vol. 1, pp. 313-318, 1991.

[9] X. H. Xia and W. B. Gao. Nonlinear Observer Design by Observer Error Linearization, *SIAM J. of Control and Optimization*, Vol. 27, pp. 199-216, 1989.

[10] D. Yu and D. N. Shields. Bilinear Fault Detection Observer and its Application to a Hydraulic System, *Int. Jnl. of Control*, Vol. 64, No. 6, pp. 1023-1047, 1996.

[11] A. N. Zhirabok. Fault Diagnosis in Nonlinear Systems with Uncertainies, *Proc.of IFAC Symp., Safeprocess '97*, Hull University, Vol. 1, pp. 528-533, 1994.

Nonlinear Observer for Signal and Parameter Fault Detection in Ship Propulsion Control

Mogens Blanke and Roozbeh Izadi-Zamanabadi

Department of Control Engineering
Aalborg University
Fredrik Bajers vej 7C
DK-9220 Aalborg, Denmark

1 Introduction

Faults in ship propulsion and their associated automation systems can cause dramatic reduction on ships' ability to propel and maneuver, and effective means are needed to prevent that faults develop into failure. The chapter analyses the control system for a propulsion plant on a ferry. It is shown how fault detection, isolation and subsequent reconfiguration can cope with many faults that would otherwise have serious consequences. The chapter emphasize analysis of re-configuration possibilities as a necessary tool to obtain fault tolerance, showing how sensor fusion and control system reconfiguration can be systematically approached. Detector design is also treated and parameter adaptation within fault detectors is shown to be needed to locate non-additive propulsion machinery fault. An adaptive observer is suggested for this purpose. est trials with a ferry are used to validate the principles.

Propulsion system availability is crucial for a ship's ability to maneuver. Nevertheless, control systems associated with propulsion required to be fail-operational or fault-tolerant. Instead, local safety systems protect machinery. They prevent continued operation or start-up if sensors inform that local shut-down. While fail-safe for each piece of machinery, the local safety approach is not globally fail-safe for the ship. The consequence has been many events where consequences vary from irregularity to major economic loss and causalities. Several events could have been prevented if automation systems had been designed to be tolerant to faults, with overall availability in mind.

Fault-tolerant control (FTC) is a methodology where analytical redundancy is employed using software that monitors the behavior of components and function blocks. Without hardware redundancy, some faults may inevitably cause a plant shut-down, but the FTC strategy is that the majority of faults, and in particular the ones with severe consequences, are accommodated. The objective is to keep plant availability but accept reduced performance as a trade-off. The first step to achieve fault tolerance is efficient detection and isolation of faults. This is a particular challenge when a system is non-linear.

In this chapter, an active solution to the FTC problem is employed where on-line fault detection and isolation can trigger a discrete event signal to a supervisor-agent when a fault is detected. The supervisor-agent will activate remedial actions. Re-configuration possibilities are analyzed for a ship propulsion system consisting of a main engine with a controllable pitch propeller. It is shown that combined parameter and output estimation is required and an adaptive observer is proposed for fault detection. A continous-time non-linear observer is shown to possess very useful features and can be used during both detection and re-configuration. Simulations on a model of a ferry [12] illustrate performance for a selected fault scenario.

2   Ship Propulsion System

This section introduces mathematical models for ship speed, propeller and prime mover, the essential propulsion system components. The purpose of the modeling is to obtain information to design fault detection and isolation (FDI) modules for essential faults and to give the prerequisites for design of re-configuration when faults occur. The block diagram in Fig. 1 illustrates the structure of the propulsion system.

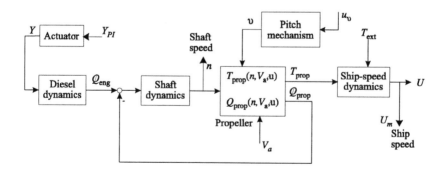

FIGURE 1. Structure of dynamic relations for CP propeller, shaft and diesel engine.

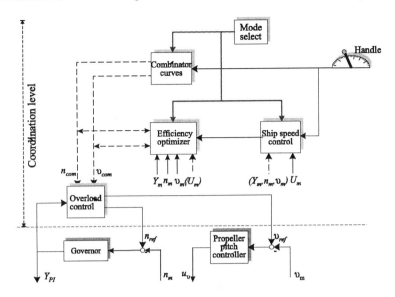

FIGURE 2. Hierarchy of controllers for propulsion system. The handle gives input to a combinator, efficiency optimizer, and ship speed control. Lower level controls are shaft speed (governor), propeller pitch and diesel overload blocks.

2.1   Propeller Thrust and Torque

Controllable pitch (CP) propellers have blade angle (pitch) controlled by a hydraulic servo system. Developed thrust and torque are functions of pitch, shaft speed and flow velocity through the propeller

$$
\begin{aligned}
T_{\text{prop}} &= f_{T\_prop}(\vartheta, n, V_a) \\
Q_{\text{prop}} &= f_{Q\_prop}(\vartheta, n, V_a)
\end{aligned}
\tag{2.1}
$$

These can be shown to approximately follow quadratic relations, for thrust

$$
T_{\text{prop}} = T_{|n|n\vartheta}\,\vartheta\,|n|\,n + T_{nv}nV_a
\tag{2.2}
$$

and for torque

$$
Q_{\text{prop}} = Q_0\,|n|\,n + Q_{nn|\vartheta|}\,|\vartheta|\,|n|\,n + Q_{nv\vartheta}\,\vartheta\,|n|\,V_a
\tag{2.3}
$$

These relations give a quite good approximation in the steady state cases whereas they are less applicable during large transients. The term $Q_0\,|n|\,n$ accounts for the torque at zero pitch.

2.2   Diesel Engine Prime Mover

Elaborate details of the dynamics [3] are not important in this context, but would be for detailed design of FDI for the engine. Here, diesel torque can be considered linearly related to the fuel index, without dynamics involved,

$$Q_{\text{eng}} = K_y Y \tag{2.4}$$

The dynamics of propeller and shaft is merely that of rotating inertias subjected to torque balance between prime mover torque and load torques,

$$\frac{d}{dt}(\frac{1}{2}I_t n^2) = n(Q_{\text{eng}} - Q_{\text{prop}} - Q_f) \tag{2.5}$$

The dynamics of the prime mover and its control system is tightly coupled to the speed dynamics of the ship through the propeller (2.3). The structure of prime mover control was also shown in Fig. 1. The measured shaft speed is compared with a reference speed and the governor (speed controller) regulates the fuel injection to the engine to obtain the desired speed. Limit curves are incorporated for shaft speed dependent torque and air pressure.

2.3   Hull Resistance

Ship's resistance to motion through the water can be described to the first order by a resistance curve, which is a third to fifth order polynomial in u. The order of the polynomial is higher the closer the ship operates into the wave making region. The resistance curve is known *a priory* but with some uncertainty. The first order equation

$$m\dot{U} = R(U) + (1 - t)T_{\text{prop}} + T_{\text{ext}}$$

is a sufficient approximation in this context.

2.4   Actuators for Fuel Injection and Propeller Pitch

The actuators can both be modeled as first order dynamic systems with limits in rate of change and in output. The electro-hydraulic pitch control system is described by the following equations:

$$\begin{aligned} u_{\dot{\vartheta}} &= k_t \left(\vartheta_{\text{ref}} - \vartheta_m \right) \\ \dot{\vartheta} &= \max(\dot{\vartheta}_{\text{min}}, \min(u_{\dot{\vartheta}}, \dot{\vartheta}_{\text{max}})) \\ \vartheta &= \max(\vartheta_{\text{min}}, \min(\vartheta, \vartheta_{\text{max}})) \end{aligned} \tag{2.6}$$

The diesel actuator is equivalent to this with command Y_c from the governor, rate limits $\dot{Y} \in [Y_{d-}, Y_{d+}]$ and output $Y \in [0, 1]$.

2.5   Sensors

Sensors for propeller pitch and fuel index are conventional angle transmitters. Shaft speed is usually measured by a set of pulse pickups. A maximum

Function blocks	Service $\begin{cases} \text{normal} \\ \text{(fault)} \end{cases}$
R_combi: $\begin{array}{c} n_d \\ \vartheta_d \end{array} = \text{RCB}(h_d)$	(n, ϑ) demand $\{$ freeze (input fault) $\}$
C_optim: $\begin{bmatrix} n_d \\ \vartheta_d \end{bmatrix} = C\_OP \begin{array}{l} h_d \\ \vartheta_m \\ n_m \\ Y_m \\ U_m \end{array}$	best efficiency $\{$ use estimate (inp.fault) $\}$ $\{$ roll-back (ref. fault) $\}$ $\{$ alter limits (diesel fault) $\}$
C_over: $\begin{bmatrix} n_r \\ \vartheta_r \end{bmatrix} = C\_OL \begin{array}{l} n_d \\ \vartheta_d \\ n_m \\ \vartheta_m \\ Y_m \end{array}$	avoid overload $\{$freeze (fault)$\}$ $\{$ use estimate (fault) $\}$
C_speed: $h_d = C\_SS \begin{bmatrix} U_r \\ U_m \\ \cdots \end{bmatrix}$	constant U $\{$ freeze $h_d(U_m$ fault) $\}$ $\{$ estimate $\text{U}_m(U_m$fault) $\}$ $\{$ roll-back$(U_r$ fault) $\}$
C_shaft: $Y_c = C\_SP \ (n_r, n_m)$	shaft control $\{$ estimate n$(n_m$fault) $\}$ $\{$ roll-back $(n_r$ fault) $\}$

TABLE 2.1. Function blocks treated as virtual components.

logic selects the higher of the two signals. This protects against drop out of one of the pick ups but not against a "high signal" fault or failure in a common processor/rate counter servicing both channels. The ship speed is measured by magnetic log, Pitot tube or Doppler log. The two former measure water speed close to the hull and are quite prone to fluctuations from the turbulence and cross flow.

3   Control Hierarchy

The control hierarchy includes controllers for: shaft speed; propeller pitch; diesel overload control; combinator curves from handle position to generate reference values of n and ϑ; efficiency optimization using n and ϑ; constant ship speed control. The signal flow between these function blocks is shown in Fig. 2. The interested reader can find details about the control functions in [12].

The input-output of each block is listed in Table 2.1. The table lists the service of the function block in normal operation and the desired function in case of specific faults. The listing of desired remedial actions is a result of a combined fault-propagation and structural analysis of the propulsion system, including the possibilities for re-configuration after serious faults

Constraint	Description
$f_1^c : n_m = n$	sensor_n
$f_2^c : \vartheta_m = \vartheta$	sensor_ϑ
$f_3^c : U_m = U$	sensor_U
$f_4^c : Y_m = Y$	sensor_Y
$f_5^c : K_y = K_{yc}$	engine gain
$f_6^c : Q_{\mathrm{eng}} = K_y Y$	engine torque
$f_7^c : -Q_{\mathrm{eng}} = Q_{\mathrm{prop}} + Q_f$	shaft balance
$f_8^c : Q_{\mathrm{prop}} = f_{Q_{\mathrm{prop}}}$	propeller torque
$f_9^c : T_{\mathrm{prop}} = f_{T_{\mathrm{prop}}}$	propeller thrust
$f_{10}^c : R(U) = f_{Ru}(U)$	hull resistance.
$f_{11}^c : R(U) = -T_{\mathrm{ext}} - (1-t)\,T_{\mathrm{prop}}$	ship speed

TABLE 2.2. Static constraints for shaft

[4]. The table list their
    input and output, faults considered, and re-configuration possibilities.
An example of this analysis is provided in the next section.

4   Structural Analysis

Structural analysis [7, 10, 17] is the study of properties which are independent of the actual values of the parameters.Constraints, here used as a
synonym for relations, between variables and parameters from the operating model are used in the analysis. The links are represented by a graph or
a table, on which the structural analysis is made.

4.1   Description of the Model

The model of the system is considered as a set of constraints, $\mathcal{F} = \{f_1^c, f_2^c, \ldots$
$, \cdots, f_m^c\}$ that are applied to a set of variables $\mathcal{Z} = X \cup \bar{X}$. X denotes the
set of unknown variables while $\bar{X}$ is the set of known variables: sensor
measurements, control variables, constants, and parameters, and reference
variables. The constraints are the relations imposed between values of the
variables, as given by the relevant physical laws. The constraints for the
propulsion system are listed in Table 2.2.

4.2   Formal Representation

The structure of the system is described by the following binary relation:

$$S : \mathcal{F} \times \mathcal{Z} \rightarrow \{0, 1\}$$

$$(f_i^c, z_j) \rightarrow \quad \begin{array}{ll} S(f_i^c, z_j) = 1 & \text{iff } f_i^c \text{ applies to } z_j, \\ S(f_i^c, z_j) = 0 & \text{otherwise.} \end{array}$$

These relations can be represented by an incidence table or the equivalent digraph. Fig. 3 a) shows the structural table for the propulsion system. Some constraints may be expressed through non-isomorphic mappings for certain variables. Such variables can not be re-constructed through an inverse mapping from knowledge about remaining variables. Elements with this property are marked by M's (for multiple), replacing the 1'es in the incidence table and unidirectional arcs in the corresponding digraph. An example of such a constraint is f_8^c: it is always possible to compute the value of Q_{prop} from f_8^c when ϑ, n, and V_a are known. However, knowing the values of Q_{prop}, n, and V_a does not enable calculation of a unique ϑ in all cases. This fact is not apparent from the equations in this chapter but is apparent when looking at the underlying propeller characteristics. The non-isomorphic problem for the Q_{prop} relation is only present in a narrow range of transient conditions (during crash stop).

4.3   Sensor Fusion for Re-configuration

In control systems, re-configuration can be obtained either by means of hardware redundancy or the use of software redundancy. In the case where hardware redundancy exists, the scope of design is FDI algorithms and hardware switching. When analytic redundancy is available, fault tolerance is obtained by means of sensor fusion: the value of the signal which is lost or corrupted due to faults, is reconstructed using known values of other signals.

The structural analysis approach is usually employed to obtain analytical redundancy relations for FDI [9]. It can, however, be used without difficulties for sensor fusion as well, since a constraint relation can be used to re-construct a signal from the other measured variables. An example for the propulsion system is a critical fault in the shaft speed measurement which can be accommodated by estimating shaft speed from other available measurements.

Fault in the Shaft Speed Measurement

A critical fault in the propulsion system is a failure in the measurement of shaft speed. The constraint f_1^c represents this device in Table 2.2. A fault occurrence means that the constraint f_1^c does not hold, e.g. the values of the variable n_m are not correctly related to the values of the variable n. Figure 3 b) shows that variable n is involved in 3 relations which are specified by the constraints f_1^c, f_8^c, and f_9^c. Since the constraint f_1^c is not valid, there are two other possible ways of calculating the values of the variable n, namely through constraints f_8^c and f_9^c. As it is shown in Figures 4 a) and 4 b), the ship speed can be described as a function of the other

	n_m	ϑ_m	U_m	Y_m	K_{yc}	n	ϑ	U	Y	K_y	Q_{eng}	Q_{prop}	T_{prop}	$R_u(U)$
f_1^c	1					1								
f_2^c		1					1							
f_3^c			1					1						
f_4^c				1					1					
f_5^c					1					1				
f_6^c								1	1	1				
f_7^c											1	1		
f_8^c						1 M	1					1		
f_9^c						1 M	1						1	
f_{10}^c							1							1
f_{11}^c													1	1

a)

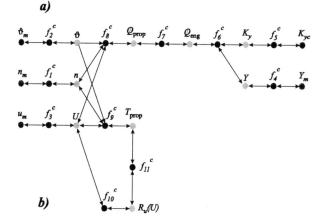

b)

FIGURE 3. a) The structural representation of the model by a (binary) table. 1's are replaced by ×'s to indicate causality (calculability) between variables. b) Corresponding digraph representation.

known variables as:

$$\hat{n}_q = f_q(\vartheta_m, Y_m, K_{yc}, U_m) \tag{4.7}$$

$$\hat{n}_t = f_t(\vartheta_m, U_m) \tag{4.8}$$

The process to apply the sensor fusion based on this approach is the following:

For the interested variable (for instance n) identify the set of related constraints (f_8^c and f_9^c) and

- choose one of the available constraints

- check the causality for the constraint in order to find out that the variable can be computed through this constraint.

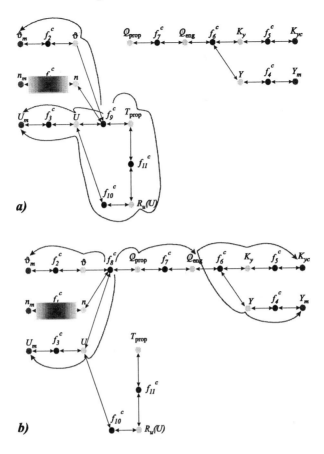

FIGURE 4. Sensor fusion methods based on structural representation: shaft speed calculation through a) propeller thrust equations f_9^c and b) propeller torque equation f_8^c.

- for all the variable connected to the chosen constraint search backward until all end variables are known variables.

Remark 2.1 *The described procedure shall find all the existing paths from the unknown variable to sets of known variables. Some of these paths may include loops, which are related to the existing control or natural loops in a system.*

By examining all the constraints, the set of equations/relation by which the variable can be calculated is identified and can be used for re-configuration purposes. For the shaft speed failure, the method is illustrated graphically in Figures 4 a) and 4 b). Grey dashed arrows show the calculation paths to the known variables.

Using quadratic representation of the propeller torque, the variable n

can be estimated from the constraint f_8^c, but estimation of n based on static relations is obviously too primitive. A non-linear observer is employed instead.

5   Isolation of Shaft Speed and Engine Faults

This section deals with the problem of detecting whether a shaft speed fault or engine fault has occurred. The relevant dynamics to be considered was described above, leading to the constraints f_7 to f_{11}. The task at hand is to estimate a signal fault in n_m and a parameter fault in K_y. The dynamic equations directly determining shaft speed are

$$I_t \dot{n} = Q_{eng} - Q_{prop} - Q_f \tag{5.9}$$
$$Q_{eng} = K_y Y_m$$

Taking ship speed U as a measured variable - a valid assumption when U_m is non-faulty,

$$Q_{prop} = Q_0 |n| n + Q_{\vartheta nn} |\vartheta| |n| n + Q_{\vartheta n V_a} \vartheta |n| (1 - w_0) U_m \tag{5.10}$$

In the sequel, we use $Q_{\vartheta nU} \equiv Q_{\vartheta n V_a}(1 - w_0)$ for brevity.

Shaft speed is positive in a controllable pitch installation, so

$$\dot{n} = \frac{1}{I_t} \left(K_y Y_m - Q_f - Q_0 n^2 - Q_{\vartheta nn} |\vartheta| n^2 - Q_{\vartheta nU} \vartheta n U_m \right) \tag{5.11}$$

Following the benchmark definition in [13], we need to consider faults in either shaft speed measurement or in the diesel torque coefficient,

$$n_m = n + n_f \quad K_y = K_{yc} - K_{yf} \tag{5.12}$$

and The detection task is hence increased from a single fault shaft speed sensor fault detection to a more complex one of simultaneous additive and non-additive faults. An adaptive observer providing simultaneous state and parameter estimates is a natural choice as a candidate for detection of the two particular faults.

5.1   Adaptive Observer

The dynamic relation (5.11) can be written in a form which is linear in the unknown parameter

$$\dot{x} = \Phi(x, u_2, u_3) + \theta u_1 \tag{5.13}$$
$$y = x$$

using

$$\Phi = \frac{1}{I_t}\left(-Q_{\vartheta nn}\,|\vartheta|\,n^2 - Q_{\vartheta nU}\vartheta nU_m - Q_0 n^2 - Q_f\right) \tag{5.14}$$

$$x = n, \ u_1 = Y_m, \ u_2 = U_m, \ u_3 = \vartheta_m, \ \theta = \frac{K_y}{I_t}$$

An adaptive observer can then be build by using the measured inputs: Y_m, U_m, ϑ_m and the measured state n_m.

It is noted that the more general case was treated in [8]. However, the detailed assessment of the Lipshitz conditions, that determine the gains in the adaptive observer, are easily made too conservative to get useful results. A few comments are thus considered appropriate.

This leads to the following theorem.

Theorem 2.1 *An adaptive observer for the problem*

$$\dot{n} = \frac{1}{I_t}\left(K_y Y_m - Q_{\vartheta nn}\,|\vartheta|\,n^2 - Q_{\vartheta nU}\vartheta nU_m - Q_0 n^2 - Q_f\right) \tag{5.15}$$

is the state estimator

$$\dot{\hat{n}} = \frac{1}{I_t}\left(-Q_{\vartheta nn}\vartheta_m\hat{n}^2 - Q_{\vartheta nU}\vartheta_m\hat{n}U_m - Q_0\hat{n}^2 - Q_f\right) + Y_m\hat{\theta} + L\left(n_m - \hat{n}\right) \tag{5.16}$$

with parameter updating

$$\dot{\hat{\theta}} = PY_m\left(n_m - \hat{n}\right) \tag{5.17}$$

The adaptive observer is semi-globally asymptotically stable with

$$Y_m > 0, \quad P > 0, \quad L > \frac{Q_{eng,max}}{I_t\,n_{max}}\left(\alpha\frac{n_{max}}{n_{min}} + \beta\right) \tag{5.18}$$

where $I_t, Q_{eng,max}, n_{max}, n_{min}, \alpha$ and β are plant specific parameters. □

The nonlinear torque function

$$\Phi(n, u) = \frac{1}{I_t}\left(-Q_{\vartheta nn}\,|\vartheta_m|\,n^2 - Q_{\vartheta nU}\,\vartheta_m n U_m - Q_0 n^2 - Q_f\right)$$

is Lipshitz

$$\|\Phi(n, u) - \Phi(\hat{n}, u)\| < \gamma\,\|n - \hat{n}\|$$

since

$$\Phi(n, u) - \Phi(\hat{n}, u) =$$

$$\frac{1}{I_t}\left(n - \hat{n}\right)\left(\left(-Q_0 - Q_{\vartheta nn}\,|\vartheta_m|\right)\left(n + \hat{n}\right) - Q_{\vartheta nU}\vartheta_m U_m\right) =$$

$$\frac{1}{I_t}\left(-\left(\left(Q_0 + Q_{\vartheta nn}\,|\vartheta_m|\right)n + Q_{\vartheta nU}\vartheta_m U_m\right) - \left(Q_0 + Q_{\vartheta nn}\,|\vartheta_m|\right)\hat{n}\right)\left(n - \hat{n}\right)$$

Practical diesel torque constraints and ship speed being dynamically related to torque lead to

$$- \left((Q_0 + Q_{\vartheta nn} |\vartheta_m|) \, n + Q_{\vartheta nU} \vartheta_m U_m \right) < \alpha \frac{Q_{eng,max}}{n_{min}}$$

$$- \left(Q_0 + Q_{\vartheta nn} \, |\vartheta_m| \right) \hat{n} < \beta \frac{Q_{eng,max}}{n_{max}}$$

hence $\| \Phi(n, u) - \Phi(\hat{n}, u) \| < \gamma \| n - \hat{n} \|$, where the Lipshitz coefficient is

$$\gamma = \frac{Q_{eng,max}}{I_t \, n_{max}} \left(\alpha \frac{n_{max}}{n_{min}} + \beta \right)$$

A Lyapunov function for the observer error

$$\tilde{n} = n - \hat{n}, \tilde{\theta} = \theta - \hat{\theta}$$

is $V = \tilde{n} P \tilde{n}^T + \tilde{\theta} \tilde{\theta}^T$. Where $P > 0$ is a scalar. Then, using the notation (5.14), and details of the proof in [8],

$$\dot{V} = 2(\Phi(n, u) - \Phi(\hat{n}, u)) P \tilde{n} + 2 u_1 \tilde{\theta} P \tilde{n} - 2 \tilde{n} L P \tilde{n} + 2 \tilde{\theta} \frac{d \tilde{\theta}}{dt} < 0$$

iff $L > \gamma$, $\quad P > 0$ and $n \subseteq [0, n_{max}] \, , \hat{n} \subseteq [0, n_{max}]$.

It is noted that to obtain $\dot{V} < 0$ it is required that $f(x, u)\theta - f(\hat{x}, u)\hat{\theta} \to 0$. This implies that $\theta \to \hat{\theta}$ iff $\hat{x} \to x$. This requires persistent excitation in $u(t)$.

Remark 2.2 *The parameters to calculate the γ value are, typically: $\alpha = 0.1, \beta = 3, n_{max} = 3 n_{min}$. The α and β values are found from (2.6) and (5.11) using observation that maximum shaft speed is limited to 1.09 n_{nom}, even during a crash stop.*

Remark 2.3 *The propeller coefficients are taken to be known parameters in the observer. With inherent parameter uncertainty, system identifica- tion is needed for practical application. Parameter convergence will require persistent excitation.*

5.2   Identification of Propeller Parameters

Direct identification of the physical parameters is conveniently done by adjusting model parameters directly until reaching the minimum of the 2-norm of the deviation between the system output and the estimate [6]. A batch processing on selected data is easily carried out, using standard methods from system identification. It must, however, be validated for each selected sequence, that the excitation in the selected data is persistent. A nice approach to this is a sensitivity analysis [14].

The functional of the output vector for the ship is predicted by the nonlinear model using measured input excitation μ as input to the model,

$$\hat{y}(k) = f(\mu_N, \hat{\theta}) \tag{5.19}$$

where $\hat{\theta}$ is a vector containing estimates of parameters, μ_N is the input vector with N samples, and $f(\mu, \theta)$ is the relevant non-linear model. The discrepancy between measurement and model prediction, when both are excited by the input signal μ, is the model error, $\varepsilon(k) = y(k) - \hat{y}(k)$. A performance function $V(\theta)$ to be minimized is then conveniently taken to be quadratic,

$$V\left(\hat{\theta}\right) = \frac{1}{2N} \sum_{k=1}^{N} \varepsilon^2\left(k, \hat{\theta}\right) \tag{5.20}$$

The parameter estimate $\hat{\theta}$ based on N input-output data points, μ_N and y_N, is the value $\hat{\theta}_N$ that minimizes $V\left(\mu_N, y_N, \hat{\theta}\right)$

$$\hat{\theta}_N = \arg\min_{\theta}\left(V(\mu_N, y_N, \hat{\theta})\right) \tag{5.21}$$

The estimate $\hat{\theta}_N$ is obtained through minimization of this criterion. Some identification methods require the Hessian H, which can be approximately determined from the model gradient $\Psi(k)$

$$\Psi(k) = \frac{\partial \hat{y}(k)}{\partial \hat{\theta}} \tag{5.22}$$

and

$$H = \frac{\partial^2 V\left(\hat{\theta}\right)}{\partial \hat{\theta}\, \partial \hat{\theta}^T} = \frac{1}{N} \sum_{k=1}^{N} \Psi(k)\ \Psi^T(k) \tag{5.23}$$

The gradient $\Psi(k)$ can be determined analytically in some cases, but is always available through numerical differentiation.

The normed root mean square output error

$$\varepsilon_{RMSn} = \frac{\varepsilon_{RMS}}{y_{RMS}} \tag{5.24}$$

is a more significant number for expressing the model fit than the not-scaled performance function $V(\theta)$. The relative normed Hessian is then

$$H_{rn} = y_{RMS}^{-2} L H L \tag{5.25}$$

where L is a diagonal matrix $L = \text{diag}(\theta)$. However, a good fit, i.e. small values of ε_{RMSn} and V, only indicates that the model structure is adequate for expressing the system behavior for a particular input signal. In [14] it is shown how characteristic sensitivity measures are very convenient for determining whether a good fit also implies accurate parameter estimates.

The minimum sensitivity of the parameter dependent part of the model error $\varepsilon_{p,RMSn}(\theta)$ with respect to one relative parameter θ_n, for arbitrary values of the remaining parameters, is found to be

$$S_{i\min} = \sqrt{\left(H_{rn}^{-1}(\theta_N)_{ii}\right)^{-1}} \tag{5.26}$$

Also, the ratio R of the maximum and minimum sensitivities in any direction in the parameter space is essential

$$R = \frac{S_{\max}}{S_{\min}}, \quad S_{\max} = \sqrt{\lambda_{\max}} \quad \text{and} \quad S_{\min} = \sqrt{\lambda_{\min}} \tag{5.27}$$

where λ denotes eigenvalues of H_{rn} and represent the sensitivity in the parameter space. These sensitivity measures are used for input design. In the sequel, we also need sensitivity measures for the individual parameters

$$S_i = \sqrt{H_{rn}(\theta_N)_{ii}} \quad \text{and} \quad R_i = \frac{S_i}{S_{i\min}} \tag{5.28}$$

Determination of Parameter Accuracy

It can be shown [15] that the estimated parameter error is inverse proportional to the sensitivity S_{imin}.

The total, relative estimated error for the i'th parameter can be determined as

$$\tilde{\theta}_{ritot} = \frac{\varepsilon_{RMSn}^x}{S_{i\min}\sqrt{N}} + \frac{\varepsilon_{RMSn}^m}{S_{i\min}} \tag{5.29}$$

where ε_{RMSn}^x and ε_{RMSn}^m are the root mean square errors caused by noise and under modeling, respectively.

The determination of parameter uncertainty for a given input signal is used in the sequel to run a batch processing identification for determination of propeller parameters, and later in assurance of identification quality when a possible diesel gain fault is suspected.

5.3   Identifiability from Usual Maneuvering Data

The parameters $Q_{\vartheta nn}$ and $Q_{\vartheta nU}$ in (2.3) are identifiable if their sensitivities are adequate, according to (5.29), given a partiqular input sequence. Figure 5 shows a sensitivity plot for the two physical parameters using data measured at sea-trials with the passenger ferry Dronning Ingrid [13] under usual maneuvering. This result is important for the practical applicability of the concept.

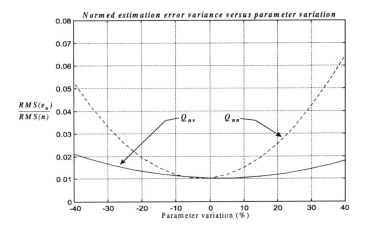

FIGURE 5. Sensitivity results: The normed variance of the estimation error calculated for variation in $Q_{\vartheta nn}$ and $Q_{\vartheta nV_a}$.

6   Fault Isolation

The combined parameter and sensor fault detection in one observer give rise to interdependency in the fault isolation. Since remedial actions are widely different to accommodate each of the two faults, fault isolation is crucial.

When used for detection and isolation, there are two observer error signals to consider

$$e_n = n_m - \hat{n} = n + \Delta n - \hat{n} \tag{6.30}$$

$$e_\theta = \frac{K_y}{I_t} + \frac{\Delta K_y}{I_t} - \frac{\hat{K}_y}{I_t} \equiv \theta + \Delta\theta - \hat{\theta} \tag{6.31}$$

where Δn and ΔK_y are the sensor and engine faults, respectively. The observer error signals are derived from from equations (5.15) to (5.17),

$$\dot{e}_n = -\left(\frac{Q_{\vartheta nn}\vartheta_m}{I_t}(n+\hat{n}) + \frac{Q_{\vartheta nU}\vartheta_m U_m}{I_t} + L\right)e_n - Y_m e_\theta - L\Delta n - Y_m\Delta\theta \tag{6.32}$$

$$\dot{e}_\theta = PY_m e_n + PY_m\Delta n \tag{6.33}$$

The linearized transfer functions in an operating point, $\{\bar{n}, \bar{\vartheta}, \bar{U}, \bar{Y}\}$, are easily determined, from the Laplace transform of the two fault signals to those of the "raw" observer error signals. Using the abbreviations

$$\alpha = 2\frac{Q_{\vartheta nn}}{I_t}\bar{\vartheta}\bar{n} + \frac{Q_{\vartheta nU}}{I_t}\bar{\vartheta}\bar{U} \tag{6.34}$$

$$N(s) = s^2 + (\alpha + L)s + P\bar{Y}^2 \tag{6.35}$$

we get

$$e_n(s) = -\frac{\left(sL + P\bar{Y}^2\right)}{N(s)}\Delta n(s) - \frac{s\bar{Y}}{N(s)}\Delta\theta(s) \tag{6.36}$$

$$e_\theta(s) = \frac{P\bar{Y}(s + \alpha)}{N(s)}\Delta n(s) - \frac{P\bar{Y}^2}{N(s)}\Delta\theta(s) \tag{6.37}$$

Since the two raw observer error signals are not linearly dependent, diagonal isolation [11] can be achieved. Nice low pass properties are obtained by defining the filtered residuals as follows,

$$r_n(s) = \frac{P\bar{Y}}{1 + s\tau}e_n(s) - \frac{s}{1 + s\tau}e_\theta(s) = -\frac{P\bar{Y}}{1 + s\tau}\Delta n(s) \tag{6.38}$$

$$r_\theta(s) = \frac{P\bar{Y}(s + \alpha)}{\left(sL + P\bar{Y}^2\right)}e_n(s) + e_\theta(s) = \frac{-PY^2}{\left(sL + PY^2\right)}\Delta\theta(s) \tag{6.39}$$

This choice assures that a steady state error gives rise to a mean value change in the residual. This is one of the fundamental properties that should be achieved to enable easy detection also in the presence of noise [1].

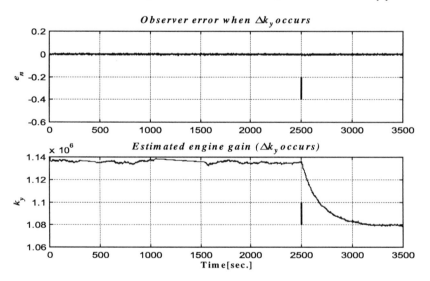

FIGURE 6. Time history for obtained state error and parameter estimate when gain fault occurs.

Simulation results

To illustrate the performance of the algorithm, test data collected on the Danish ferry Dr. Ingrid, were used. Fig. 6 shows a gain fault in K_y occurring at time 2500 second. The gain changes stepwise down by 5%, corresponding to the partial failure of a single cylinder. The raw observer error signals are plotted. Balancing of gain factors for P and L can be made such that further filtering is not necessary. The e_n signal is well below a reasonable threshold value.

The second fault considered is the critical failure of the shaft speed measurement. A negative fault in shaft speed measurement is shown in Fig. 7. The simultaneous discrepancy in both signals make it necessary to use further filtering as proposed to achieve isolation.

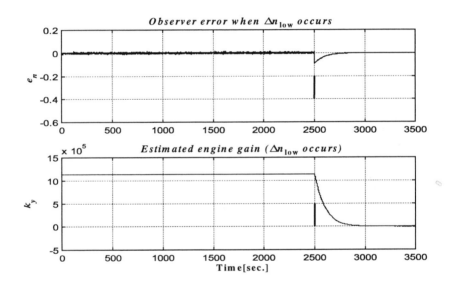

FIGURE 7. Time history for obtained state error and parameter estimate when a negative fault in shaft speed measurement.

A positive fault in shaft speed was simulated and plotted in Fig. 7. The response of the K_y deserves some comments. What happens is that the n_m high fault is immediately reacted to by the shaft speed governor, which drives the fuel index rapidly to zero. Y being zero is a violation of the conditions for convergence of the non-linear observer. This is quite obvious, since a zero gain is the present in the parameter update, (5.17). This phenomenon is not accounted for by the linear filter analysis above that should provide fault isolation. It is noted that none of the above simulations were made with re-configuration activated. This is treated below.

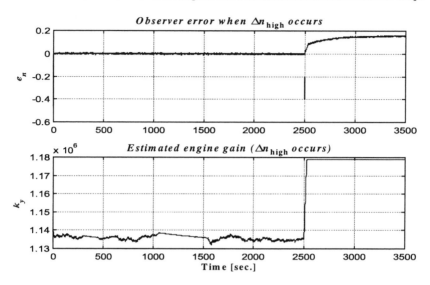

FIGURE 8. Time history for obtained state error and parameter estimate when a positive fault in shaft speed measurement.

6.1   Re-configuration

Re-configuration actions depend on the actual fault that has occurred. The key issue in this context is that there is difference in the time it takes to verify which of the possible faults have occurred. Simultaneously there is a dramatic difference in the requirements of how fast a fault must be detected, isolated and accommodated. If the n-fault has happened, the diesel engine can reach and over-speed condition within fractions of a second. If the diesel engine gain has deteriorated, action within a minute span of time will suffice. Since the n-fault is the most severe, and the time to isolate the actual fault is longer than the required time to re-configure for the n-fault, the supervisor will need to choose the worst case assumption until final isolation can be made.

If necessary, re-configuration will thus first assume that an n-fault has happened, and accommodate to this situation. Having re-configured the n-measurement, isolation might run until finished, and the final remedial action take place. Continued observation for correct isolation of the cause of the fault is not possible when only shaft speed is monitored. In an actual application, ship speed, pitch angle and fuel throttle sensors would all be supervised, and more elaborate schemes could be employed to detect possible faults. In the solution chosen, all signals used are available by the governor. Limiting fault detection and accommodation to signals available by the governor, would enhance the integrity of this device.

Re-configuration to accommodate the n fault is then to switch the controller to use the estimate of n instead of the faulty measurement n_m. When a gain fault of the engine has occurred, the remedial action is to change the overload limits within the governor to the reduced capability available from the engine. Since the two reactions are entirely different, proper isolation of the two faults is crucial.

Detection and isolation of a change is done using standard methods for change detection. This is not immediately possible with only one residual, redundant information in the system is needed. The possibilities can be derived from the structure diagrams. They show that observation of ship speed through the thrust constraints is indeed feasible.

The slow dynamics of ship speed, and the fast reactions of a diesel engine make it necessary, however, to assume a worst case fault in the shaft speed measurement. When a discrepancy in n is observed, diesel control is maintained using the open loop nonlinear observer. In due time, other measures will show whether the fault was the less serious cylinder defect of the engine, and the re-configurated sensor signal could be switched to normal, while other appropriate steps are taken to accommodation the fault now isolated.

The simulations show the performance of the adaptive observer with re-configuration when a shaft speed sensor failure occurs.

Symbol	Unit	Explanation
f_{T_prop}	N	thrust function
f_{Q_prop}	Nm	torque function
I_t	kgm^2	total inertia
K_y	Nm	torque coefficeint
n	$rads^{-1}$	shaft speed
$R(U)$	N	hull resistance
T_{prop}	N	propeller thrust
T_{ext}	N	external force
$1-t$	-	thrust deduction factor
U	ms^{-1}	ship speed
V_a	ms^{-1}	flow at propeller
1-w	-	wake fraction
Q_{eng}	Nm	diesel torque
Q_f	Nm	shaft friction
Q_{prop}	Nm	propeller torque
Y_d	$0..1$	fuel index
ϑ	$-1..1$	propeller pitch

TABLE 2.3. List of Symbols

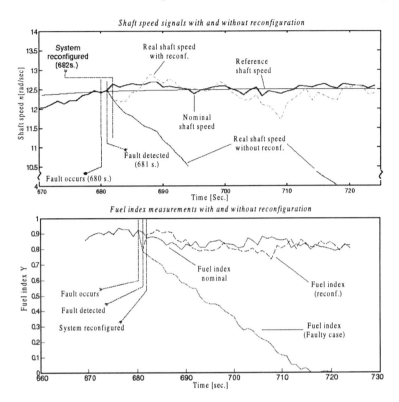

FIGURE 9. A zoom in on the shaft speed and fuel index values in the worst case, using the observer generated shaft speed.

7   Simulation Results

The methods are applied to the ship propulsion benchmark where the following scenario is simulated. A shaft sensor failure occurs at $t = 1000$(s). A statistical fault detection method (CUSUM) has detected the sensor failing high at $t = 1001$(s). The supervisor accommodates the fault at $t = 1002$(s). by activating a dedicated procedure that estimates the variable n. The calculated variable replaces the measured one in the shaft speed control loop.

The upper part of Fig. 9 shows the real shaft speed for nominal, faulty and re-configured cases. The non-linear observer is used and when the Δn fault is detected, observer gains L and P are immediately set to zero. The K_y parameter is taken as the value it had a few seconds before the fault was detected. The estimated shaft speed is immediately used by the shaft speed controller as a substitute for the faulty sensor signal.

The lower part of the figure shows fuel index. It is seen that switching from the faulty n_m to the estimated $\hat{n}_q$ results in an overshoot of less than 5% in shaft speed. The failure has been simulated to happen exactly during a transient command to obtain a worst case condition. It is essential that the 9% critical limit of over-speed shut down of the main engine is not reached. The resulting overshoot is thus well below the critical over-speed shut down limit.

For all figures, the curves represent: normal case (solid), faulty case (dash dotted), and re-configured case (dashed), reference signal (dotted).

8   Conclusions

This chapter has analyzed fault detection and isolation, and re-configuration possibilities for a ship propulsion system with a main engine and a controllable pitch propeller: It was demonstrated how fault-tolerance could be achieved against critical sensor failure and cylinder malfunction of the prime mover engine. A non-linear adaptive observer was designed for fault detection and re-configuration, and filters for efficient isolation of faults was found. A structural analysis was used to obtain a scheme of consistent re-configuration possibilities when the goal was to achieve uninterrupted prime propulsion of the ship. Simulations of a model of a ferry illustrated how a critical failure of the shaft speed measurement could be accommodated by the controller. The unavoidable penalty in control quality was found to be quite small and certainly acceptable against the alternative, which was a temporal loss of main propulsion of the ship.

The essential contributions of the chapter were combining structural analysis and re-configuration design, and applying this approach to a realistic case using a nonlinea, adaptive observer.

Acknowledgments

This research was partly funded by the European DISC II project under ECC DG7. Support of collaboration between the Universities of Lille (France) and Aalborg (Denmark) was received from the European Science Foundation under the COSY program.

9   REFERENCES

[1] M. Basseville and I. Nikiforov. *Statistical Change Detection*, Prentice Hall Inc., 1994.

[2] M. Blanke. Consistent design of dependable control systems. *Control Engineering Practice* 4(9): 1305–1312, 1996.

[3] M. Blanke and J. S. Andersen. On dynamics of large two stroke diesel engines: New results from identification. *Proceedings 9th IFAC World*

Conference. Budapest, Hungary, 1984.

[4] M. Blanke and R. Izadi-Zamanabadi. Reconfigurable control of a ship propulsion plant. *Proceedings IFAC Conference CAMS'98*, Fukuoka, Japan. pp. 51–58, October 1998.

[5] M. Blanke and R. Izadi-Zamanabadi and T. F. Loostma. Fault Monitoring and Re-configurable Control for a Ship Propulsion Plant. In: *Journal of Adaptive Control and Signal Processing* , 1999.

[6] M. Blanke and R. Knudsen. A sensitivity approach to identification of ship dynamics from sea trial data. *Proceedings IFAC Conference CAMS'98*, Fukuoka, Japan. pp. 261–269, October 1998.

[7] J. Ph. Cassar, M. Staroswiecki and P. Declerck. Structural decomposition of large scale systems for the design of failure detection and isolation procedures. *International Journal of System Science* **20**(1): 31–42, 1994.

[8] Y. M. Cho and R. Rajamani A systematic approach to adaptive observer synthesis for nonlinear systems. *IEEE Transactions of Automatic Control* **42**(4), 534–537, 1977.

[9] V. Cocquempot, R. Izadi-Zamanabadi, M. Staroswiecki and M. Blanke. Residual generation for the ship benchmarck using structural approach. *Proceedings CONTROL'98*. Swansea, U.K., 1998.

[10] P. Declerck and M. Staroswiecki. Characterisation of the canonical components of a structural graph for fault detection in large scale industrial plants. ECC'91. Grenoble, France. pp. 298–303, 1991.

[11] Janos J. Gertler. *Fault Detection and Diagnosis in Engineering Systems*, Marcel Dekker Inc., 1998.

[12] R. Izadi-Zamanabadi and M. Blanke. Ship propulsion system as a benchmark for fault-tolerant control. *Control Engineering Practice (to appear)*, 1999.

[13] R. Izadi-Zamanabadi and M. Blanke. A ship propulsion system model for fault-tolerant control. Tech. report, Department of Control Engineering, Aalborg University, 1998.

[14] M. Knudsen. A sensitivity approach for estimation of physical parameters. *Proceedings 10th IFAC Symposium on System Identification*. Vol. 2 p.231, Copenhagen, Denmark, 1994.

[15] M. Knudsen. Determination of parameter estimation errors due to noise and under modelling. *Proceedings IEEE Instrumentation and Measurement Technology Conference*, Brussels, 1996.

[16] L. Ljung. *System Identification - Theory for the User.* Prentice Hall Int., 1987.

[17] M. Staroswiecki and P. Declerck. Analytical redundancy in nonlinear interconnected systems by means of structural analysis. *IFAC-AIPAC'89*, Vol. II. Nancy, France. pp. 23–27, 1989.

[18] A. Tiano and M. Blanke. Multivariable identification of ship steering and roll motions. *Proceedings Institute of Measurement and Control (UK). Special Issue on Marine Systems Modelling and Control,* 1997.

Nonlinear Observers for Fault Detection and Isolation

P. M. Frank, G. Schreier and E. Alcorta García

Department of Measurement and Control
University of Duisburg
47048 Duisburg, Germany

1   Introduction

One of the essential requirements of fault tolerant control (FTC) is fault detection and isolation (FDI) [9, 44]. The main task of fault detection and isolation can be roughly described as the early determination (detection) and localization (isolation) of faulty elements of a dynamic system, as well as the time of the appearance of the faults. Because of the great relevance of FDI in industrial plants as well as the availability of appropriate methods, this subject has become a fundamental issue of research within the control community during the last 25 years [4, 14, 22, 27, 31, 34, 46].

Among the well-established concepts of fault detection and isolation using analytical redundancy, i.e. based on mathematical models, are the parity space approach, the parameter estimation approach and the observer-based approach, where the parity space approach can be considered as a special version of the observer-based approach. It has also been shown that the parameter estimation approach has some interesting relationships to the observer-based approach [2, 11, 23, 30]. This is why the observer-based approach has become a most relevant subject of research in connection with FDI.

The traditional observer-based approach to FDI makes use of linear models in the observer even if the plant under consideration is (as is usual) nonlinear [14, 16, 31, 33, 34]. Only in the last decade is there a trend to design nonlinear diagnostic observers for FDI in nonlinear systems [1, 7, 12, 13, 20, 26, 28, 38, 39, 47]. It is worth noting that the use of observers for FDI requires some special design efforts, because diagnostic observers are output observers with different goals from the state observers used for control. This point has often been overseen in the FDI literature in the past.

As mentioned above, design methods for diagnostic observers for nonlinear systems found in the literature are often based on the assumption that the system works during normal operation in the neighborhood of a certain operating point. Clearly, in many such cases linearisation is possible, however the linearisation errors can cause some difficulties in the FDI algorithm such that these errors can be misinterpreted as faults and hence lead to false alarms. This is why approaches using nonlinear observer design for FDI tasks are becoming more and more important in order to augment the performance of the resulting fault detection systems. During the research of the last decade on nonlinear diagnostic observers, some results have been achieved for certain classes of nonlinear systems not only for fault detection but also for fault isolation [20, 29, 38, 41]. Nevertheless, a general theory of nonlinear FDI as well as the design of nonlinear diagnostic observers is still missing. The main reason for this is that the estimation of the set or of sub-sets of the state or the measurement vector of a nonlinear system is not well solved even if there are no disturbances or the nonlinearity of the system is independent of the disturbances. At present, there are attempts to overcome the difficulty of analytical treatment of the nonlinearity by using non-analytical (qualitative and knowledge-based) methodologies such as neural networks or fuzzy techniques. In this chapter we will limit our consideration to the analytical approaches to the design of nonlinear diagnostic observers [17, 18, 50, 51].

The purpose of this chapter is to give a brief survey of the state of the art of nonlinear diagnostic observers for FDI in deterministic nonlinear dynamic systems, where quantitative models are available [3]. This survey is mainly covered by the first part of the chapter. The methods are discussed in terms of robustness and similarities between the different approaches, and open problems are pointed out. The second part of the chapter introduces the basic ideas of the design of a nonlinear diagnostic observer based on the explicit use of the Lipschitz constant.

For simplicity of notation, the time dependence of the functions will be dropped throughout the chapter. The brackets ( ) will be used to express the functions' arguments and the brackets [ ] to signify Lie brackets (in addition to their standard use for matrices and to indicate priority of operations). As usual, $\| \ \|$ means the Euclidean norm of a vector.

2   Preliminaries

Consider a nonlinear system, described by the equations:

$$\begin{aligned}
\dot{x} &= f(x, u, \theta_f, \theta_d), \quad x(0) = x_0 \\
y &= h(x, u, \theta_{fs}).
\end{aligned} \tag{2.1}$$

where $x \in \mathbb{R}^n$ is the state vector, $u \in \mathbb{R}^m$ is the input vector, $y \in \mathbb{R}^p$ is the output vector of the system, $\theta_f \in \mathbb{R}^l$ represents the actual system parameters, i.e., when no faults are present in the system, $\theta_f = \theta_{f0}$ where θ_{f0}

is the nominal parameter vector (understanding "fault" as an unpermitted parameter deviation in the system), $\theta_{fs} \in \mathbb{R}^{l_s}$ represents the parameters in the output equation (if a sensor fault occurs $\theta_{fs} \neq \theta_{fs0}$, where θ_{fs0} represent the nominal parameters in the output equation) and $\theta_d \in \mathbb{R}^{l_d}$ represents modelling mismatches (if the model of the system is perfectly known, $\theta_d = 0$).

In an ideal case, a residual $r(t)$ will be zero if no faults are present, different from zero when a fault is present (fault detection) and only manifest its ith element if a fault in the ith element is present (fault isolation). The procedure for evaluating the redundancy given by the mathematical model of a system is frequently divided into the following two steps [14]: **Residual generation** is the construction of signals that are accentuated by the changes in the parameter vector (faults) and **Residual evaluation** is the decision and isolation of the occurrence of changes in the parameter vector (faults).

3 Observer-Based Residual Generation

In this section, we briefly review the principal observer-based residual-generation methods for fault diagnosis which have been developed in recent years for special classes of nonlinear dynamic systems.

3.1 Nonlinear Identity Observer Approach

This approach to fault diagnosis was first proposed in [26] for the detection and isolation of component faults (see also [13]); for a more general class of faults in [14] and for further design considerations in [1]. The starting point is the nonlinear model (2.1) and the following observer structure

$$
\begin{aligned}
\dot{z} &= f(z, u, \theta_{f0}, 0) + K(z, u)[y - \hat{y}] \\
r &= y - h(z, u, \theta_{fs0}).
\end{aligned}
\tag{3.2}
$$

The design of the observer (3.2) is carried out on the assumption that no faults ($\theta_f = \theta_{f0}$ and $\theta_{fs} = \theta_{fs0}$) and no modelling mismatches ($\theta_d = 0$) are present in the system. Defining the estimation error $e \overset{\triangle}{=} x - z$, the differential equations governing its dynamics could be written by

$$
\begin{aligned}
\dot{e} &= F(z, u, \theta_{f0}, 0)e - K(z, u)H(z, u, \theta_{fs0}, 0)e \\
&\quad + O_1(e^j, t) \\
r &= H(z, u, \theta_{fs0}, 0)e + O_2(e^j, t)
\end{aligned}
\tag{3.3}
$$

where

$$F(z, u, \theta_{f0}, 0) = \left. \frac{\partial f(x, u, \theta_{f0}, 0)}{\partial x} \right|_{x=z} \tag{3.4}$$

and

$$H(z, u, \theta_{fs0}) = \left. \frac{\partial h(x, u, \theta_{fs0})}{\partial x} \right|_{x=z} \tag{3.5}$$

$O_1(e^j, t)$ and $O_2(e^j, t)$ represent the second- and higher-order terms with respect to e. The terms $O_1(e^j, t)$ and $O_2(e^j, t)$ will be neglected here.

The remaining problem is to design the matrix $K(z, u)$, in such a way that the equilibrium $e = 0$ of (3.3) is asymptotically stable.

A solution to this problem, with the assumption that $h(x, u) = Cx$ and $\ker[C] \neq 0$, was proposed in [1]. Following [1], the matrix $K(z, u)$ takes the form

$$K(z, u) = P^{-1} \hat{F}(z, u) C^T Q \tag{3.6}$$

where the matrix $P = P^T > 0$ is positive definite, and should be assigned such that

$$\bar{K}^T P \left. \frac{\partial f(x, u, \theta_{f0}, 0)}{\partial x} \right|_{x=z} \bar{K}$$

is a negative definite matrix, where the matrix $\bar{K}$ is the highest rank right orthogonal matrix to C. The matrix valued function $\hat{F}(z, u)$ is given by

$$\hat{F}(z, u) = diag \left\{ \frac{1}{2} \sum_{j=1}^{n} |\psi_{ij} + \psi_{ji}| \right\} \quad i = 1, \cdots, n, \tag{3.7}$$

where ψ_{ij} is the ij^{th} element of the matrix $P \left. \frac{\partial f(x, u)}{\partial x} \right|_{x=z}$, and Q is a matrix satisfying $C^T Q C - I \geq 0$. With this selection of the matrix $K(z, u)$, the equilibrium $e = 0$ is asymptotically stable for the first order approximation [1].

In some cases the use of a constant matrix $K(z, u)$ will be sufficient to guarantee stability of (3.3), as pointed out in [13]. It could be the case, for example, when $f(x, u)$ has a special form i.e. $f(x, u) = Ax + g(x)u$ and the function $g(x)u$ satisfing a Lipschitz condition. In this case the design problem becomes similar to that for Thau's observer [43]. If $f(x, u) = f(x) + g(x)u$ and if the vector-valued functions $f(x)$ and $g(x)$ satisfy some technical conditions, the observer design with constant $K(z, u) = K$ can be achieved in a similar way to the one given in [21].

3.2 Nonlinear Unknown Input Observer Approach

A direct extension of the unknown input observer (UIO) results in linear systems to the nonlinear case was considered in [47]. The approach takes advantage of the structure of the system model, which is assumed to be in observable canonical form [6]. In this case, a constant state transformation could be used (as in the linear case), and a complete design procedure can be achieved. The class of systems considered in [47] can be described as follows

$$
\begin{aligned}
\dot{x} &= Ax + B(y, u) + Ed + K(x, u)f_f \\
y &= Cx + K_s(x, u)f_s
\end{aligned}
\tag{3.8}
$$

where $d = d(x, u, \theta_d)$ represents the unknown inputs, $f_f \overset{\triangle}{=} \theta_f - \theta_{f0}$ represents a component or an actuator fault to be detected and $f_s \overset{\triangle}{=} \theta_{fs} - \theta_{fs0}$ the sensor faults. The question of how to obtain a system representation according to (3.8) from a more general nonlinear system representation by a transformation of the state vector is not of concern in this section. Existance conditions and design considerations of the required transformations have been studied by Birk and Zeitz [8].

A fault-detection observer for the system (3.8) is given by

$$
\begin{aligned}
\dot{z} &= Fz + J(y, u) + Gy \\
r &= L_1 z + L_2 y.
\end{aligned}
\tag{3.9}
$$

The following conditions on the observer matrices are necessary in order to provide total decoupling from the unknown input d and sensitivity to the fault vector f_f.

$$
\begin{aligned}
TA - FT &= GC \quad F \text{ stable} \\
J(y, u) &= TB(y, u) \\
L_1 T + L_2 C = 0 \quad&, \quad TE = 0 \\
\text{rank}\{TK(x, u)\} &= \text{rank}\{K(x, u)\} \\
\text{rank}\left\{ \begin{bmatrix} G \\ L_2 \end{bmatrix} K_s(x, u) \right\} &= \text{rank}\{K_s(x, u)\}
\end{aligned}
$$

If these conditions are fulfilled, the residuals obey the equations

$$
\begin{aligned}
\dot{e} &= Fe - GK(x, u)f_f + TK_s(x, u)f_s \\
r &= L_1 e + L_2 K_s(x, u)f_s
\end{aligned}
\tag{3.10}
$$

where $e \triangleq Tx - z$. A drawback of the nonlinear UIO (NUIO) approach as formulated in [47] is the difficulty of transforming a general system (2.1) into the required form (3.8). As for the linear case, the existence conditions of the NUIO are also very restrictive.

3.3 The Disturbance Decoupling Nonlinear Observer Approach

An alternative to the NUIO approach, considering a more general class of systems, was proposed in [38, 39]. The basic idea was the same as for the NUIO, but a nonlinear state transformation instead of a linear one is used. The class of systems that can be treated with this approach is described by

$$\begin{aligned} \dot{x} &= A(x,u) + E(x,u)\theta_d + K(x,u)f_f \\ y &= C(x,u) \end{aligned} \qquad (3.11)$$

where the matrices

$$A(x,u) \triangleq f(x,u,\theta_{f0},0)$$

$$E(x,u) \triangleq \frac{\partial f(x,u,f_f+\theta_{f0},\theta_d,)}{\partial \theta_d}\bigg|_{f_f=\theta_d=0}$$

$$K(x,u) \triangleq \frac{\partial f(x,u,f_f+\theta_{f0},\theta_d,)}{\partial f_f}\bigg|_{f_f=\theta_d=0}$$

$f_f \triangleq \theta_f - \theta_{f0}$. The model (3.11) corresponds to a first- order approximation of (2.1) at the set point $f_f = 0$, $\theta_d = 0$. Here, the second- and higher-order terms with respect to f_f and θ_d are not considered.

The first step is to find a nonlinear state transformation $z = T(x)$ in order to decouple the faults from the disturbances. This can be met if the condition

$$\frac{\partial T(x)}{\partial x}E(x) = 0 \qquad (3.12)$$

is satisfied, where it is assumed that $E(x,u) = E(x)$, which implies that the required transformation depends only on the state x $(T(x))$. Note that $T(x)$ is not a similarity transformation, i.e. the transformed states span only a subspace of the one spaned by x.

If the matrix $E(x,u)$ has an explicit dependence on $u(t)$, a transformation depending on the input $u(t)$ $(T(x,u))$ will be required. Consequently, the transformed system will depend on the derivative of $u(t)$. Some ways of avoiding this difficulty are pointed out in [38, 39].

The necessary and sufficient condition to solve (3.12) is given by the Frobenius theorem [38, 39] and can be formulated as follows: Assume that the rank of $E(x)$ is equal to q for all x and $T(x)$ is a $(n - q)$ vector. There exists a solution $T(x)$ for the equation (3.12) if and only if

$$rank\,(E(x)\;\;[e_i(x),\;e_j(x)]) = q \tag{3.13}$$

for $i, j = 1, \cdots, q$ and all x, where $e_i(x)$ denotes the ith column of the matrix $E(x)$ and $[e_i(x), e_j(x)]$ is the so-called Lie-bracket (w.r.t.x).

If the existence conditions of the nonlinear state transformation are fulfilled, the transformed system can be described as

$$\dot{z} = \frac{\partial T(x)}{\partial x}\,[A(x, u) + K(x, u)f_f]$$
$$y^* = c^*(z, u, y) \tag{3.14}$$

where the output has been transformed in order to obtain a new output, depending only on the state z, the input u and the original output y. The second step is the design of an observer for the reduced system (3.14). Some approaches were considered in [38, 39]:

i) A design based on the nonlinear identity observer as described in Section 3.1.

ii) If the transformed system (3.14) is in an observable canonical form, the observer can be designed directly as in Section 3.2 with $d = 0$.

iii) The use of a parallel model of the system (i.e., observer without feedback).

Remark 3.1 *The decoupling condition (3.12) was formulated independently using a different approach, the so-called "algebra of functions" [41].*

Remark 3.2 *A similar approach to the one considered in i) was also introduced in [41]. The observer matrix H is selected in order to stabilize the sensitivity equations of the observer with respect to parameter variations. An optimal index is also given for the case when the decoupling condition (3.12) is not satisfied. Unfortunately, the result is extremely complex, as pointed out by Shumsky [41].*

Note that the approach in i) could be considered as a robustification of the nonlinear identity observer (NIO) presented in Section 3.1.

3.4   Adaptive Nonlinear Observer Approach

One problem of the observer-based methods for fault diagnosis is their weakness in detecting slowly developing faults, especially when model uncertainties are present [15]. An adaptive observer-based residual generator approach to overcome this difficulty has been proposed independently in [7] and in [12].

In [7] a nonlinear observer is used in order to detect a leak in a pipeline, and the nonlinear open-loop observer's performance is improved with the adaptation of a friction coefficient. In [12] a more general case is considered. Based on a modified version of the observer proposed in [5], an adaptive residual generator is developed. Following [12], consider a nonlinear system described by

$$\begin{aligned}
\dot{x} &= a(x) + q_0(x, u) + Q(x, u)\theta_d + G(x, u)f + g(t) \\
y &= c(x) \qquad x_0 = x(0).
\end{aligned} \qquad (3.15)$$

where the output y is considered scalar for sake of simplicity, $a : \mathbb{R}^n \to \mathbb{R}^n$, $Q : \mathbb{R}^n \times \mathbb{R}^m \to \mathbb{R}^{n \times l_d}$, $G : \mathbb{R}^n \times \mathbb{R}^m \to \mathbb{R}^{n \times l}$, $g : \mathbb{R} \to \mathbb{R}^n$ and $c : \mathbb{R}^n \to \mathbb{R}$ are assumed to be known and smooth enough, f represents abrupt changes, and $\theta_d \in \mathbb{R}^{l_d}$ is an unknown vector which represents, for example, unknown time-varying parameters, slowly varying faults or part of the nonlinearities of the system. Assume that

$$0 < ||\dot{\theta}|| \leq M << \infty$$

The adaptive residual generator is given in two steps: first, a transformation $\xi = T(x) \in \mathbb{R}^k$, $k \leq n$ has to be found (if possible), defined on a neighbourhood of the initial state x_0 such that:

$$\begin{aligned}
\dot{\xi} &= F\xi + \psi_o(y, u) + \Psi(y, u)\theta + \Xi(x, u)f \\
y &= \begin{bmatrix} 0 & \cdots & 0 & 1 \end{bmatrix} \xi
\end{aligned} \qquad (3.16)$$

and rank$[\Xi(y, n)] = l$. Necessary and sufficient conditions for the existence of the transformation $\xi = T(x)$ can be found in [12]. The second step consists of designing the adaptive residual as

$$\dot{z} = Fz + \psi_o(y, u) + \Psi(y, u)\hat{\theta} + L_1 y$$
$$+ \begin{bmatrix} V(t) \\ 0 \end{bmatrix} \dot{\hat{\theta}}$$
$$r(t) = y(t) - z_k(t)$$
$$\dot{\hat{\theta}}(t) = \Gamma \phi^T(t) r(t)$$
$$\dot{V}(t) = RV(t) + \bar{\Psi}_k(y, u), \quad V(0) = 0$$
$$\phi(t) = k^T V(t) + \psi_k(y, u)$$

where Γ is a positive definite matrix, R is a stable matrix, the elements l_j of L_1 are assigned such that $s^n + l_n s^{n-1} + \cdots + l_2 s + l_1 = 0$ is Hurwitz, $k^T = [\, 0 \quad \cdots \quad 0 \quad 1 \,]$. Let a matrix Q satisfy

$$QFQ^{-1} = \begin{bmatrix} R & \times \\ k^T & \times \end{bmatrix} \tag{3.17}$$

then

$$Q\Psi(y, u) \triangleq \begin{bmatrix} \bar{\Psi}_n(y, u) \\ \Psi_n(y, u) \end{bmatrix} \in \mathbb{R}^{n \times l} \tag{3.18}$$

and the conditions

1. $\phi(t)$ is bounded, $\dot{\phi}(T)$ is bounded except possibly at a countable number of points

2. $\exists \, \alpha, \beta$ such that $0 < \alpha I \leq \int_t^{t+\beta} \phi(\tau)\phi^T(\tau)d\tau$

3. $\exists \, M_1$ such that $|V(t)\dot{\theta}| \leq M_1 < \infty$

are required in order to guarantee that $|y - \hat{z}_k| < K < \infty \;\; \forall t$.

It is usually assumed that $\|\dot{\theta}\|$ is small. Thus, it is reasonable to suppose that the estimation errors are restricted to a small range around zero [12].

A different approach is considered in [48]. It is based on a modified version of the nonlinear observer proposed recently in [21]. The systems considered are all observable (in the sense of [21]), and transformable into the following form

$$\dot{x} = F'(x) + G'(x)u + T'(y, u)\theta_d + p(x)f$$
$$y = Cx \tag{3.19}$$

where

$$
\begin{aligned}
F'(x) &= [x_2, \cdots, x_n, \phi(x)]^T, \\
G'(x) &= [g_1(x_1), \cdots, g_n(x_1 \cdots x_n)]^T, \\
C &= [1\ 0\ \cdots\ 0] \\
T'(y, u) &= S_\infty^{-1}
\begin{bmatrix}
\tau_1 & \cdots & \tau_s \\
0 & \cdots & 0 \\
\vdots & \vdots & \vdots \\
0 & \cdots & 0
\end{bmatrix}
\end{aligned}
$$

$\theta_d \in \mathbb{R}^S$, S_∞ will be defined later. If $\theta_d = f = 0$, the observability guarantees the existence of the transformation [21]. For sake of simplicity, the SISO case is considered here; the reader is referred to [48] for the MIMO case. Under the assumption that the functions g_i are Lipschitz, an adaptive observer for the system (3.19) is given by

$$
\begin{aligned}
\dot{z} &= F'(z) + G'(z)u + T'(y, u)\hat{\theta}_d + S_\infty^{-1}C^T[z_1 - y] \\
r &= z_1 - y \qquad\qquad\qquad\qquad (3.20) \\
\dot{\hat{\theta}}_d &= -[\tau_1\ \cdots\ \tau_s]^T[z_1 - y]
\end{aligned}
$$

where S_∞ is the solution of $0 = -\gamma S_\infty - A^T S_\infty - S_\infty A + C^T C$ and A is the anti-shift operator $A : \mathbb{R}^n \to \mathbb{R}^n$, $A_{i,j} = \delta_{i,j-1}$. Unlike [12], in [48] no mention is made concerning to the behaviour of the adaptive observer (3.20) if θ_d is time-dependent.

3.5   The Nonlinear Fault Detection Filter Approach

In this section, an extension of the fault-detection filter for linear systems to a class of nonlinear systems is considered. As for the linear case, detection filters are output observers that produce residuals with directional properties [20]. Based on a Thau-type nonlinear observer, the results given in [45] for the fault-detection filter for linear systems are generalized in [20] for a class of nonlinear systems. Assume that a nonlinear system is represented as

$$
\begin{aligned}
\dot{x} &= Ax + f(x) + Bu + \sum_{i=1}^{s} F_i\phi_i(x, u)f_i \\
y &= Cx \qquad\qquad\qquad\qquad\qquad (3.21)
\end{aligned}
$$

where $F_i \in \mathbb{R}^n$, s is the number of parameter changes (faults $f_i = \theta_{fi} - \theta_{f0i}$) considered, ϕ_i is a function of $x(t)$ and/or $u(t)$ depending of the parameter considered, and $x(t)$, $y(t)$ and $u(t)$ are as in (2.1). $F_i\phi_i(x, u)$ is defined as follows:

$$F_i\phi_i(x, u) = \frac{\partial[Ax + f(x) + Bu]}{\partial\theta_{fi}}\bigg|_{\theta_{fi}=\theta_{f0i}} \qquad (3.22)$$

The representation used here is slightly different from that used in [20] in order to maintain uniformity of presentation in this chapter.

Defining the matrix F as:

$$F = [F_1 \ F_2 \ \cdots \ F_s] \qquad (3.23)$$

and following [20], some assumptions are required:

1. The nonlinear term $f(x)$ is Lipschitz, i.e.
   $||f(x_1) - f(x_2)|| \leq \gamma||x_1 - x_2||$

2. (A, C) is observable

3. CF is of rank s. This means that the relative degree of the fault vector with respect to the output is equal to one.

Consider the next Thau-type observer for the system (3.21)

$$\dot{z} = Az + f(z) + Bu + H[y - Cz]$$
$$r = y - Cz \qquad (3.24)$$

where the matrices

$$H = Q_h[CF]^\dagger + H_F[I - CF[CF]^\dagger]C$$

$$Q_h = \left[AF_1 - \sum_{j=1}^{n}\alpha_j^1\lambda_j^1\nu_j^1 \ \cdots \ AF_s - \sum_{j=1}^{n}\alpha_j^s\lambda_j^s\nu_j^s\right]$$

$$F_i = \sum_{j=1}^{n}\alpha_j^i\nu_j^i$$

$$H_F = \frac{1}{2\epsilon}PC^T[I - CF[CF]^\dagger]^TC \qquad (3.25)$$

and λ_i, ν_i are the eigenvalues and eigenvectors of the matrix $A_H = A - HC$ respectively, $\epsilon > 0$ is a positive constant such that there exists a symmetric and positive definite matrix P that the modified Riccati equation

$$0 = A_H P + P A_H^T + P[I\gamma^2 - \frac{1}{\epsilon}C_F^T C_F]P + I + I\epsilon \qquad (3.26)$$

is satisfied, α_i are constants such that F_i can be represented by the summation $\sum_{j=1}^{n} \alpha_j \nu_j$, $C_F = [I - CF[CF]^\dagger]C$, and the superscript $\dagger$ means the pseudo-inverse of a matrix. Note that the solutions of (3.25) and (3.26) can only be obtained by an iterative algorithm.

Note that (3.26) given in [20] has a misprint.

Remark 3.3 *The Thau-type observer (3.24) for sensor fault detection and isolation was presented in [29]. In this case the selection of the matrix H is simplified because no directionality is required.*

A different approach, including a more general class of systems, was considered in [42]. This approach is based on a generalization of the notion of (h,f)-inva riance using the so-called "algebra of functions" [41]. The same idea, but utilizing the disturbance decoupling approach, was proposed in [38]. Based on the invariance principle (or on the disturbance decoupling approach), a set of state transformations for the system (2.1) is defined. Each transformation maps the state of (2.1) (or the system (3.11)) into a subsystem that depends only on one fault or on a set of faults, and is robust to the rest of them. At this point, nonlinear identity observers are used to build a bank of residuals that have the desired fault directionality.

3.6   Observer for Fault Diagnosis in Bilinear Systems

Fault-diagnostic methods for bilinear systems have been studied only in recent years, maybe because sometimes the linear approaches (such as the UIO) can be extended to bilinear systems, or because bilinear systems are a special case of more general nonlinear systems.

The study of bilinear systems is, however, important, because a set of physical systems (nuclear reactor systems, suspension systems, fermentation processes, hydraulic drives, heat exchange systems, etc.) **could be modeled** by bilinear equations [32]. Further, it is possible to take advantage of the special model structure in order to improve the design of the residuals.

Different approaches have been proposed [25, 40, 48, 49]. In [40, 48, 49] the unknown input fault detection observer approach (in different versions) is extended to bilinear systems.

In [25] a more general (and maybe a more realistic) class of bilinear systems was considered. The approach includes systems represented by

$$\dot{x} = A_0 x + \sum_{j=1}^{p} A_j u_j x + \Psi(u, y) + \sum_{i=1}^{2}(E_i x + F_i) f_i$$

$$y = Cx. \tag{3.27}$$

For sake of simplicity, like in [25], only two possible faults are considered here. The approach given in [25] is reviewed. A fault-detection filter for the system (3.27) is given by

$$\dot{z} = \hat{A}_0 z + \sum_{j=1}^{p}[\hat{A}_j u_j z + \hat{B}_j u_j y] + P\Psi(u, y) + \hat{B}_0 y + R^{-1} L_2^T [L_1 y - L_2 z]$$

$$r = L_1 y - L_2 z$$

$$\dot{R} = -\theta R - \hat{A}_u^T R - R\hat{A}_u + L_2^T L_2$$

$$\hat{A}_u \triangleq \hat{A}_0 + \sum_{j=i}^{p} \hat{A}_j u_j$$

if the conditions

$$\hat{A}_i P - P A_i + \hat{B}_i C = 0$$

$$L_1 C - L_2 P = 0$$

$$P \begin{bmatrix} E_2 & F_2 \end{bmatrix} = 0$$

with L_1, L_2 and P non-zero, and $u(t)$ a θ-strictly persistent input [25] for the system

$$\dot{\eta} = \hat{A}_0 \eta + \sum_{j=1}^{p} \hat{A}_j u_j \eta$$

$$q = L_2 \eta, \tag{3.28}$$

are satisfied.

Here $u(t)$ is said to be a θ-strictly persistent input for the system (3.28) if:

$$\exists t_0 > 0, \ \exists \alpha > 0, \text{ such that for any } t > t_0$$

$$\int_0^t e^{-\theta(t-s)} \phi_u(t-s)^T C^T C \phi_u(t-s) ds \geq \alpha I$$

where I is the nxn identity matrix, ϕ_u is the transition matrix of (3.28) and θ must be positive real.

4   Nonlinear Observer Design via Lipschitz Condition

Consider a class of nonlinear systems described by

$$E\dot{x} = Ax + Bu + f(x, u) \tag{4.29}$$

$$y = Cx \tag{4.30}$$

where $x \in \mathbb{R}^n$, $y \in \mathbb{R}^p$, $u \in \mathbb{R}^m$ and $f(x, u) \in \mathbb{R}^q$ and the matrices A, B, C and E have appropiate dimensions. The matrice A, B, C and E are known.

In this section, each component of the nonlinearity can be a nonlinear function on the state and the input too. First, an observer is designed for the system (4.29), (4.30). Then its stability is discussed. It is shown in which cases this observer can be applicable. Finally the residual generation for fault detection and isolation is presented.

A nonlinear observer for a class of nonlinear regular systems was presented by [24]. [36] discussed the same observer as [24], but better results of the upper bound of the Lipschitz constant were obtained by [36]. Generalizing the Lyapunov-like equation [37], the upper bound of the Lipschitz constant can be augmented/greater. The observer design was discussed also for nonlinear singular systems. [10] presented a reduced order observer for nonlinear system, which is independant on the control variable. A method to reconstruct the whole state of a class of nonlinear singular systems is given in [19]. In this section, the gain matrix of the observer presented by [19] will be obtained with a more general Lyapunov-like equation. So a better solution for the upper bound of the Lipschitz constant can be found.

4.1   Observer Presentation

So that the observer can be designed, the following three hypotheses must be satisfied.

- The row vectors of the matrices C and E must be a basis of the n-dimensional vector space:

$$rank \begin{pmatrix} E \\ C \end{pmatrix} = n \tag{H1}$$

- The linear part of the system has to be observable:

$$rank \begin{pmatrix} sE - A \\ C \end{pmatrix} = n \tag{H2}$$

- The nonlinearity $f(x, u)$ satisfies a Lipschitz condition, which requires that there exists a positive constant, ϵ, such that

$$||f(\hat{x}, u) - f(x, u)|| \leq \epsilon ||\hat{y} - y|| \tag{H3}$$

Moreover, if the measurement matrix has full row rank, i.e.:

$$rank(C) = p \tag{4.31}$$

is satisfied, the matrix computation is much easier. Under these hypotheses the following procedure can be used to design an observer for FDI.

Proposition 3.1 *The parameterized system equation* (α, κ)

$$\dot{z} = Nz + Ly + Gu + Rf(\hat{x}, u) - P^{-1}C^T(\hat{y} - y) \tag{4.32}$$

$$\hat{x} = z + Ky \tag{4.33}$$

$$\hat{y} = C\hat{x} \tag{4.34}$$

where $\hat{x}, z \in \mathbb{R}^n$, $\hat{y} \in \mathbb{R}^p$ and $f(\hat{x}, u) \in \mathbb{R}^q$ and the matrices N, L, G, R, K and P have appropriate dimensions, is a stable observer of the nonlinear system (4.29), (4.30), where the Lipschitz constant ϵ must hold the following inequality:

$$\epsilon \leq \epsilon_0(\alpha, \kappa) \tag{4.35}$$

with

$$\epsilon_0(\alpha, \kappa) = \frac{\lambda_{min}((2 - \alpha)C^T C + \kappa P)}{2\sigma_{max}(PR)} \tag{4.36}$$

and the matrices satisfy the following conditions:

$$N - P^{-1}C^T C \qquad stable \tag{4.37}$$

$$G - RB = 0 \tag{4.38}$$

$$NRE + LC - RA = 0 \tag{4.39}$$

$$RE + KC = I_n \tag{4.40}$$

and the matrix P, depending on the parameters α and κ, is the solution of the Lyapunov equation

$$N^T P + PN - \alpha C^T C + \kappa P = 0 \tag{4.41}$$

Note that the parameters (α, κ) have to be chosen so that the matrix P is positive definite and the condition (4.35) is satisfied.$\square$

Proof. The estimation error is defined as:

$$e = \hat{x} - x \tag{4.42}$$

With the estimated state (4.33) and the condition (4.40), the estimation error becomes

$$e = z - REx \tag{4.43}$$

Taking into account (4.32) and (4.29), the time derivative of the estimation error becomes

$$\begin{aligned}
\dot{e} \;=\; & Ne + (LC - RA - NRE)x + (G - RB)u \\
& + R(f(\hat{x}, u) - f(x, u)) - P^{-1}C^T(\hat{y} - y)
\end{aligned} \tag{4.44}$$

Using the matrix conditions (4.38) and (4.39), the error dynamics are governed by the following equation

$$\dot{e} = (N - P^{-1}C^T C)e + R(f(\hat{x}, u) - f(x, u)) \tag{4.45}$$

To discuss the stability, the direct method of Lyapunov is applied. Consider the following Lyapunov function

$$V = e^T P e \tag{4.46}$$

This function V is positive definite if and only if the time constant matrix P is positive definite, i.e., if the eigenvalues of P are positive. So the second step is the discussion of the negative definiteness of the time derivative. Taking into account the error dynamic (4.45) and the Lyapunov equation (4.41), the time derivative becomes

$$\dot{V} = -e^T((2 - \alpha)CC + \kappa P)e + 2e^T PR(f(\hat{x}, u) - f(x, u)) \tag{4.47}$$

The second term must be overestimated by the greatest singular value of PR, because R can only be a square matrix if the system is regular, i.e., if E is a square matrix with full rank. The first term can be overestimated by the smallest eigenvalue of the matrix:

$$\begin{aligned}
\dot{V} \;=\; & -\lambda_{min}((2 - \alpha)CC + \kappa P)\|e\|^2 \\
& + 2\sigma_{max}PR\|e\|\|f(\hat{x}, u) - f(x, u)\|
\end{aligned} \tag{4.48}$$

Applying the Lipschitz condition (H3), the time derivative of the Lyapunov function can be overestimated as follows:

$$\dot{V} = (-\lambda_{min}((2-\alpha)CC + \kappa P) + 2\epsilon\sigma_{max}PR)\|e\|^2 \qquad (4.49)$$

Now it can be concluded that the time derivative of the Lyapunov function is positive definite, if the condition (4.36) is satisfied. □

An observer for the above nonlinear systems can be designed, if the nonlinearity satisfies locally the Lipschitz condition. It has been proved above that the asymptotic stability holds if the Lipschitz condition (H3) is satisfied. There are two parameters to design the observer. The main problem is the stability, but if there are different pairs of the parameters α and κ which satisfy the stability condition, the dynamics of the observer can be a second criterion to choose the parameters.

A further interesting point will be the conditions of the parameters α and κ so that the positive definiteness of the Lyapunov matrix P can be guaranteed.

Note that the matrix computation of the proposed observer satisfying the conditions (4.37)-(4.40) is presented in [19]. Normally, a system is not singular. But it can be said that a model of a system is not fully known. In the next section details the kind of systems to which the proposed observer can be applied.

4.2 Contribution of this Observer

A nonlinear observer for nonlinear, singular systems was presented in section 4.1. This observer is a generalization of the observer presented by [19, 24, 36, 37]. The proposed observer can be applied for failure diagnosis of nonlinear, singular systems. It can be shown that this observer is also applicable for nonlinear systems with unknown inputs, which can be described by the following equations:

$$\dot{x} = A_k x + B_k u + f_k(x, u) + D_k d \qquad (4.50)$$
$$y = Cx \qquad (4.51)$$

where $d \in \mathbb{R}^{n-q}$ is the vector of unknown inputs and q has to be smaller than n. D is a known matrix, whose rank is equal to $(n-q)$. If there exists a matrix E_k, so that $E_k D_k = 0$ and the hypotheses of the presented observer for the nonlinear singular system are all satisfied, the observer can also be used to reconstruct the state of the nonlinear system under unknown inputs.

The analytical equation of the nonlinear system with unknown inputs can be transformed into the equations of the nonlinear singular system by multiplying (4.50) from the left by the matrix E_k

$$E_k \dot{x} = E_k A_k x + E_k B_k u + E_k f_k(x, u) + E_k D_k d \tag{4.52}$$

Taking into account that $E_k D_k = 0$, (4.52) can be compared with the differential equation of the singular system (4.29). The following relations can be obtained

$$E = E_k \tag{4.53}$$
$$A = E_k A_k \tag{4.54}$$
$$B = E_k B_k \tag{4.55}$$
$$f(x, u) = E_k f_k(x, u) \tag{4.56}$$

If the condition $E_k D_k = 0$ is satisfied, it can be concluded, that the proposed observer can be well applied for a class of nonlinear regular systems, nonlinear singular systems and nonlinear systems with unknown inputs. Now it will be shown that the state can be well reconstructed applying the proposed observer for a class of nonlinear uncertain systems. Consider the nonlinear uncertain system described by the following equations:

$$\dot{x} = (A_u + \Delta A_u)x + (B_u + \Delta B_u)u + f_u(x, u) \tag{4.57}$$
$$y = Cx \tag{4.58}$$

Under the constraint that the uncertain matrices ΔA and ΔB are parameterized as in [35]:

$$\Delta A_u = D_u \nabla F_1 \tag{4.59}$$
$$\Delta B_u = D_u \nabla F_2 \tag{4.60}$$

Taking into account the parameterization of the uncertain matrices (4.59) and (4.60), the differential equation of the nonlinear uncertain system can be written in the form:

$$\dot{x} = A_u x + B_u u + f_u(x, u) + D_u(\nabla F_1 x + \nabla F_2 u) \tag{4.61}$$

Comparing this differential equation with the differential equation of the nonlinear system with unknown inputs (4.50), it can be concluded that these two equations are equal if the unknown inputs are defined as follows:

$$d = (\nabla F_1 x + \nabla F_2 u) \tag{4.62}$$

This shows that the proposed observer can be applied for the above class of nonlinear systems, even if the system is uncertain or if it has unknown inputs.

4.3 Residual Generation

In the last paragraph, an observer for a class of nonlinear singular systems was proposed and different cases were presented of systems to which this observer can be applied. Now the aim is to give an approach for generating the residuals for nonlinear singular systems with faults and/or parameter uncertainties. Consider a nonlinear singular system with faults θ_f and uncertainties θ_d of the form:

$$
\begin{aligned}
E\dot{x} &= Ax + Bu + f(x,u) + \Psi(x,u)\theta_d + \theta_f & (4.63)\\
y &= Cx & (4.64)
\end{aligned}
$$

where the matrices A, B, C and the nonlinear matrix $\Psi(x,u)$ are well known. The residual of this observer can be defined as follows:

$$r = C(\hat{x} - x) \tag{4.65}$$

where $\hat{x} - x$ is the state estimation error e, and the residual is equal to the estimation error of the output. The dynamic of the state estimation error becomes:

$$\dot{e} = \dot{z} - RE\dot{x} \tag{4.66}$$

Taking into account the system dynamics with faults (4.63), the observer dynamics (4.32) and the matrices conditions (4.37)-(4.40), the state estimation error becomes:

$$\dot{e} = (N - P^{-1}C^T C)e + R(f(\hat{x},u) - f(x,u)) - R\Psi(x,u)\theta_d - R\theta_f \tag{4.67}$$

Determining this proposed residual, faults and/or uncertainties can be detected.

5 Conclusions

In this chapter the different approaches to the design of nonlinear observers for residual generation for FDI in nonlinear systems have been briefly reviewed. The survey also incorperates some recent results obtained with a nonlinear observer that has been designed for a class of nonlinear singular systems. This observer is designed with a Lyapunov-like equation with two degrees of freedom. This allows the determination of an upper bound of the Lipschitz constant better than in [19]. The whole state of the system can be reconstructed if the three hypotheses given in the chapter are satisfied and if only part of the process is modeled. This observer design can also be applied to FDI of nonlinear systems with unknown inputs or for a class of nonlinear systems with uncertainties. Note however, that the relationship of of the fault sensitivity with the degrees of freedom is still an open problem.

As can be seen, the fault detection problem for nonlinear systems is still neither generally nor completely solved. This chapter presents ideas for residual generation with nonlinear observers under the restriction to certain classes of nonlinear systems. Also, the fault isolation problem is a further interesting issue, in which there are still many open questions, because of the well known difficulties associated with the design of nonlinear observers not only for feedback control but also for fault diagnosis.

6 REFERENCES

[1] H. K. Adjallah, D. Maquin and J. Ragot. Non-linear Observer-Based Fault Detection. 3rd IEEE Conference on Control Applications, United Kingdom, pp. 1115-1120, 1994.

[2] E. Alcorta García and P. M. Frank. On the Relationship Between Observer and Parameter Identification Based Approaches to Fault Detection. 13th World Congres of IFAC, San Francisco, USA, Vol. N, pp. 25-29, 1996.

[3] E. Alcorta García and P. M. Frank. Deterministic Nonlinear Observer-based Approaches to Fault Diagnosis: A Survey. Control Engineering Practice, Vol. 5, pp. 663-670, 1997.

[4] M. Basseville. Detecting Changes in Signals and Systems - A Survey. Automatica, Vol. 24, pp. 309-326, 1988.

[5] G. Bastin and M. R. Gevers. Stable Adaptive Observers for Nonlinear Time-varying Systems. IEEE Transactions on Automatic Control, Vol. 33, pp. 650-657, 1988.

[6] D. Bestle and M. Zeitz. Canonical Form Observer Design for Nonlinear

Time- invariant Systems. *International Journal of Control, Vol. 38, pp. 419-431, 1983.*

[7] L. Billmann and R. Isermann. Leak Detection Methods for Pipelines. *Automatica, Vol. 23, pp. 381-385, 1987.*

[8] J. Birk and M. Zeitz. Extended Luenberger Observer for Non-Linear Multivariable Systems. *International Journal of Control, Vol. 47, pp. 1823-1836, 1988.*

[9] M. Blanke, R. Izadi-Zamanabadi, S. A. Bogh and C. P. Lunau. Fault Tolerant Control Systems - a Holistic View. *Control Engineering Practice, Vol., pp., 1997.*

[10] M. Boutayeb and M. Darouach. Observer Design for Non Linear Descriptor Systems. 34th *Conference on Decision and Control, New Orleans, USA, pp. 2369-2374, 1995.*

[11] G. Delmaire, J.-P. Cassar and M. Staroswiecki. Identification and Parity Space Techniques for Failure Detection in SISO Systems Including Modelling Error. 33rd *Conference On Decision and Control, Florida, USA, pp. 2279-2285, 1994.*

[12] X. Ding and P. M. Frank. Fault Diagnosis Using Adaptive Observers. *SICICI'93, Singapore, 1992.*

[13] P. M. Frank. Advanced Fault Detection and Isolation Schemes Using Nonlinear and Robust Observers. 10th *World Congress on Automatic Control IFAC'87, Vol. 3, pp. 63-68, 1987.*

[14] P. M. Frank. Fault Diagnosis in Dynamic Systems Using Analytical and Knowledge- based Redundancy - A Survey and some new Results. *Automatica, Vol. 26, pp. 459-474, 1990.*

[15] P. M. Frank. On-line Fault Detection in Uncertain Nonlinear Systems Using Diagnostic Observer: A Survey. *International Journal of Systems Science, Vol. 25, pp. 2129- 2154, 1994.*

[16] P. M. Frank and X. Ding. Frequency Domain Approach to Optimally Robust Residual Generation and Evaluation for Model Based Fault Diagnosis. *Automatica, Vol. 30, pp. 789- 804, 1994.*

[17] P. M. Frank. Application of Fuzzy Logic to Process Supervision and Fault Diagnosis. *IFAC Safeprocess, Finland, pp. 631-538, 1994.*

[18] P. M. Frank. Analytical and Qualitative Model-based Fault Diagnosis - A Survey and Some New Results. *European Journal of Control, Vol. 2, pp. 6-28, 1996.*

[19] B. Gaddouna Ouladsine, G. Schreier and J. Ragot. Asymtotic Observer for a Nonlinear Descriptor System. *CESA IMACS, Symposium on Control, Optimization and Supervision, France, pp. 374-379, 1996.*

[20] V. Garg and J. K. Hedrick. Fault Detection Filters for a Class of Nonlinear Systems. *American Control Conference, Seattle, USA, pp. 1647-1651, 1995.*

[21] J. P. Gauthier, H. Hammouri and S. Othman. A Simple Observer for Nonlinear Systems, Applications to Bioreactor. *IEEE Transactions on Automatic Control, Vol. 37, pp. 875-880, 1992.*

[22] J. Gertler. Model Based Fault Diagnosis. *Control-Theory and Advanced Technology, Vol. 9, pp. 259-285, 1993.*

[23] J. Gertler. Disgnosing Parametric Faults: Form Parameter Estimation to Parity Space. *American Control Conference, Seatle, Washington, USA, pp. 1615-1620, 1995.*

[24] M. A. Hammami. Stabilization of a Class of Nonlinear Systems Using an Observer Design. 32nd *Conference on Decicion and Control, pp. 1954-1959, 1993.*

[25] H. Hammouri, M. Kinnaert and E. H. El Yaagoubi. Fault Detection and Isolation for State Affine Systems. *European Journal of Control, Vol.4, pp. 2-16, 1998.*

[26] D. Hengy and P. M. Frank. Component Failure Detection Using Local Second-Order Observers. *IFAC Workshop, Kyoto, Japan, 1986.*

[27] R. Isermann. Process Fault Detection Based on Modeling and Estimation Methods A Survey. *Automatica, Vol. 20, pp. 387-404, 1984.* tection and Isolation. *European Control Conference, pp. 1970-1974, 1993.*

[28] M. Kinnaert, Y. Peng and H. Hammouri. The Fundamental Problem of Residual Generation for Bilinear Systems up to Output Injection. *European Control Conference, Italy, pp. 3777-3782, 1995.*

[29] V. Krishnaswami and G. Rizzoni. A Survey of Observer-Based Residual generation for FDI. *IFAC Safeprocess, Finland, pp. 34-39, 1994.*

[30] J.-F. Magni. On Continuous Time Parameter Identification by using Observers. *IEEE Transactions on Automatic control, Vol. 40, pp. 1789-1792, 1995.*

[31] L. A. Mironovskii. Functional Diagnosis of Dynamic Systems. *Automation and Remote Control, pp. 1122-1143, 1980.*

[32] R. R. Mohler and W. J. Kolodziej. An Overview of Bilinear System Theory and Applications. *IEEE Transactions on Systems, Man and Cybernetics, Vol. SMC-10, pp. 683-688, 1980.*

[33] W. Nuninger, F. Kratz and J. Ragot. Structural Equivalence Between Direct Residuals Based on Parity Space and Indirect Residuals Based on Unknown Input Observer. *IFAC Safeprocess, United Kingdom, pp. 462-467, 1997.*

[34] R. J. Patton. Robust Model-based Fault Diagnosis: The State of the Art. *IFAC Safeprocess, Finland, pp. 1-24, 1994.*

[35] I. R. Petersen. A Stabilization Algorithm for a Class of Uncertain Linear Systems. *Systems and Control Letters, Vol. 8, pp. 181-188, 1987.*

[36] G. Schreier, J. Ragot, R. J. Patton and P. M. Frank. Observer Design for a Class of Nonlinear Systems. *IFAC Safeprocess, United Kingdom, pp. 498-503, 1997.*

[37] G. Schreier, P. M. Frank and F. Kratz. Stability Discussion of an Observer for a Class of Nonlinear Systems. *IAR Annual Conference, France, pp. 68-75, 1998.*

[38] R. Seliger and P. M. Frank. Fault Diagnosis by Disturbance Decoupled Nonlinear Observers. 30th *Conference on Decision and Control, England, pp. 2248-2253, 1991.*

[39] R. Seliger and P. M. Frank. Robust Component Fault Detection and Isolation in Nonlinear Dynamic Systems. *IFAC Safeprocess, Germany, pp. 313-318, 1991.*

[40] D. N. Shields. Quantitative Approaches for Fault Diagnosis Based in Bilinear Systems. *IFAC, 13th Triennial World Congress, pp. 151-156, 1996.*

[41] A. Ye Shumsky. Failure Detection and Isolation in Nonlinear Systems Based on Robust Observer Approach. *TOOLDIAG, France, pp. 524-530, 1993.*

[42] A. Ye Shumsky. Failure Detection Filter for Diagnosis of Nonlinear Dynamic Systems. *IFAC Safeprocess, Finland, pp. 335-340, 1994.*

[43] F. E. Thau. Observing the State of Non-linear Dynamic Systems. *International Journal of Control, Vol. 17, pp. 471-479, 1973.*

[44] C. Thybo and M. Blanke. Industial Cost-Benefit Assessment for Fault Tolerant Control Systems. *International Conference on Control, Wales Swansea, United Kingdom, pp. 1151-1156, 1998.*

[45] J. E. White and J. L. Speyer. Detection Filter Design: Spectral Theory and Algorithms. *IEEE Tranactions on Automatic Control, Vol. 32, pp. 593-603, 1987.*

[46] A. S. Willsky. A Survey of Design Methods for Failure Detection in Dynamic Systems. *Automatica, Vol. 12, pp. 601-611, 1976.*

[47] J. Wünnenberg. Observer-Based Fault Detection in Dynamic Systems. *VDI-Fortschrittsbericht, VDI-Verlag, Reihe 8, Nr. 222, Germany.*

[48] H. Yang and M. Saif. Nonlinear Adaptive Observer Design for Fault Detection. *American Control Conference, Seattle, USA, pp. 1136-1139, 1995.*

[49] D. Yu and D. N. Shields. Fault Diagnosis in Bilinear Systems - A Survey. *European Control Conference, Italy, pp. 360-365, 1995.* Systems Research. 12th *IFAC Congress, Sydney, Australia, Vol. 3, pp. 485-488, 1993.*

[50] Z. Zhuang and P. M. Frank. Qualitative Observer and its Application to Fault Detection and Isolation Systems. *Journal of Systems and Control Engineering, I MECH E, Vol. 211, Part I, pp. 253-262, 1997.*

[51] Z. Zhuang, G. Schreier and P. M. Frank. A Qualitative-Observer Approach to Generating and Evaluating Residuals. 37th *Conference on Decision and Control, Florida, USA, pp. 102-107, 1998.*

Application of Nonlinear Observers to Fault Detection and Isolation

H. Hammouri[1], M. Kinnaert[2] and E.H. El Yaagoubi[3]

[1]LAGEP, University of Lyon 1, Lyon, France
[2]Department of Control Engineering, Université Libre de Bruxelles, Brussels, Belgium
[3]LCPI, ENSEM, Casablanca, Morocco

1   Introduction

Fault detection and isolation (FDI) systems differ from classical alarm systems by the fact that they give early warning of faults. Alarm systems essentially process measured signals separately by comparing them to thresholds or by computing their trend. FDI systems take into account the correlation existing between those signals thanks to the use of a mathematical model of the supervised process.

A typical FDI system is made of two parts, a residual generator and a decision module. The residual generator is a filter with the actuator commands and the measured plant outputs as inputs, which generates a set of signals called residuals. The latter have zero mean in the absence of fault (after the filter transient has vanished), and the mean of some of them becomes distinguishably different from zero upon occurrence of specific faults. The decision module processes the residuals in order to decide whether some of them have a mean significantly different from zero (fault detection). Then, by analysing the pattern of non-zero mean residuals, it decides what is(are) the most likely faulty component(s) . This operation is called fault isolation.

In this text, only the design of residual generators is considered. There is a vast literature on this topic, and our aim is not to provide a survey but rather to stress the basic principle behind one approach to residual generation, namely one type of observer based methods. For linear systems, observer based residual generation dates back to the work of Beard [2] and Jones [11]. In their approach, the residuals are the output error of a Luenberger observer of which the gain is tuned in a very specific way. Indeed, the particular choice of the observer gain ensures that the residuals take a fixed direction or lie in a specific plane upon occurrence of a given fault. This problem can be seen as a simultaneous assignment of eigenvalues and eigenvectors [21]. Eigensystem assignment has also been used to tackle robustness issues in detection filters [18]. Massoumnia [14] has considered the same problem in a geometric framework.

Another approach to the synthesis of observer-based residual generators was developed in [15], [22], [4]. The basic idea on which it relies in order to design a residual which is only sensitive to a given fault is the following. One has to determine from the initial model of the plant a detectable subsystem of which the state is not affected by unknown inputs or by faults except for the specific fault to be detected. Next, a Luenberger observer can be designed for this particular subsystem and the output error of the observer is a suitable residual.

The latter approach is reviewed here for linear systems. Next its extension to nonlinear systems is considered. In particular, the application of high gain nonlinear observers for residual generation is investigated. The theory is illustrated by a simulation study on a hydraulic process.

2   Residual Generation for Linear Systems

2.1   Problem Statement

We consider the class of continuous time-invariant linear systems described by the following state space model :

$$\dot{x}(t) = Ax(t) + Bu(t) + F_1 v_1(t) + F_2 v_2(t) \qquad (2.1)$$

$$y(t) = Cx(t) \qquad (2.2)$$

where $x(t) \in \mathcal{X} \subset R^n, u(t) \in \mathcal{U} \subset R^m, y(t) \in \mathcal{Y} \subset R^p, \; v_i(t) \in \mathcal{V}_i \subset R^{n_{v_i}}, i = 1, 2.$

$\mathcal{X}, \mathcal{U}, \mathcal{Y}, \mathcal{V}_i, i = 1, 2$ denote linear vector spaces.

In (2.1),(2.2), $x(t)$ denotes the state of the system, $u(t)$, the known input signals, $y(t)$, the measured output signals, $v_1(t)$ and $v_2(t)$ are unknown functions of time which we call failure modes. In the ith failure mode, the following relations hold : $v_i(t) \neq 0, t \geq t_0$ and $v_j(.) = 0, j \neq i$. A, B, F_1, F_2 and C are known matrices, and we assume without loss of generality that F_1 and F_2 have full column rank.

Different types of faults can be modelled in the framework of (2.1),(2.2). If the dynamics of the actuators are negligible with respect to the process time constants, an actuator failure such as a valve sticking can be described as follows. The jamming of the first actuator can be modelled with $F_1 = B_{.,1}, v_1(t) = \bar{u}_1 - u_1(t)$ where $B_{.,1}$ denotes the first column of matrix $B, u_1(t)$, the first component of vector $u(t)$, and $\bar{u}_1$ is the value at which the control signal is stuck.

A leak in an hydraulic system can also be modelled by an additive signal, namely the flow of the leaking fluid. Even a change in the dynamics of the plant could be considered as a fault of the type indicated in (2.1), (2.2), by choosing adequately F_i and $v_i(t)$.

The simplest problem of residual generator design, called the fundamental problem of residual generation (FPRG) can be stated as follows, for system (2.1), (2.2):

(FPRG) Determine a linear time-invariant system with inputs $u(t)$ and $y(t)$, and output $r(t) \in R^q$ such that :

1) In the absence of fault (i.e. when $v_i(t) = 0, i = 1,2$),$r(t)$ asymptotically decays to zero.

2)In the second failure mode (i.e. when $v_2(t) \neq 0, t \geq t_0, t_0$ being the fault occurrence time), $r(t)$ asymptotically decays to zero.

3) In the first failure mode (i.e. when $v_1(t) \neq 0, t \geq t_0, t_0$ being the fault occurrence time), $r(t)$ does not asymptotically decay to zero.

A restatement of this problem using the terminology of linear system theory is instrumental in the determination of a solution, especially in the framework of geometric system theory. This is the object of the next subsection.

2.2 Second Problem Formulation

The most general form of linear time-invariant (LTI) system with inputs $u(t)$ and $y(t)$ and output $r(t)$ is :

$$\dot{w}(t) = A_r w(t) + B_r u(t) + M_r y(t) \qquad (2.3)$$
$$r(t) = C_r w(t) + D_r u(t) + N_r y(t) \qquad (2.4)$$

where $w(t) \in \mathcal{W}$. Subsequently, n_r and q denote the dimension of $w(t)$ and $r(t)$ respectively.

Combining (2.1),(2.2) and (2.3), (2.4) yields :

$$\begin{bmatrix} \dot{x}(t) \\ \dot{w}(t) \end{bmatrix} = \begin{bmatrix} A & 0 \\ M_r C & A_r \end{bmatrix} \begin{bmatrix} x(t) \\ w(t) \end{bmatrix}$$
$$+ \begin{bmatrix} B & F_2 \\ B_r & 0 \end{bmatrix} \begin{bmatrix} u(t) \\ v_2(t) \end{bmatrix} + \begin{bmatrix} F_1 \\ 0 \end{bmatrix} v_1(t)$$

$$(2.5)$$

$$r(t) = \begin{bmatrix} N_r C & C_r \end{bmatrix} \begin{bmatrix} x(t) \\ w(t) \end{bmatrix} + \begin{bmatrix} D_r & 0 \end{bmatrix} \begin{bmatrix} u(t) \\ v_2(t) \end{bmatrix}$$

$$(2.6)$$

Introducing the extended state $x_e(t) = [x^T(t), w^T(t)]^T$, which belongs to $\mathcal{X}_e = \mathcal{X} \oplus \mathcal{W}$, and the extended control signal $u_e(t) = [u^T(t), v_2^T(t)]^T$, which belongs to $\mathcal{U}_e = \mathcal{U} \oplus \mathcal{V}_2$, (2.5), (2.6) can be written :

$$\dot{x}_e\,(t) \;=\; A_e x_e(t) + B_e u_e(t) + F_e v_1(t) \qquad (2.7)$$

$$r(t) \;=\; C_e x_e(t) + D_e u_e(t) \qquad\qquad\quad (2.8)$$

The definition of the different matrices is obvious from (2.5), (2.6).

We now restate the FPRG as a set of conditions to be fulfilled by (2.7), (2.8).

Clearly 1) and 2) in the definition of the FPRG are equivalent to :

1') the map $u_e(t) \to r(t)$ is zero

2') the observable modes of the pair (C_e, A_e) are asymptotically stable.

Several criteria can be considered for condition 3) in the FPRG, as discussed in [15]. As in the latter reference, the requirement that the system relating $v_1(t)$ to $r(t)$ be input observable is imposed here. Remember that the map $v_1(t) \to r(t)$ is input observable if the magnitude v_1, of a step like fault $v_1(t)$ can be determined uniquely from $r(t), t \geq 0$ when $x_e(0) = 0$.

Subsequently, we consider that, input observability of the map $v_1(t) \to r(t)$ is sufficient to guarantee condition (3) of the FPRG in practice.

This yields a new statement for the FPRG :

(FPRG1) Determine a system of the form (2.3), (2.4) such that :

1') the map $u_e \longmapsto r$ is zero,

2') the observable modes of the pair (C_e, A_e) are asymptotically stable,

3') the map $f \longmapsto r$ is input observable.

2.3 Principle of the Solution

As already announced in the introduction, the solution relies on the determination of a detectable system with v_1 as only unknown input, from the original state space model of the plant,(2.1),(2.2).

To this end, an output injection map $L : \mathcal{Y} \to \mathcal{X}$, and an output mixing map $H : \mathcal{Y} \to \mathcal{Y}$ are introduced in order to define the following system class :

$$\dot{x}\,(t) \;=\; (A + LC)\,x(t) - Ly(t) + Bu(t) + F_1 v_1(t) + F_2 v_2(t) \quad (2.9)$$

$$z(t) \;=\; HCx(t)$$

The major part of the design consists in determining the matrices L and H so that the pair $(HC, A + LC)$ is unobservable, and $Im F_2$ is included in the unobservable subspace of the pair $(HC, A + LC)$. Let h denote the dimension of this subspace. Once such matrices are obtained, there exists a linear change of coordinates $x = T\bar{x}$ such that system (2.9) can be written in the standard form for a nonobservable system [12] :

$$\dot{\bar{x}}_1(t) = \overline{A}_{11}\bar{x}_1(t) - \overline{L}_1 y(t) + \overline{B}_1 u(t) + \overline{F}_{11} v_1(t) \tag{2.10}$$

$$\dot{\bar{x}}_2(t) = \overline{A}_{21}\bar{x}_1(t) + \overline{A}_{22}\bar{x}_2(t) - \overline{L}_2 y(t) + \overline{B}_2 u(t) + \overline{F}_{21} v_1(t) + \overline{F}_{22} v_2(t)$$

$$z(t) = \overline{C}_1 \bar{x}_1(t) \tag{2.11}$$

where $\bar{x}_1 \in R^{(n-h)}$, $\bar{x}_2 \in R^h$, and

$$T^{-1}(A+LC)T = \left(\begin{array}{cc} \overline{A}_{11} & 0 \\ \overline{A}_{21} & \overline{A}_{22} \end{array} \right) ; T^{-1}B = \left(\begin{array}{c} \overline{B}_1 \\ \overline{B}_2 \end{array} \right) ; T^{-1}F_1 = \left(\begin{array}{c} \overline{F}_{11} \\ \overline{F}_{21} \end{array} \right) ;$$

$$T^{-1}F_2 = \left(\begin{array}{c} 0 \\ \overline{F}_{22} \end{array} \right) ; T^{-1}L = \left(\begin{array}{c} \overline{L}_1 \\ \overline{L}_2 \end{array} \right) ; HCT = \left(\overline{C}_1 \quad 0 \right)$$

The first $(n-h)$ rows of $T^{-1}F_2$ are null since ImF_2 lies in the unobservable subspace of the pair $(HC, A+LC)$. Moreover, the pair $(\overline{C}_1, \overline{A}_{11})$ is observable by construction.

This last remark implies that we can build a linear observer for estimating $\bar{x}_1(t)$ from (2.10),(2.11) when $v_1(t) = 0, t \geq 0$:

$$\dot{w}(t) = \overline{A}_{11} w(t) - \overline{L}_1 y(t) + \overline{B}_1 u(t) \tag{2.12}$$
$$+ K(z(t) - \overline{C}_1 w(t)),$$

We claim that the output reconstruction error :

$$r(t) = z(t) - \overline{C}_1 w(t) = \overline{C}_1 \varepsilon(t) \tag{2.13}$$

is a suitable residual provided some additional condition are fulfilled to ensure 3') in FPRG1.

Indeed, notice that $\varepsilon(t)$ is governed by:

$$\dot{\varepsilon}(t) = \left(\overline{A}_{11} - K\overline{C}_1 \right) \varepsilon(t) + \overline{F}_{11} v_1(t) \tag{2.14}$$

Hence $r(t)$ asymptotically decays to zero when $v_1 \equiv 0$.

To fulfil FPRG1, the map $v_1 \to r$ must be input observable. It can be shown that this condition is verified provided,

$$ImF_1 \cap S(HC, A+LC|ImF_2) = 0$$

where $S(HC, A+LC|ImF_2)$ denotes the unobservable subspace of $(HC, A+LC)$ (containing ImF_2).

Notice that $S(HC, A+LC|ImF_2)$ is a (C, A) unobservability subspace (u.o.s.) containing ImF_2. Indeed, a subspace R is a (C, A) u.o.s. if it is the unobservable subspace of a pair $(GC, A+MC)$ for some $p \times p$ and $n \times p$ matrices G and M[14]. It can be shown that the set of u.o.s. containing ImF_2 has an infimal element, S^*. It turns out that necessary and sufficient conditions for FPRG1 to have a solution can be expressed in terms of this subspace, namely :

Theorem 4.1 *[15] FPRG1 has a solution if and only if*

$$\mathcal{S}^* \cap ImF_1 = 0 \qquad (2.15)$$

where $\mathcal{S}^* := infS(C, A; ImF_2)$ *is the smallest* (C, A)*-unobservability subspace containing* ImF_2. *Moreover, if (2.15) holds the dynamics of the residual generator, i.e. the eigenvalues of* $\overline{A}_{11} - K\overline{C}_1$ *in (2.12), can be assigned arbitrarily.*

3 Residual Generation for Nonlinear Systems

3.1 Introduction

The basic idea behind the design of residual generators for linear systems can be extended to nonlinear systems, provided the appropriate nonlinear notions are used. One of the problems that arises in the extension is the design of an asymptotic observer for the nonlinear system from which the residual is deduced as the output reconstruction error. One has to restrict the considered systems to a specific class to ensure the existence of an asymptotic observer. Here observers for uniformly observable nonlinear systems will be used. Observers with linear error dynamics have been used by Seliger and Frank [19]. Other classes of nonlinear systems have been considered elsewhere, such as bilinear systems [23], [13], and state affine systems [7].

The remaining part of this section is organised as follows. A review of some basic notions from observability theory for nonlinear systems is presented. Next, nonlinear observers for uniformly observable systems are described. Finally those preriquisites are applied to design nonlinear residual generators, and the theory is illustrated by a simulation on a hydraulic process.

3.2 Basic Notions

For the sake of simplicity, we only consider control affine nonlinear systems :

$$\begin{cases} \dot{x} & = & f(x) + \sum_{i=1}^{m} g_i(x)u_i \\ y & = & h(x) & = & (h_1(x), \dots, h_p(x)) \end{cases} \qquad (3.16)$$

where $x(t) \in I\!\!R^n, u(t) = (u_1(t), \dots, u_m(t))^T \in \mathcal{U}$ a mesurable subset of $I\!\!R^m$, $y(t) \in I\!\!R^p$ are respectively the state, the input and the output of the dynamical system (3.16).

System (3.16) is said to be observable if and only if, for every pair of initial states, $(x, \bar{x}), x \neq \bar{x}$, there exist an admissible control $u : [0, T] \rightarrow$

$\mathcal{U}$ and a time instant $t \in [0, T]$ such that $y(x, u, t) \neq y(\bar{x}, u, t)$, where $y(x, u, t) = h(x_u(t))$, and $x_u(t)$ is the unique trajectory of (3.16) such that $x_u(0) = x$. If such an input u exists, we say that u distinguishes $(x, \bar{x})$. An input $u : [0, T] \to \mathcal{U}$, which distinguishes every $(x, \bar{x})$, $x \neq \bar{x}$ is said to be universal on $[0, T]$. System (3.16) is said to be uniformly observable if, for every $T > 0$, every admissible control $u : [0, T] \to \mathcal{U}$ is a universal input on $[0, T]$.

The observation space $\mathcal{O}(h)$ of system (3.16) is defined as the smallest vector space containing $h_1, \ldots, h_p$ and closed under the Lie derivative L_X, where X stands for the vector fields $f, g_1, \ldots, g_m$. This space allows to define a geometric notion of observability, namely the rank observability condition. System (3.16) is observable in the sense of rank at a fixed $x \in \mathbb{R}^n$ if $\dim \, d\mathcal{O}(h)(x) = n$ where $d\mathcal{O}(h)(x) = \{d\tau(x); \tau \in \mathcal{O}(h)\}$ (d is the classical differential operator). This notion extends to nonlinear systems the Kalman rank observability condition for linear systems. For more details on this topic see [9].

3.3 High Gain Observers for Uniformly Observable Systems

In [3] (for a short proof see [5]), single-output nonlinear systems which are uniformly observable are characterized. To describe this result, let $L_f(h)(x)$ denote the Lie derivative of a scalar function h w.r.t. the vector field f, as already mentioned above, and let $L_f^i(h)(x) = L_f(L_f^{i-1}(h)(x))$.

If $\Phi : x \to (h(x), L_f(h)(x), \cdots, L_f^{n-1}(h)(x))^T = z$ is a local diffeomorphism, and if system (3.16), with $y \in \mathbb{R}$, is uniformly observable, then Φ transforms locally system (3.16) into the following canonical form :

$$\begin{cases} \dot{z} & = & A\, z + \gamma(z) + \sum_{i=1}^m u_i \Psi_i(z) \\ y & = & C\, z \end{cases} \tag{3.17}$$

where

$$A = \begin{pmatrix} 0 & 1 & 0 & \cdots & 0 \\ \vdots & \ddots & \ddots & \ddots & 0 \\ \vdots & & \ddots & & 1 \\ 0 & & & \cdots & 0 \end{pmatrix}, \; C = (1, 0, \cdots, 0), \; \gamma(z) = \begin{pmatrix} 0 \\ \vdots \\ 0 \\ \gamma_n(z) \end{pmatrix} \tag{3.18}$$

and $\Psi_i(z) = [\Psi_{i1}(z_1), \Psi_{i2}(z_1, z_2), \cdots, \Psi_{in}(z)]^T$ (i.e. $\Psi_{ij}(z) = \Psi_{ij}(z_1, \cdots, z_j)$).

Under the hypothesis that γ and the Ψ_i's are global Lipschitz, an observer for (3.17) can take the form :

$$\dot{\hat{z}} = A\hat{z} + \gamma(\hat{z}) + \sum_{i=1}^m u_i \Psi_i(\hat{z}) - S_\theta^{-1} C^T (C\hat{z} - y) \tag{3.19}$$

where S_θ is the unique solution of the algebraic Lyapunov algebraic equation :

$$\theta S_\theta + A^T S_\theta + S_\theta A = C^T C \qquad (3.20)$$

More generally, consider the triangular form :

$$\begin{cases} \dot{z} &= Az + \Psi(t, u, z) \\ y &= C\,z \end{cases} \qquad (3.21)$$

where A and C are defined by (3.18), and the ith component $\Psi_i(t, u, z)$ of $\Psi(t, u, z)$ is such that $\Psi_i(t, u, z) = \Psi_i(t, u, z_1, \dots, z_i)$. Moreover, assume that Ψ fulfils hypothesis H1) below:

H1) Ψ is global Lipschitz w.r.t. z, locally w.r.t. u and globally w.r.t. t, i.e. $\forall \alpha > 0; \exists \gamma > 0; \forall z, z' \in \mathbb{R}^n; \forall t \geq 0; \forall u \in \mathbb{R}^m, \|u\| \leq \alpha$, the following inequality holds: $\|\Psi(t, u, z) - \Psi(t, u, z')\| \leq \gamma \|z - z'\|$

Then, an observer for (3.21) can take the form :

$$\dot{\hat{z}} = A\hat{z} + \Psi(t, u, \hat{z}) - S_\theta^{-1} C^T (C\hat{z} - y) \qquad (3.22)$$

where S_θ is given by (3.20).

More precisely, the following result, which is a slight extension of the work reported in [5], holds :

Theorem 4.2 $\forall \alpha > 0; \exists \theta_0 > 0; \forall \theta \geq \theta_0; \exists \lambda_\theta > 0; \exists \mu_\theta > 0 s.t.$

$$\|\hat{z}(t) - z(t)\| \leq \lambda_\theta e^{-\mu_\theta t} \|\hat{z}(0) - z(0)\|$$

for every bounded admissible control with upper bound α. Moreover, $\mu_\theta \to +\infty$ as $\theta \to +\infty$.

This result can be extended to the more general triangular form [8]:

$$\begin{cases} \dot{z} &= A(t)z + \Psi(t, u, z) \\ y &= Cz \end{cases} \qquad (3.23)$$

where

$$A(t) = \begin{pmatrix} 0 & a_1(t) & 0 & \dots & 0 \\ \vdots & \ddots & \ddots & \ddots & 0 \\ \vdots & & \ddots & & a_{n-1}(t) \\ 0 & & \dots & & 0 \end{pmatrix}, \qquad C = (1, 0, \cdots, 0),$$

and the ith component $\Psi_i(t, u, z)$ of $\Psi(t, u, z)$ is such that $\Psi_i(t, u, z) = \Psi_i(t, u, z_1, \dots, z_i)$. To achieve this goal, let us introduce the following hypothesis :

H2) $a_i, i = 1, \ldots, n-1$ are known differentiable functions with unknown derivatives, and there exist $\epsilon > 0, M > 0, M' > 0$ such that, for every $t \geq 0$, $\epsilon \leq |a_i(t)| \leq M$ and $|\frac{d}{dt}a_i(t)| \leq M'$ for $i = 1, \ldots, n-1$.

Under hypotheses H1) and H2), our candidate observer for (3.23) takes the form ;

$$\dot{\hat{z}} = A(t)\hat{z} + \Psi(t, u, \hat{z}) - \Lambda^{-1}(t)S_\theta^{-1}C^T(C\hat{z} - y) \qquad (3.24)$$

where S_θ is solution of (3.20) and $\Lambda(t)$ is the $n \times n$ diagonal matrix $diag(1, a_1(t), a_1(t)a_2(t), \ldots, a_1(t) \ldots a_{n-1}(t))$.

The convergence property of the estimation error for observer (3.24) can be stated as follows.

Theorem 4.3 *Assume that H1) and H2) hold, and consider an arbitrary constant $\alpha > 0$, then $\exists\theta_0 > 0; \forall\theta \geq \theta_0; \exists\lambda_\theta > 0; \exists\mu_\theta > 0 s.t.*$

$$\|\hat{z}(t) - z(t)\| \leq \lambda_\theta e^{-\mu_\theta t}\|\hat{z}(0) - z(0)\|$$

for every admissible control u with upper bound $\|u\|_\infty \leq \alpha$

Proof See [8]

3.4 The Fundamental Problem of Residual Generation for Nonlinear Systems

We consider the following class of nonlinear systems:

$$(\Sigma_{NL}) \qquad \begin{cases} \dot{x} & = f(x) + \sum_{i=1}^m g_i(x)u_i + e_1(x)v_1 + e_2(x)v_2 \\ y & = h(x) \end{cases} \qquad (3.25)$$

where $x(t) \in I\!\!R^n$, $u(t) = (u_1(t), \ldots, u_m(t)) \subset \mathcal{U}$, an open set of $I\!\!R^m$, $v_i(t) \in I\!\!R, i = 1, 2$, $y(t) \in I\!\!R^p$. $f, g_i, i = 1, \ldots, m, e_j, j = 1, 2$ and h are of class $\mathcal{C}^\infty$. $u(t)$ and $y(t)$ are the system input and output signals. They are assumed to be known, while the functions $v_i(t), i = 1, 2$, are arbitrary unknown functions of time. When the first (second) failure mode occurs, $v_1(t)$ $(v_2(t))$ becomes non zero while $v_2(t)(v_1(t))$, remains equal to zero. Sensor failures can also possibly be modelled by $v_i(t)$ if the sensor dynamics is included in the model Σ_{NL} [6]. We restrict our developments to the case where v_1 and v_2 are scalar signals for the sake of simplicity.

In order to define a residual generator for the system Σ_{NL}, we consider the class of smooth systems :

$$\begin{cases} \dot{z} & = f_r(z, u, y) \\ r & = h_r(z, u, y) \end{cases} \qquad (3.26)$$

with inputs u and y, state $z \in I\!\!R^{n_r}$ and $I\!\!R^q$-valued output $r(t)$; f_r and h_r are of class $\mathcal{C}^\infty$.

Now, let $L^\infty([0,T], I\!\!R^n)$ (or simply $L^\infty([0,T])$) denote the space of bounded measurable functions, and $\|\ \|_\infty$ denote its classical norm. In the sequel, for the sake of simplicity, the notation $L^\infty([0,T])$ is used whatever the dimension, N, of the considered signal may be. Set $\mathcal{B}_T^\infty(v_0, \epsilon) = \{v \in L^\infty([0,T]); \|v - v_0\|_\infty < \epsilon\}$ the ball of $L^\infty([0,T])$. Using the fact that $f, g_i, e_j, f_r,$ and h_r are of class $\mathcal{C}^\infty$ w.r.t. their arguments, and noticing that $r(t) = r(x(0), z(0), u, v_1, v_2, t)$, we deduce the following statement.

For every fixed $x(0) \in I\!\!R^n, z(0) \in I\!\!R^{n_r}, u, v_1, v_2$ in $L^\infty([0,\tau]), \tau > 0$; there exist $T \in]0, \tau], \epsilon > 0$ such that :

$$r : \mathcal{B}_T^\infty(v_1, \epsilon) \to L^\infty([0,T]) : \tilde{v}_1 \to r(x(0), z(0), u, \tilde{v}_1, v_2, \bullet)$$

is Frechet differentiable at v_1. We let $(D_F r)(v_1)$ denote the Frechet derivative of r at v_1.

In a similar way, $y(\bullet) = y(x(0), u, v_1, v_2, \bullet)$ is Frechet differentiable w.r.t. v_1, and we let $(D_F y)(v_1)$ denote its Frechet derivative at v_1.

Definition 4.1 *(3.26) is a residual generator for the detection and isolation of failure v_1 in system (3.25) if there exists $\mathcal{U} \subset L_{loc}^\infty$, the space of locally bounded measurable functions such that :*
1) $\forall u \in \mathcal{U}; \forall x(0) \in I\!\!R^n; \forall z(0) \in I\!\!R^{n'}; \forall v_2 \in L_{loc}^\infty$:

$$r(x(0), z(0), u, 0, v_2)(t) \to 0 \quad as \quad t \to +\infty$$

2) $\exists u \in \mathcal{U}; \exists x(0) \in I\!\!R^n; \exists z(0) \in I\!\!R^{n'}; \exists T > 0; \exists v_1, v_2 \in L^\infty([0,T])$ such that $(D_F r)(v_1) \neq 0$

The determination of a residual signal which fulfils 1) and 2) above is called the fundamental problem of residual generation (FPRG) for system (3.25). Condition 1 in definition 4.1 is equivalent to 1) and 2) in FPRG1, for the linear case.

Remark 4.1 *In the linear case, if we denote (C, A, B) a realization of the map $v_1 \to r$, the above Frechet derivative is given by :*

$$(D_F r)(v_1) : \xi \to C \int_0^\bullet e^{(\bullet - s)A} B\xi(s) ds$$

It is independent of $u, x(0), z(0), v_1$ and v_2. Clearly, in this case, condition 2 of definition 4.1 is equivalent to the existence of a left inverse for the transfer function between v_1 and r. The latter requirement also appears in the statement of the FPRG for linear systems when only scalar failure modes are considered, as is the case here (see Section 2.2).

Before giving sufficient conditions for the existence of a residual generator for (3.25), we introduce a few known notions [17].

$\mathcal{O}(h)$ denotes the observation space of system (3.25), with $v_1 = v_2 = 0$. $d\mathcal{O}(h)$ defines a codistribution on $I\!\!R^n$ denoted by the same letters.

$Kerd\mathcal{O}(h)$ is the distribution spanned by all vector fields X such that $L_X(\tau) = 0$ for every $\tau \in \mathcal{O}(h)$. It is invariant under $f, g_1, \ldots, g_m$ (i.e. for each $Y \in Kerd\mathcal{O}(h)$ and each $X \in \{f, g_1, \ldots, g_m\}$, the Lie bracket $[X, Y]$ belongs to $Kerd\mathcal{O}(h)$). Moreover, it is obviously involutive (i.e. $\forall X, Y \in Kerd\mathcal{O}(h), [X, Y] \in Kerd\mathcal{O}(h)$). Finally, notice that $Kerd\mathcal{O}(h)$ can be seen as the largest distribution invariant under $f, g_1, \ldots, g_m$ and contained in $Kerdh$.

Now assume that $dim\, d\mathcal{O}(h)(x) = dim\, d\mathcal{O}(h)(x_0) = k < n$ for every x in some neighbourhood of x_0. Then it is well known that Σ_{NL}, restricted to some neighbourhood V_{x_0} of x_0, can be transformed by a change of coordinates into the following form:

$$\dot{\xi}^1 = \bar{f}_1(\xi^1) + \sum_{i=1}^{m} \bar{g}_{1i}(\xi^1)u_i + \bar{e}_{11}(\xi)v_1 + \bar{e}_{12}(\xi)v_2 \qquad (3.27)$$

$$\dot{\xi}^2 = \bar{f}_2(\xi) + \sum_{i=1}^{m} \bar{g}_{2i}(\xi)u_i + \bar{e}_{21}(\xi)v_1 + \bar{e}_{22}(\xi)v_2 \qquad (3.28)$$

$$y = \bar{h}(\xi^1) \qquad (3.29)$$

where $\xi^1 \in \mathbb{R}^k, \xi^2 \in \mathbb{R}^{n-k}, \xi = (\xi^{1T}\ \xi^{2T})^T, \xi^1 = (\xi_1, \ldots, \xi_k)^T$ and $\xi^2 = (\xi_{k+1}, \ldots, \xi_n)^T$,
$(d\xi_1, \ldots, d\xi_k)$ spans $d\mathcal{O}(h)$ and $(\partial/\partial\xi_{k+1}, \ldots, \partial/\partial\xi_n)$ spans $Kerd\mathcal{O}(h)$.

In the sequel, the same type of diffeomorphism will be used on a different system. For the sake of simplicity, we shall only consider the case where the diffeomorphism $x \to \xi(x)$ is a global one (When this situation does not hold, only initial conditions and inputs for which the associated trajectories lie into V_{x_0} should be considered). Under this assumption we state the following proposition.

Proposition 4.1 *The FPRG for Σ_{NL} has a solution if there exists an output map $\Psi = \varphi \circ h$, with*
$\varphi : \mathbb{R}^p \to \mathbb{R}^{p'}$ *$(p' \le p)$ such that :*
1)

$$e_2 \in Kerd\mathcal{O}(\Psi) \qquad (3.30)$$

2) There exists a set $U \subset L_{loc}^\infty$ such that :

- *2.i) The system (3.27) with output map $\bar{\Psi}(\xi^1) = \Psi(x)$ admits an asymptotic observer for $v_1 = v_2 = 0$ and $u \in U$, of the classical form :*

$$\begin{cases} \dot{\hat{\xi}}^1 = \bar{f}_1(\hat{\xi}^1) + \sum_{i=1}^{m} \bar{g}_{1i}(\hat{\xi}^1)u_i + p(\hat{\xi}^1, G)(\bar{\Psi}(\hat{\xi}_1) - \bar{\Psi}(\xi_1)) \\ \dot{G} = K(u, G, \hat{\xi}_1) \end{cases}$$

$$(3.31)$$

where $\hat{\xi}^1 \in \mathbb{R}^k$, $k = \dim d\mathcal{O}(\Psi)$, p and K are smooth functions

- *2-ii)* $\exists u \in U; \exists x(0) \in \mathbb{R}^n; \exists T > 0; \exists v_1, v_2 \in L^\infty([0,T])$ *s.t.*
  $D_F \bar{\Psi}(\xi^1)(v_1) \neq 0$.

Remark 4.2 : *In the case where U is such that $U|_{[0,T]} = L^\infty([0,T], \mathcal{U})$ for some $T > 0$, $e_1 \notin \ker d\mathcal{O}(\Psi)$ implies condition 2-ii) (see proposition 4.14 [17]).*

Proof. Under hypotheses 1) and 2), there exists a residual generator of the form (3.31), with output $r = \bar{\Psi}(\hat{\xi}^1) - \bar{\Psi}(\xi^1)$ that solves the FPRG. Indeed, we show that conditions 1), 2) of definition 1 are satisfied for the output r. First notice that (3.30) implies $\bar{e}_{12}(\xi) = 0$ in (3.27). Hence, by assumption 2, $r(t) \to 0$ as $t \to \infty$ for every $v_2 \in L^\infty_{loc}$ and $v_1 = 0$, and thus conditiion 1 of definition 4.1 holds. It remains to show condition 2) of the same definition. Assume that $\forall u \in \mathcal{U}, \forall v_1 \in L^\infty_{loc}, \forall v_2 \in L^\infty_{loc}, \forall \xi \in \mathbb{R}^n, \forall \hat{\xi}^1 \in \mathbb{R}^k, D_F r(\xi(0), \hat{\xi}^1(0), u, \bullet, v_2)(v_1) = 0$. This means that r does not depend on v_1. Hence the controlled dynamical system (3.31) does not depend of v_1, and neither does $\bar{\Psi}(\hat{\xi}^1)$. Thus $\bar{\Psi}(\xi^1)$ does not depend of v_1 or, equivalently, $D_F \bar{\Psi}(\xi^1)(v_1) = 0$. This is in contradiction with 2. ii). ∎

3.5   Application of Nonlinear Observers to the FPRG

We now show how the observer (3.19) can be applied for fault detection and isolation. Consider again system (3.25), and suppose that there exists a C^∞ function $\varphi : \mathbb{R}^p \to \mathbb{R}$ satisfying the following assumptions:

- A1) There exists an integer $k \geq 1$ such that the Jacobian of $[\varphi \circ h, \ldots, L_f^{k-1}(\varphi \circ h)]^T$ is of rank k at each $x \in V$, where V denotes some open set of $\mathbb{R}^n$.

- A2) $dL_f^k(\varphi \circ h) \wedge dL_f^{k-1}(\varphi \circ h) \wedge \cdots \wedge d(\varphi \circ h) = 0$
  $dL_{g_i} L_f^j(\varphi \circ h) \wedge dL_f^j(\varphi \circ h) \wedge \cdots \wedge d(\varphi \circ h) = 0$ for $i = 1, \cdots, m$, $j = 0, \cdots, k - 1$,
  where $\wedge$ means the exterior product of differential forms, and $L_f^0(\varphi \circ h) = \varphi \circ h$.

Proposition 4.2 *Under the assumptions A1) and A2), consider system (3.25), with output $\bar{y} = (\varphi \circ h)(x)$. There exists an infinite choice of local systems of coordinates $(\xi_1, \cdots, \xi_n)$ in which system (3.25) takes the form (3.27),(3.28), (3.29) where*

$$\bar{f}_1(\xi^1) = [\xi_2 + \eta_1(\xi_1), \xi_3 + \eta_2(\xi_1, \xi_2), \cdots, \xi_k + \eta_{k-1}(\xi_1, \ldots, \xi_{k-1}), \eta_k(\xi^1)]^T \tag{3.32}$$

$$\bar{g}_{1i}(\xi^1) = \left[\ \bar{g}_{1i}^1(\xi^1)\ \ \cdots\ \ \bar{g}_{1i}^k(\xi^1)\ \right]^T \tag{3.33}$$

with

$$\bar{g}_{1i}^j(\xi^1) = \bar{g}_{1i}^j(\xi_1,\cdots,\xi_j)$$

and

$$\bar{h}(\xi^1) = \xi_1 \tag{3.34}$$

Proof. Using assumption A1), we can construct a diffeomorphism $\Phi = [\Phi_1,\cdots,\Phi_n]^T$ from an open subset $W \subset V$ such that $\Phi_j = L_f^j(\varphi \circ h)$ for $j = 0,\cdots,k-1$. Now set $(\xi_1,\cdots,\xi_n) = (\Phi_1(x),\cdots,\Phi_n(x))$, $\xi^1 = (\xi_1,\cdots\xi_k)^T$ and $\xi^2 = (\xi_{k+1},\cdots,\xi_n)^T$.

Taking the derivative of ξ_j along trajectories of system (3.25), for $j = 1,\cdots,k-1$, yields :

$$\dot{\xi}_j(t) = \xi_{j+1}(t) + \sum_{i=1}^m u_i(t)L_{g_i}(\Phi_j)(x(t)) + v_1 L_{e_1}(\Phi_j)(x(t)) + v_2 L_{e_2}(\Phi_j)(x(t)) \tag{3.35}$$

Assumption A2) implies that $dL_{g_i}(\xi_j) \wedge d\xi_j \wedge \cdots \wedge d\xi_1 = 0$, which means that $L_{g_i}(\xi_j) = \bar{g}_{1i}^j(\xi_1,\cdots,\xi_j)$ for $i = 1,\cdots,m$.

Hence, (3.35) becomes:

$$\dot{\xi}_j(t) = \xi_{j+1}(t) + \sum_{i=1}^m u_i(t)\bar{g}_{1i}^j(\xi_1,\cdots\xi_j) + v_1 \bar{e}_{11}^j(\xi) + v_2 \bar{e}_{12}^j(\xi) \tag{3.36}$$

for $j = 1,\cdots,k-1$.

For the kth component, ξ_k, we obtain :

$$\begin{aligned}
\dot{\xi}_k &= L_f(\Phi_k)(x(t)) \\
&+ \sum_{i=1}^m u_i(t)L_{g_i}(\Phi_k)(x(t)) + v_1 L_{e_1}(\Phi_k)(x(t)) + v_2 L_{e_2}(\Phi_k)(x(t))
\end{aligned}$$

Using again assumption A2), we get :

$$\begin{cases} dL_f(\xi_k) \wedge d\xi_k \wedge \cdots \wedge d\xi_1 &= 0 \\ dL_{g_i}(\xi_k) \wedge d\xi_k \wedge \cdots \wedge d\xi_1 &= 0 \end{cases}$$

Hence,

$$\begin{cases} L_f(\xi_k) &= \eta(\xi_1,\cdots,\xi_k) = \eta_k(\xi^1) \\ L_{g_i}(\xi_k) &= \bar{g}_{1i}^k(\xi_1,\cdots,\xi_k) = \bar{g}_{1i}^k(\xi^1) \end{cases} \tag{3.37}$$

Other choices of coordinates which bring system (3.25) into the form (3.27), (3.28), (3.29) with the particular structure (3.32),(3.33), (3.34) can be obtained as follows :
$\xi_1' = \xi_1$, $\xi_i' = \xi_i + \mu_i(\xi_1, \dots, \xi_{i-1})$ for $i = 2, \dots, k$, where $\mu_i : I\!\!R^{i-1} \to I\!\!R$ are arbitrary C^∞ functions. ∎

Now consider the reduced controlled system :

$$
\begin{cases}
\dot{\xi}^1 &= \bar{f}_1(\xi^1) + \displaystyle\sum_{i=1}^{m} \bar{g}_{1i}(\xi^1)u_i \\
\bar{y} &= \bar{h}(\xi^1)
\end{cases}
\tag{3.38}
$$

where $\bar{f}_1, \bar{g}_{1i}$, and $\bar{h}$ are given respectively by (3.32), (3.33), and (3.34).
Let W_1 denote the open set :

$$
\{\xi^1 \in I\!\!R^k; \exists \xi^2 \in I\!\!R^{n-k} s.t. \begin{pmatrix} \xi^1 \\ \xi^2 \end{pmatrix} \in \Phi(W)\}
$$

$W \subset V$, where W and V are given above.

Without loss of generality, we assume that we are only concerned with a set of initial states and a class U of bounded admissible controls u such that $\|u\|_\infty \leq M$ (M is a given constant) for which the trajectories of (3.38) lie into a domain $W_1' \subset W_1$, and such that $\eta_j, \bar{g}_{1i}, j = 1, \cdots, k, i = 1, \cdots, m$ can be extended to global Lipschitz functions $\tilde{\eta}_j, \tilde{g}_{1i}, j = 1, \cdots, k, i = 1, \cdots, m$ on $I\!\!R^k$ (i.e. $\tilde{\eta}_j|_{W_1'} = \eta_j, \tilde{g}_{1i}|_{W_1'} = \bar{g}_{1i}$ and $\tilde{\eta}, \tilde{g}_{1i}$ are global Lipschitz functions on $I\!\!R^k$). Under this assumption, an observer of the form (3.22) can be constructed in order to estimate exponentially the concerned unknown trajectories of (3.38). More precisely, this observer can be written as follows (see Section 2):

$$
\dot{\hat{\xi}}^1 = \tilde{f}_1(\hat{\xi}^1) + \sum_{i=1}^{m} \tilde{g}_{1i}(\hat{\xi}^1)u_i - S_\theta^{-1}C^TC(\hat{\xi}^1 - \xi^1)
\tag{3.39}
$$

where $\tilde{f}_1(\hat{\xi}^1) = A\hat{\xi}^1 + \begin{pmatrix} \tilde{\eta}_1(\hat{\xi}_1) \\ \vdots \\ \tilde{\eta}_{k-1}(\hat{\xi}_1, \dots, \hat{\xi}_{k-1}) \\ \tilde{\eta}_k(\hat{\xi}^1) \end{pmatrix}$, with A, C as in (3.18), and

S_θ given by (3.20).

In order to design a residual generator, the following corollary to proposition 4.1 can be used.

Corollary 4.1 *Consider system (3.25), and suppose there exists a C_∞ function $\varphi : I\!\!R^p \to I\!\!R$ satisfying assumptions A1) and A2). Let U be as above, and such that $U|_{[0,T]} = L^\infty([0,T], \mathcal{U})$ for some $T > 0$. Moreover, assume that :*

a) $L_{e_2} L_f^i(\Psi) = 0$ *for* $i = 0, \ldots, k - 1$.

b) $\exists i \in \{0, \ldots, k - 1\}$ *s.t.* $L_{e_1} L_f^i(\Psi) \neq 0$.

where $\Psi = \varphi \circ h$.

Then system (3.39) with output $r(t) = \hat{\xi}_1(t) - \bar{y}(t)$ *($\bar{y}(t) = \Psi(x(t))$) is a residual generator which detects and isolates* v_1.

Proof. It suffices to check conditions 1, 2-i) and 2-ii) of proposition 4.1. 2-i) is satisfied since (3.39) is an observer which converges for every $u \in U$. Assumption a) is nothing but condition 1). Finally, assumption b) implies $e_1 \notin ker d\mathcal{O}(\Psi)$, and by remark 4.2, U satisfies condition 2-ii). ∎

4 Hydraulic System

The considered system consists of a spool valve and a single rod piston acting on an inertial load (see figure 1). The external force F_e controls the flow entering the head side chamber of the piston from a pressure supply P_a. The rod side chamber is always connected to the return pressure P_r.

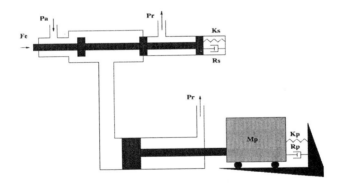

FIGURE 1. Hydraulic system.

Our aim is to detect and isolate two faults in this system : a drop of the spool control force F_e, and an increase of the internal leakage of the piston (which is normally assumed to be negligible).

4.1 Modelling of the System

The following notations will be used:x_1,displacement of the spool, x_2,velocity of the spool, x_3, displacement of the piston, x_4, velocity of the piston, x_5, pressure in the head side chamber, v_1, failure mode corresponding to the control force, v_2, failure mode corresponding to the internal leakage of the piston, A_P, area of the piston, D, diameter of the spool, B, bulk modulus, C_d, discharge coefficient, ρ, density of the fluid, K_S and R_S, respectively spring and damping coefficients associated to the spool, K_P and R_P, respectively spring and damping coefficients associated to the load, M_S and M_P, respectively mass of the spool and mass of the piston together with the load.

The model of the process can now be derived :

$$\dot{x}_1 = x_2 \tag{4.40}$$

$$\dot{x}_2 = -\left(K_S x_1 + R_S x_2\right)/M_S + \left(F_e - F_F - v_1\right)/M_S \tag{4.41}$$

$$\dot{x}_3 = x_4 \tag{4.42}$$

$$\dot{x}_4 = \left(-K_P x_3 - R_P x_4 + A_P x_5\right)/M_P \tag{4.43}$$

$$\dot{x}_5 = \frac{B}{A_P x_3} C_d \Pi D x_1 \sqrt{\frac{2}{\rho}\left(P_a - x_5\right)} - B\frac{x_4}{x_3} - \frac{x_5 x_4}{x_3} - \frac{B}{A_P}\frac{v_2}{x_3} \tag{4.44}$$

where $F_F = \frac{2C_d\Pi D}{\rho} x_1\left(P_a - x_5\right)$ represents the resultant flow force acting on the spool. We assume that the available measurements are $y = \begin{bmatrix} y_1 & y_2 \end{bmatrix}^T = \begin{bmatrix} x_1 & x_3 \end{bmatrix}^T$.

All the state variables $x_i, i = 1,\dots,5$ take values in closed intervals $[a_i, b_i], i = 1,\dots,5$. The position measurements are calibrated so that the lower bound of the interval is positive, and thus the division by x_3 in (4.44) does not cause any problem.

4.2   Design of a Residual Generator

First a residual generator to detect and isolate v_2 is obtained by noting that proposition 4.2 can be used with $(\varphi \circ h)(x) = x_3 = y_2$, if we consider $x_1 = y_1$ as a known input in (4.44). Indeed, it is easily checked that conditions A1) and A2) are fulfilled with $k = 3$. Besides, the change of coordinates :

$$\begin{bmatrix} \xi_1 & \xi_2 & \xi_3 & \xi_4 & \xi_5 \end{bmatrix}^T = \begin{bmatrix} x_3 & x_4 & A_P x_5/M_P & x_1 & x_2 \end{bmatrix}^T$$

transforms equations (4.42), (4.43) and (4.44) into :

$$\begin{cases} \dot{\xi}_1 = \xi_2 \\ \dot{\xi}_2 = \xi_3 - K_P\xi_1/M_P - R_P\xi_2/M_P \\ \dot{\xi}_3 = -\frac{\xi_2}{\xi_1}\left(\frac{A_P B}{M_P} + \xi_3\right) + \frac{B}{M_P}C_d\Pi D y_1\frac{1}{\xi_1}\sqrt{\frac{2}{\rho}\left(P_a - \frac{M_P}{A_P}\xi_3\right)} - \frac{B}{M_P}\frac{v_2}{\xi_1} \end{cases} \tag{4.45}$$

which is in the form (3.27), with $\overline{f}_1\left(\xi^1\right)$ and $\overline{g}_{11}\left(\xi^1\right)$ respectively given by (3.32) and (3.33). Besides

$$\overline{h}\left(\xi^1\right) = \xi_1 = (\varphi \circ h)(x) = y_2 \qquad (4.46)$$

The functions $\eta_2(\xi) = -K_P\xi_1/M_P - R_P\xi_2/M_P, \eta_3(\xi) = \frac{\xi_2}{\xi_1}\left(\frac{A_P B}{M_P} + \xi_3\right)$, and

$\overline{g}_{13}(\xi) = \frac{B}{M_P}C_d\Pi D\frac{1}{\xi_1}\sqrt{\frac{2}{\rho}\left(P_a - \frac{M_P}{A_P}\xi_3\right)}$ have bounded derivatives for any

ξ in

$W_1' = \{\xi s.t.\xi_1 \in [a_3, b_3], \xi_2 \in [a_4, b_4], \xi_3 \in [\frac{A_P}{M_P}a_5, \frac{A_P}{M_P}b_5 - \epsilon]\}$. The arbitrarily small number ϵ appearing in the upper bound for ξ_3 is introduced to take care of the fact that $\xi_3 = \frac{A_P}{M_P}b_5$ zeros the argument of the square root in $\overline{g}_{13}(\xi)$ (indeed $b_5 = P_a$). This phenomenon cannot occur on the actual system, as there will always be a small leak around the piston which will prevent ξ_3 from reaching this extreme value. It is now straighforward to extend $\eta_2(\xi), \eta_3(\xi)$ and $\overline{g}_{13}(\xi)$ to global Lipschitz functions on $\mathbb{R}^3$.

Conditions a) and b) of corollary 4.1 (in which v_1 and v_2 are interchanged as our aim is to detect v_2 here) are fulfilled for any input signal that keeps the state trajectory of (4.45) into W_1'. Hence a residual generator of the form (3.39) with output $r_1 = \widehat{\xi}_1 - y_2$ can be designed on the basis of model (4.45).

This was verified by simulating the process described by equations (4.40) to (4.44) together with the residual generator. The numerical values were taken as described in Appendix A. $v_1(t)$ was chosen as a step-like failure signal between 10 s and 20 s, while $v_2(t)$ was non zero between 30 s and 40 s. Figure 2 shows that the resulting residual is indeed unaffected by v_1 and that it allows to detect and isolate v_2.

We now illustrate the performance of an observer of the form (3.24) by considering the system made of equations (4.40),(4.41),(4.43), (4.44), with output $y_1 = x_1$. It turns out that the output reconstruction error for this observer, r_2, is a signal which is sensitive to v_1 and v_2. Hence the combined monitoring of r_1 and r_2 allows one to detect and isolate both failure modes provided they do not occur simultaneously.

The equations mentioned in the previous paragraph can be put in the form of model (3.23) thanks to the following obvious change of coordinates:

$$\left[\begin{array}{ccccc} z_1 & z_2 & z_3 & z_4 & z_5 \end{array}\right]^T = \left[\begin{array}{ccccc} x_1 & x_2 & x_5 & x_4(1 + x_5/B) & x_3 \end{array}\right]^T$$

which is indeed a diffeomorphism for any x in $W = \{x \in \mathbb{R}^5 : x_i \in [a_i, b_i], i = 1, \ldots, 5\}$.

When $v_1 = v_2 = 0$, $A(t)$ and $\Psi(t, F_e, z)$ can be written:

$$A = \left[\begin{array}{cccc} 0 & 1 & 0 & 0 \\ 0 & 0 & \frac{2C_d\Pi D}{\rho M_S}y_1(t) & 0 \\ 0 & 0 & 0 & -\frac{B}{y_2(t)} \\ 0 & 0 & 0 & 0 \end{array}\right]$$

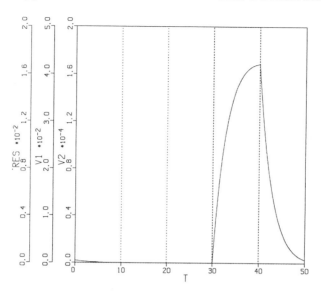

FIGURE 2. Residual r_1 and failure modes v_1 and v_2.

and $\Psi(t, F_e, z) =$

$$
\begin{bmatrix}
0 \\
-\frac{2C_d \Pi D}{\rho M_S} y_1 P_a - (K_S y_1 + R_S z_2)/M_S + F_e/M_S \\
f_\Psi(y_1, y_2, z_3) \\
(-K_P y_2 + A_P z_3)(1 + \frac{z_3}{B})\frac{1}{M_P} - \frac{R_P z_4}{M_P} + \frac{z_4}{1 + z_3/B}(-\frac{B}{y_2} z_4 + f_\Psi(y_1, y_2, z_3))
\end{bmatrix}
$$

with $f_\Psi(y_1, y_2, z_3) = \frac{B}{A_P y_2} C_d \Pi D y_1 \sqrt{\frac{2}{\rho}(P_a - z_3)})$

Notice that $y_1(t)$ and $y_2(t)$ belong to closed intervals with a positive lower bound, and the components of Ψ have bounded derivatives with respect to z on the domain $W' = \{z \in \mathbb{R}^5 : z_1 \in [a_1, b_1], z_2 \in [a_2, b_2], z_3 \in [a_5, b_5 - \epsilon], z_4 \in [a_4(1 + a_5/B), b_4(1 + b_5/B)], z_5 \in [a_3, b_3]\}$, where ϵ is an arbitrarily small real introduced for the same reason as above. Hence hypotheses H1) and H2) can be fulfilled by extending $\Psi(t, F_e, z)$ to a global Lipschitz function w.r.t. z and t.

An observer of the form (3.24) can thus be designed, and the second residual is defined by $r_2(t) = \hat{z}_1 - y_1$. Figure 3 shows the evolution of $r_2(t)$ for faults $v_1(t)$ and $v_2(t)$ with relatively realistic values. The timing of the faults, and the numerical values used for the simulation are the same as for figure 2. The influence of v_2 on r_2 does not appear on this plot, because it is low in comparison with the effect of v_1.

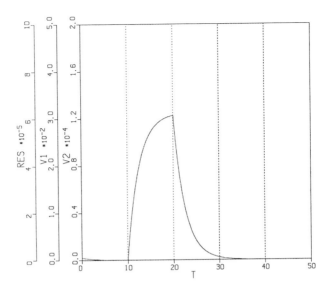

FIGURE 3. Residual r_2 and failure modes v_1 and v_2.

5 Conclusions

The fundamental problem of residual generation for linear systems has been reviewed, and the principle behind its solution has been described. Next, a formulation of the the same problem for nonlinear systems which are affine in the control signals and the failure modes was derived. Sufficient conditions for the existence of a solution have been presented. The possibility to use high gain nonlinear observers in the design of a residual generator has been investigated. The theoretical results have been illustrated by simulations on a hydraulic process.

The considered observers are based on a single measurement. Extension of the results to the multi-output situation is a topic for further research.

6 References

[1] P. Alexandre and M. Kinnaert. Numerically reliable algorithm for the synthesis of linear fault detection and isolation filters, based on the geometric approach. Proc. of the 1993 IEEE Conference on Systems, Man and Cybernetics, vol.5, pp 359-364, 1993.

[2] R. V. Beard. Failure accommodation in linear systems through self-reorganization. Ph.D. Dissertation. Dep. Aeronautics and Astronautics. Mass. Inst. Technol., Cambridge, MA, 1971.

[3] J. P. Gauthier and G. Bornard. Observability for any $u(t)$ of a class of bilinear systems. IEEE Trans. Automatic Control. 26, pp 922-926,

1981.

[4] P. M. Frank. Fault diagnosis in dynamic systems using analytical and knowledge based redundancy. A survey and some new results. Automatica, 26(3), pp 459-474, 1990.

[5] J. P. Gauthier, H. Hammouri and S. Othman. A simple observer for nonlinear systems, application to bioreactors. IEEE Trans. Automatic Control, 37 (6), pp 875-880, 1992.

[6] J. P. Gauthier and I. Kupka. Observability and observers for nonlinear systems. SIAM J. Control and Optimization, 32(4), pp 975-994, 1994.

[7] H. Hammouri, M. Kinnaert and E.H. El Yaagoubi. Fault detection and isolation for state affine systems. European Journal of Control, 4, pp 2-16, 1998.

[8] H. Hammouri, M. Kinnaert and E.H. El Yaagoubi. Observer based approach to fault detection and isolation for nonlinear systems. IEEE Trans. Automatic Control, 1998.

[9] R. Hermann and A.J. Krener. Nonlinear controllability and observability. IEEE Trans. on Automatic Control, 22(5), pp 728-740, 1977.

[10] M. Hou and P.C. Muller. Fault detection and isolation observers. Int. J. Control, vol.60 (5), pp 827-846, 1994.

[11] H. L. Jones. Failure detection in linear systems. Ph.D. Dissertation. Dep. Aeronautics and Astronautics, Mass. Inst. Technol., Cambridge, MA, 1973.

[12] T. Kailath. Linear Systems. Prentice-Hall, Englewood Cliffs, N.J, 1980.

[13] M. Kinnaert. Innovation generation for bilinear systems : application to robust fault detection. Proc. of the 1998 American Control Conference, Philadelphia, pp 1595-1599, 1998.

[14] M. Massoumnia. Geometric approach to the synthesis of failure detection filters. IEEE Trans. Automatic Control, AC-31, pp 839-846, 1986.

[15] M. Massoumnia, G.C. Verghese and A.S. Willsky. Failure detection and identification. IEEE Trans. Automatic Control, AC-34, pp 316-321, 1989.

[16] R. Nikoukhah. Innovation generation in the presence of unknown inputs : application to robust failure detection. Automatica, 30(12), pp 1851-1867, 1994.

[17] H. Nijmeijer and A.J. van der Schaft. Nonlinear Dynamical Control Systems. Springer-Verlag, 1990.

[18] R. J. Patton and J. Chen. Robust fault detection using eigenstructure assignment: a tutorial consideration and some new results. Proc. of the 30-th IEEE Conference on Decision and Control, pp 2242-2246, 1991.

[19] R. Seliger and P.M. Frank. Robust component fault detection and isolation in nonlinear dynamic system using nonlinear unknown input observers. Proc. of the IFAC/IMACS Symposium SAFEPROCESS, pp 313-318, 1991.

[20] P. M. Van Dooren. The generalized eigenstructure problem in linear system theory. IEEE Trans. Automat. Control, vol. AC-26, pp 111-129, 1981.

[21] E. White and J.L. Speyer. Detection filter design: spectral theory and algorithms. IEEE Trans. Automatic Control, AC-32, 7, pp 593-603.

[22] J. Wunnenberg. Observer based fault detection in dynamic systems. VDI-Fortschrittsberichte. Reihe 8, Nr. 222, 1990.

[23] D. Yu and D.N. Shields. A bilinear fault detection observer. Automatica, 32(11), pp 1597-1602, 1996.

Appendix A: Numerical Values used for the Simulation of the Hydraulic System

$M_S = 0.1$ kg, $R_S = 2.10$ Ns/m, $K_s = 10^3$ N/m, $D_P = 0.2$ m (Diameter of the piston), $D = 0.01$ m, $\rho = 840$ kg/m^3, $B = 10^9$ N/m^2, $C_d = 0.7$ kg/m^3, $P_a = 220 \quad 10^5$N/m^2, $M_P = 5 \quad 10^3$kg, $R_P = 10^4$ Ns/m, $K_P = 5 \quad 10^5$ N/m

Innovation Generation for Bilinear Systems with Unknown Inputs

M. Kinnaert and L. El Bahir

Department of Control Engineering
Université Libre de Bruxelles
Brussels, Belgium

1 Introduction

Fault detection systems are typically made of two parts : a residual generator and a decision system. The first module generates sequences, called residuals, from the sampled input and output signals of the supervised process. These sequences have nominally zero mean in the absence of fault (after a possible transient has vanished), and their mean becomes distinguishably different from zero upon occurrence of a fault. The decision system evaluates the residuals in order to determine whether their mean differs significantly from zero. This task can be efficiently performed by appropriate statistical tests [2]. However, such tests are often based on standard hypotheses, such as whiteness of the evaluated sequence. To be able to know the properties of the distribution of the residual, the problem of residual generation has to be stated in a stochastic setting.

For linear models with additive faults, the innovation sequence (associated with a Kalman filter) has been used as a residual. Indeed, it is a sufficient statistics, which means that the information about the faults contained in the measurements is retained in the innovation. Moreover, it is zero-mean in the absence of fault, and it is white with known covariance. Hence, it can be evaluated, for instance, by cumulative sum (CUSUM) or generalized likelihood ratio (GLR) tests, depending on whether the fault magnitude is known or not [2]. Yet such a signal can only be obtained from a standard linear model without unknown inputs. The latter often arise due to the presence of unmeasured signals or unknown parameters which cannot be modelled as random processes with known statistics. To handle this problem, Nikhoukhah [16] has developed, for linear time invariant (LTI) systems, a method to design a filter which, from the observed quantities (measured plant inputs and outputs), generates a signal with the following properties. It is zero mean and white in the absence of fault; it is decoupled from the unknown inputs, and it preserves, as much as possible, information on the faults. The resulting residual is called innovation because, in the absence of unknown inputs, it coincides with the innovation of a Kalman filter designed for the plant model.

Our aim in this chapter is to extend the known results for LTI systems to the class of bilinear systems. There are several reasons for the choice of this particular class of systems. First, when modelling systems using physical laws, one often obtains bilinear models. This is for instance the case for processes involving heat transfer for which energy balance equations yield products of temperature (state) and flow (input). Secondly, bilinear models can give a significantly better approximation of the behaviour of a nonlinear system over its working range, than what can be achieved using a linearized model around the nominal set point. This will be illustrated by an example below. Finally, observer theory, which is a fundamental ingredient for the design of residual generators, is well developed for bilinear systems [11], [5] [4]. The latter can actually be seen as a particular class of linear time-varying systems.

Observer-based residual generators for bilinear systems have been developed and studied in a deterministic setting in [20], [21] and [15]. In those papers the state equation describing the residual generator is linear and time-invariant up to an output injection. This choice of structure restricts the set of systems for which the problem admits a solution [13]. A more general structure is considered in [19]. However the authors do not deal with robustness issues in that paper, since there are no unknown inputs beside the faults in the problem they consider. Moreover, they obtain an estimate of the fault signals, which is a nice feature, but requires stronger hypotheses on the class of monitored systems than what is needed in our work.

Our formulation of the problem of innovation generation for bilinear systems is directly inspired by Nikoukhah's work [16]. However, the proposed solution does not resort to transfer function manipulation, like in [16], since this approach cannot be extended to time-varying systems. First a uniformly completely observable subsystem without unknown inputs is extracted from the original system. Next a Kalman filter is designed for the bilinear system resulting from the first step.

The work reported here is built on previous results obtained in [9] and [14]. The first paper describes a geometric approach to the synthesis of residual generators for deterministic bilinear systems. No constructive design procedure is obtained in that work. In the second paper, an algebraic approach is considered to solve the same problem. The algorithm derived in that paper is, however, limited to a smaller class of bilinear systems than the results presented here, if we translate them for deterministic systems.

The chapter is organized as follows. The problem of innovation generator design for bilinear systems is stated in Section 2. A solution is presented in section 3. Next the issue of innovation monitoring by GLR test is addressed in section 4. Finally the theory is applied to design and validate a fault detection and isolation system for a three-tank process.

2   Problem Statement

The following class of discrete-time bilinear systems is considered :

$$\begin{cases} x(k+1) = A_0x(k) + \sum_{i=1}^{m} u_i(k)A_ix(k) + \sum_{i=1}^{n_d}(E_i^dx(k) + F_i^d)d_i(k) \\ \qquad\qquad +Bu(k) + \sum_{i=1}^{n_f}\left(E_i^f x(k) + F_i^f\right)f_i(k) + Gw(k) \\ y(k) = Cx(k) + Du(k) + Hw(k) \end{cases}$$

$$(2.1)$$

where $x(k) \in R^n, u(k) \in R^m, y(k) \in R^p, d(k) \in R^{n_d}, f(k) \in R^{n_f}, w(k) \in R^{n_w}$ are respectively the state, the known inputs, the measured outputs, the unknown inputs, the failure modes, and a zero mean Gaussian white noise sequence with covariance matrix $E(w(k)w(\tau)^T) = I_{n_w}\delta_{k\tau}$ (where I_{n_w} denotes the n_w-by-n_w identity matrix and $\delta_{k\tau} = 1$ if $k = \tau$ and it is null otherwise). $u_i(k)$ denotes the ith component of $u(k)$, and similar notations are used for the other variables. With $f(k)$ set to zero, (2.1) describes the normal (fault free) behaviour of the monitored system. We assume that H has full row rank. Moreover, we let $\mathcal{U}$ denote the set of admissible input sequences.

Consider the following class of systems :

$$\begin{cases} \hat{z}(k+1) = \bar{A}(k)\hat{z}(k) + \bar{B}_y(k)y(k) + \bar{B}_u(k)u(k) \\ r(k) = \bar{C}\hat{z}(k) + \bar{D}_yy(k) + \bar{D}_uu(k) \end{cases} \qquad (2.2)$$

where $\bar{A}(k), \bar{B}_y(k)$ and $\bar{B}_u(k)$ are time-varying matrices of appropriate dimensions, while $\bar{C}, \bar{D}_y$ and $\bar{D}_u$ have constant entries.

Definition 5.1 *A system of the form (2.2) is called an innovation filter (or an innovation generator) for system (2.1) if there exists a set $\bar{\mathcal{U}}$ in $\mathcal{U}$ such that, in the absence of fault, for all $u \in \bar{\mathcal{U}}$, (2.2) is uniformly asymptotically stable, and the output r is a zero mean white noise sequence which is invariant under u and d, once the transient due to initial conditions has vanished.*

Notice that, in the linear case, an extra condition is added in the definition in order to ensure that no useful information on f, contained in y, is lost. This notion is difficult to translate in a nonlinear framework. However the innovation filter resulting from our algorithm will be seen to offer certain guarantees on this point.

3   Design Procedure

The algorithm is based on the following two theorems of which the proofs yield a constructive design procedure.

Theorem 5.1 *There exists an innovation filter of the form (2.2) for system (2.1) if the following two statements hold true :*

1. *There exist constant matrices $P, \widehat{A}_i, \widehat{B}_i, i = 0, ..., m, L_1$ and L_2, with P, L_1 and L_2 different from zero, such that :*

$$PA_i - \widehat{A}_i P = \widehat{B}_i C \qquad i = 0, ..., m \qquad (3.3)$$

$$P\left[E_i^d \vdots F_i^d\right] = 0 \qquad i = 1, ..., n_d \qquad (3.4)$$

$$L_1 C - L_2 P = 0 \qquad (3.5)$$

2. *There exists a set $\bar{U}$ in $\mathcal{U}$ such that, for all $u \in \bar{U}$ the following system is uniformly completely observable and uniformly completely controllable :*

$$\begin{cases} \tilde{z}(k+1) = \tilde{A}(k)\tilde{z}(k) + \tilde{G}(k)w(k) \\ \tilde{q}(k) = L_2\tilde{z}(k) \end{cases} \qquad (3.6)$$

where $\tilde{A}(k) = \hat{A}(u(k)) - S(k)R^{-1}L_2$, $\hat{A}(u(k)) = \hat{A}_0 + \sum_{i=1}^m u_i(k)\hat{A}_i$, $R = L_1 HH^T L_1^T$, $S(k) = (PG - \hat{B}(u(k))H)H^T L_1^T$, $\hat{B}(u(k)) = \hat{B}_0 + \sum_{i=1}^m u_i(k)\hat{B}_i$, and $\tilde{G}(k) = PG - \hat{B}(u(k))H - S(k)R^{-1}L_1 H$.

Proof. As there is no need to consider the term in $f(k)$ for this proof, we let $f(k) = 0$ in (2.1).

Define $z(k) = Px(k)$. Then, from (2.1) and (3.4)

$$z(k+1) = PA_0 x(k) + \sum_{i=1}^m u_i(k)PA_i x(k) + PBu(k) + PGw(k) \qquad (3.7)$$

Now substituting (3.3) for $PA_i, i = 0, ..., m$ in (3.7), and taking the output equation of model (2.1) into account, (3.7) yields :

$$z(k+1) = \widehat{A}(u(k))z(k) + \widehat{B}(u(k))(y(k) - Du(k)) + PBu(k)$$

$$+ (PG - \widehat{B}(u(k))H)w(k) \qquad (3.8)$$

where $\widehat{A}(u(k)) = \widehat{A}_0 + \sum_{i=1}^m u_i(k)\widehat{A}_i$ and $\widehat{B}(u(k)) = \widehat{B}_0 + \sum_{i=1}^m u_i(k)\widehat{B}_i$.

(3.8) is a bilinear system up to output injection which is only affected by d via y. In order to build an innovation filter, we shall design a Kalman filter for estimating z. To this end, we have to determine that part of the measurement y which depends on z, u and w only. This is achieved by defining the signal q as :

$$q(k) = L_1 y(k) = L_1 C x(k) + L_1 D u(k) + L_1 H w(k) \qquad (3.9)$$

and, from (3.5),

$$q(k) = L_2 z(k) + L_1 D u(k) + L_1 H w(k) \qquad (3.10)$$

The noise terms are correlated in (3.8) and (3.10). Yet one can easily transform (3.8), (3.10) into a system for which the state excitation noise and the observation noise are not correlated, by adding and subtracting $S(k)R^{-1}q(k)$ in (3.8), with $S(k)$ and R as defined in the theorem statement [8]. This yields :

$$z(k + 1) = \tilde{A}(k)z(k) + \tilde{B}_y(k)y(k) + \tilde{B}_u(k)u(k) + \tilde{G}(k)w(k) \qquad (3.11)$$

where $\tilde{B}_y(k) = \hat{B}(u(k)) + S(k)R^{-1}L_1$, $\tilde{B}_u(k) = PB - \hat{B}(u(k))D - S(k)R^{-1}L_1 D$, and the other notations are as in the theorem statement.

The Kalman filter for (3.11), (3.10) is :

$$\hat{z}(k + 1) = \tilde{A}(k)\hat{z}(k) + \tilde{B}_y(k)y(k) + \tilde{B}_u(k)u(k)$$
$$+ \Gamma(k)(L_1 y(k) - L_2 \hat{z}(k) - L_1 D u(k)) \qquad (3.12)$$

where the gain is obtained from

$$\Gamma(k) = \tilde{A}(k)\Pi(k)L_2^T (L_2\Pi(k)L_2^T + R)^{-1} \qquad (3.13)$$

with $\Pi(k)$ given by

$$\Pi(k + 1) = -\tilde{A}(k)\Pi(k)L_2^T (L_2\Pi(k)L_2^T + R)^{-1}L_2\Pi(k)\tilde{A}(k)^T$$
$$+ \tilde{A}(k)\Pi(k)\tilde{A}(k)^T + Q(k) \quad with \quad \Pi(0) = \Pi_0 \qquad (3.14)$$

where Π_0 is a positive definite matrix and $Q(k) = \tilde{G}(k)\tilde{G}(k)^T$.

The filter (3.12) is of the form of the first equation in (2.2) with: $\bar{A}(k) = \tilde{A}(k) - \Gamma(k)L_2$, $\bar{B}_y(k) = \tilde{B}_y(k) + \Gamma(k)L_1$, and $\bar{B}_u(k) = \tilde{B}_u(k) - \Gamma(k)L_1 D$. By the second condition in the theorem statement, the system (3.12), (3.13), (3.14) is uniformly asymptotically stable for all $u \in \tilde{\mathcal{U}}$ [12]. Now set

$$r(k) = q(k) - L_2\hat{z}(k) - L_1 D u(k) = -L_2\hat{z}(k) + L_1 y(k) - L_1 D u(k) \quad (3.15)$$

(3.15) is of the form of the second equation in (2.2), and $r(\cdot)$ is a white noise sequence once the transient due to initial conditions has vanished, since it is the innovation of a Kalman filter. ∎

The next theorem gives necessary and sufficient conditions for the existence of a solution to the equations (3.3), (3.4), (3.5). The proof of the result is constructive and directly yields a solution for $P, L_1, L_2, \widehat{A}_i, \widehat{B}_i, i = 0, \cdots, m$.

For the clarity of the developments, before stating the theorem, we rewrite (3.4) as follows :

$$PK = 0 \qquad (3.16)$$

where $K = \begin{bmatrix} E_1^d ... E_{n_d}^d & F_1^d ... F_{n_d}^d \end{bmatrix}$.

The singular value decomposition (SVD) of several matrices will be used. For a matrix M^γ (the exponent indicates the iteration number in the recursive method resulting from the proof below), with rank r_{M^γ}, the different factors of this decomposition are denoted in the following way:

$$M^\gamma = U_M^\gamma \overline{\Sigma}_M^\gamma V_M^{\gamma T} = [U_{M1}^\gamma \quad U_{M2}^\gamma] \begin{bmatrix} \Sigma_M^\gamma & 0 \\ 0 & 0 \end{bmatrix} \begin{bmatrix} V_{M1}^{\gamma T} \\ V_{M2}^{\gamma T} \end{bmatrix} \qquad (3.17)$$

where Σ_M^γ is a r_{M^γ}-by-r_{M^γ} diagonal matrix containing the singular values of M^γ, and U_M^γ, V_M^γ are orthogonal matrices.

Theorem 5.2 : *Equations (3.3)-(3.5) and (3.16) have a solution such that $P \neq 0$, $L_1 \neq 0$ and both matrices have full row rank if and only if there exists an integer $\alpha > 0$ such that $K^\alpha = 0$ or $Q^{(\alpha-1)}$ has full column rank, and none of the matrices $K^0, K^1, \ldots, K^{(\alpha-1)}, Q^0, Q^1, \ldots, Q^{(\alpha-1)}$ has full row rank. The matrices K^j and Q^j, $j = 0, \ldots, \alpha$ are given by the following recursive formulas :*

- *When $Q^j \neq 0$:*

$$A_i^{(j+1)} = U_{K2}^{jT} A_i^j (U_{K2}^j - U_{K1}^j V_{Q1}^j (\Sigma_Q^j)^{-1} U_{Q1}^{jT} C^j U_{K2}^j) \qquad i = 0, \ldots, m \qquad (3.18)$$

$$C^{(j+1)} = U_{Q2}^{jT} C^j U_{K2}^j \qquad (3.19)$$

$$K^{(j+1)} = U_{K2}^{jT} [A_0^j U_{K1}^j V_{Q2}^j \quad A_1^j U_{K1}^j V_{Q2}^j \cdots A_m^j U_{K1}^j V_{Q2}^j] \qquad (3.20)$$

$$Q^{(j+1)} = C^{(j+1)} U_{K1}^{(j+1)} \qquad (3.21)$$

- *When $Q^j = 0$:*

$$A_i^{(j+1)} = U_{K2}^{jT} A_i^j U_{K2}^j \qquad i = 0, \ldots, m \qquad (3.22)$$

$$C^{(j+1)} = C^j U_{K2}^j \qquad (3.23)$$

$$K^{(j+1)} = U_{K2}^{jT} [A_0^j U_{K1}^j \quad A_1^j U_{K1}^j \cdots A_m^j U_{K1}^j] \qquad (3.24)$$

$$Q^{(j+1)} = C^{(j+1)} U_{K1}^{(j+1)} \qquad (3.25)$$

where $A_i^0 = A_i, i = 0, \ldots, m, K^0 = K, C^0 = C, Q^0 = C^0 U_{K1}^0 = C U_{K1}$.

Proof.

If part

The sufficiency part of the proof is constructive and it can be divided into two sections : first the computation of $P, \widehat{A}_i, \widehat{B}_i, i = 0, \ldots, m$ that fulfil (3.3) and (3.4), next the computation of L_1 and L_2 that fulfil (3.5) for the matrix P obtained in the first step.

Computation of $P, \widehat{A}_i, \widehat{B}_i, i = 0, \ldots, m$

Set $P^0 = P$ and $\widehat{B}_i^0 = \widehat{B}_i$. Then (3.3) and (3.16) can be written as follows, given the notations introduced for the initialization of the recursive formulas :

$$P^0 A_i^0 - \widehat{A}_i P^0 = \widehat{B}_i^0 C^0 \qquad i = 0, ..., m \tag{3.26}$$

$$P^0 K^0 = 0 \tag{3.27}$$

Equation (3.27) is fulfilled if and only if

$$P^0 = P^1 U_{K2}^{0T} \tag{3.28}$$

for some appropriate matrix P^1. (3.28) can only be written when $rank K^0 < n$ (and hence K^0 has not full row rank), which holds by hypothesis. Substituting (3.28) into (3.26), and multiplying the resulting expression on the right by $U_K^0 = [U_{K1}^0 \quad U_{K2}^0]$ yields:

$$P^1 U_{K2}^{0T} A_i^0 U_{K1}^0 = \widehat{B}_i^0 C^0 U_{K1}^0 \tag{3.29}$$

$$P^1 U_{K2}^{0T} A_i^0 U_{K2}^0 - \widehat{A}_i P^1 = \widehat{B}_i^0 C^0 U_{K2}^0 \tag{3.30}$$

for $i = 0, \ldots, m$. Define Q^0 as in the theorem statement, and introduce its SVD in (3.29). Three situations are considered successively:

- $Q^0 \neq 0$ and Q^0 has not full column rank
  Equation (3.29) yields, after multiplication on the right by $[V_{Q1}^0 \quad V_{Q2}^0]$:

$$P^1 U_{K2}^{0T} A_i^0 U_{K1}^0 V_{Q1}^0 = \widehat{B}_i^0 U_{Q1}^0 \Sigma_Q^0 \tag{3.31}$$

$$i = 0, \ldots, m$$

$$P^1 U_{K2}^{0T} A_i^0 U_{K1}^0 V_{Q2}^0 = 0 \tag{3.32}$$

For a fixed P^1, any $\widehat{B}_i^0$ that solves (3.31) has the form :

$$\widehat{B}_i^0 = P^1 U_{K2}^{0T} A_i^0 U_{K1}^0 V_{Q1}^0 (\Sigma_Q^0)^{-1} U_{Q1}^{0T} + \widehat{B}_i^1 U_{Q2}^{0T} \tag{3.33}$$

for some matrix $\widehat{B}_i^1$ of appropriate dimension, and for $i = 0, \ldots, m$. Indeed, U_{Q2}^0 exists since Q^0 has not full row rank by hypothesis. Substituting (3.33) for $\widehat{B}_i^0$ into (3.30) yields :

$$P^1 U_{K2}^{0T} A_i^0 (U_{K2}^0 - U_{K1}^0 V_{Q1}^0 (\Sigma_Q^0)^{-1} U_{Q1}^{0T} C^0 U_{K2}^0) - \widehat{A}_i P^1$$

$$= \widehat{B}_i^1 U_{Q2}^{0T} C^0 U_{K2}^0 \qquad i = 0, \ldots, m, \quad (3.34)$$

which can be written as follows using the recursive formulas (3.18), (3.19) :

$$P^1 A_i^1 - \widehat{A}_i P^1 = \widehat{B}_i^1 C^1 \qquad i = 0, ..., m. \qquad (3.35)$$

Moreover, the set of equations (3.32) can be written :

$$P^1 K^1 = 0 \qquad (3.36)$$

thanks to (3.20). Notice that (3.35), (3.36) have the same form as (3.26), (3.27). If $K^1 = 0$, one directly finds a solution for (3.35), for instance $P^1 = I, \widehat{A}_i = A_i^1, \widehat{B}_i^1 = 0, i = 0, \ldots, m$. If $K^1 \neq 0$, one repeats the above procedure, starting from (3.28), with adequate changes of exponents.

- $Q^0 = 0$
  (3.29) yields :

$$P^1 U_{K2}^{0T} A_i^0 U_{K1}^0 = 0 \qquad (3.37)$$

Hence defining K^1, A_i^1 and C^1 respectively as in (3.24), (3.22) and (3.23), and setting $\widehat{B}_i^1 = \widehat{B}_i^0, i = 0, \ldots, m$, (3.30) and (3.37) take the form of (3.35) and (3.36). The iterative procedure continues if $K^1 \neq 0$; otherwise it stops since the solution for $P^1, \widehat{A}_i^1, \widehat{B}_i^1, i = 0, \ldots m$ can be obtained as above .

- Q^0 has full column rank
  In this case, $V_Q^0 = V_{Q1}^0$ and (3.33) is obtained from (3.31) as a solution for $\widehat{B}_i^0$. Substituting this expression in (3.30) yields (3.35). The difference with the first case ($Q^0 \neq 0$ and Q^0 has not full column rank) is that no equation like (3.32) arises. Hence one can directly solve (3.35) ($P^1 = I, \widehat{A}_i = A_i^1, \widehat{B}_i^1 = 0, i = 0, \ldots, m$).

The combination of the above three cases defines a recursive procedure which stops either when $K^\alpha = 0$ or $Q^{(\alpha-1)}$ has full column rank, for some integer $\alpha > 0$. At that stage, one is left with the equations :

$$P^\alpha A_i^\alpha - \widehat{A}_i P^\alpha = \widehat{B}_i^\alpha C^\alpha \qquad i = 0, ..., m \qquad (3.38)$$

for which $P^\alpha = I$, $\widehat{A}_i = A_i^\alpha$ and $\widehat{B}_i^\alpha = 0, i = 0, \ldots, m$ is a solution. For $P^\alpha = I$, the matrix P resulting from the recurrence based on (3.28), say P^*, is easily seen to be:

$$P^* = U_{K2}^{(\alpha-1)T} \ldots U_{K2}^{1T} U_{K2}^{0T} \tag{3.39}$$

Moreover, $\widehat{B}_i, i = 0, \ldots, m$ can be computed by applying backward, from $j = \alpha - 1$ to $j = 0$ the following recursive formulas deduced from (3.33) :

$$\begin{cases} \widehat{B}_i^j = P^{(j+1)} U_{K2}^{jT} A_i^j U_{K1}^j V_{Q1}^j (\Sigma_Q^j)^{-1} U_{Q1}^{jT} + \widehat{B}_i^{(j+1)} U_{Q2}^{jT} & \text{when} \quad Q^j \neq 0 \\ \widehat{B}_i^j = \widehat{B}_i^{j+1} & \text{when} \quad Q^j = 0 \end{cases} \tag{3.40}$$

with $\widehat{B}_i^\alpha = 0, i = 0, \ldots, m$, and by remembering that $\widehat{B}_i^0 = \widehat{B}_i, i = 0, \ldots, m$.

Computation of L_1 and L_2

It now remains to solve (3.5). A particular solution corresponding to P^* is :

$$L_1^* = U_{Q2}^{(\alpha-1)T} \ldots U_{Q2}^{1T} U_{Q2}^{0T} \tag{3.41}$$

$$L_2^* = U_{Q2}^{(\alpha-1)T} \ldots U_{Q2}^{0T} C U_{K2}^0 \ldots U_{K2}^{(\alpha-1)} \tag{3.42}$$

One can check that (3.5) is fulfilled as follows. Let us compute $L_2^* P^*$, with P^* given by (3.39) :

$$L_2^* P^* = U_{Q2}^{(\alpha-1)T} \ldots U_{Q2}^{0T} C U_{K2}^0 \ldots U_{K2}^{(\alpha-1)} U_{K2}^{(\alpha-1)T} \ldots U_{K2}^{0T} \tag{3.43}$$

Noticing that $U_{K2}^i U_{K2}^{iT} = I - U_{K1}^i U_{K1}^{iT}, i = 0, \ldots, \alpha - 1$, and substituting successively $U_{K2}^{(\alpha-1)} U_{K2}^{(\alpha-1)T}, U_{K2}^{(\alpha-2)} U_{K2}^{(\alpha-2)T}, \ldots, U_{K2}^0 U_{K2}^{0T}$ in terms of these expressions in (3.43), one gets :

$$\begin{aligned} L_2^* P^* = {}&U_{Q2}^{(\alpha-1)T} \ldots U_{Q2}^{0T} C - U_{Q2}^{(\alpha-1)T} \ldots U_{Q2}^{0T} C U_{K1}^0 U_{K1}^{0T} \\ &- \sum_{i=0}^{\alpha-2} U_{Q2}^{(\alpha-1)T} \ldots U_{Q2}^{0T} C U_{K2}^0 \ldots U_{K2}^i U_{K1}^{(i+1)} U_{K1}^{(i+1)T} U_{K2}^{iT} \ldots U_{K2}^{0T} \end{aligned} \tag{3.44}$$

The first term on the right hand side of (3.44) is nothing but $L_1^* C$. Hence, to conclude this part of the proof, it suffices to show that all the other terms are equal to zero. Let us consider an arbitrary term of the sum (the same reasoning also applies to the second term of the right hand side of (3.44)). Using the recursive formula (3.19), one easily deduces :

$$\begin{aligned} U_{Q2}^{(\alpha-1)T} &\ldots U_{Q2}^{0T} C U_{K2}^0 \ldots U_{K2}^i U_{K1}^{(i+1)} U_{K1}^{(i+1)T} U_{K2}^{iT} \ldots U_{K2}^{0T} \\ &= U_{Q2}^{(\alpha-1)T} \ldots U_{Q2}^{(i+1)T} C^{(i+1)} U_{K1}^{(i+1)} U_{K1}^{(i+1)T} U_{K2}^{iT} \ldots U_{K2}^{0T} \\ &= U_{Q2}^{(\alpha-1)T} \ldots U_{Q2}^{(i+1)T} Q^{(i+1)} U_{K1}^{(i+1)T} U_{K2}^{iT} \ldots U_{K2}^{0T} \end{aligned} \tag{3.45}$$

where the definition of $Q^{(i+1)}$, (3.21), was used to obtain the last expression.

Finally notice that, in the right hand side of (3.45), $U_{Q2}^{(i+1)T}Q^{(i+1)} = U_{Q2}^{(i+1)T}U_{Q1}^{(i+1)}\Sigma_{Q}^{(i+1)}V_{Q1}^{(i+1)T} = 0$, and hence any term in the sum is null in (3.44). This also holds for the second term of the right hand side of (3.44) as already mentioned.

In the expressions for L_1^* and L_2^*, one should set $U_{Q2}^j = I$ when $Q^j = 0$, which corresponds to what is done in the definition of $C^{(j+1)}$ (compare (3.23) and (3.19)).

It remains to verify that P^* and L_1^* are non-zero. As all the matrices U_{K2}^{iT}, $i = 0,\ldots,(\alpha-1)$ have full row rank, one easily checks that $rankP^* = rankU_{K2}^{(\alpha-1)T}$. The latter is non-zero as $K^{(\alpha-1)}$ has not full row rank by hypothesis. Hence $P^* \neq 0$, and P^* has full row rank. On the other hand, as $Q^i, i = 0,\ldots,\alpha - 1$ have not full row rank, $U_{Q2}^{iT}, i = 0,\ldots\alpha - 1$ exist. By a similar argument as for P^* one concludes that L_1^* has full row rank and is different from zero.

Only if part

Assume that equations (3.3)-(3.5) and (3.16) have a solution with $P \neq 0$ and $L_1 \neq 0$, both matrices having full row rank. Let the dimensions of P and L_1 be denoted $\ell \times n$ and $s \times p$ respectively. From (3.16), K cannot have full row rank. Hence $P = P^0$ has the form (3.28) for some matrix P^1 that fulfils (3.29), (3.30). Equivalently, P^1 must fulfil (3.35) for some matrices $\widehat{B}_i^1, i = 0,\ldots,m$, linked to $\widehat{B}_i$ via (3.40) and it must also fulfil (3.36) when Q^0 has not full column rank. P^1 has dimensions $\ell \times (n - r_K)$ where r_K denotes the rank of matrix K. Now, two situations can be distinguished :

1. if Q^0 has full column rank or $K^1 = 0$, then $\alpha = 1$, and $P^1, \widehat{A}_i, i = 0,\ldots,m$ must be a solution of (3.35) for some $\widehat{B}_i^1, i = 0,\ldots,m$.

2. if none of the conditions in 1. hold, K^1 cannot have full row rank by (3.36), and P^1 must be of the form $P^1 = P^2 U_{K2}^{1T}$ where P^2 is an $\ell \times (n - rankK^1 - rankK)$ matrix. P_2 must fulfil equations of the form (3.29), (3.30) with all the exponents increased by 1. One can repeat for P^2 the same procedure as for P^1, and so on.

This corresponds to an iterative procedure. One realizes that there must exist a finite integer α, for which either $Q^{(\alpha-1)}$ has full column rank or $K^\alpha = 0$. Indeed, as the number of columns of $P^j, j = 0, 1,\ldots$ keeps decreasing when the number of iterations increases, this iterative procedure must stop, otherwise P cannot have full row rank.

It remains to show that the existence of a solution implies that $Q^0, Q^1, \ldots, Q^{(\alpha-1)}$ cannot have full row rank. To this end, let us consider equation (3.5). Introduce (3.28) into (3.5), and multiply the resulting equation on the right by $[U_{K1}^0 \quad U_{K2}^0]$. This yields :

$$L_1 C U_{K2}^0 - L_2 P^1 = 0 \qquad (3.46)$$

$$L_1 C U_{K1}^0 = L_1 Q^0 = 0 \tag{3.47}$$

(3.47) implies that Q^0 cannot have full row rank, as $L_1 \neq 0$. Now, substituting P^1 for its value in terms of P^2 in (3.46), and multiplying the resulting expression on the right by $[U_{K1}^1 \ U_{K2}^1]$, one deduces :

$$L_1 C U_{K2}^0 U_{K2}^1 - L_2 P^2 = 0 \tag{3.48}$$

$$L_1 C U_{K2}^0 U_{K1}^1 = 0 \tag{3.49}$$

The latter equation can be written :

$$L_1 U_{Q1}^0 U_{Q1}^{0T} C U_{K2}^0 U_{K1}^1 + L_1 U_{Q2}^0 U_{Q2}^{0T} C U_{K2}^0 U_{K1}^1 = 0 \tag{3.50}$$

By (3.47), L_1 is necessarily of the form $L_1 = L_1^1 U_{Q2}^{0T}$ for some non zero $\ell \times (p - rankQ^0)$ matrix L_1^1. Substituting this value for L_1 in (3.50), and taking (3.19) and (3.21) into account, one obtains : $L_1^1 Q^1 = 0$. Hence Q^1 cannot have full row rank. Proceeding in the same way as above, one gets $L_1^j Q^j = 0$, $0 \leq j \leq \alpha - 1$, where $L_1^0 = L_1$, $L_1^{j-1} = L_1^j U_{Q2}^{(j-1)T}$ and $L_1^j \neq 0$. Hence $Q^j, 0 \leq j \leq \alpha - 1$ have not full row rank. ∎

The matrices P^* and L_1^* resulting from Theorem 2 have full row rank, and moreover P^* and L_1^* have the largest possible rank among the set of solutions to (3.3)-(3.5) . Indeed, from the necessity part of the proof of Theorem 1, one notices that any pair of matrices P and L_1 that make a solution of (3.3)-(3.5) (together with adequate matrices $L_2, \hat{A}_i, \hat{B}_i, i = 1, \ldots, m$) must be of the form $P = P^\alpha P^*, L_1 = L_1^\alpha L_1^*$ for some matrices P^α, L_1^α of appropriate dimensions. Hence $z = P^*x$ and $q = L_1^*y$ have the largest possible dimension, which is a normal requirement for avoiding loss of information on f.

The design method resulting from Theorems 1 and 2 can be summarized as follows:

1. determine a solution to (3.3),(3.4) by applying the recursive formulas (3.18)-(3.21) or (3.22)-(3.25) for $j = 0, \ldots, \alpha$. If one of the matrices $K^0, K^1, \ldots, K^{(\alpha-1)}, Q^0, Q^1, \ldots, Q^{(\alpha-1)}$ has not full row rank, the procedure stops : the algorithm does not give a solution. Otherwise, compute P^* according to (3.39); set $\hat{A}_i = A_i, i = 0, \ldots, m$ and compute $\hat{B}_i, i = 0, \ldots, m$ from (3.40).

2. Solve (3.5), with $P = P^*$. A solution is given by (3.41), (3.42).

3. Implement the innovation filter (3.12)-(3.14).

To be able to use the innovation filter described above for fault detection, one should monitor on-line its output, or a function of its output, by adequate statistical tests. This issue is discussed in the next section.

4 Innovation Monitoring

4.1 Introductory Remark

Two situations must be distinguished depending on whether $E_i^f = 0, i = 1, \ldots, n_f$ or not in (2.1). In the first case, the faults are additive, namely they only change the mean of the innovation. Then, the latter is known to be a sufficient statistic for the faults f, and it can be monitored by the generalized likelihood ratio (GLR) test, for instance [18], [2]. If some of the matrices $E_i^f, i = 1, \ldots, n_f$ are non zero, the faults are not additive, and the innovation is not a sufficient statistic for f anymore [2]. Monitoring the innovation could still allow one to detect the faults, but it is not the best solution. One potential approach could be to work with the least-squares-score associated to the innovation filter [1]. However, some issues still have to be clarified for that method. Therefore, only additive faults will be considered in the remaining part of Section 4.

The distribution of the residual in the absence and in the presence of faults is first determined, before presenting a review of the GLR test.

4.2 Innovation in the Presence of Additive Faults

To be able to apply the GLR test, step-like faults will be considered, namely $f(k) = \mu 1_{\{k \geq t_0\}}$, where μ is a constant vector, and $1_{\{k \geq t_0\}}$ is equal to 1 when $k \geq t_0$ and it is null otherwise. It is straightforward to compute the signature of the fault on the innovation (also called the dynamic profile of the fault). Indeed, with $f(k)$ non zero, (3.11) can be written :

$$
\begin{aligned}
z(k+1) = {}& \tilde{A}(k)z(k) + \tilde{B}_y(k)y(k) + \tilde{B}_u(k)u(k) + PF^f\mu 1_{\{k \geq t_0\}} \\
& + \tilde{G}(k)w(k)
\end{aligned}
\tag{4.51}
$$

By subtracting (3.12) from (4.51), and by substituting (3.10) for $q(k)$ in (3.15), one deduces :

$$
\begin{aligned}
\epsilon_z(k+1) = {}& (\tilde{A}(k) - \Gamma(k)L_2)\epsilon_z(k) + PF^f\mu 1_{\{k \geq t_0\}} \\
& + (\tilde{G}(k) - \Gamma(k)L_1 H)w(k)
\end{aligned}
\tag{4.52}
$$

$$
r(k) = L_2\epsilon_z(k) + L_1 H w(k)
\tag{4.53}
$$

where $\epsilon_z(k) = z(k) - \hat{z}(k)$.

Hence the innovation can be written :

$$
r(k) = r_0(k) + \rho(k, t_0)\mu
\tag{4.54}
$$

where $r_0(k)$ is the innovation for the fault free system, and $\rho(k, t_0)\mu$ is the signature of the fault. The latter is null for $k < t_0$ and it can be obtained by simulating (4.52),(4.53) with $w \equiv 0$ and $\epsilon_z(t_0) = 0$ for $k \geq t_0$. Since the noise $w(k)$ is assumed to have a normal distribution (see model (2.1)), $r(k)$ is also Gaussian. More precisely,

$$\mathcal{L}(r(k)) = \mathcal{N}(0, \Sigma(k)) \quad \text{when no fault has occurred} \tag{4.55}$$

$$\mathcal{L}(r(k)) = \mathcal{N}(\rho(k, t_0)\mu, \Sigma(k)) \quad \text{after occurrence of a fault} \tag{4.56}$$

where $\mathcal{N}(\xi, \Theta)$ denotes the normal distribution with mean ξ and variance Θ, $\Sigma(k) = L_2 \Pi(k) L_2^T + L_1 H H^T L_1^T$, and $\Pi(k)$ is given by (3.14).

4.3 Generalized Likelihood Ratio Test

For the sake of simplicity, μ is assumed to be a scalar (see [18] for the non-scalar case). The GLR test is aimed at choosing between two hypotheses:

- H_0 : no fault has occurred

- H_1 : a fault of unknown magnitude, μ, has occurred at an unknown time instant, $t_0 \leq k$, where k denotes the present time instant.

To explain the idea behind this test, let us first assume that μ is known. Classical tests between both hypotheses rely on the log-likelihood ratio of H_1 versus H_0 for the residual sequence, namely:

$$s(k) = \ln \frac{p_\mu(r(k))}{p_0(r(k))} \tag{4.57}$$

where $p_\mu(r(k))$ $(p_0(r(k)))$ is the probability density of $r(k)$ under hypothesis H_1 (H_0). This quantity has the following fundamental property:

$$E_\mu(s(k)) > 0 \qquad\qquad E_0(s(k)) < 0$$

where E_μ and E_0 denote expectation under the distributions associated to $p_\mu(\cdot)$ and $p_0(\cdot)$ respectively. Therefore, the cumulative sum of log-likelihood ratios, $S_k = \sum_{i=1}^{k} s(i)$ has a negative drift in the absence of fault, and a positive drift when a fault has occurred. Hence the maximum likelihood estimate of the fault occurrence time, $\hat{t}_0$, can be computed by maximizing, w.r.t. t_0, the log-likelihood ratio of H_1 versus H_0 for the residual samples from time t_0 to k, namely :

$$S_{t_0}^k = \sum_{i=t_0}^{k} \ln \frac{p_\mu(r(i))}{p_0(r(i))} \tag{4.58}$$

This yields the cumulative sum (CUSUM) test which amounts to computing the function:

$$g_{\text{CUSUM}}(k) = \max_{1 \leq t_0 \leq k} S_{t_0}^k$$

and to generating an alarm when $g_{\text{CUSUM}}(k) > \epsilon$, where ϵ is a user defined threshold.

Now, when μ is not known, a similar reasoning still holds. Yet the log-likelihood ratio is a function of μ, $S_{t_0}^k(\mu)$. In order to evaluate this function, μ is replaced by its maximum likelihood estimate, $\hat{\mu}$; hence the name generalized likelihood ratio. The maximum likelihood estimates of t_0 and μ are obtained by solving the following double maximization problem:

$$(\hat{t}_0, \hat{\mu}) = arg\{\max_{1 \le t_0 \le k} \sup_{\mu} S_{t_0}^k(\mu)\} \tag{4.59}$$

When the probability density functions appearing in (4.57) correspond to Gaussian distributions, it is possible to find an explicit expression for $\hat{\mu}_k(t_0)$, the maximum likelihood estimate of μ at time k assuming the fault occurred at time t_0. This expression is then used for computing the log-likelihood ratio of H_1 versus H_0 given $r(1), \cdots, r(k)$, and it yields the following GLR test function:

$$g_{\text{GLR}}(k) = \max_{1 \le t_0 \le k} \sup_{\mu} S_{t_0}^k(\mu) = \frac{1}{2} \max_{1 \le t_0 \le k} \frac{(\sum_{i=t_0}^k \rho(i, t_0)^T \Sigma(i)^{-1} r(i))^2}{\sum_{i=t_0}^k \rho(i, t_0)^T \Sigma(i)^{-1} \rho(i, t_0)} \tag{4.60}$$

As previously, when $g_{\text{GLR}}(k)$ exceeds a suitable threshold, hypothesis H_1 is considered to be true, and an alarm is generated.

The maximization problem in (4.60) is performed over all possible past time instants. This increasing time span yields an increasing search duration for finding the optimum. To avoid that problem, the fault occurrence time is estimated using a window of fixed size, M, in practice. Thus (4.60) becomes :

$$g_{\text{GLR}}(k) = \max_{k-M \le t_0 \le k} \sup_{\mu} S_{t_0}^k(\mu) \tag{4.61}$$

Upon occurrence of a fault, the GLR test provides an estimate of the fault magnitude and the fault occurrence time. This allows one to keep monitoring the system in its faulty behaviour, to detect whether further changes take place. The procedure used to update the innovation generator upon occurrence of a fault is described in [7]. It requires the introduction of an additional tuning parameter, namely a tolerance on the estimated fault magnitude, μ_{tol}. The system is considered to be in faulty mode when the GLR test has been triggered and the estimated fault magnitude is greater than μ_{tol}.

5   Design and Validation of a FDI System for a three Tank Process

5.1   Process Description

The flowsheet of the three tank process is given in Figure 1. It consists of three tanks, R1, R2 and R3, which serve to supply water from the reservoir R0. The capacity of the latter reservoir is much greater than the capacity of R1, R2 and R3, so that its level remains practically constant during the operation. In the study reported here, tanks R1 and R3 play the role of buffers for supplying R2. The level in tank R1 is controlled by manipulating the speed of pump P1, while the level in R3 is adjusted by manipulating the aperture of valve V5. Pump P2 runs at constant speed. Two PI controllers are used to drive the levels in R1 and R3 at their desired reference value.

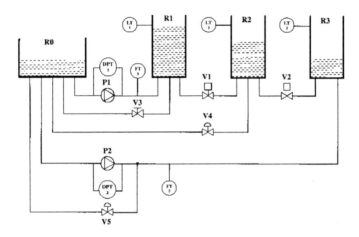

FIGURE 1. Flowsheet of the three-tank process.

The plant can be modelled by a nonlinear state space model of the form:

$$\dot{X}(t) = f(X(t), U(t)) \tag{5.62}$$

$$Y(t) = X(t) + N(t) \tag{5.63}$$

where $X(t) = [\ h_1(t) \quad h_2(t) \quad h_3(t)\ ]^T$ is the state vector consisting of the levels in tanks R1, R2 and R3. $U(t) = [\ \Omega_1(t) \quad S_5(t)\ ]^T$ is the command input vector made of the speed of rotation of pump P1 and the position of the stem of the continuous valve V5. $Y(t)$ are the measured outputs, and $N(t)$ is a white noise process. The explicit form of (5.62) can be found in [6], where the model parameters have been identified from experimental data taken on the actual plant. In the same reference, a simulator that reproduces the process behaviour under faulty and fault free situations is described. Here, only three faults will be considered, namely a leak in tank R1, emulated by opening valve V3, a clog in the branch with pump P2, and a bias on the measurement of the level in R3, h_3.

5.2 Design and Validation of the Innovation Generator

A bilinear model of the following form has been identified from data obtained by running the simulator around a set point $(w_1^0, s_5^0, h_1^0, h_2^0, h_3^0)$, without introducing any noise in the simulations:

$$x(k+1) = A_0 x(k) + w_1(k)A_1 x(k) + s_5(k)A_2 x(k) + Bu(k)$$
$$+ f_1(k)E_1 x(k) + F_1 f_1(k) + f_2(k)E_2 x(k) + F_2 f_2(k) \qquad (5.64)$$

$$y(k) = x(k) + E_3 f_3(k) \qquad (5.65)$$

where $x(k), y(k), w_1(k)$ and $s_5(k)$ denote deviations from the nominal set point, and $u(k) = [\ w_1(k) \quad s_5(k)\ ]^T$. $f_1(k), f_2(k)$ and $f_3(k)$ correspond to the faults mentioned in the previous section, in the same order (leak, clog, bias).

The aim is to detect and isolate the faults. To this end, one approach consists in generating three innovation signals, each of them being only sensitive to a single fault. Due to a lack of space, only the design of the innovation filter aimed at detecting fault f_3 will be considered here. The reader is referred to [7] for the complete design of the FDI system.

To generate an innovation sequence sensitive to the sensor fault f_3 only, the two other faults must be considered as unknown inputs (vector d in (2.1)). Thus, equation (3.4) can be written $P[\ E_1 \quad F_1 \quad E_2 \quad F_2\ ]$ $= PK_{12} = 0$. In the present case, K_{12} has full row rank, and thus this equation has no solution. To alleviate this problem, an approximate decoupling with respect to f_1 and f_2 will be achieved using the method of [17]. The procedure consists in computing a matrix K_{12}^* which has not full row rank and which is close to K_{12} in the sense that the following Frobenius norm is minimum: $||K_{12} - K_{12}^*||_F$. Then, one substitutes K_{12}^* for K_{12} in (3.4), and the design is pursued by following the procedure presented at the end of Section 3. The variance of the measurement noise was evaluated from experimental data, and the variance of the state excitation noise was chosen in such a way that the transient of the innovation upon occurrence of a fault is sufficiently fast while keeping a reasonable sensitivity to the fault.

A linear time-invariant innovation filter has also been designed for the same application. It is based on a linearized model of the plant (around the same set point that was used for model (5.64), (5.65). The innovation sequences obtained with both types of filters are now compared by inputting simulation data in faulty and fault free conditions in the innovation generators. Constant faults f_3, f_1 and f_2 are simulated respectively in the time intervals $[600s, 800s]$, $[1850s, 2050s]$, and $[4600s, 4800s]$. The simulations reported here were performed in open-loop with input signals covering a large working range. The command signal deviations from their nominal value are plotted in Figure 2. Figure 3 represents the corresponding output signal deviations. The effect of the three faults is indicated. The innovation sequences aimed at detecting fault f_3, and obtained from the linear and the bilinear models, are plotted in Figures 4. The following points can be

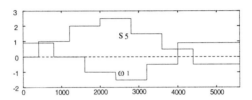

FIGURE 2. Command signal deviations from their nominal value.

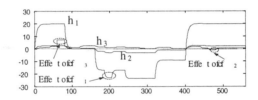

FIGURE 3. Output signal deviations from their nominal value.

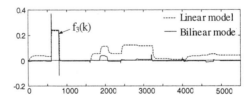

FIGURE 4. Innovation sequences obtained from linear and bilinear models

noticed :

- The innovation filter based on the bilinear model generates a signal with lower magnitude than its counterpart based on the linear model, when the command signal deviations from their nominal value are large, in the absence of faults.

- The error introduced by the approximate decoupling procedure of [17] is acceptable since each innovation sequence is significantly more affected by the fault to which it must be sensitive, namely f_3, than by the other faults.

The same conclusions remain valid when the system operates in closed-loop.

5.3 Evaluation of the Innovation Sequence

To tune the GLR test, the following heuristic rules might be helpful :

- The test window M should be approximately equal to the duration of the transient in the innovation upon occurrence of a fault.

- The test threshold should be fixed beyond the typical test output in the absence of fault in order to avoid most false alarms. Indeed the innovation is not perfectly zero mean due to modelling imperfections, and to the effect of Kalman filter transients caused by abrupt changes in the command signals.

The innovation filter for detection and isolation of f_3, followed by a GLR test, was applied on actual plant data. The obtained results are plotted in figure 5, were the innovation sequence, the test function, the estimated fault magnitude, and the actual and estimated fault occurrence times are successively presented. The fault occurrence and disappearance times are quite well estimated. One notices that the residual mean is not exactly zero after the initialization stage, which lasts for 150 seconds. This is due to modelling uncertainties. Yet, reasonable results are obtained although the innovation generator and GLR test were designed from model (5.64), (5.65), which was identified from plant simulator data. The robustness of

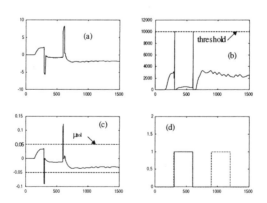

FIGURE 5. (a)Innovation for detection of f_3, (b) GLR test function, (c) Estimated fault magnitude, (d) True fault durations for f_3 and f_1 (dashed line) and estimated fault duration for f_3 (solid line)

the GLR test with respect to modelling uncertainties (including errors in the state excitation noise variance) is illustrated by the above results. It was already noticed in several other applications, as reported in [3] for instance.

6   Conclusions

A method to design an innovation filter for bilinear systems with unknown inputs has been presented. Further work is needed to ensure that no information on the fault is lost in the innovation generation. The issue of sensitivity of the residual to the fault has not been addressed here. Some results on this point were obtained in [10] in a deterministic framework. The innovation monitoring by the GLR test has been reviewed for additive faults. In the case of non-additive faults further work is needed to combine innovation filters with adequate statistical detection methods. Simulation results have shown that, for some nonlinear systems, the use of a bilinear model instead of a linear model, for innovation generation, can make the residual magnitude small over a larger working range.

7   REFERENCES

[1] M. Basseville. On-board component fault detection and isolation using the statistical local approach. *Automatica*, **34(11)**, pp. 1391–1415, 1998.

[2] M. Basseville and I.V. Nikiforov. *Detection of Abrupt Changes, Theory and Application*, Prentice-Hall, Englewood Cliffs, N.J., 1993.

[3] M. Basseville and A. Benveniste. Design and comparative study of some sequential jump detection algorithms for digital signals. *IEEE Trans. Acoustics, Speech and Signal Processing*, **ASSP-31(3)**, pp. 521–535, 1983.

[4] G. Besançon and H. Hammouri. On uniform observation of non-uniformly observable systems. *Systems and Control Letters*, **29(1)**, pp. 9–19, 1996.

[5] G. Bornard, N. Couenne and F. Celle. Regularly persistent observers for bilinear systems. in *New Trends in Nonlinear Control Theory*, Lecture Notes in Control and Information Science, **122**, Springer-Verlag, pp. 130–140, 1988.

[6] Dolanc G., D. Juricic, A. Rakar, J. Petrovcic and D. Vrancic. Three-tank benchmark test. *Report COPL007R for Copernicus Project CT94-0237*, 1997.

[7] L. El Bahir and M. Kinnaert. Fault detection and isolation for a three tank system based on a bilinear model of the supervised process. *Proceedings of the UKACC International Conference on CONTROL'98*, pp. 1486–1491, 1998.

[8] G. C. Goodwin and K.S. Sin. *Adaptive Filtering Prediction and Control*, Prentice-Hall, 1994, page 251.

[9] H. Hammouri, M. Kinnaert and E.H. El Yaagoubi. Residual generator synthesis for bilinear systems up to output injection. *Proceedings of the 33rd IEEE Conference on Decision and Control*, Orlando, December 14-16, pp. 1548–1553, 1994.

[10] H. Hammouri, M. Kinnaert and E.H. El Yaagoubi. Fault detection and isolation for state affine systems. *European Journal of Control*, **4**, pp. 2–16, 1998.

[11] S. Hara and K. Furuta. Minimal order state observers for bilinear systems. *International Journal of Control*, **24(5)**, pp. 705–718, 1976.

[12] A. H. Jazwinski. *Stochastic Processes and Filtering Theory*, Academic Press, 1970, page 240.

[13] M. Kinnaert. Robust fault detection based on observers for bilinear systems. Internal Report. Université Libre de Bruxelles, 1998.

[14] M. Kinnaert, Y. Peng and H. Hammouri. The fundamantal problem of residual generation for bilinear systems up to output injection. *Proceedings of the 3rd European Control Conference*, pp. 3777–3782, 1995.

[15] C. Mechmeche, S. Nowakowski and M. Darouach. A failure detection procedure for bilinear systems based on a new formulation of unknown input bilinear observers.*Proc. of SAFEPROCESS'94*, pp. 64–68, 1994.

[16] R. Nikoukhah. Innovation generation in the presence of unknown inputs: application to robust failure detection. *Automatica*, **30(12)**, pp. 1851–1867, 1994.

[17] R. J. Patton and Jie Chen. Optimal unknown input distribution matrix selection in robust fault diagnosis. *Automatica*, **29(4)**, pp. 837–841, 1993.

[18] A. S. Willsky and H.L. Jones. A generalized likelihood ratio approach to the detection and estimation of jumps in linear systems. *IEEE Trans. Automatic Control*, **AC-21**, pp. 108–112, 1976.

[19] H. Yang and M. Saif. State observation, failure detection and isolation (FDI) in bilinear systems. *Proc. of the 34th IEEE Conference on Decision and Control*, pp. 2391–2396, 1995.

[20] D. Yu and D.N. Shields. A bilinear fault detection observer. *Automatica*, **32**(11), pp. 1597–1602, 1996.

[21] D. Yu, D.N. Shields and S. Daley. A bilinear fault detection observer and its application to a hydraulic drive system. *Int. J. Control*, **64**(6), pp. 1023–1047, 1996.

Part IV

Synchronization and Observers

In this last part of the book the notion of observers is combined with synchronization. Synchronization has attracted in the last decade a tremendous interest from researchers within physics, dynamical systems, circuit theory and more lately control theory. Indeed, depending from the specific setting, synchronization often refers to the ability of asymptotic reconstruction of the state of a system on the basis of a suitable (drive) signal. Formulated in this way, the synchronization problem fits into the usual formulation of an observer problem, and numerous examples, either at simulation level or even experimentally, illustrate successful synchronizing systems. However, it should be remarked that most of the existing physical examples are much more on a "try-out" basis than that a systematic procedure for the designing of a synchronized system. Since synchronization may be viewed as an observer problem it is important to see how nonlinear observers can be used for typical classes of nonlinear systems which can be synchronized. Most of these systems are highly nonlinear, or even complex/chaotic, and it is not at all evident why some of the simplest "try-out schemes" work. This question is at the heart of the last part of the book, where one also can find a number of interesting illustrative examples of synchronized systems.

Synchronization Through Extended Kalman Filtering

César Cruz[1] and Henk Nijmeijer[2]

[1]Department of Electronics & Telecom., Scientific Research and
Advanced Studies Center of Ensenada (CICESE), México
[2]Faculty of Mathematical Sciences, University of Twente, Enschede and
Faculty of Mechanical Engineering, Eindhoven University of Technology,
The Netherlands

1 Introduction

Synchronization is a fascinating phenomenon and has been observed in many diverse systems. Synchronization in chaotic systems may bring many interesting possibilities in practical applications. For example, it is believed that synchronization plays a crucial role in information processing in living organisms and could lead to important applications in speech and image processing (Ogorzalek [17]). Another area where synchronization may play an important role, is (secure) communication. Due to the fact that chaotic signals are noise-like and unpredictable in nature, such signals can possibly be used as potential carriers for secure communication ([20], [12], [18], [6], [10]). Moreover, the synchronization property of chaotic circuits has revealed potential applications to secure communications (see, e.g., [5], Kocarev et al., [12]; Parlitz et al., [18]; Cuomo et al., [6]; Halle et al., [10]).

However, the problem of obtaining two chaotic systems oscillating in a synchronized way is a nontrivial question. This is because any small difference in initial conditions would be exponentially amplified and thus the motions of the systems rapidly become uncorrelated (Ogorzalek [17]). Despite this fact, Pecora and Carroll discovered (see [5], [19]), that in particular cases it is very well possible that synchronization between a chaotic transmitter and driven receiver is possible. Much further work about synchronization can be found in the special issue [24].

In this paper we will concentrate on synchronization for systems in discrete time. Mostly, synchronization has been studied for systems in continuous time but many ideas go through in discrete time. On the other hand, the theory is less far developed for discrete time systems -even despite the fact that in many cases the actual implementation is done in discrete time.

Consider an autonomous, discrete-time, dynamical system

$$x\left(k+1\right) = f\left(x\left(k\right)\right), \tag{1.1}$$

where $x(k)$ is a n-dimensional vector. The Pecora-Carroll scheme for synchronization can be described as follows. Divide the system into two subsystems via $x(k) = (x_1(k), x_2(k))^T$, where $x_1(k)$ is n_1-dimensional and $x_2(k)$ is n_2-dimensional, with $n_1 + n_2 = n$ (often $n_1 = 1$, see [5], [6], [19]). We refer to $x_1(k)$ as the *drive signal*. So, the transmitter system (1.1) can be written as

$$x_1(k+1) = f_1(x_1(k), x_2(k)), \qquad (1.2)$$
$$x_2(k+1) = f_2(x_1(k), x_2(k)) \qquad (1.3)$$

where $f(x) = (f_1(x_1, x_2), f_2(x_1, x_2))^T$. The driven replica subsystem is described by

$$x_2^R(k+1) = f_2\left(x_1(k), x_2^R(k)\right), \qquad (1.4)$$

the so-called receiver dynamics. In (1.4), $x_2^R(k)$ is the response variable and (1.4) is known as the response subsystem.

The receiver system (1.4) *synchronizes* with the transmitter system (1.2) and (1.3), if

$$\lim_{k \to \infty} \| x_2(k) - x_2^R(k) \| = 0, \qquad (1.5)$$

no matter which initial values $x_1(0)$, $x_2(0)$ and $x_2^R(0)$ have.

In the context of communication the drive signal is transmitted from the transmitter system to the receiver system. The full state $x(k)$ of the transmitter is unknown at the receiver. By driving with the known signal $x_1(k)$ to a replica response subsystem (1.4), we can then obtain $x_2(k)$, if the copy synchronizes with the full system (1.2)-(1.3) (according to (1.5)).

In general, no matter if f_i is chaotic or complex, no complete general answer exists to the problem whether a replica system (1.4) would achieve synchronization according to (1.5). For that reason several attempts for achieving synchronization of signals like $x_2(k)$ and $x_2^R(k)$ have been proposed. The above idea of synchronization by decomposition into subsystems (by using a replica (1.4)) has first been given in Pecora and Carroll [19]; the ocurrence of such synchronization is conditioned on whether the conditional Lyapunov exponents for (1.4) are negative. In such case, the system (1.1) is said to possess a self-synchronizing property. Note however, the negativity of the conditional Lyapunov exponents is not a guarantee for the successful synchronization, cf. Badola *et al.* [2]. However, recent research has shown, that one is not necessarily constrained to the use of a replica (1.4) when choosing a system to achieve synchronization (at least when one is partly free to choose the receiver dynamics-driven by the transmitter's drive signal), i.e., it is also possible if, instead of (1.4), we utilize a different response system

$$x_2^R(k+1) = f_2^R\left(x_1(k), x_2^R(k)\right). \qquad (1.6)$$

This may leads to considerably more flexibility in applications like secure communication. This added flexibility may facilitate potential improvement in synchronization, e.g., Ding and Ott [8] have obtained synchronization using (1.6) in case a replica (1.4) does not synchronize. Another adventage of using (1.6) is to improve the convergence of the synchronization. In particular, we like to recall the (reduced) observer viewpoint advocated by Nijmeijer and Mareels in [15] which basically admits -under suitable assumptions- the construction of dynamics (1.6) such that (1.5) holds, whatever initial conditions (1.2), (1.3) and (1.6) have. A possible extension of [15] in discrete time has been given by Huijberts *et al.* [11].

On the other hand, for communication purposes (specifically, additive signal masking and recovery), it is required that the synchronization is not affected by some noise in the synchronizing drive signal. In other words, the use of synchronized chaotic systems for secure communications relies on the robustness of the synchronization to perturbations in the drive signal and in the system dynamics, [6], see also [20]. For certain synchronized chaotic systems, the ability to synchronize is robust. In the Lorenz system, for example, one can observe such property (Cuomo *et al.* [6]) for an exact replica (1.4).

On the basis of these considerations, we propose a system of the form

$$\hat{x}(k+1) = f^R(\bar{x}_1(k), \hat{x}(k)), \tag{1.7}$$

where $\bar{x}_1(k)$ is the drive signal corrupted by noise from the transmitter dynamics i.e., $\bar{x}_1(k) = x_1(k) + v(k)$, with $v(k)$ the noise signal, and $\hat{x}(k)$ is the state estimate for the original system (1.1), provided by an extended Kalman filter (EKF). In the next section, we give the particular equations for the EKF and further background on it. Among some advantages of using (1.7) as receiver/estimator for (1.1) are:

- The EKF possesses some natural robustness property to additive Gaussian noise in the drive signal (see Cuomo *et al.* [6]).

- The EKF is easily implemented.

- Flexibility in applications since a decomposition into two subsystems (1.2)-(1.3) is not necessary.

The idea of using an EKF to estimate the states of a chaotic system is given perhaps for the first time by Fowler [9]. Cuomo *et al.* [6] shown that the EKF estimates of a Lorenz system approach its true states. Recently in [21], Sobiski and Thorp describe a method of synchronizing two chaotic systems by implementing an EKF for continuous-time systems.

Motivated by these papers and by its implication in secure communications, we study synchronization between the discrete-time systems (1.1) and (1.7) when the EKF is used as receiver driven by a noisy drive signal. Besides synchronization per se, we also study the utility of using the EKF for reconstruction of a binary message. The idea is that some parameter in the transmitter can be used as the (binary signal) message carrier. In the EKF we include in this case as extra state a parameter-estimator, see also [21] where a similar idea was used in continuous time. At the same time, we mention some differences with [21]:

- The context we use here is discrete-time. This on the one hand makes the presentation even simpler than in continuous time. On the other hand, to the best of our (knowledge) using the EKF for synchronization in discrete time is new.

- By means of extensive simulations we make an attempt to give an evaluation of the performance of the EKF by changing filter-parameters (covariance noise, initial error, etc.).

- For the given examples a rigorous result that ensures the *local* (optimal) convergence of the EKF is given.

The organization of this paper is as follows: in Section 2 we present our approach to achieve synchronization of discrete-time systems via an extended Kalman filter which is driven by a noisy drive signal from the transmitter. By using computer simulations, the approach used in this study is illustrated with two examples in Section 3.1, while in Section 3.2 an application of these results to secure communication is given. Finally, Section 4 contains some concluding remarks.

2   An Extended Kalman Filter as Receiver

Briefly, given a stochastic (linear) model description, the Kalman filtering problem is to produce an estimate $\hat{x}(k)$ of the state model $x(k)$, using measurements till time k so as to minimize the mean-square error between estimate and state. From this viewpoint, the Kalman filter is the *optimal* linear filter since it produces an estimate minimizing a mean-square error. For application to nonlinear models, the so-called extended Kalman filter is often used in practice, but, in general, no guarantee in this case of producing a good (optimal) state estimate can be given. In this case the nonlinear system is linearized by employing the best estimates of the state as the reference values used at each stage for the linearization, i.e., the EKF consists of using the classical Kalman filter equations for the first-order approximation of the nonlinear system about the last estimate. As a direct consequence of taking this approximation, the EKF is no longer linear or optimal. For further details we refer the reader to Anderson and Moore [1] and references therein.

We consider transmitter dynamics of the form

$$x(k+1) = f(x(k)) + w(k), \quad x(0) = x_0, \tag{2.8}$$

with transmitted signal

$$y(k) = h(x(k)) + v(k).\tag{2.9}$$

Typically, $x(k)$ is an n-dimensional vector and $y(k)$ is a scalar signal (although much of what follows can be extended to a vector signal $y(k)$).

In (2.8) $w(k)$ represents the noise in the dynamics of the transmitter which is assumed to be a zero mean noise process with $E\left[w(k)w^T(l)\right] = Q\delta_{kl} > 0$, with δ_{kl} the Kronecker delta function. Also $v(k)$ is a zero mean noise process with $E[v(k)v(l)] = R\delta_{kl} > 0$; $v(k)$ and $w(k)$ are assumed to be independent.

Remark 1.1 *Although it may not be necessary to introduce the dynamics noise $w(k)$ in (2.8), we find it convenient and more flexible to do so. In fact from certain perspective it can be argued that an error free dynamics would be an over-idealization, which may prohibit a successful synchronization through the EKF, see Sorenson [23]. In the theoretic developments to follow one might think about the covariance matix Q as being reasonably small, mimicking at least a very accurate modeling. For convenience we have assumed in the examples of Section 3 that $w(k) = 0$ for all k.*

The receiver dynamics we propose is a filter that will produce an estimate for the state $x(k)$ given the measurements $y(k)$ according to (2.9). The EKF that we use here as the receiver dynamics for (2.8) and (2.9) is given as follows, cf. [1].

1) Measurement update equations:

$$\hat{x}(k) = \hat{x}(k/k-1) + K_{\hat{x}}(k)\left[y(k) - h(\hat{x}(k/k-1))\right],\tag{2.10}$$

the vector $\hat{x}(k)$ is referred to as the *filtered* estimate of the state $x(k)$. The covariance of the error in $\hat{x}(k)$ is given by

$$P_{\hat{x}}(k) = \left[I - K_{\hat{x}}(k)H_{\hat{x}}(k)\right]P_{\hat{x}}(k/k-1).\tag{2.11}$$

2) Time update equations:
The (one-step ahead) *predictor* of $x(k+1)$ is given by

$$\hat{x}(k+1/k) = f(\hat{x}(k)),\tag{2.12}$$

the covariance matrix of the prediction error is

$$P_{\hat{x}}(k+1/k) = F_{\hat{x}}(k)P_{\hat{x}}(k)F_{\hat{x}}^T(k) + Q,\tag{2.13}$$

where

$$K_{\hat{x}}(k) = P_{\hat{x}}(k/k-1)H_{\hat{x}}^T(k)\left[H_{\hat{x}}(k)P_{\hat{x}}(k/k-1)H_{\hat{x}}^T(k) + R\right]^{-1}\tag{2.14}$$

is known as the Kalman gain matrix, and

$$F_{\hat{x}}(k) = \left.\frac{\partial f(x(k))}{\partial x(k)}\right|_{x(k)=\hat{x}(k)}, \tag{2.15}$$

$$H_{\hat{x}}(k) = \left.\frac{\partial h(x(k))}{\partial x(k)}\right|_{x(k)=\hat{x}(k/k-1)}. \tag{2.16}$$

Remark 1.2 *In practice it might be impossible to determine $x(0)$ exactly. In this case, $x(0)$ is assumed to be a Gaussian random variable of known mean value $E\{x(0)\} = \bar{x}(0)$ and known covariance matrix $E\left\{[x(0) - \bar{x}(0)][x(0) - \bar{x}(0)]^T\right\} = P(0)$, and it is independent of $w(k)$ and $v(k)$.*

The filter is initialized by setting $\hat{x}(0) = \bar{x}_0$ and $P(0) = P_0 = P_0^T > 0$. Thus, $x(0)$ is given and we choose arbitrarily $\bar{x}_0$, $P_0 = P_0^T > 0$.

The definition of synchronization given in the introduction, i.e., equation (1.5), can be extended to include approximate or noisy synchronization to accommodate inaccurate parameters and non-ideal signal transmission. In this case, the receiver (2.10)-(2.13) can not synchronize with the transmitter (2.8) in the way that condition (1.5) is fulfilled so we need to replace it by a weaker condition

$$\lim_{k\to\infty} \| x(k) - \hat{x}(k) \| \leq \rho, \quad \forall k \geq \tau, \tag{2.17}$$

where ρ should be related to R and is a constant of the synchronization error. If for a given ρ there exists a time instant τ (to be called the synchronization time) such that condition (2.17) is fulfilled, then the transmitter (2.8) and the EKF receiver (2.10)-(2.13) are *approximately synchronized.*

Remark 1.3 *Also one might consider as an adequate condition for approximate synchronization in the noisy context*

$$\lim_{k\to\infty} \| E(x(k) - \hat{x}(k)) \| \leq \rho, \quad \forall k \geq \tau.$$

In particular this may be a more relevant requirement if $w(k)$ is not necessarily bounded. Since we will later on assume that both $v(k)$ and $w(k)$ are bounded it suffices to take (2.17) as the definition for approximate synchronization.

We define for $e_i(k) = x_i(k) - \hat{x}_i(k)$, $i = 1, 2, \ldots, n$

$$\tau_i(\rho) = \min_\tau \{|e_i(k)| < \rho, \quad k = \tau, \tau+1, \ldots\}, \tag{2.18}$$

and the synchronization time by

$$\tau = \max\left(\tau_i\right), \quad i = 1, 2, \ldots, n. \tag{2.19}$$

Remark 1.4 *From (2.17) note that there exists a compromise between the quantities ρ and τ, since if ρ increases then τ decreases, and vice versa.*

The EKF is often used to design observers (to deal with state estimation) for forced or non forced nonlinear systems. In spite of the fact that only local convergence is ensured, this method is widely used in practice and often gives convincing results (for a summary of the theory and applications see, e.g. Boutayeb *et al.* [4], Baras *et al.* [3], La Scala *et al.* [13], Ljung [14], Song and Grizzle [22] and references therein). The convergence aspects of the EKF when it is used as a deterministic observer for discrete-time system, are analyzed through a Lyapunov approach in Boutayeb *et al.* [4], and Song and Grizzle [22]. We follow the approach proposed in La Scala *et al.* [13], for the establishing the convergence of the EKF when applied to a stochastic, discrete-time nonlinear system with a linear output map. To this end, define the error in the filtered state as

$$e\left(k\right) \triangleq x\left(k\right) - \hat{x}\left(k\right). \tag{2.20}$$

From (2.10) we have

$$e\left(k\right) = \left[I - K_{\hat{x}}\left(k\right) H_{\hat{x}}\left(k\right)\right] e\left(k/k - 1\right) - K_{\hat{x}}\left(k\right) v\left(k\right),$$

where $e\left(k/k - 1\right) \triangleq x\left(k\right) - \hat{x}\left(k/k - 1\right)$ is the error in the predicted state estimate, thus,

$$e\left(k + 1/k\right) = \frac{\partial f}{\partial x}\left(x\left(k\right)\right) \cdot e\left(k\right) - \mathcal{O}_f\left(x\left(k\right), -e\left(k\right)\right) + w\left(k\right)$$

$\mathcal{O}_f$ is the remainder term from the Taylor series expansion of f, i.e.,

$$\mathcal{O}_f\left(a, b\right) = f\left(a + b\right) - f\left(a\right) - \frac{\partial f}{\partial a}\left(a\right) \cdot b.$$

So, for the (corrupted) drive signal (2.9), we have the error dynamics equation

$$
\begin{aligned}
e\left(k\right) = {} & \left[I - K_{\hat{x}}\left(k\right) H_{\hat{x}}\left(k\right)\right] F_x\left(k - 1\right) e\left(k - 1\right) \tag{2.21} \\
& - \left[I - K_{\hat{x}}\left(k\right) H_{\hat{x}}\left(k\right)\right] \mathcal{O}_f\left(x\left(k - 1\right), -e\left(k - 1\right)\right) \\
& + \left[I - K_{\hat{x}}\left(k\right) H_{\hat{x}}\left(k\right)\right] w\left(k - 1\right) - K_{\hat{x}}\left(k\right) v\left(k\right).
\end{aligned}
$$

From the last equation, we see that the dynamics for the filtering error of the EKF driven by a noisy drive signal from the transmitter, is composed by the sum of the error dynamics for the deterministic case (neglecting linearization errors), and nonlinear perturbation terms driven by the noise processes and remainder term from the Taylor series expansion of f.

Consider the time-varying linear system

$$\begin{aligned} \xi\,(k+1) &= F_z\,(k)\,\xi\,(k) + w\,(k)\,, &\qquad(2.22)\\ Y\,(k) &= H\xi\,(k) + v\,(k)\,, \end{aligned}$$

with $F_z\,(k) = (\partial f/\partial x)\,(\psi\,(k) - \phi\,(k))$. Define the observability Gramian of $\left[F_z,\,R^{-\frac{1}{2}}H\right]$ along a trajectory $\{z\,(k)\}$ of (2.22) as

$$\mathcal{O}\,(k,N) = \sum_{i=k-N}^{k} \Phi^T\,(i,k)\,H^T\,(i)\,R^{-1}H\,(i)\,\Phi\,(i,k)$$

for some $N \geq 0$ and for all $k \geq N$, where $\Phi\,(k_2, k_1) = F_z\,(k_2 - 1)\,F_z\,(k_2 - 2)\cdots F_z\,(k_1)$. Similarly, the controllability Gramian of $[F_z, Q]$ along a trajectory $\{z\,(k)\}$ of (2.22) as

$$\mathcal{C}\,(k,N) = \sum_{i=k-N}^{k-1} \Phi\,(k,i+1)\,Q\Phi^T\,(k,i+1)\,.$$

A system is said to be controllable (observable) along a trajectory $\{z\,(k)\}$ if there exists N such that for all $R_x > 0$ there exist $0 < \varepsilon_r < R_x$, $a_i\,(R_x, \varepsilon_r, N)$ and $b_i\,(R_x, \varepsilon_r, N)$, $i = 1, 2$, such that for some arbitrary sequence $\{\psi\,(k)\}$, $\|\psi\,(k)\| \leq R_x$, and for all $\{\phi\,(k)\}$ such that $\|\phi\,(k)\| \leq \varepsilon_r$

$$a_1 I \geq \mathcal{C}\,(k,N) \geq a_2 I, \qquad 0 < a_2 \leq a_1 < \infty, \qquad(2.23)$$

$$b_1 I \leq \mathcal{O}\,(k,N) \leq b_2 I, \qquad 0 < b_1 \leq b_2 < \infty \qquad(2.24)$$

where these Gramians are evaluated along the trajectory $z\,(k) = \psi\,(k) - \phi\,(k)$ of (2.22).

The following assumptions are needed, cf. [13]:

1) The transmitted signal (2.9) is a linear in x, i.e.,

$$y\,(k) = Hx\,(k) + v\,(k)\,. \qquad(2.25)$$

2) $f \in C^3\,(\mathbb{R}^n, \mathbb{R}^n)$, $(\partial f/\partial x)\,(x)$ is invertible for all $x \in \mathbb{R}^n$, and
3) for all k (see Remark 1.3)

$$\|x\,(k)\| \leq R_x, \qquad(2.26)$$

$$\|w\,(k)\| \leq \|w\| < \infty \quad \text{and} \quad \|v\,(k)\| \leq \|v\| < \infty. \qquad(2.27)$$

From Assumption 2), we can find $\rho_i > 0$, $i = 1, 2, 3$ such that

$$\left\|\frac{\partial f}{\partial x}(x)\right\| \le \rho_1, \quad \left\|\frac{\partial^2 f}{\partial x^2}(x)\right\| \le \rho_2, \quad \left\|\frac{\partial^3 f}{\partial x^3}(x)\right\| \le \rho_3$$

for all $\|x\| \le R_x + \varepsilon_r$. Furthermore, by the continuity of f, there exists a $\rho_4 > 0$ such that

$$\left\|\frac{\partial f}{\partial x}(x_1) - \frac{\partial f}{\partial x}(x_2)\right\| \le \rho_4 \|x_1 - x_2\|$$

for all $\|x_1\|, \|x_2\| \le R_x$, and

$$\left\|\frac{\partial f}{\partial x}(x_1) - \frac{\partial f}{\partial x}(x_2) - \frac{\partial^2 f}{\partial x^2}(x_2) \cdot (x_1 - x_2)\right\| \le \frac{1}{2}\rho_4 \|x_1 - x_2\|^2$$

(see, La Scala *et al.* [13]). Let

$$p = a_1 + \frac{1}{b_1}, \quad q = \frac{1}{a_2} + b_2, \quad s = \frac{1}{a_2} + b_2 + \rho_1^2 \delta_1^{-1}.$$

Since $P_z(k)$ and $P_z(k+1/k)$ are defined by means of the linear system (2.22), we find, using [7] the bounds (depending on ε_r, R_x and N)

$$q^{-1}I \le P_z(k) \le pI,$$
$$q^{-1}I \le P_z(k+1/k) \le sI.$$

Let

$$\varepsilon_z = \min\left\{\varepsilon_r, \frac{\sqrt{2}}{\sqrt{\rho_4}}\left(-\rho_1 + \sqrt{\rho_1^2 + \frac{1}{q}\left(\frac{1}{sp^2} - \gamma\right)}\right)^{\frac{1}{2}}\right\}, \qquad (2.28)$$

where $0 < \gamma < 1/sp^2$, provided that $-1/sp^2 + \rho_1\rho_4 q \|z(k)\|^2 + \frac{1}{4}\rho_4^2 q \|z(k)\|^4 \le -\gamma$. Define $\alpha, \beta > 0$ via

$$\beta\alpha^k = (pq)^{\frac{1}{2}}\left(1 - \frac{\gamma}{q}\right)^{\frac{k}{2}} \qquad (2.29)$$

and consider

$$\bar{z}(k+1) = A(k)\bar{z}(k) + \bar{f}_2(k, \bar{z}(k)) \qquad (2.30)$$

where

$$A(k) = [I - K_{\bar{z}}(k+1)H]F_x(k),$$

$$\bar{f}_2(k, \bar{z}(k)) = -\frac{\partial K_{\bar{z}}(k+1)}{\partial z(k)}HF_x(k)z(k)\bar{z}(k)$$

and

$$\tilde{z}(k) = x(k) - z(k).$$

Note that (2.30) is the linearized, undriven component of the EKF error dynamics, neglecting linearization errors. In La Scala $et~al.$ [13] an explicit expression for an upper bound on $\bar{f}_2$ is given as

$$\left\| \bar{f}_2 (k, \bar{z}(k)) \right\| \leq \zeta_{\bar{z}} = \delta_k \rho_1 \rho_5 \varepsilon_z \tag{2.31}$$

where $\delta_k = \delta_p \rho_5 \delta_2^{-1} \left(1 + s\rho_5^2 \delta_2^{-1}\right)$, $\|H\| \leq \rho_5$ and $\delta_p = 2\rho_1 \rho_2 p$.

Let

$$\eta \triangleq \max \left\{ pq \left(\|w\| + \frac{1}{2}\rho_4 \varepsilon_r^2 \right) + p\rho_5 \delta_2^{-1} \|v\|, \tag{2.32}$$
$$\delta_k \left(\rho_5 \|w\| + \|v\| + \frac{1}{2}\rho_4 \rho_5 \varepsilon_r^2 \right) + 2pq\rho_1 \right\},$$

and

$$\zeta = 2\rho_1 \rho_5 \delta_k + \rho_1 \rho_5 \delta_{k2} \varepsilon_r \tag{2.33}$$

with $\delta_{k2} (\rho_1, \rho_2, \rho_3, \rho_5, \delta_2, p) > 0$ such that

$$\left\| \frac{\partial^2 K_{\bar{z}}(k+1)}{\partial z^2} \right\| \leq \delta_{k2}.$$

Assume there exist η, $\zeta > 0$ as defined in (2.32) and (2.33) such that

$$\left\| [I - K_{\bar{z}}(k+1)H] [w(k) - \mathcal{O}_f (x(k), e(k))] - K_{\bar{z}}(k) v(k) \right\| \leq \eta \varepsilon_r,$$

and

$$\left\| \frac{\partial^2}{\partial z^2} [I - K_{\bar{z}}(k+1)H] F_x(k) \cdot z(k) \right\| \leq \zeta$$

for all $\|z(k)\| \leq \varepsilon_r$ and for all $k \geq 0$.

Theorem 1.1 *[EKF Stability] (La Scala et al. [13]) Consider the error dynamics of the EKF given in (2.21) when the EKF is applied to a signal model (2.8)-(2.25) which satisfies the standing assumptions 1)-3). Select N and $0 < \varepsilon_r < R_x$ such that the controllability and observability conditions (2.23) and (2.24) are satisfied. Then if*

$$\beta (\zeta + \eta) + \alpha < 1, \tag{2.34}$$

$$\beta \zeta_{\bar{z}} + \alpha < 1, \tag{2.35}$$

$$\|e\left(0\right)\| < \varepsilon_z \left(pq\right)^{-\frac{1}{2}}, \tag{2.36}$$

where ε_z, α and β, $\zeta_{\bar{z}}$, η , and ζ are given in (2.28), (2.29), (2.31), and (2.32), and (2.33), respectively, we have that the error dynamics satisfies

$$\|e\left(k\right)\| \leq \beta \left(\alpha^k \left(\alpha + \zeta\beta\right)^k\right) \|e_0\| \tag{2.37}$$

for all $0 \leq k < \tau$, and

$$\|e\left(k\right)\| \leq \frac{\beta\eta\left(R_x - \beta\varepsilon_z\right)}{1 - \left(\alpha + \zeta\beta\right)} \leq \varepsilon_r \tag{2.38}$$

for all $k \geq \tau$.

It is interesting to observe that the above theorem implies that under suitable technical conditions on the system dynamics f we obtain the 'practical stability' condition (2.37)-(2.38). It is clear that in the given noisy context this is the best we can hope far; convergence to zero is obviously impossible. The constant ε_r will play in the next section the approximate synchronization constant, see (2.17). Similarly the synchronization time τ in (2.19) is related to the integer N.

3   Examples

3.1   Synchronization

Example 1 (see Badola *et al.* [2])
   Consider two coupled logistic maps as the transmitter dynamics

$$
\begin{aligned}
x_1\left(k+1\right) &= \left(1 - \epsilon\right)\mu x_1\left(k\right)\left(1 - x_1\left(k\right)\right) + \epsilon x_2\left(k\right), &\quad (3.39)\\
x_2\left(k+1\right) &= \left(1 - \epsilon\right)\mu x_2\left(k\right)\left(1 - x_2\left(k\right)\right) + \epsilon x_1\left(k\right).
\end{aligned}
$$

Treating $y\left(k\right) = x_2\left(k\right)$ as the (ideal) drive signal ($n_1 = n_2 = 1$), Badola *et al.*, in [2] investigated the synchronization of $x_1\left(k\right)$ and the receiver signal $x_1^R\left(k\right)$ of which the dynamics were taken as

$$x_1^R\left(k+1\right) = \left(1 - \epsilon\right)\mu x_1^R\left(k\right)\left(1 - x_1^R\left(k\right)\right) + \epsilon x_2\left(k\right). \tag{3.40}$$

In Badola *et al.* [2] it turned out that only for particular initial conditions synchronization between (3.39) and (3.40) occurs. We therefore reconsider (3.39) in the frame of Section 2. To do this, we use an EKF presented in the previous section as receiver dynamics for the noisy transmitter

$$
\begin{aligned}
x_1\left(k+1\right) &= \left(1 - \epsilon\right)\mu x_1\left(k\right)\left(1 - x_1\left(k\right)\right) + \epsilon x_2\left(k\right) + w_1\left(k\right), &\quad (3.41)\\
x_2\left(k+1\right) &= \left(1 - \epsilon\right)\mu x_2\left(k\right)\left(1 - x_2\left(k\right)\right) + \epsilon x_1\left(k\right) + w_2\left(k\right)
\end{aligned}
$$

and the (corrupted) drive signal

$$y(k) = x_2(k) + v(k). \tag{3.42}$$

The EKF will yield the state estimates $\hat{x}_1(k)$ and $\hat{x}_2(k)$ for the signals $x_1(k)$ and $x_2(k)$. The structure of the EKF is given by

$$
\begin{aligned}
\hat{x}_1(k+1) &= (1-\epsilon)\,\mu\hat{x}_1(k)(1-\hat{x}_1(k)) + \epsilon\hat{x}_2(k) \tag{3.43}\\
&\quad + k_1(k)\,[y(k) - \hat{x}_2(k)],\\
\hat{x}_2(k+1) &= (1-\epsilon)\,\mu\hat{x}_2(k)(1-\hat{x}_2(k)) + \epsilon\hat{x}_1(k)\\
&\quad + k_2(k)\,[y(k) - \hat{x}_2(k)]
\end{aligned}
$$

where the gain vector $(k_1(k), k_2(k))^T$ is given via equations (2.10)-(2.14).

We investigate the evolution of the estimation process created by the EKF with the assumption that the initial matrix P_0 is of the form $P_0 = \mathrm{diag}\{p_{0i}\}$, $i = 1, 2$. Also, the variance of the noise R was fixed at 0.00005 and $p_{01} = p_{02} = 100$. To simplify the presentation we realized the dynamics noise as being identical zero or, which is equivalent, the covariance Q of $(w_1, w_2)^T$ was supposed to be extremely small of the form $Q = \mathrm{diag}\{q_i\}$, $i = 1, 2$, with $q_i = 10^{-8}$.

For the parameter values of $\epsilon = 0.09$ and $\mu = 3.7$ and initial conditions in $[0, 1] \times [0, 1]$, we have that $\|x(k)\| \le R_x = \sqrt{1.9}$ for all $k \ge 0$. Although $\partial f(x)/\partial x$ is not invertible everywhere, it turns out that when initializing (3.41) at $x(0) = (0.4, 0.7)$ the Jacobian remained nonsingular along the trajectory. The controllability and observability conditions hold for all $k \ge 1$. Finally, we take $\varepsilon_r = 1$, $\rho_1 = 4.76$, $\rho_2 = \rho_3 = \rho_4 = 3.36$ and $\rho_5 = 1$ to satisfy the Theorem 1.1.

Initial conditions $x(0) = (0.4, 0.7)$ have been used for the subsequent simulations. For the above parameter values of μ and ϵ, the transmitter (3.41) is apparently chaotic. Following [2], $x_1(k)$ and $x_1^R(k)$ do not synchronize for these parameter values and initial conditions $x_1^R(0) = \hat{x}_1(0) = 0.65$ while we obtain synchronization (according to (2.17)) using the EKF as receiver. Figure 1 shows the synchronization error between transmitter and receiver dynamics. We see that, after some transient behavior, the approximate synchronization is clearly visible; according to (2.19) it is obtained when $\tau = 3$ when $\rho = 0.04$ is considered (see Figure 2).

To evaluate the performance of the EKF from the point of view of sensitivity to initial errors, twenty different Monte Carlo runs were taken in order to obtain root-mean-square error statistics. The results are summarized in Table 1.1; where sd_i $(i = 1, 2)$ is the sum of square errors given by

$$sd_i = \sum_{i=1}^{N} (x_i(k) - \hat{x}_i(k))^2, \quad k = 0, 1, \ldots, N \tag{3.44}$$

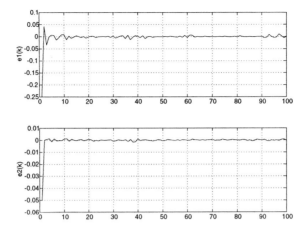

FIGURE 1. Synchronization errors $e_i(k) = x_i(k) - \hat{x}_i(k)$ $(i = 1, 2)$ for transmitter (3.39) and EKF receiver (3.43) for Example 1: $e(0) = (-0.25, -0.05)$, $R = 0.00005$, $\mu = 3.7$, and $\epsilon = 0.09$.

where $x_i(k)$ and $\hat{x}_i(k)$ are the true and estimated states, respectively, and N the number of time steps. Thus, the mean square error (MSE) is obtained by $\frac{sd_i}{N}$. With the purpose to know the same statistics, when the transient has died out we define the truncated mean-square error as

$$TMSE = \frac{1}{N+1-\tau} \sum_{i=\tau}^{N} (x(i) - \hat{x}(i))^2. \tag{3.45}$$

The results in Table 1.1 show the good performance of the EKF for the system (3.41). In particular, it should be observed that even with larger initial errors, the truncated mean-square errors remain within similar ranges.

Example 2 (see Huijberts *et al.* [11]) Consider the third order transmitter

$$
\begin{aligned}
x_1(k+1) &= (1-\epsilon)\mu x_1(k)(1 - x_1(k)) + \epsilon x_2(k), & (3.46)\\
x_2(k+1) &= (1-\epsilon)\mu x_2(k)(1 - x_2(k)) + \epsilon x_3(k),\\
x_3(k+1) &= (1-\epsilon)\mu x_3(k)(1 - x_3(k)) + \epsilon x_1(k)
\end{aligned}
$$

as an extension of the system (3.39). Similarly that the last example we consider the noisy transmitter

$$
\begin{aligned}
x_1(k+1) &= (1-\epsilon)\mu x_1(k)(1 - x_1(k)) + \epsilon x_2(k) + w_1(k), & (3.47)\\
x_2(k+1) &= (1-\epsilon)\mu x_2(k)(1 - x_2(k)) + \epsilon x_3(k) + w_2(k),\\
x_3(k+1) &= (1-\epsilon)\mu x_3(k)(1 - x_3(k)) + \epsilon x_1(k) + w_3(k),
\end{aligned}
$$

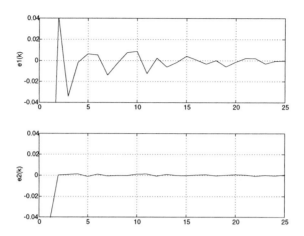

FIGURE 2. Approximate synchronization when $\rho = 0.04$ is considered for Example 1: $R = 0.00005$, $\mu = 3.7$, and $\epsilon = 0.09$.

in this case the (corrupted) drive signal is

$$y(k) = x_2(k) + v(k). \tag{3.48}$$

The equations for the EKF (receiver system for (3.46)) are

$$
\begin{aligned}
\hat{x}_1(k+1) &= (1-\epsilon)\,\mu\hat{x}_1(k)\,(1-\hat{x}_1(k)) + \epsilon\hat{x}_2(k) \\
&\quad + k_1(k)\,[y(k) - \hat{x}_2(k)], \\
\hat{x}_2(k+1) &= (1-\epsilon)\,\mu\hat{x}_2(k)\,(1-\hat{x}_2(k)) + \epsilon\hat{x}_3(k) \\
&\quad + k_2(k)\,[y(k) - \hat{x}_2(k)], \\
\hat{x}_3(k+1) &= (1-\epsilon)\,\mu\hat{x}_3(k)\,(1-\hat{x}_3(k)) + \epsilon\hat{x}_1(k) \\
&\quad + k_3(k)\,[y(k) - \hat{x}_2(k)]
\end{aligned}
\tag{3.49}
$$

with gain vector $(k_1(k), k_2(k), k_3(k))^T$ given via equations (2.10)-(2.14).

In the following simulations we take $x(0) = (0.2, 0.4, 0.6)$, $R = 0.0001$, $P_0 = \text{diag}\{p_{0i}\}$, $p_{0i} = 100$, $i = 1, 2, 3$, $\mu = 3.7$, and $\epsilon = 0.35$ (in this case, we take this value since estimating the signals $x_1(k)$ and $x_3(k)$ is more difficult because $x_1(k)$ is only indirectly influenced via $x_3(k)$) have been used. Again, to simplify the presentation we realized the dynamics noise as being identical zero or, which is equivalent, the covariance Q of $(w_1, w_2, w_3)^T$ was supposed to be extremely small of the form $Q = \text{diag}\{q_i\}$, $i = 1, 2, 3$, $q_i = 10^{-8}$.

For the above parameter values of ϵ and μ and initial condition in $[0, 1.2] \times [0, 1.2] \times [0, 1.2]$, we have that $\|x(k)\| \le R_x = \sqrt{3.15}$ for all $k \ge 0$. Again, we have that $f(x)/\partial x$ is not invertible everywhere, it turns out that when

$x(0)$	$\hat{x}(0)$	$x(0)-\hat{x}(0)$	sd_1	sd_2	τ	tsd_1	tsd_2
(0.4,0.7)	(0.4,0.7)	(0,0)	0.0019	0.0002	0	0.0019	0.0002
-	(0.39,0.69)	(0.01,0.01)	0.0021	0.0003	0	0.0021	0.0003
-	(0.37,0.67)	(0.03,0.03)	0.0022	0.0003	0	0.0022	0.0003
-	(0.35,0.85)	(0.05,-0.15)	0.0052	0.0225	2	0.0027	2.4625e-05
-	(0.35,0.90)	(0.05,-0.20)	0.0057	0.0400	2	0.0030	2.7650e-05
-	(0.35,0.95)	(0.05,-0.25)	0.0062	0.0625	4	0.0032	2.8954e-05
-	(0.35,1)	(0.05,-0.30)	0.0082	0.0900	6	0.0035	3.3452e-05
-	(0.30,0.85)	(0.10,-0.15)	0.0214	0.0225	3	0.0028	2.7802e-05
-	(0.25,0.85)	(0.15,-0.15)	0.0745	0.0226	8	0.0036	4.3795e-05
-	(0.20,0.85)	(0.20,-0.15)	0.1697	0.0226	8	0.0046	5.0678e-05
-	(0.65,0.75)	(-0.25,-0.05)	0.0666	0.0025	3	0.0023	2.6513e-05
-	(0.70,0.75)	(-0.30,-0.05)	0.1026	0.0025	3	0.0024	4.1756e-05
-	(0.75,0.75)	(-0.35,-0.05)	0.1870	0.0025	8	0.0056	7.0948e-05

TABLE 1.1. Dependence of the synchronization time on the initial condition and truncated mean-square error according to (3.45) for Example 1: $p_{0i} = 100$, $i = 1, 2$, $R = 0.00005$, $\mu = 3.7$, $\epsilon = 0.09$, $\rho = 0.04$, $N = 100$.

initializing (3.47) at $(0.2, 0.4, 0.6)$ the Jacobian remained nonsingular along the trajectory. The controllability and observability conditions hold for all $k \geq 1$. Finally, we take $\epsilon_r = 1.2$, $\rho_1 = 5.94$, $\rho_2 = \rho_3 = \rho_4 = 4.2$ and $\rho_5 = 1$ in order to satisfy the Theorem 1.1.

Figure 3 shows the synchronization error evolution between (3.47) and (3.49) for $\hat{x}(0) = (1.3, 6.4, 0)$. Notice that we have assumed here a rather large initial error as to see the clear effect that the EKF needs more time for approximate synchronization. Again, we can see, after some transient behavior, that approximate synchronization is achieved; according to (2.19) it is obtained at $\tau = 8$ when $\rho = 0.04$ is considered (see Figure 4).

Twenty different Monte Carlo runs were taken in order to obtain root-mean-square error statistics. The results are summarized in Table 1.2 (sd_i and τ) and Table 1.3 (tsd_i).

3.2   Secure Communication

Finally, in this subsection, we want to present an illustration of the potential use of synchronized systems through the EKF in secure communications.

Parameter switching is the simplest form of chaotic parameter modulation. In this method the message $s(k)$ is supposed to be binary, and is used to modulate one or more parameters of the (switching) transmitter, i.e., $s(k)$ controls a switch whose action changes the parameter values of the transmitter. Thus, according to the value of $s(k)$ at any given instant k, the transmitter has either the parameter set value p or the parameter set value $\bar{p}$.

At the receiver, $s(k)$ is decoded by using the synchronization error to

$x(0)$	$\hat{x}(0)$	$x(0)-\hat{x}(0)$	sd_1	sd_2	sd_3	τ
(0.2,0.2,0.6)	(0.2,0.2,0.6)	(0,0,0)	0.0025	0.0010	0.0037	0
-	(0.21,0.21,0.61)	(-0.01,-0.01,-0.01)	0.0028	0.0012	0.0040	0
-	(0.22,0.22,0.62)	(-0.02,-0.02,-0.02)	0.0030	0.0014	0.0044	0
-	(0.23,0.23,0.63)	(-0.03,-0.03,-0.03)	0.0035	0.0020	0.0049	0
-	(0.24,0.24,0.64)	(-0.04,-0.04,-0.04)	0.0051	0.0026	0.0057	3
-	(0.25,0.25,0.65)	(-0.05,-0.05,-0.05)	0.0078	0.0035	0.0089	3
-	(0.26,0.26,0.66)	(-0.06,-0.06,-0.06)	0.0112	0.0044	0.0107	5
-	(0.28,0.28,0.68)	(-0.08,-0.08,-0.08)	0.0191	0.0072	0.0126	5
-	(0.3,0.3,0.7)	(-0.1,-0.1,-0.1)	0.0277	0.0108	0.0158	5
-	(0.4,0.4,0.8)	(-0.2,-0.2,-0.2)	0.1130	0.0439	0.1377	10
-	(-0.05,-0.05,0.35)	(0.25,0.25,0.25)	0.6244	0.0825	0.3338	17
(0.2,0.4,0.6)	(1.1,2.2,0.1)	(-0.9,-1.8,0.5)	0.7098	3.2463	0.3684	6
-	(1.1,3.9,0)	(-0.9,-3.5,0.6)	0.8112	12.2512	0.3984	6
-	(1.3,6.4,0)	(-1.1,-6,0.6)	1.3486	36.0032	0.4523	8
-	(1.3,7.4,-0.1)	(-1.1,-7,0.7)	1.4776	49.0023	0.7087	11

TABLE 1.2. Dependence of the synchronization time on the initial condition for Example 2: $p_{0i} = 100$, $i = 1,2,3$, $R = 0.0001$, $\mu = 3.7$, $\epsilon = 0.35$, $\rho = 0.04$, $N = 100$.

$x(0)$	$\hat{x}(0)$	$x(0)-\hat{x}(0)$	tsd_1	tsd_2	tsd_3
(0.2,0.2,0.6)	(0.2,0.2,0.6)	(0,0,0)	0.0025	0.0010	0.0037
-	(0.21,0.21,0.61)	(-0.01,-0.01,-0.01)	0.0028	0.0012	0.0040
-	(0.22,0.22,0.62)	(-0.02,-0.02,-0.02)	0.0030	0.0014	0.0044
-	(0.23,0.23,0.63)	(-.03,-.03,-.03)	0.0035	0.0020	0.0049
-	(0.24,0.24,0.64)	(-0.04,-0.04,-0.04)	0.0010	0.0004	0.0017
-	(0.25,0.25,0.65)	(-0.05,-0.05,-0.05)	0.0013	0.0005	0.0020
-	(0.26,0.26,0.66)	(-0.06,-0.06,-0.06)	0.0017	0.0007	0.0022
-	(0.28,0.28,0.68)	(-0.08,-0.08,-0.08)	0.0019	0.0009	0.0024
-	(0.3,0.3,0.7)	(-0.1,-0.1,-0.1)	0.0030	0.0011	0.0025
-	(0.4,0.4,0.8)	(-0.2,-0.2,-0.2)	0.0017	0.0016	0.0028
-	(-0.05,-0.05,0.35)	(0.25,0.25,0.25)	0.0100	0.0059	0.0062
(0.2,0.4,0.6)	(1.1,2.2,0.1)	(-0.9,-1.8,0.5)	0.0005	0.0005	0.0010
-	(1.1,3.9,0)	(-0.9,-3.5,0.6)	0.0005	0.0007	0.0011
-	(1.3,6.4,0)	(-1.1,-6,0.6)	0.0018	0.0014	0.0020
-	(1.3,7.4,-0.1)	(-1.1,-7,0.7)	0.0024	0.0015	0.0026

TABLE 1.3. Truncated mean-square error according to (3.45) for Example 2: $p_{0i} = 100$, $i = 1,2,3$, $R = 0.0001$, $\mu = 3.7$, $\epsilon = 0.35$, $\rho = 0.04$, $N = 100$.

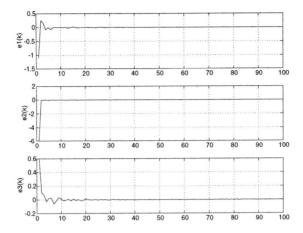

FIGURE 3. Synchronization errors $e_i(k) = x_i(k) - \hat{x}_i(k)$ $(i = 1, 2, 3)$ for transmitter (3.46) and EKF receiver (3.49) for Example 2: $e(0) = (-1.1, -6, 0.6)$, $R = 0.0001$, $\mu = 3.7$, and $\epsilon = 0.35$.

decide whether the received signal corresponds to one parameter value, or the other (it can be interpreted as a zero or one).

The usefulness of this simple idea has been demonstrated by Parlitz *et al.* [18] and Cuomo *et al.* [6] for a replica.

In our case, to transmit $s(k)$ via parameter modulation scheme, the EKF is modified to estimate the value of this parameter. Thus, a combined state and parameter estimation is made by the extension of the state vector with the unknown parameter. Let μ be the parameter to be modulated in the transmitter dynamics (3.41) and (3.47); in both examples, the parameter value ϵ was fixed. If no other apriori information is available, an additional state $\mu(k)$ is used to extend the original state vector by treating μ as a function of time according to $\mu(k+1) = \mu(k)$. So, the noisy transmitter dynamics (3.41) and (3.47) become for Example 1:

$$x_1(k+1) = (1-\epsilon)\mu(k)x_1(k)(1-x_1(k)) + \epsilon x_2(k) + w_1(k), \quad (3.50)$$
$$x_2(k+1) = (1-\epsilon)\mu(k)x_2(k)(1-x_2(k)) + \epsilon x_1(k) + w_2(k),$$
$$\mu(k+1) = \mu(k) + w_3(k),$$

and for Example 2:

$$x_1(k+1) = (1-\epsilon)\mu(k)x_1(k)(1-x_1(k)) + \epsilon x_2(k) + w_1(k), \quad (3.51)$$
$$x_2(k+1) = (1-\epsilon)\mu(k)x_2(k)(1-x_2(k)) + \epsilon x_3(k) + w_2(k),$$
$$x_3(k+1) = (1-\epsilon)\mu(k)x_3(k)(1-x_3(k)) + \epsilon x_1(k) + w_3(k),$$
$$\mu(k+1) = \mu(k) + w_4(k),$$

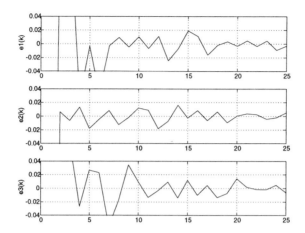

FIGURE 4. Approximate synchronization when $\rho = 0.04$ is considered for Example 2: $R = 0.0001$, $\mu = 3.7$, and $\epsilon = 0.35$.

respectively, with the (corrupted) transmitted signal

$$y(k) = x_2(k) + v(k). \tag{3.52}$$

We note that both examples satisfy the conditions of Theorem 1.1; for communication purposes we take values of R_x smaller than $\sqrt{1.9}$ and $\sqrt{3.15}$ for Examples 1 and 2, respectively. We use a 'modulation rule' to modulate $s(k)$ in the parameter μ of the transmitter (3.50) and (3.51). Then the EKF used as receiver maintains synchronization by estimating the changes in the modulated parameter μ (while the parameter ϵ is fixed at the same value as in the transmitter). So, $s(k)$ can be recovered by the estimation given through the EKF. The modulation rule is given by

$$\mu(k) = \mu + a \cdot s(k), \quad \hat{\mu}(k) = \mu + a \cdot \tilde{s}(k) \tag{3.53}$$

where a is a constant and $\tilde{s}(k)$ is the recovered message. The message is defined as follows

$$s(k) = \begin{cases} 0, & 0 \le k < 200, \\ 1, & 200 \le k < 400, \\ 0, & 400 \le k < 600, \\ 1, & 600 \le k < 800, \\ 0, & 800 \le k < 1000. \end{cases}$$

An illustration for the binary communication of Example 1, via modulation and estimation of parameter μ with $a = 0.08$, i.e., when μ is switched between $\mu(0) = 3.7$ and $\mu(1) = 3.78$ is shown in Figure 5. Figure 6 shows

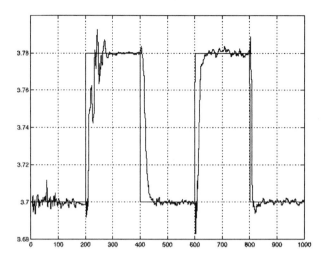

FIGURE 5. Estimated and true value μ for Example 1: $x(0) = \hat{x}(0) = (0.4, 0.7, 3.7)$, R=0.00005, and $a = 0.08$.

binary communication for Example 2, via modulation and estimation of parameter μ with $a = 0.1$, i.e., when μ is switched between $\mu(0) = 3.7$ and $\mu(1) = 3.8$.

4 Concluding Remarks

We have discussed the use (in discrete-time) of an extended Kalman filter (EKF) as receiver system for chaotic synchronization purposes. Synchronization is obtained between transmitter and receiver dynamics when the EKF is driven by a noisy drive signal from the transmitter.

The computer simulation results presented, show that our chaotic synchronization approach is robust to additive Gaussian noise. Besides synchronization per se, we have presented the utility of using the EKF for reconstruction of a binary message. In this case, by modulating a parameter in the transmitter and estimating this parameter via a modified EKF. Thus, we expect that it can be possibly applied to experimental systems, especially, for secure communication systems based on signal masking and parameter modulation.

Complementary simulations showed that synchronization and binary communication are also possible in case $x(0) \neq \hat{x}(0)$. Although, the synchronization time depends on the initial conditions and is different to the case $x(0) = \hat{x}(0)$. Obviously, the synchronization time is smaller for $x(0) \approx \hat{x}(0)$ and for smaller noise variance.

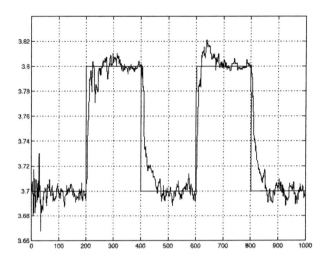

FIGURE 6. Estimated and true value μ for Example 2: $x(0) = \hat{x}(0) = (0.2, 0.4, 0.6, 3.7)$, R=0.0001, and $a = 0.1$.

Acknowledgements

This chapter was realized during a postdoctoral stay of the first author at the Univ. of Twente, supported by CONACYT (México) under Grant 973093.

5 REFERENCES

[1] B.D.O. Anderson and J.B. Moore, *Optimal Filtering*, Prentice-Hall, INC. Englewood Cliffs, New Jersey, 1979.

[2] P. Badola, S.S. Tambe, and B.D. Kulkarni, "Driving systems with chaotic signals", *Physical Review A*, **46**(10), pp. 6735-6737, 1992.

[3] J.S. Baras, A. Bensoussan and M.R. James, "Dynamic observers as asymptotic limits of recursive filters: special cases", *SIAM Journal on Applied Mathematics*, **48**(5), pp. 1147-1158, 1988.

[4] M. Boutayeb, H. Rafaralahy and M. Darouach, "Convergence Analysis of the Extended Kalman Filter Used as an Observer for Nonlinear Deterministic Discrete-Time Systems", *IEEE Trans. Automat. Contr.*, **42**(4), pp. 581-586, 1997.

[5] T.L. Carroll and L.M. Pecora, "Synchronizing chaotic circuits", *IEEE Trans. Circuits Syst.*, **38**, pp. 453-456, 1991.

[6] K.M. Cuomo, A.V. Oppenheim and S.H. Strogratz, "Synchronization of Lorenz-based chaotic circuits with application to communication", *IEEE Trans. Circuits Syst. II*, **40**(10), pp. 626-633, 1993.

[7] J.J. Deyst, Jr. and C.R. Price, "Conditions for asymptotic stability of the dicrete minimum-variance linear estimator", *IEEE Trans. Automat. Contr*, **13**(6), pp. 702-705, 1968.

[8] M. Ding and E. Ott, "Enhancing synchronism of chaotic systems", *Physical Review E*, **49**(2), pp. 945-948, 1994.

[9] T.B. Fowler, "Application of Stochastic Control Techniques to Chaotic Nonlinear Systems", *IEEE Trans. Automat. Contr.*, **34**(2), pp. 201-205, 1989.

[10] K.S. Halle, C.W. Wu, M.Itoh and L.O. Chua, "Spread spectrum communication through modulation of chaos", *International Journal of Bifurcation and Chaos*, **3**(2), pp. 469-477, 1993.

[11] H.J.C Huijberts, T. Lilge and H. Nijmeijer, 1999 "A control perspective on synchronization and the Takens-Aeyels-Sauer Reconstruction Theorem", To appear in *Phys. Rev. E*.

[12] Lj. Kocarev, K.S. Halle, K. Eckert and L.O. Chua, "Experimental demostration of secure communications via chaotic synchronization", *International Journal of Bifurcation and Chaos*, **2**(3), pp. 709-713, 1992.

[13] B.F. La Scala, R.R. Bitmead and M.R. James, "Conditions for Stability of the Extended Kalman Filter and Their Application to the Frequency Tracking Problem", *Mathematics of Control, Signals, and Systems*, **8**, pp. 1-26, 1995.

[14] L. Ljung, "Asymptotic behavior of the extended Kalman filter as a parameter estimator for linear systems", *IEEE Trans. Automat. Contr.*, **24**, pp. 36-50, 1979.

[15] H. Nijmeijer and I.M.Y. Mareels, "An Observer Looks at Synchronization", *IEEE Trans. Circ. Syst. I*, **44**(10), pp. 882-890, 1997.

[16] H. Nijmeijer, "On Synchronization of Chaotic Systems", *Proc. 36th IEEE Conf. on Decision and Control*, San Diego, USA, pp. 384-388, 1997.

[17] M.J. Ogorzalek, "Taming Chaos-Part I: Synchronization", *IEEE Trans. on Circuits and Systems-I: Fundamental Theory and Applications*, **40**(10), pp. 693-699, 1993.

[18] U. Parlitz, L.O. Chua, Lj. Kocarev, K.S. Halle and A. Shang, "Transmission of digital signals by chaotic synchronization", *International Journal of Bifurcation and Chaos*, **2**(4), pp. 973-977, 1992.

[19] L.M. Pecora and T.L. Carroll, "Synchronization in chaotic systems", *Phys. Rev. Let.*, **64**, pp. 821-824, 1990.

[20] G. Pérez and H.A. Cerdeira, "Extracting Messages Masked by Chaos", *Physical Review Letters*, **74**(11), pp. 1970-1973, 1995.

[21] D.J. Sobiski and J.S Thorp, "PDMA-1: Chaotic Communication via the Extended Kalman Filter", *IEEE Trans. Circuits and Systems-I: Fundamental Theory and Applications*, **45**(2), pp. 194-197, 1998.

[22] Y. Song and J.W. Grizzle, "The Extended Kalman Filter as a Local Asymptotic Observer for Discrete-time Nonlinear systems", *Journal of Mathematical Systems Estimation and Control*, **5**(1), pp. 59-78, 1995.

[23] H.W. Sorenson, "Least-squares estimation: from Gauss to Kalman", *IEEE Spectrum*, **7**, pp. 63-68, 1970.

[24] Special Issue on Chaos Synchronization and Control: Theory and Applications of the *IEEE Trans. Circuits and Systems I*, **44**(10), 1997.

Nonlinear Discrete-Time Observers for Synchronization Problems

T. Lilge

Institut für Regelungstechnik
University of Hannover
Hannover, Germany

1 Introduction

In recent years there has been an increasing interest in the synchronization of two coupled dynamical systems. Pecora and Carroll presented in their fundamental work [15] the synchronization of two identical Lorenz systems with different initial conditions if the second systems was driven by a state variable of the first system. In the following, many other works considered the synchronization of continuous-time systems, e.g. [14, Chapter 15], [18], [17] or [16], but also of discrete-time systems, e.g. [1], [11] or [19]. One possible use of synchronizing two systems can be found in secure communications where the receiver has to reconstruct the desired information from the transmitted state variable(s) of the transmitter system regardless of the initial conditions of transmitter and receiver.

In many examples found in the literature, the corresponding receiver system is an identical copy of the transmitter driven by the transmitted signals. There is often no theoretical proof given for the synchronization of transmitter and receiver. Moreover, the time until the systems synchronize can not be influenced and can be very long. Finally, synchronization of some systems can not be achieved for arbitrary initial conditions, cf. [1].

>From a control perspective, the problem of synchronization can be regarded as an observer problem, cf. [13] for continuous-time and [5] for discrete-time systems. This work focuses on a design of nonlinear observers for discrete-time systems (transmitters) of the form

$$
\begin{aligned}
x_{k+1} &= f(x_k), & x_k &\in \mathbb{R}^n \\
y_k &= h(x_k), & y_k &\in \mathbb{R}
\end{aligned}
\tag{1.1}
$$

where the output y can be interpreted as the transmitted signal and $f : \mathbb{R}^n \to \mathbb{R}^n$, $h : \mathbb{R}^n \to \mathbb{R}$ are smooth functions with $f(0) = 0$ and

$h(0) = 0$. For reasons of readability the step variable k is written as an index, i.e. $x_k = x(k)$. A design procedure for receiver systems (observers) of the form

$$\hat{x}_{k+1} = \hat{f}(\hat{x}_k, y_k, y_{k-1}, ..., y_{k-n+1}), \qquad \hat{x}_k \in I\!R^n \qquad (1.2)$$

is presented. System (1.2) exploits the transmitted signal y_k and $n-1$ buffered past measurements $y_{k-1}, ..., y_{k-n+1}$ in order to compute an estimate for the states of system (1.1) such that

$$\lim_{k \to \infty} \|\hat{x}_k - x_k\| = 0 \qquad (1.3)$$

and the convergence rate can be determined by an appropriate choice of the observer parameters.

An observer design via nonlinear observer form with linearizable error dynamics allows to choose the eigenvalues of the error dynamics in observer form as known from linear observers. The discrete-time *nonlinear observer form* is given by

$$z_{k+1} = A z_k + f_z(y_k), \qquad y_k = h_z(z_{n,k}), \qquad (1.4)$$

where $z \in I\!R^n$, $z_{n,k}$ is the n-th component of z_k, $f_z : I\!R \to I\!R^n$, $h_z : I\!R \to I\!R$ is the invertible output function and

$$A = \begin{pmatrix} 0 & \cdots & 0 & 0 \\ 1 & \cdots & 0 & 0 \\ \vdots & \ddots & \vdots & \vdots \\ 0 & \cdots & 1 & 0 \end{pmatrix}.$$

This observer form has been obtained by Lee and Nam in [8] but also by Chung and Grizzle in [3]. The main drawbacks of their approaches are that the dynamics f in (1.1) has to be a diffeomorphism and that a system of partial differential equations has to be solved in order to obtain the coordinate transformation into canonical form. In addition, the output function h_z has to be linear ($y_k = z_{n,k}$). An extension to a nonlinear output function h_z and some simplification for calculating the transformation is given by Ingenbleek in [6] and [7].

Brodmann in [2] and Lin and Byrnes in [10] presented a transformation of system (1.1) into nonlinear observer form (1.4) via the so called nonlinear observability form. In a first step, the system is transformed into observability form. The calculation of the transformation map is straightforward. A crucial point is computing the inverse map in order to obtain the system representation in observability form. However, this inversion problem also appears when system (1.1) is directly transformed into observer form as presented in [8, 3, 7]. In the second step, the system is transformed from observability form into observer form, which is only possible if the Hessian matrix of the nonlinear function showing up in observability form is diagonal. Since the transformation map between these two forms is always invertible, every system in observer form also has a representation in observability form. This implies that an existing representation in observer form can always be found from observability form. In addition, there are no restrictions for the system function f and the calculation of the transformation does not require solving a set of partial differential equations. However, the necessity that the Hessian matrix of a nonlinear function is diagonal is very restrictive.

In this chapter, an extended observer form depending on past measurements of the system output y_k is considered (see also [9] for systems with input). This considerably enlarges the class of systems for which a nonlinear observer with linearizable error dynamics can be designed. Once, a system is given in observability form, the extended observer form is also available. Furthermore, the transformation into this extended observer form offers some degrees of freedom which allow to design several observers with different characteristics.

Necessary and sufficient conditions are derived for (1.1) to be equivalent to a system in *extended observer form* (EOF)

$$
\begin{aligned}
z_{k+1} &= \begin{pmatrix} 0 & \cdots & 0 & 0 \\ 1 & \cdots & 0 & 0 \\ \vdots & \ddots & \vdots & \vdots \\ 0 & \cdots & 1 & 0 \end{pmatrix} z_k + \begin{pmatrix} f_{z,0}(y_k) \\ f_{z,1}(y_{k-1}, y_k) \\ \vdots \\ f_{z,n-1}(y_{k-n+1}, \dots, y_k) \end{pmatrix}, \quad (1.5) \\
y_k &= h_z(z_{n,k}),
\end{aligned}
$$

with $z \in \mathbb{R}^n$, $f_{z,\mu} : \mathbb{R}^{\mu+1} \to \mathbb{R}$, and $h_z : \mathbb{R} \to \mathbb{R}$ is the invertible output function of the system in EOF, i.e. $z_{n,k}$ can be calculated by $z_{n,k} = h_z^{-1}(y_k)$. This canonical form is obtained by state and output transformation and results in a linear structure with additive nonlinearities just depending on the actual and $n - 1$ buffered past output measurements. Therefore, an observer design based on (1.5) with linear observer error dynamics is possible.

The rest of this chapter is organized as follows. In the next section, necessary and sufficient conditions are given for system (1.1) being equivalent to a system representation in extended observer form. Section 3 focuses on the observer design via observer form and includes a discussion of some degrees of freedom in the coordinate transformation. Three alternative observers via EOF with different structure and characteristics are presented in Section 4. Finally, the observer design is applied to a synchronization problem in the field of communication in discrete-time and to a synchronization problem of the continuous-time Rössler system.

2 State Equivalence to a System in Extended Observer Form

After a short review of results concerning the observability of system (1.1) and its equivalence to a representation in observer form as presented in [2] and [10], the necessary and sufficient condition for system (1.1) being equivalent to a system in EOF is presented in this section.

Theorem 2.1 ([12, 2]) *If the so-called observability map*

$$
\psi(x) = \begin{pmatrix} h \circ f^0(x) \\ h \circ f^1(x) \\ \vdots \\ h \circ f^{n-1}(x) \end{pmatrix} \tag{2.6}
$$

has full rank at $x = 0$, then the system (1.1) is strongly locally observable at $x = 0$.

Since the observability map $\psi(x)$ of a strongly locally observable system (1.1) has full rank at $x = 0$, the state transformation $s = \psi(x)$, with $s_k = (y_k, y_{k+1}, ..., y_{k+n-1})^T$, leads to a new local representation of system (1.1) around $x = 0$, the so called nonlinear observability form.

Definition 2.1 (Nonlinear observability form ([12, 2]))   *The local representation in nonlinear observability form of a strongly locally observable system (1.1) at $x = 0$ has the form*

$$
s_{k+1} = \begin{pmatrix} y_{k+1} \\ \vdots \\ y_{k+n-1} \\ y_{k+n} \end{pmatrix} = \begin{pmatrix} s_{2,k} \\ \vdots \\ s_{n,k} \\ f_s(s_k) \end{pmatrix}, \quad y_k = s_{1,k}, \tag{2.7}
$$

with $\quad s_k = \psi(x_k) \Leftrightarrow x_k = \psi^{-1}(s_k)$

and $\quad f_s(s_k) = h \circ f^n \circ \psi^{-1}(s_k)$

If ψ is a global diffeomorphism the state transformation $s = \psi(x)$ leads to a global system representation in observability form.

As shown in [2] and [10], system (1.1) is equivalent to a system in non-linear observer form (1.4) if and only if the observability form (2.7) exists and there exists analytic functions $f_{z,\mu} : I\!R \to I\!R$, $\mu = 0, 1, ..., n-1$, and $h_z : I\!R \to I\!R$ such that the Hessian matrix of $h_z^{-1}(f_s(s_k))$ is diagonal, i.e.

$$h_z^{-1}(f_s(s_k)) = \sum_{\mu=0}^{n-1} f_{z,\mu}(s_{\mu+1,k}). \qquad (2.8)$$

Then, the observer form (1.4) contains the functions $f_{z,\mu}$ as the n components of f_z and h_z as the output function. Equation (2.8) represents a very restrictive condition for a system being equivalent to a representation in nonlinear observer form. As presented in the following theorem, the conditions for a system (1.1) being equivalent to a representation in extended observer form (1.5) are less restrictive and the class of systems for which an observer with linearizable error dynamics can be designed, is considerably enlarged.

Theorem 2.2 *A discrete-time system (1.1) with single output is state equivalent to a system in extended observer form (1.5) via state and output transformation if and only if the observability map (2.6) is a global diffeomorphism, i.e. a global system representation in observability form (2.7) exists.*

Proof.
   The transformation from observability form (2.7) into EOF (1.5) can be found by considering the structure of the EOF and taking into account $s_k = (y_k, ..., y_{k+n-1})^T$. Then, $z_{n,k} = h_z^{-1}(y_k)$ where y_k is replaced by $s_{1,k}$. From the last component of the system function in EOF, it follows $z_{n-1,k} = z_{n,k+1} - f_{z,n-1}(y_{k-n+1}, ..., y_k)$ where $z_{n,k+1}$ can be replaced by $h_z^{-1}(y_{k+1})$ and $y_{k+\mu}$, $\mu = 0, 1$, by $s_{\mu+1,k}$. Continuing in this way, the state variables $z_{i,k}$, $i = 1, 2, ..., n$ in EOF result from $s_{i,k}$ and $y_{k-n+1}, ..., y_{k-1}$ in the form

$$
\begin{aligned}
z_{n,k} &= h_z^{-1}(s_{1,k}) \\
z_{n-1,k} &= h_z^{-1}(s_{2,k}) - f_{z,n-1}(y_{k-n+1}, ..., y_{k-1}, s_{1,k}) \\
z_{n-2,k} &= z_{n-1,k+1} - f_{z,n-2}(y_{k-n+2}, ..., y_k) \\
&= h_z^{-1}(s_{3,k}) - f_{z,n-1}(y_{k-n+2}, ..., y_{k-1}, s_{1,k}, s_{2,k}) \\
&\quad - f_{z,n-2}(y_{k-n+2}, ..., y_{k-1}, s_{1,k}) \\
&\vdots \\
z_{1,k} &= z_{2,k+1} - f_{z,1}(y_{k-1}, y_k) \\
&= h_z^{-1}(s_{n,k}) - \sum_{\mu=1}^{n-1} f_{z,\mu}(y_{k-1}, s_{1,k}, ..., s_{\mu,k}).
\end{aligned}
\qquad (2.9)
$$

Since h_z is invertible, the inverse transformation exists and can easily be calculated:

$$
\begin{aligned}
s_{1,k} &= h_z(z_{n,k}) \\
s_{2,k} &= h_z\left(z_{n-1,k} + f_{z,n-1}(y_{k-n+1}, ..., y_{k-1}, s_{1,k})\right) \\
&\vdots \\
s_{n,k} &= h_z\left(z_{1,k} + \sum_{\mu=1}^{n-1} f_{z,\mu}(y_{k-1}, s_{1,k}, ..., s_{\mu,k})\right).
\end{aligned}
\tag{2.10}
$$

For the rest of the proof it has to be shown how the functions h_z and $f_{z,\mu}$ with $\mu = 0, 1, ..., n-1$ in EOF depend on the function f_s in observability form. Equation (2.10) and $y_{k+i-1} = s_{i,k}$ for $i = 1, 2, ..., n$ lead to the last component of the system representation in observability form

$$
\begin{aligned}
s_{n,k+1} &= h_z\left(z_{1,k+1} + \sum_{\mu=1}^{n-1} f_{z,\mu}(y_k, s_{1,k+1}, ..., s_{\mu,k+1})\right) \\
&= h_z\left(f_{z,0}(s_{1,k}) + \sum_{\mu=1}^{n-1} f_{z,\mu}(s_{1,k}, s_{2,k}, ..., s_{\mu+1,k})\right) \\
&= h_z\left(\sum_{\mu=0}^{n-1} f_{z,\mu}(s_{1,k}, s_{2,k}, ..., s_{\mu+1,k})\right).
\end{aligned}
$$

Since $s_{n,k+1} = f_s(s_k)$, it follows

$$
h_z^{-1}\left(f_s(s_k)\right) = \sum_{\mu=0}^{n-1} f_{z,\mu}(s_{1,k}, s_{2,k}, ..., s_{\mu+1,k}),
\tag{2.11}
$$

which is the counterpart of (2.8). The functions $f_{z,\mu}$ and h_z are showing up in the system representation (1.5) in EOF. In contrast to (2.8), which is very restrictive, (2.11) can always be fulfilled. Setting $h_z^{-1}\left(f_s(s_k)\right) = \tilde{f}_s(s_k)$ leads to

$$
\tilde{f}_s(s_k) = \sum_{\mu=0}^{n-1} f_{z,\mu}(s_{1,k}, s_{2,k}, ..., s_{\mu+1,k}).
\tag{2.12}
$$

Since $\tilde{f}_s$ and $f_{z,n-1}$ depend on the same arguments, it is always possible to fulfil (2.11) by an appropriate choice of $f_{z,n-1}$.

The transformation between observability form (2.7) and extended observer form (1.5) is always invertible and the inverse transformation can easily be calculated. Therefore, each system representation in EOF can be transformed into observability form which implies that the transformation of a system into EOF via observability form includes all existing solutions. □

As a consequence, there are no more conditions for the system representation in observability form (2.7). A remaining drawback is finding the inverse $x = \psi^{-1}(s)$ of the observability map.

Remark 2.1 *If the observability map (2.6) is a local diffeomorphism at* $x = 0$ *system (1.1) is locally equivalent to a system in extended observer form (1.5) and only a local observer can be obtained.*

Equation (2.11), which describes the relation between observability and extended observer form, leads to some freedom for the system representation in EOF because the selection of functions $f_{z,\mu}$ and h_z which are matching this equation is not unique. The influence of these functions on the observer design is considered in the next section.

3   Observer Design via Extended Observer Form

The dynamical system

$$
\begin{aligned}
\hat{z}_{k+1} &= \begin{pmatrix} 0 & \cdots & 0 & 0 \\ 1 & \cdots & 0 & 0 \\ \vdots & \ddots & \vdots & \vdots \\ 0 & \cdots & 1 & 0 \end{pmatrix} \hat{z}_k + \begin{pmatrix} f_{z,0}(y_k) \\ f_{z,1}(y_{k-1}, y_k) \\ \vdots \\ f_{z,n-1}(y_{k-n+1}, ..., y_k) \end{pmatrix} \\
&\quad + \begin{pmatrix} q_0 & q_1 & \cdots & q_{n-1} \end{pmatrix}^T \Big(\underbrace{h_z^{-1}(y_k)}_{z_{n,k}} - \hat{z}_{n,k} \Big), \\
\hat{y}_k &= h_z(\hat{z}_{n,k}), \qquad k \geq n-1,
\end{aligned}
$$

(3.13)

is a nonlinear observer for a system in EOF (1.5). It is easy to check that, for $k \geq n-1$, this observer leads to the linear error dynamics in EOF

$$
e_{k+1} = \hat{z}_{k+1} - z_{k+1} = \begin{pmatrix} 0 & \cdots & 0 & -q_0 \\ 1 & \cdots & 0 & -q_1 \\ \vdots & \ddots & \vdots & \vdots \\ 0 & \cdots & 1 & -q_{n-1} \end{pmatrix} e_k.
$$
(3.14)

The parameters $q_0, q_1, ..., q_{n-1}$, are the coefficients of the characteristic polynomial which allow to assign the eigenvalues of the error dynamics in EOF.

Since system and observer equations in EOF are depending on the selection of functions h_z and $f_{z,\mu}$, it is of interest to know how the observer equations in x-coordinates depend on these functions. As shown in [9], the observer equation (3.13) in EOF transformed back into x-coordinates, in which the system (1.1) is described, does not depend on the selection of functions $f_{z,0}, f_{z,1}, ..., f_{z,n-1}$. As a consequence, the selection process of $f_{z,\mu}$ for a given h_z is not suitable to affect the behavior of the observer in x-coordinates and the simple choice

$$f_{z,\mu}(s_{1,k}, ..., s_{\mu+1,k}) = 0 \qquad \text{for} \quad \mu = 0, 1, ..., n-2 \qquad (3.15)$$

$$f_{z,n-1}(s_{1,k}, ..., s_{n,k}) = h_z^{-1}(f_s(s_k)) \qquad (3.16)$$

does not require any computational effort and leads to the same observer equations in x-coordinates like other selections.

However, it is often more suitable to calculate the observer equations in EOF and afterwards transform the observer state $\hat{z}_k$ back into x-coordinates because this leads in general to shorter computation times. Since in this case the calculation of $\hat{z}_k$ requires all past measurements appearing in the EOF the selection of functions $f_{z,\mu}$ has an influence on the number of past measurements having to be buffered for the online calculation of the observer.

In [2], the nonlinear output function h_z in observer form could help in some cases to fulfill (2.8). Here, the nonlinear output function h_z in extended observer form can therefore help in some cases to reduce the number of necessary past measurements.

In addition, h_z considerably affects the transformation between original form and EOF (x- and z-coordinates). The transformation from x-coordinates into EOF is given by substituting $s_{i,k}$ by $h(f^{i-1}(x_k))$ in the components of the transformation map (2.9) from observability form into EOF. This leads to expressions of the form $h_z^{-1}(h(\cdot))$ which in turn are added to the functions $f_{z,\mu}$. The structure of $f_{z,\mu}$ follows from (2.11) which left hand side is of the form $h_z^{-1}(f_s(\cdot))$. Substituting f_s by $h(f^n(x_k))$ also results in an expression of the form $h_z^{-1}(h(\cdot))$. Therefore, the selection of h_z^{-1} or h_z respectively is rather important for the structure and the nonlinearities of the transformation from x-coordinates into EOF. Especially, a choice of h_z being of the same structure as h is mostly beneficial because in this case, cancellations of nonlinear expressions may occur in $h_z^{-1}(h(\cdot))$ leading to a simpler transformation rule. In what follows, this general requirement is explained in more detail for a special form of the system output function h.

If h depends on a linear combination of state variables, i.e. $h(x_k) = \bar{h}(v^T x_k)$ with v^T being a constant n-dimensional row vector, the transformation into EOF leads to $h_z(z_{n,k}) = \bar{h}(v^T x_k)$. It is advisable to choose $h_z(z_{n,k}) = \bar{h}(z_{n,k})$ which leads to a linear component $z_{n,k} = v^T x_k$ of the transformation from x-coordinates into EOF. Nonlinearities resulting from h and h_z in the other $n-1$ components are also cancelled which can be seen by inserting (2.6) in (2.9). This ensures, that the dynamic of the observer error in x-coordinates does not considerably differ from the one assigned in EOF.

4   Alternative Observer Structures via EOF

4.1   Observer Equations

The results presented in this subsection are based on the choice (3.15) with $f_{z,\mu} = 0$ for $\mu = 0, 1, ..., n-2$. This choice represents a very special case because iterating system (1.5) in EOF with arbitrary initial conditions leads to

$$z_{i,k} = 0 \quad \text{for} \quad k \geq i \quad \text{and} \quad i = 1, 2, ..., n-1. \tag{4.17}$$

Since $z_{n-1,k} = 0$ for $k \geq n-1$, the last component of the system function in EOF results in $z_{n,k+1} = f_{z,n-1}(y_{k-n+1}, ..., y_k)$ and from $h_z(z_{n,k+1}) = y_{k+1}$, it follows

$$f_{z,n-1}(y_{k-n+1}, ..., y_k) = h_z^{-1}(y_{k+1}) \quad \text{for} \quad k \geq n-1. \tag{4.18}$$

Besides observer (3.13) (in the following called **observer 1**), these results offer a design of three other observers in EOF.

Replacing $f_{z,n-1}(y_{k-n+1}, ..., y_k)$ by $h_z^{-1}(y_{k+1})$ in (3.13) with $f_{z,\mu} = 0$ for $\mu = 0, 1, ..., n-2$ leads to **observer 2**. Using (4.17), the resulting **observer 3** is given by

$$
\hat{z}_{k+1} = \begin{pmatrix} 0 \\ \vdots \\ 0 \\ f_{z,n-1}(y_{k-n+1}, ..., y_k) \end{pmatrix} + \begin{pmatrix} \lambda_1 \hat{z}_{1,k} \\ \vdots \\ \lambda_{n-1} \hat{z}_{n-1,k} \\ \lambda_n(\hat{z}_{n,k} - h_z^{-1}(y_k)) \end{pmatrix},
$$
$$
\hat{y}_k = h_z(\hat{z}_{n,k}), \quad k \geq n-1,
$$

$$\tag{4.19}$$

where λ_i are the eigenvalues of the observer error dynamics in EOF for $k \geq n-1$

$$e_{k+1} = \hat{z}_{k+1} - z_{k+1} = \begin{pmatrix} \lambda_1 & \cdots & 0 \\ \vdots & \ddots & \vdots \\ 0 & \cdots & \lambda_n \end{pmatrix} e_k. \tag{4.20}$$

The convergence rate for the i-th component of the observer error can be assigned by λ_i. Finally, replacing $f_{z,n-1}(y_{k-n+1}, ..., y_k)$ in (4.19) by $h_z^{-1}(y_{k+1})$ leads to **observer 4** with error dynamics (4.20). This observer is similar to the one presented by Ciccarella et al. in [4].

Since the selection of the functions $f_{z,\mu}$ according to (3.15) represents a necessary condition for the three observers presented in this section, the only remaining degree of freedom in the transformation into EOF is the choice of the output function h_z.

4.2   Main Characteristics of the Observers

A comparison of the different observers derived in the previous section is done for the special case $h_z(z_{n,k}) = z_{n,k}$ because it leads to a straightforward investigation of the observer characteristics. It is only necessary to consider the observer equations and error dynamics in EOF: Observers 2, 3 and 4 are based on the same system equations in EOF because the transformation from x-to z-coordinates is always identical (the degrees of freedom in the transformation from observability form into EOF are chosen following (3.15)). Since observer 1 in x-coordinates does not depend on the choice of the functions $f_{z,\mu}$, it is also possible to choose the transformation from observability form into EOF in compliance to (3.15) without affecting the characteristics of this observer. Then, the transformation between x-and z-coordinates is identical for all four observers which implies that differences can only have their cause in EOF.

Taking into consideration that $h_z(z_{n,k}) = z_{n,k}$, the following characteristics for the observers in EOF can be found:

- Since the observer equations of observers 2 and 4 do not depend on system parameters, these observers are robust to parameter uncertainties. Observers 1 and 3 are depending on parameters.

- Since the observer equations of observers 2 and 4 depend on y_{k+1}, the observer state $\hat{z}_{k+1}$ can not be calculated before step $k+1$. Using observers 1 or 3, the estimation $\hat{z}_{k+1}$ can already be calculated within step k.

- Using observers 3 and 4, it is easy to check from error dynamics (4.20) that $\|e_{i,k+1}\| < \|e_{i,k}\|$, $i = 1, 2, ..., n$, if the eigenvalues of the error dynamics are chosen within the unit circle. This does not hold for the error dynamics (3.14) of observers 1 or 2.

- Assuming the measured output signal is $\tilde{y}_k = y_k + r_{y,k}$ where $r_{y,k}$ is an additive noise, the resulting error dynamics for observer 4 and $k \geq n - 1$ are given by

$$
e_{k+1} = \begin{pmatrix} 0 \\ \vdots \\ 0 \\ r_{y,k+1} - \lambda_n r_{y,k} \end{pmatrix} + \begin{pmatrix} \lambda_1 & \cdots & 0 \\ \vdots & \ddots & \vdots \\ 0 & \cdots & \lambda_n \end{pmatrix} e_k. \qquad (4.21)
$$

For $e_{n-1} = 0$ and $k \geq n-1$, $e_k = \begin{pmatrix} 0 & \cdots & 0 & r_{y,k} \end{pmatrix}^T$ is a solution of this difference equation. This also holds for observer 2. Therefore, using observer 2 or 4, noise is not filtered regardless of the chosen eigenvalues whereas observers 1 and 3 are able to filter noise. This especially holds for observer 1, which, designed for a linear system

$$x_{k+1} = A x_k, \qquad y_k = c^T x_k \qquad\qquad (4.22)$$

with A a matrix and c a vector of appropriate dimensions, leads to the well known linear observer

$$\hat{x}_{k+1} = A \hat{x}_k + g(c^T \hat{x}_k - y_k) \qquad\qquad (4.23)$$

with the gain vector g. Choosing all eigenvalues of the observer error near the system eigenvalues (if possible) the gain and therefore the influence of noise at the output measurements on the observer is small. It even vanishes if system and observer eigenvalues are identical. Assuming small noise and system states near the operating point, the results from the linear case can be adopted to nonlinear systems. Therefore, good robustness of observer 1 to noise at the output measurements designed for a nonlinear systems is achieved by eigenvalues in EOF near to those of the linearized system.

To summarize, table 2.1 shows the main characteristics of the observers in EOF with $h_z(z_{n,k}) = z_{n,k}$. Supposing a smooth transformation map between EOF and x-coordinates, it is probable that the characteristics of the observers are also found in x-coordinates.

TABLE 2.1. Characteristics of the observers in EOF $(h_z(z_{n,k}) = z_{n,k})$.

Characteristic	Observer			
	1	2	3	4
Filtering of measurement noise	$+$	$-$	$+$	$-$
Robustness to model uncertainties	$-$	$+$	$-$	$+$
Transient behavior	$-$	$-$	$+$	$+$
Computation for step $k+1$ at step	k	$k+1$	k	$k+1$

5 An Example in the Field of Communication

As an example for reconstructing a desired information from the transmitted signal, the presented observer design is applied to the second order system

$$x_{k+1} = \begin{pmatrix} (1-\epsilon)\mu x_{1,k}(1-x_{1,k}) + \epsilon x_{2,k} \\ (1-\epsilon)\mu x_{2,k}(1-x_{2,k}) + w_k x_{2,k} \end{pmatrix}, \qquad (5.24)$$

$$y_k = x_{1,k}$$

where μ and ϵ are constants, $y_k = x_{1,k}$ is the transmitted signal and w_k contains the desired information. The signal w_k is a discrete-time signal with $0.06 \le w_k \le 0.12$ and a step width of 5, i.e. w takes a new value for $k = 0, 5, 10, 15, \ldots$ and remains constant for other values of k. For $w = \epsilon$, system (5.24) is identical to the one presented in [1]. An observer design via EOF for $w = \epsilon$ was already considered in [5].

The observer design via EOF for system (5.24) is based on the third order model

$$
\begin{pmatrix} x_{1,k+1} \\ x_{2,k+1} \\ w_{k+1} \end{pmatrix} = \begin{pmatrix} (1-\epsilon)\mu x_{1,k}(1-x_{1,k}) + \epsilon x_{2,k} \\ (1-\epsilon)\mu x_{2,k}(1-x_{2,k}) + w_k x_{1,k} \\ w_k \end{pmatrix},
$$

$$
y_k = x_{1,k}.
$$
(5.25)

The signal w is assumed to be constant. Since this only holds for the duration of five steps, this assumption requires an observer that converges fast enough. For $x_{1,k} > 0$, the representation of system (5.25) in observability form exists and an observer design via EOF is possible. The observer equations are omitted for reasons of space.

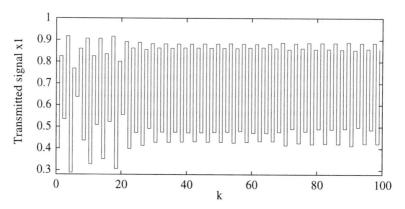

FIGURE 1. Transmitted signal $y_k = x_{1,k}$

Figure 1 shows the transmitted signal $y_k = x_{1,k}$ for initial conditions $x_{1,0} = 0.4$, $x_{2,0} = 0.2$ and w_k as presented in Figure 2, which also shows the reconstructed signal $w_{e,k}$ using observers 1, 3 and 4 via EOF with $\hat{x}_{1,0} = 0.5$, $\hat{x}_{2,0} = 0.5$ and $w_{e,0} = 0$. The eigenvalues of the observer error dynamics in EOF were chosen to $\lambda_1 = \lambda_2 = \lambda_3 = 0.1$.

The reconstruction of w_k using observer 1 has a delay of 3 steps and reaches satisfactory accuracy just before the observer starts to converge to the next value of w_k (A dead-beat design with $\lambda_1 = \lambda_2 = \lambda_3 = 0$ considerably improves the behaviour). Since observer 3 shows better transient behaviour than observer 1 (see Section 4.2), the reconstruction of w_k is quite

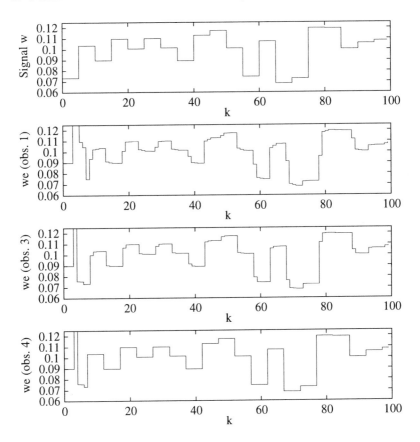

FIGURE 2. Information signal w_k and the estimated signal $w_{e,k}$ using observers 1, 3 and 4 with $\lambda_1 = \lambda_2 = \lambda_3 = 0.1$

good although it also has a delay of 3 steps. Best results can be achieved using observer 4. For $k > 6$ the reconstruction of w_k is exact except for a delay of 2 steps, i.e. $w_{e,k} = w_{k-2}$. This observer exploits the actual measurement y_k for estimating w_k which obviously leads to this good transient behaviour when w_k takes a new value.

6   Observer Design for the Rössler System

This section focuses on the design and realisation of an observer for the continuous-time Rössler system which has the form

$$\dot{x}(t) \;=\; \begin{pmatrix} -x_2(t) - x_3(t) \\ x_1(t) + a x_2(t) \\ c + x_3(t)\big(x_1(t) - b\big) \end{pmatrix}, \quad y(t) = x_3(t) \qquad (6.26)$$

with the coefficients $a, b, c > 0$. As shown in [13] the initial condition $x_3(0) > 0$ leads to $y(t) = x_3(t) > 0$ for all $t \geq 0$ and the observer problem is well posed.

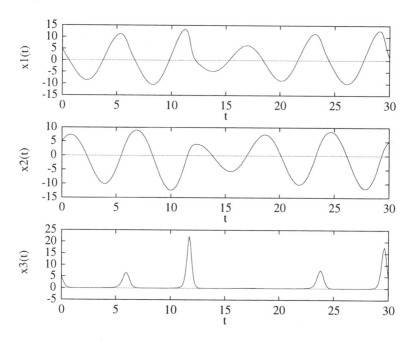

FIGURE 3. State variables of the Rössler system

Figure 3 shows the trajectories of the system states for initial conditions $x(0) = \begin{pmatrix} 5 & 5 & 5 \end{pmatrix}^T$ and coefficients $a = c = 0.2$ and $b = 7.5$.

There exists two approaches for the realisation of an observer for a continuous-time system:

1. The observer is designed in continuous-time. The computation of this observer in a real-time program with constant sampling time requires the discretization of the observer equations. Since the output signal y between the samplings is unknown between two samplings, the discretization can not be exact even in the linear case. In addition, the sampling time has to be chosen with respect to the systems and the observer dynamics.

2. The system can be, at least approximately, discretized which allows an observer design for the discrete-time model. The obtained observer can directly be implemented in a real-time program. Whereas a linear system can always be exactly discretized, the discretization of a nonlinear system is in general an approximation.

In what follows, a discretized continuous-time observer and the discrete-time observer 4 via EOF are designed for the Rössler system and implemented in a real-time program which reads all $T = \frac{1}{128}$s the output y of the simulated Rössler system. The two observers are compared for initial conditions $\hat{x}(0) = (\ 2\ \ 2\ \ 2\ )^T$ and different error dynamics.

6.1   Observer Design in Continuous-Time

The transformation

$$\begin{pmatrix} z_1(t) \\ z_2(t) \\ z_3(t) \end{pmatrix} = \begin{pmatrix} x_1(t) \\ x_2(t) \\ ln(x_3(t)) \end{pmatrix} \tag{6.27}$$

leads to a system representation in new coordinates (cf. [13])

$$\begin{aligned} \dot{z}(t) &= \begin{pmatrix} 0 & -1 & 0 \\ 1 & a & 0 \\ 1 & 0 & 0 \end{pmatrix} z(t) + \begin{pmatrix} -e^{z_3(t)} \\ 0 \\ ce^{-z_3(t)} - b \end{pmatrix}, \\ y(t) &= e^{z_3(t)} \end{aligned} \tag{6.28}$$

which allows to design an observer with linearizable error dynamics. The resulting observer has the form

$$\dot{\hat{x}}(t) = \begin{pmatrix} -\hat{x}_2(t) - y(t) + o_1\big(ln(\hat{x}_3(t)) - ln(y(t))\big) \\ \hat{x}_1(t) + a\hat{x}_2(t) + o_2\big(ln(\hat{x}_3(t)) - ln(y(t))\big) \\ \hat{x}_3(t)\big(\hat{x}_1(t) + \frac{c}{y(t)} - b + o_3\big(ln(\hat{x}_3(t)) - ln(y(t))\big)\big) \end{pmatrix} \tag{6.29}$$

with

$$o_3 = -(a + q_2), \quad o_1 = 1 + o_3 a - q_1, \quad o_2 = q_0 + o_3 - o_1 a \tag{6.30}$$

and q_0, q_1, q_2 the coefficients of the desired characteristic polynomial of the observer error dynamics. For the realisation in a real-time program, the differential equations are numerically computed using the Runge-Kutta algorithm.

6.2 Observer Design in Discrete-Time

A continuous-time system of the form

$$\dot{x}(t) = f\big(x(t)\big), \quad y(t) = h\big(x(t)\big) \tag{6.31}$$

can be discretized by a Taylor-Series expansion

$$
\begin{aligned}
x(t+T) &= x(t) + T\,\dot{x}(t) + \tfrac{T^2}{2}\ddot{x}(t) + \ldots \\
&= x(t) + T\,f(x(t)) + \tfrac{T^2}{2}\tfrac{\partial f(x(t))}{\partial x(t)}f(x(t)) + \ldots
\end{aligned} \tag{6.32}
$$

with sampling time T. Setting $x(t) = x(kT) =: x_k$, $x(t+T) = x((k+1)T) =: x_{k+1}$ leads to the discrete-time system representation

$$
\begin{aligned}
x_{k+1} &= x_k + T\,f(x_k) + \tfrac{T^2}{2}\tfrac{\partial f(x_k)}{\partial x_k}f(x_k) + \ldots \\
y_k &= h(x_k).
\end{aligned} \tag{6.33}
$$

In practice it is necessary to neglect higher order terms of the Taylor-series expansion. The higher the order of the first neglected term, the more accurate is the discretization. However, the complexity of the obtained system representation also increases with the number of considered terms. For the Rössler system, all terms with order higher than one must be neglected. Otherwise, the inverse of the observability map is not found. Therefore, the observer design is based on the discrete-time system model

$$
\begin{aligned}
x_{k+1} &= \begin{pmatrix} x_{1,k} - T(x_{2,k} + x_{3,k}) \\ x_{2,k} + T(x_{1,k} + a x_{2,k}) \\ x_{3,k} + T(c + x_{3,k}(x_{1,k} - b)) \end{pmatrix}, \\
y_k &= x_{3,k}.
\end{aligned} \tag{6.34}
$$

For this system, observer 4 via EOF with linearizable error dynamics was designed and compared to the continuous-time observer.

6.3 Observer Errors for Slow Error Dynamics

First the eigenvalues of the error dynamics were chosen to $\lambda_{i,c} = -10$ (continuous-time) and $\lambda_{i,d} = e^{T\lambda_{i,c}} \approx 0.925$ (discrete-time) with $i = 1, 2, 3$. Figure 4 shows the observer errors $e_{i,k} = \hat{x}_{i,k} - x_{i,k}$ for observer 4 via EOF. During the peaks of $x_3(t)$, there are considerable observer errors. Especially $|e_{2,k}|$ reaches values up to 1 whereas the observer errors of the continuous-time observer are always smaller than 0.003 for $t > 1.5$s which can be seen in Figure 5.

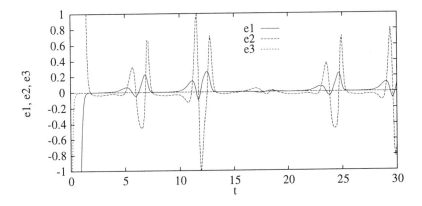

FIGURE 4. Observer error using discrete-time observer 4, $\lambda_{i,d} = e^{-10T}$

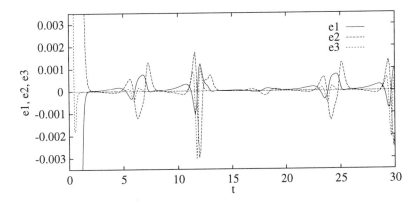

FIGURE 5. Observer error using discretized continuous-time observer, $\lambda_{i,c} = -10$

6.4 Observer Errors for Fast Error Dynamics

Whereas the eigenvalues of a discrete-time observer can be chosen to $\lambda_{i,d} = 0$ (dead-beat), the error dynamics of the continuous-time observer can not be chosen arbitrarily fast for a given sampling time. Otherwise, the errors of the numerical computation of the differential equations considerably increase.

The observer errors using observer 4 with dead-beat design are shown in Figure 6. For $t > 2$, they are nearly identical to the case with slow dynamics whereas the continuous-time observer with $\lambda_{i,c} = -120$ leads to higher observer errors (see Figure 7). In addition, the corresponding discrete-time eigenvalue is $e^{-120T} = 0.391$ and therefore the convergence rate is smaller compared to a dead-beat design.

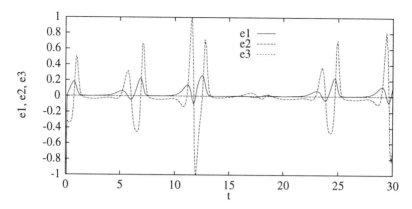

FIGURE 6. Observer error using discrete-time observer 4 with $\lambda_{i,d} = 0$

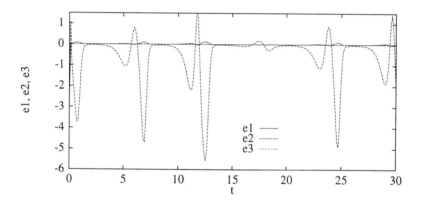

FIGURE 7. Observer error using discretized continuous-time observer with $\lambda_{i,c} = -120$

6.5   Concluding Remarks

The simulations have shown that the continuous-time observer leads to better results if the error dynamics are 'slow'. However, since the sampling time has to be chosen with respect to the observer dynamics, the eigenvalues of the observer dynamics must lie within a certain range or the sampling time has to be decreased.

A discrete-time observer even allows a dead-beat design because the sampling time has to be chosen only with respect to the system dynamics. This point is very important if high convergence rates are desired. However, a crucial point is finding a discrete-time representation of the system with satisfying accuracy. In the considered case, only a coarse discretization was possible.

Further simulations have shown that the discrete-time observer for this example is much more sensitive to noise but less sensitive to parameter uncertainties.

7 Discussion and Conclusions

In this chapter a discrete-time observer design via nonlinear observability form and extended nonlinear observer form using additional past output values is presented. Like the design procedure presented in [2] and [10], it does not require a diffeomorphic system function and the transformation can be calculated without solving a system of partial differential equations. In addition, the main problem of [2] and [10], the restrictive condition that the Hessian matrix of the nonlinear function appearing in observability form has to be diagonal, does not occur. Every strongly locally observable system is state equivalent to a system in EOF.

The observer design was applied to a problem in the field of communication in discrete-time and to a synchronization problem in the continuous-time context. For the latter, the performance of the discrete-time observer depends on the accuracy of the system discretization.

8 REFERENCES

[1] P. Badola, S. S. Tambe, and B. D. Kulkarni. Driving systems with chaotic signals. *Physical Review A*, 46(10):6735–6737, 1992.

[2] M. Brodmann. *Beobachterentwurf für nichtlineare zeitdiskrete Systeme*. Number 416 of line 8 in VDI-Fortschrittberichte. VDI-Verlag, Düsseldorf, 1994. Dissertation, Universität Hannover.

[3] S.-T. Chung and J. W. Grizzle. Sampled-data observer error linearization. *Automatica*, 26(6):997–1007, 1990.

[4] G. Ciccarella, M. Dalla Mora, and A. Germani. A robust observer for discrete time nonlinear systems. *Systems & Control Letters*, 24:291–300, 1995.

[5] H. Huijberts, T. Lilge, and H. Nijmeijer. A control perspective on synchronization and the takens-aeyels-sauer recontruction theorem. *Accepted for publication in Physical Review E*, 1998.

[6] R. Ingenbleek. *Beobachtbarkeit und Beobachterentwurf für zeitdiskrete nichtlineare Systeme*. Number 03/93 in Forschungsberichte. Universität Duisburg, Duisburg, 1993.

[7] R. Ingenbleek. *Zustandsbeobachter für zeitdiskrete nichtlineare Systeme - Geometrische Analyse und Synthese*. Number 527 of line 8 in

VDI-Fortschrittberichte. VDI-Verlag, Düsseldorf, 1996. Dissertation, Universität Duisburg.

[8] W. Lee and K. Nam. Observer design for autonomous discrete-time nonlinear systems. *Systems & Control Letters*, 17:49–58, 1991.

[9] T. Lilge. On observer design for nonlinear discrete-time systems. *European Journal of Control*, (4):306–319, 1998.

[10] W. Lin and C. I. Byrnes. Remarks on linearization of discrete-time autonomous systems and nonlinear observer design. *Systems & Control Letters*, 25:31–40, 1995.

[11] M. Loecher and E. R. Hunt. Control of high-dimensional chaos in systems with symmetry. *Physical Review Letters*, 79(1):63–66, 1997.

[12] H. Nijmeijer. Observability of autonomous discrete time nonlinear systems: A geometric approach. *Int. Journal of Control*, 36(5):867–74, Nov. 1982.

[13] H. Nimeijer and I. Mareels. An observers look to synchronization. *IEEE Transactions on Circuits and Systems*, 44(10):882–90, 1997.

[14] E. Ott, T. Sauer, and J. A. Yorke. *Coping with Chaos*. John Wiley & Sons, Inc., New York, 1994.

[15] L. M. Pecora and T. L. Carroll. Synchronization in chaotic systems. *Physical Review Letters*, 64(8):821–824, 1990.

[16] K. Pyragas. Generalized synchronization of chaos in directionally coupled chaotic systems. *Physical Review E*, 51(2):980–994, 1995.

[17] K. Pyragas. Weak and strong synchronization of chaos. *Physical Review E*, 54(5):R4508–R4511, 1996.

[18] T. Stojanovski, U. Parlitz, L. K., and R. Harris. Exploiting delay reconstruction for chaos synchronization. *Physics Letters A*, 233:355–360, 1997.

[19] Y. Zhang, M. Dai, W. Hua, Y. Ni, and G. Du. Digital communication by active-passive-decomposition synchronization in hyperchaotic systems. *Physical Review E*, 58(3):3022–3027, 1998.

Chaos Synchronization

Ulrich Parlitz[1], Lutz Junge[1] and Ljupco Kocarev[2]

[1]Drittes Physikalisches Institut, Universität Göttingen, Bürgerstr. 42-44,
D-37073 Göttingen, Germany
[2]Department of Electrical Engineering, St Cyril and Methodius
University, Skopje, PO Box 574, Macedonia

1 Introduction

Synchronization is a phenomenon of interest in many scientific areas ranging from celestial mechanics to laser physics, from electronics to communications, and from biophysics to neuroscience [1]. In particular, synchronization of chaotic dynamics [7, 4, 24] has attracted much attention during the last years because of its role in understanding the basic features of man-made and natural systems. Thus, for example, optical communication using chaotic waveforms demonstrated experimentally [8, 37, 38] and theoretically [3], is possible because of chaos synchronization between transmitter and receiver. On the other hand, the evidence of chaotic behavior in the brain [34] and the importance of synchronization in perceptive processes of mammals [33] indicate a possible role of chaos synchronization in neural ensembles [32] as well.

The phenomenon of synchronization also occurs for uni-directionally coupled systems and in this case the driven system (or response system) may be viewed as a nonlinear observer of the driving system. Or, conversely, nonlinear observer theory may be used to construct pairs of uni-directionally coupled synchronizing systems. Such pairs may then be used for system and parameter estimation or for potential applications in communication systems, see [1] and [31]. In all these cases, the two coupled systems are (almost) identical and therefore *identical synchronization* occurs which means that the difference of drive and response state vectors converges to zero for $t \to \infty$. If two different systems are coupled, more sophisticated types of synchronization [5] may occur like generalized synchronization or phase synchronization that will be discussed in Sections 3 and 4.

We shall begin with a presentation of synchronization phenomena of spatially extended systems that are given in terms of (chaotic) partial differential equations. For more information about chaos synchronization and related problems the reader is refered to the article collections in [31].

2 Synchronization of Spatially Extended Systems

In the following we shall discuss identical synchronization of uni-directionally coupled spatially extended systems that are described by a partial differential equation in the form

$$\frac{\partial u}{\partial t} = F(u, \frac{\partial u}{\partial x}, \frac{\partial^2 u}{\partial x^2}, \dots), \quad x \in [0, L] \tag{2.1}$$

with spatial length L. In this chapter we consider only one dimensional PDE's. Generalizations for higher dimensional systems are straightforward.

For this class of systems we define:

Two spatially extended systems are called synchronized, if their states $\mathbf{u}(x, t)$ *and* $\mathbf{v}(x, t)$ *converge to each other in the whole spatial domain, i.e. if* $\forall x \in [0, L] : \lim_{t \to \infty} \|\mathbf{u}(x, t) - \mathbf{v}(x, t)\| = 0$

As in the case of low dimensional systems there exists an invariant manifold $\mathbf{u} \equiv \mathbf{v}$ (also called synchronization manifold), whose stability properties determine the occurence of stable synchronization. If the transverse system $\mathbf{w}_\perp = \frac{1}{2}(\mathbf{u} - \mathbf{v})$ has an asymptotically stable fixed point at the origin, then this manifold is asymptotically stable and synchronization occurs. Indeed, all known techniques (see [1], [31]) for verifying synchronization such as *necessary* criteria like negative conditional Lyapunov exponents or *sufficient* criteria like Lyapunov functions and stability of unstable periodic orbits, can be generalized and can in principle be applied to spatially extended systems, too.

On the other hand, the generalization of the coupling techniques used for low dimensional systems is not so straightforward. A coupling along the whole spatial axis is possible for numerical simulations but may turn out to be impractical or even impossible for experiments. A similar argument holds for local pinning coupling schemes that are used for synchronizing coupled map lattices (CMLs)[19, 15, 16, 11, 9]. These schemes use coupling in points which is not only practically impossible but also in some sense useless for PDE's.

An alternative is the *sensor* coupling scheme, introduced in [20], which generalizes the pinning schemes to systems with continuous space variables. The idea is, that typical experimental measurement devices have a finite resolution l and measure local spatial averages of the desired quantity. The left plot of Figure 1 shows the concept of the sensor coupling scheme and the right plot illustrates the notion of a measured sensor signal. According to [20] we want to call these elements *sensors*. Each sensor measures a scalar time-series of the form

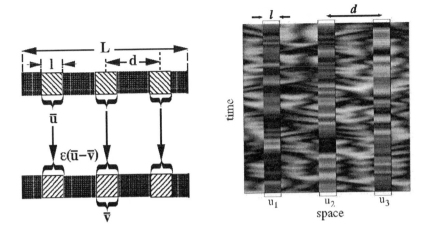

FIGURE 1. Principle of the sensor coupling scheme. Left: sketch, right: visualization of three sensor time series measured from spatio-temporal chaos.

$$\overline{u}_n(t) = \frac{1}{l} \int_{nd-l/2}^{nd+l/2} u(x,t)dx, \quad n = 1, \ldots, N \qquad (2.2)$$

which is averaged over a width l.

Because of the exponential decrease of spatial correlations in spatially extended chaotic systems, we need several but a finite number N of coupling signals that contain all the necessary information to reconstruct the whole state in the synchronization process. Therefore the N sensors are distributed with equal distance $d = \frac{L}{N}$ (for periodic boundary conditions) along the spatial axis. Numerical investigations have shown that the equidistant arrangement is nearly optimal for systems with extensive chaos [13]. Now we have to choose a coupling scheme that will be applied locally using the sensor signals as driving forces. To do this we measure in the driven system N sensor signals at the same positions and apply a diffusive coupling term with coupling strength ϵ

$$f(u,v) = \begin{cases} \epsilon(\overline{u}_n - \overline{v}_n) & : \quad nd - l/2 \le x \le nd + l/2 \\ 0 & : \quad \text{else} \end{cases} \qquad (2.3)$$

at each sensor position. As an example we shall examine now the one dimensional complex Ginzburg-Landau model

$$\frac{\partial u}{\partial t} = \mu u - (1 - i\alpha)|u|^2 u + (1 + i\beta)\Delta u, \quad u \in [0, L] \qquad (2.4)$$

with periodic boundary conditions. This equation possesses uniform travelling wave solutions. For $1 - \alpha\beta < 0$ they become unstable and different

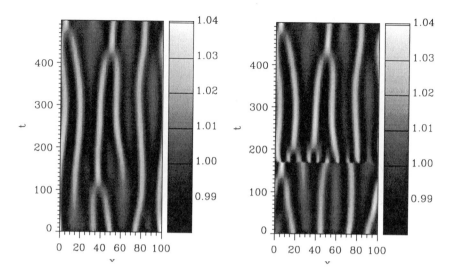

FIGURE 2. Synchronization of two Ginzburg-Landau equations in the phase turbulent regime (Left: drive, right: response) using $N = 15$ sensors with width $l = 3$ and coupling strength $\epsilon = 0.2$. The amplitudes of drive and response PDE are grey scaled.

types of turbulence occur. In the following we will consider two parameter sets, $\mu = 1.0, \alpha = 2.0, \beta = 0.7$ corresponding to phase turbulence and $\mu = 1.0, \alpha = 2.0, \beta = 1.2$ which yields defect turbulence. In both cases *extensive chaos* is observed and the Lyapunov dimension D_L of the underlying attractor increases with the system size L like $D_L \sim 0.102L$ for phase turbulence and with $D_L \sim 0.332L$ for defect turbulence.

In order to achieve synchronization we drive an identical copy of (2.4) using N sensor signals (2.3) that are applied in intervals of width l.

$$\frac{\partial v}{\partial t} = \mu v - (1 - i\alpha)|v|^2 v + (1 + i\beta)\Delta v + f(u, v) \qquad (2.5)$$

Note that this is a local control technique and the driven system (2.5) evolves freely between the sensor locations. Figure 2 shows the synchronization of drive (left) and response (right) in the phase turbulent regime. For this example we used $N = 15$ equally spaced sensors with width $l = 3$ to synchronize two Ginzburg-Landau equations with length $L = 100$. At $t = 170$ the coupling is switched on and the response system quickly converges to the synchronized state. In the beginning of the coupling the perturbation introduced through the sensors signals induces a periodic pattern, which decays very fast due to the synchronization.

Similar results have been obtained for defect turbulence. If the sensor coupling is applied only locally one may observe *local synchronization* as it is shown in Fig. 3. Replacing the driving sensor signals by vanishing sig-

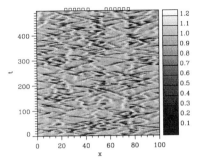 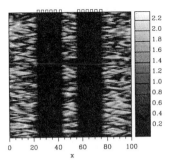

FIGURE 3. Local synchronization of defect turbulence of the Ginzburg-Landau equations. The left figure shows the dynamics of the response system and the right figure the synchronization error between drive and response that vanishes (dark areas) in those intervals where sensors are placed ($N = 2*6, l = 2, \varepsilon = 2.0$).

nals one may also suppress (locally) the chaotic oscillations of the response system and stabilize the homogeneous state. Furthermore, the sensor coupling has also been succesfully applied to a pair of Kuramoto-Sivashinski equations and provides nonlinear observers that can be used for estimating parameters of PDEs from time series [13].

3   Generalized Synchronization

If a pair of very similar or even identical systems is coupled one may observe *identical synchronization* (IS) where the difference of the state vectors of both systems converges to zero, even in the case of chaotic dynamics. This kind of synchronization, however, cannot be expected for coupled systems that are of completely different origin (e.g., an electrical circuit coupled to a mechanical system). What does "synchronization" mean in such a more general case? Periodic systems are usually called synchronized if either their phases or frequencies are locked. For chaotic systems, however, the notions of "frequency" or "phase" are in general not well defined and can thus not be used for characterizing synchronization (except for some class of chaotic systems where a phase variable can be introduced to quantify *chaotic phase synchronization* that will be discussed in the next section). In this Section we present different notions of *generalized synchronization* (GS) that have been proposed during the last years [4, 29, 2, 17, 26, 6, 12, 35, 21].

Basically two types of generalized synchronization of uni-directionally coupled systems have been investigated in the literature so far. In its strongest form GS leads to the existence of a function that maps (asymptotically for $t \to \infty$) states of the drive system to states of the response system. In this case the chaotic dynamics of the response system can be predicted from the drive system. Whether such a function exists and whether it is continuous or even smooth depends on the features of the drive and response system [6, 12, 35]. As an example consider a m-dimensional (chaotic) dynamical system

$$\mathbf{x}^{n+1} = \mathbf{f}(\mathbf{x}^n) \tag{3.6}$$

that drives the following one-dimensional system

$$y^{n+1} = by^n + \cos(2\pi x_1^n) \tag{3.7}$$

with $b < 1$. Is is easy to see that

$$y^n = \frac{1}{b} \sum_{i=1}^{\infty} b^i \cos(2\pi x_1^{n-i}). \tag{3.8}$$

For invertible drive dynamics x_1^{n-i} can be replace by $f_1^{-i}(\mathbf{x}^n)$ and since the cosine function is bounded and continuous, the state of the response system y^n is a continuous function of the state of the drive system:

$$y^n = g(\mathbf{x}^n) = \frac{1}{b} \sum_{i=1}^{\infty} b^i \cos(2\pi f_1^{-i}(\mathbf{x}^n)). \tag{3.9}$$

In order to study the smoothness feature of this function we compute its first derivative

$$\frac{\partial g}{\partial x_j}(\mathbf{x}) = -\frac{2\pi}{b} \sum_{i=1}^{\infty} b^i \frac{\partial f_1^{-i}}{\partial x_j}(\mathbf{x}) \sin(2\pi f_1^{-i}(\mathbf{x})). \tag{3.10}$$

For g being differentiable this sum has to converge, or in other words the coefficients $c_i = b^i \frac{\partial f_1^{-i}}{\partial x_j}(\mathbf{x})$ have to be less than 1 for large i. This is the case when the contraction of the response system is stronger than that of the drive, or in terms of Lyapunov exponents, if $\ln b < \lambda_m$ where λ_m denotes the smallest Lyapunov exponent of the drive. To illustrate this point we shall use as drive system Arnold's cat-map

$$x_1^{n+1} = (x_1^n + x_2^n) \mod 1 \tag{3.11}$$
$$x_2^{n+1} = (x_1^n + 2x_2^n) \mod 1 \tag{3.12}$$

which is an ergodic map on the torus with a constant Jacobian matrix. The eigenvalues of this matrix are $\mu_1 = (3 + \sqrt{5})/2$ and $\mu_2 = (3 - \sqrt{5})/2$. The Jacobian matrix $Df^{-i}(\mathbf{x})$ of the iterated inverse map is given by $A^{-i} = U \cdot diag(\mu_1^{-i}, \mu_2^{-i}) \cdot U^{tr}$ where U is an orthogonal 2×2 matrix. Since $\mu_1 > 1 > \mu_2$ the coefficients c_i can for large i by approximated by $c_i \approx const. \cdot (b/\mu_2)^i$. Therefore, the first derivative of g diverges for $b > \mu_2$ (or in terms of

(a)
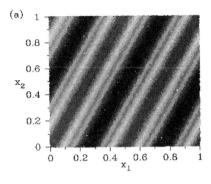

(b)
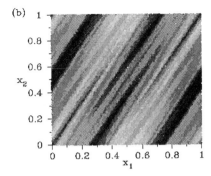

FIGURE 4. Grey scaled plot of the response state y vs. the (x_1, x_2) coordinates of the driving cat map. (a) $b = 0.01$ (b) $b = 0.4$. Both figures have been computed by iterating the dynamical systems (3.7), (3.11) and (3.12) and transients have been discarded.

Lyapunov exponents $\ln(b) > \ln(\mu_2) = \lambda_2$). In this case the function g is essentially a Weierstrass function. Figure 4 shows how the function g looses its smoothness when b is increased from $0.01 = b < \mu_2 = (3-\sqrt{5})/2 = 0.382$ to $0.4 = b > \mu_2$.

Typically a function exists if the response system is asymptotically stable when driven by the coupling signal *and* no subharmonic entrainment occurs [21]. If, for example, a periodic orbit of the drive entrains a stable periodic orbit of the response with twice the period (i.e. $T_D : T_R = 1 : 2$) then any point on the attractor of the drive is mapped to *two* points on the response orbit and in this case there exists a relation but *not* a function. This multivaluedness always occurs for *subharmonic periodic entrainment* with $T_D < T_R$.

Note that identical synchronization implies GS in any diffeomorphic equivalent coordinate system. On the other hand, if GS is observed between two dynamical systems with a diffeomorphic function this diffeomorphism can be used to perform a change of the response coordinate system such that in the new coordinate system the response system synchronizes identically with the drive system.

To find evidence for the existence of a (continuous) function relating states of the drive to states of the response one may apply nearest neighbors statistics [29]. This approach for identifying generalized synchronization can be applied to uni- and bi-directionally coupled systems if the original (physical) state spaces of drive and response are accessible. If only (scalar) time series from the drive and the response system can be sampled, then delay embedding [30] may be used to investigate neighbourhood relations in the corresponding reconstructed state spaces. In this case, however, only generalized synchronization of *uni*-directionally coupled systems can be detected by predicting the (reconstructed) state of the response system using a time series from the drive system. A prediction of the evolution of the drive system based on data from the response system is *always* possible (i.e. with and without generalized synchronization), because (almost) any time series measured at the response system may also be viewed as a time series from the combined systems *drive & response* and may thus be used to reconstruct and predict the dynamics of drive and response. In this sense a time series based test provides *no* information about GS in the case of *bi-directionally* coupled systems.

In such a case where drive and response are not related by a function, a second, weaker notion of GS may apply that assumes only asymptotic stability of the response system but not the existence of a function mapping states of the drive to states of the response system [2, 21]. This type of GS can be verified using the so-called auxiliary system method where two identical copies of the response system are driven by the same driving signal. Identical synchronization of both response systems indicates GS in the weaker sense. Note that using the auxiliary systems approach one may also observe *nonidentical* (i.e. generalized) synchronization of *identical* systems that fail to synchronize identically.

Current research in the field of generalized synchronization focuses on the question whether the different phenomena and approaches for characterizing (generalized) synchronization can be unified in a mathematically rigorous sense using the notion of *normally hyperbolic invariant manifolds* that are smooth and persistent under perturbations of the system(s) [39, 10].

4 Phase Synchronization

Another generalization of the notion of identical synchronization is the phenomenon of *phase synchronization* (PS) [36, 27, 22, 25, 18].

It can easily be observed when a well defined phase variable can be identified in both coupled systems. This can be done heuristically for strange attractors that spiral around some particular point (or "hole") in a two-dimensional projection of the attractor, like the Rössler attractor shown in Fig. 5. In such a case, a phase angle $\phi(t)$ can be defined that de- or increases monotonically. Phase synchronization of two coupled systems occurs if the difference $|\phi_1(t) - \phi_2(t)|$ between the corresponding phases is bounded by some constant. A more general definition includes rational relations $|n\phi_1 - m\phi_2| < const$ for arbitrary integers n and m.

Using the phase angle $\phi(t)$ one may define a mean rotation frequency $\Omega = \lim_{t\to\infty} \phi(t)/t$ and in the case of PS, this mean rotation frequencies of the drive and the response system coincide, i.e., also for chaotic systems PS leads to the frequency entrainment known from coupled periodic oscillations. The amplitudes of both systems remain in this case completely uncorrelated [27].

This phenomenon may be used in technical or experimental applications where a coherent superposition of several output channels is desired.

In more abstract terms PS occurs when a zero Lyapunov exponent of the response system becomes negative. This leads to a reduction of the degree of freedom of the response system in the direction of the flow. For systems where a phase variable can be defined the direction of the flow coincides in general with the coordinate that is described by the phase variable. A zero LE that becomes negative reflects in this sense a restriction that is imposed on the motion of the phase variable. If the zero LE that decreases is the largest LE of the response system then phase synchronization occurs together with GS [22]. If there exist, however, in addition to the formerly zero LE, other LEs which are and remain positive, PS occurs but no GS.

This scenario for the onset of PS may be observed for a sinusoidally driven Rössler system [36]:

$$\begin{aligned} \dot{x}_1 &= 0.4 + x_1(x_2 - 8.5) \\ \dot{x}_2 &= -x_1 - x_3 + a\cos(t) \\ \dot{x}_3 &= x_2 + 0.15x_3. \end{aligned} \tag{4.13}$$

Figure 5 shows the onset of PS when the driving amplitude a exceeds some critical value of $a_c \approx 0.4$. The solid gray lines belong to the chaotic attractor of the driven system (4.13) and the black dots are plotted at times $t_n = n2\pi$ yielding a stroboscopic phase portrait. As can be seen in Fig. 5a these dots are scattered on the chaotic attractor if the driving force is too weak, indicating no fixed phase relation of the chaotic oscillation with respect to the driving signal. Figure 5b shows the distribution near the onset of PS where the dots already start to form a cluster.

If the amplitude a is sufficiently high, phase sychronization occurs as can be seen for an amplitude of $a = 0.7$ in Fig. 5c.

This transition can also be studied in terms of the Lyapunov exponents of the response system. Figure 6a shows the two largest exponents λ_1 and λ_2 of the Rössler system (4.13) plotted in dependence on the coupling parameter a. For $a > 0.4$ the zero exponent starts decreasing while λ_1 remains positive. The driven system thus looses a degree of freedom although it stays chaotic.

This degree of freedom is associated with the zero Lyapunov exponent λ_2, i.e., with the (tangential) direction of the trajectories. This direction, however, is exactly the direction of the spiraling motion around the "hole" in the attractor that was used for introducing a phase variable.

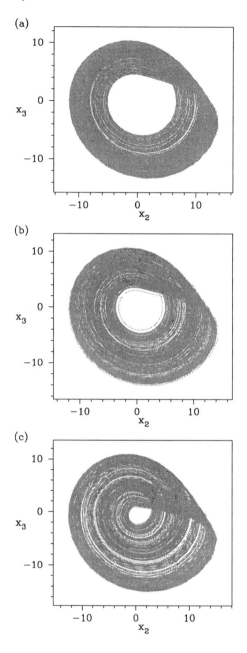

FIGURE 5. Phase synchronization of a periodically driven Rössler system (4.13) (a) $a = 0.1$, no PS; (b) $a = 0.5$, onset of PS; (c) $a = 0.7$, full PS.

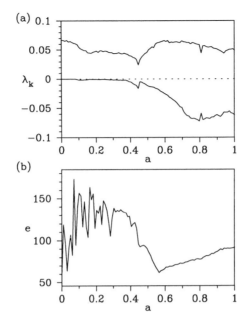

FIGURE 6. (a) The two largest Lyapunov exponents λ_1 and λ_2, and (b) the mean synchronization error e (4.14) of a pair of identical Rössler response systems (4.13) vs. the driving amplitude a.

Figure 6b shows the mutual averaged synchronization error e

$$e = \sqrt{\frac{1}{N}\sum_{n=1}^{N}\|\mathbf{x}(t_n) - \tilde{\mathbf{x}}(t_n)\|^2} \qquad (4.14)$$

of two identical Rössler systems that are driven by the same sinusoidal signal and sampled with $t_n = n \cdot 2\pi/25$. For GS such a comparison with an auxiliary system would result in an asymptotically vanishing error, but here both response systems are chaotic and the PS leads only to a decrease of e by a factor of about two. In this sense PS leads to a constructive interference of chaotic response signals that has also been observed in mean field variables of arrays of slightly different response systems which were driven by a common signal [25, 18].

Another phenomenon that is closely related to PS is *lag synchronization* that was observed recently by Rosenblum et al. [28] and leads to synchronization with some time delay between drive and response.

5   Conclusions

In this chapter we have addressed specific topics and examples of "chaos synchronization": synchronization of spatially extended systems (PDEs), generalized synchronization and phase synchronization. In particular synchronization phenomena of uni-directionally coupled *identical* systems (here: pairs of PDEs) are very closely related to questions of observability and observer design. If the underlying dynamics is chaotic, intermittent breakdown of synchronization may occur if not *not* all of the unstable orbits which form the skeleton of the chaotic attractor are synchronized (for details see [1] and references cited therein). This phenomenon can be excluded if proper (global) stability conditions can be established (for example, in terms of Lyapunov functions). Another topic that is worth mentioning are potential applications of (chaos) synchronization like system identification and data encryption. For system identification or parameter estimation a model equation is varied until it synchronizes with a given time series. This approach has been successfully applied to maps, coupled ODE's and coupled PDEs in order to estimate some free parameters [23, 14]. The advantage of chaotic dynamics for this task is the fact that a larger part of the state space is explored compared to periodic solutions. The second potential application mentioned above, "chaos communication", was actually for many researchers the main motivation to study synchronization mechanisms of uni-directionally coupled systems. The basic idea is to transmit a modulated chaotic signal and to use synchronization for recovering the message at the receiver (see [37, 38] for a recent fast optical implementation and [1] for other examples). Whether this approach can really compete with standard cryptographie is still an open question but some special applications seem possible.

Acknowledgments

This work was supported by the German Science Foundation (DFG grant Pa 643/1-1) and a binational German-Macedonian grant (MAK-004-96).

6   REFERENCES

[1] *Chaos* **7**(4), pp. 509-826, *IEEE Trans. Circuits and Systems, part I* **44**, 1997.

[2] H.D.I. Abarbanel, N.F. Rulkov and M.M. Sushchik. Generalized synchronization of chaos: The auxiliary system approach, *Phys. Rev. E* **53**(5), pp. 4528-4535, 1996.

[3] H.D.I Abarbanel and M. Kennel. Synchronizing High-Dimensional

Chaotic Optical Ring Dynamics, *Phys. Rev. Lett.* **80**(14), pp. 3153-3156, 1998.

[4] V.S. Afraimovich, N.N. Verichev and M.I. Rabinovich. Stochastic synchronization of oscillations in dissipative systems, *Radiophys. Quantum Electron.* **29**, pp. 795-803, 1986.

[5] I.I. Blekhman, A.L. Fradkov, H. Nijmeijer and A.Yu. Progromsky. On self-synchronization and controlled synchronization, *Syst. Contr. Lett.* **31**(5), pp. 299-306, 1997.

[6] M.E. Davies and K.M. Campbell. Linear recursive filters and nonlinear dynamics, *Nonlinearity* **9**, pp. 487-499, 1996.

[7] H. Fujisaka and T. Yamada. Stability Theory of Synchronized Motion in Coupled-Oscillator Systems, *Progr. Theor. Phys.* **69**(1), pp. 32-47, 1983.

[8] J. Goedgebuer, L. Larger and H. Porte. Optical Cryptosystem Based on Synchronization of Hyperchaos Generated by a Delayed Feedback Tunable Laser Diode, *Phys. Rev. Lett.* **80**(10), pp. 2249-2252, 1998.

[9] R. O. Grigoriev, M. C. Cross and H. G. Schuster. Pinning Control of Spatiotemporal Chaos, *Phys. Rev. Lett.* **79**(15), pp. 2795-2798, 1997.

[10] M. Hirsch and C. Pugh. *Stable Manifolds and Hyperbolic Sets*, In *Global Analysis*, AMS Proc. Symp. Pure Math. **14**, 1970.

[11] G. Hu and Z. Qu. Controlling Spatiotemporal Chaos in Coupled Map Lattice Systems, *Phys. Rev. Lett.* **72**(1), pp. 68-71, 1994.

[12] B.R. Hunt, E. Ott and J.A. Yorke. Differentiable generalized synchronization of chaos, *Phys. Rev. E* **55**(4), pp. 4029-4034, 1997.

[13] L. Junge, U. Parlitz, Z. Tasev and L. Kocarev. Synchronization and control of spatially extended systems using sensor coupling, *To appear in the Int. J. Bif. Chaos*, 1999.

[14] L. Junge and. U. Parlitz. Control and synchronization of spatially extended systems, Proceedings of the *1998 International Symposium on Nonlinear Theory and its Applications - NOLTA'98*, Le Régent, Crans-Montana, Switzerland, Sept. 14-17, pp. 303-306, 1998.

[15] L. Kocarev, Z. Tasev, T. Stojanovsk and U. Parlitz. Synchronizing spatiotemporal chaos, *Chaos* **7**(4), pp. 635-643, 1997.

[16] L. Kocarev, Z. Tasev and U. Parlitz. Synchronizing Spatiotemporal Chaos of Partial Differential Equations, *Phys. Rev. Lett.* **79**(1), pp. 51-54, 1997.

[17] L. Kocarev and U. Parlitz. Generalized Synchronization, Predictability, and Equivalence of Unidirectionelly Coupled Dynamical Systems, *Phys. Rev. Lett.* **76**(11), pp. 1816-1819, 1996.

[18] G. Osipov, A. Pikovsky, M. Rosenblum and J. Kurths. Phase synchronization effects in a lattice of nonidentical Rössler oscillators *Phys. Rev. E* **55**(3), pp. 2353-2361, 1997.

[19] N. Parekh, S. Parthasarathy and S. Sinha. Global and Local Control of Spatitemporal Chaos in Coupled Map Lattices, *Phys. Rev. Lett.* **81**(7), pp. 1401-1404, 1998.

[20] U. Parlitz and L. Kocarev. Synchronization of Chaotic Systems, In *H.-G. Schuster (Ed.) Handbook of Chaos Control*, Wiley-VCH, 1998.

[21] U. Parlitz, L. Junge and L. Kocarev. Subharmonic entrainment of Unstable Periodic Orbits and Generalized Synchronization, *Phys. Rev. Lett.* **79**(17), pp. 3158-3161, 1997.

[22] U. Parlitz, L. Junge, W. Lauterborn and L. Kocarev. Experimental observation of phase synchronization, *Phys. Rev. E* **54**(2), pp. 2115-2117, 1996.

[23] U. Parlitz, L. Junge and L. Kocarev. Synchronization-based parameter estimation from time series, *Phys. Rev. E*, **54**(6), pp. 6253-6260, 1996.

[24] L.M. Pecora and T.L. Carroll. Synchronization in chaotic systems, *Phys. Rev. Lett.* **64**(8), pp. 821-824, 1990.

[25] A.S. Pikovsky, M.G. Rosenblum and J. Kurths. Synchronization in a population of globally coupled oscillators, *Europhys. Letters* **34**(3), pp. 165-170, 1996.

[26] K. Pyragas. Weak and strong synchronization of chaos, *Phys. Rev. E* **54**(5), pp. R4508-R4511, 1996.

[27] M.G. Rosenblum, A.S. Pikovsky and J. Kurths. Phase Synchronization of Chaotic Oscillators, *Phys. Rev. Lett.* **76**(11), pp. 1804-1807, 1996.

[28] M.G. Rosenblum, A.S. Pikovsky and J. Kurths. From phase to lag synchronization in coupled chaotic oscillators, *Phys. Rev. Lett.* **78**(22), pp. 4193-4196, 1997.

[29] N. Rulkov, M. Sushchik, L. Tsimring and H.D.I. Abarbanel. Generalized synchronization of chaos in directionally coupled chaotic systems, *Phys. Rev. E* **51**(2), pp. 980-994, 1995.

[30] T. Sauer, J.A. Yorke and M. Casdagli. Embedology, *J. Stat. Phys.* **65**(3,4), pp. 579-616, 1991.

[31] H.-G. Schuster (Ed.). *Handbook of Chaos Control*, Wiley-VCH, 1998.

[32] S. J. Schiff, P. So, T. Chang, R.E. Burke and T. Sauer. Detecting dynamical interdependence and generalized synchrony through mutual prediction in a neural ensemble, *Phys. Rev. E* **54**(6), pp. 6708-6724, 1996.

[33] W. Singer. Synchronization of cortical activity and its putative role in information processing and learning, *Ann. Rev. Physiol.* **55**, pp.349-374, 1993.

[34] C. Skarda and W. J. Freeman. How brains make chaos to order to make sense of the world *Behav. Brain Sci.* **10**, pp. 161-195, 1987.

[35] J. Stark. Invariant graphs for forced systems, *Physica D* **109**, pp. 163-179, 1997.

[36] E.-F. Stone. Frequency entrainment of a phase coherent attractor, *Phys. Lett. A* **163**, pp.367-374, 1992.

[37] G.D. Van Wiggeren and R. Roy. Communication with Chaotic Lasers, *Science* **279**, 1998.

[38] G.D. Van Wiggeren and R. Roy. Optical Communication with Chaotic Waveforms, *Phys. Rev. Lett.* **81**(16), pp. 3547-3550, 1998.

[39] S. Wiggins. *Normally Hyperbolic Invariant Manifolds in Dynamical Systems*, Springer-Verlag, New York, 1994.

Lecture Notes in Control and Information Sciences

Edited by M. Thoma

1993–1999 Published Titles:

Vol. 224: Magni, J.-F.; Bennani, S.; Terlouw, J. (Eds)
Robust Flight Control: A Design Challenge
664 pp. 1997 [3-540-76151-9]

Vol. 225: Poznyak, A.S.; Najim, K.
Learning Automata and Stochastic Optimization
219 pp. 1997 [3-540-76154-3]

Vol. 226: Cooperman, G.; Michler, G.; Vinck, H. (Eds)
Workshop on High Performance Computing and Gigabit Local Area Networks
248 pp. 1997 [3-540-76169-1]

Vol. 227: Tarbouriech, S.; Garcia, G. (Eds)
Control of Uncertain Systems with Bounded Inputs
203 pp. 1997 [3-540-76183-7]

Vol. 228: Dugard, L.; Verriest, E.I. (Eds)
Stability and Control of Time-delay Systems
344 pp. 1998 [3-540-76193-4]

Vol. 229: Laumond, J.-P. (Ed.)
Robot Motion Planning and Control
360 pp. 1998 [3-540-76219-1]

Vol. 230: Siciliano, B.; Valavanis, K.P. (Eds)
Control Problems in Robotics and Automation
328 pp. 1998 [3-540-76220-5]

Vol. 231: Emel'yanov, S.V.; Burovoi, I.A.; Levada, F.Yu.
Control of Indefinite Nonlinear Dynamic Systems
196 pp. 1998 [3-540-76245-0]

Vol. 232: Casals, A.; de Almeida, A.T. (Eds)
Experimental Robotics V: The Fifth International Symposium Barcelona, Catalonia, June 15-18, 1997
190 pp. 1998 [3-540-76218-3]

Vol. 233: Chiacchio, P.; Chiaverini, S. (Eds)
Complex Robotic Systems
189 pp. 1998 [3-540-76265-5]

Vol. 234: Arena, P.; Fortuna, L.; Muscato, G.; Xibilia, M.G.
Neural Networks in Multidimensional Domains: Fundamentals and New Trends in Modelling and Control
179 pp. 1998 [1-85233-006-6]

Vol. 235: Chen, B.M.
H∞ Control and Its Applications
361 pp. 1998 [1-85233-026-0]

Vol. 236: de Almeida, A.T.; Khatib, O. (Eds)
Autonomous Robotic Systems
283 pp. 1998 [1-85233-036-8]

Vol. 237: Kreigman, D.J.; Hagar, G.D.; Morse, A.S. (Eds)
The Confluence of Vision and Control
304 pp. 1998 [1-85233-025-2]

Vol. 238: Elia, N. ; Dahleh, M.A.
Computational Methods for Controller Design
200 pp. 1998 [1-85233-075-9]

Vol. 239: Wang, Q.G.; Lee, T.H.; Tan, K.K.
Finite Spectrum Assignment for Time-Delay Systems
200 pp. 1998 [1-85233-065-1]

Vol. 240: Lin, Z.
Low Gain Feedback
376 pp. 1999 [1-85233-081-3]

Vol. 241: Yamamoto, Y.; Hara S.
Learning, Control and Hybrid Systems
472 pp. 1999 [1-85233-076-7]

Vol. 242: Conte, G.; Moog, C.H.; Perdon A.M.
Nonlinear Control Systems
192 pp. 1999 [1-85233-151-8]

Vol. 243: Tzafestas, S.G.; Schmidt, G. (Eds)
Progress in Systems and Robot Analysis and Control Design
624 pp. 1999 [1-85233-123-2]